Communications
in Computer and Information Science 487

Alexander Dudin   Anatoly Nazarov
Rafael Yakupov   Alexander Gortsev (Eds.)

Information Technologies and Mathematical Modelling

13th International Scientific Conference, ITMM 2014
named after A.F. Terpugov
Anzhero-Sudzhensk, Russia, November 20-22, 2014
Proceedings

 Springer

Volume Editors

Alexander Dudin
Belarusian State University
Minsk, Belarus
E-mail: dudin@bsu.by

Anatoly Nazarov
Tomsk State University, Russia
E-mail: nazarov.tsu@gmail.com

Rafael Yakupov
Kemerovo State University
Anzhero-Sudzhensk, Russia
E-mail: yrt@asf.ru

Alexander Gortsev
National Research Tomsk State University, Russia
E-mail: gam@fpmk.tsu.ru

ISSN 1865-0929   e-ISSN 1865-0937
ISBN 978-3-319-13670-7   e-ISBN 978-3-319-13671-4
DOI 10.1007/978-3-319-13671-4
Springer Cham Heidelberg New York Dordrecht London

Library of Congress Control Number: 2014955253

Typesetting: Camera-ready by author, data conversion by Scientific Publishing Services, Chennai, India

Printed on acid-free paper

Springer is part of Springer Science+Business Media (www.springer.com)

Preface

The series of scientific conferences "Information Technologies and Mathematical Modelling" (ITMM) started in 2002. In the beginning, it had the status of a national conference or national conference with international participation. In 2012, it was named after A. F. Terpugov, an outstanding scientist of the Tomsk State University, a leader of the famous Siberian school on applied probability, who was one of the first organizers of the conference.

Today we observe the process of globalization of the sciences involving Russian scientists into global science scene. This is why the ITMM conference was given an international status.

Traditionally, the conference has from eight to 12 sections in various fields of mathematical modelling and information technologies. A strong focus is on applied problems in education, economics, technology and management. Throughout the years, the sections on probabilistic methods and models, queueing theory, telecommunication systems, and software engineering have been the most popular at the conference.

This volume presents new results in the theory of random processes, methods of study of queueing systems, probabilistic methods and models, analysis of telecommunication systems and networks, software engineering, and others. It is targeted at specialists in probabilistic theory, random processes, mathematical modeling, as well as engineers engaged in logical and technical design and operational management of telecommunication and computer networks, contact centers, databases, software design, etc.

November 2014   Anatoly Nazarov

Organization

The ITMM conferences are organized by the Anzhero-Sudzhensk Branch of Kemerovo State University together with Kemerovo State University, National Research Tomsk State University, Kemerovo Scientific Centre of SB RAS (the Siberian Branch of Russian Academy of Sciences), and the Institute of Computational Technologies of SB RAS since 2002.

In 2014 the XIII International Scientific Research and Practice Conference was organized named after A. F. Terpugov "Information Technologies and Mathematical Modelling."

International Program Committee

A. Dudin, Belarus, Chair
A. Nazarov, Russia, Co-chair
R. Yakupov, Russia, Co-chair
I. Atencia, Spain
S. Chakravarthy, USA
B. Choi, Korea
T. Czachorski, Poland
N. Danilov, Russia
D. Efrosinin, Austria
A. Gortsev, Russia
K. Al-Begain, UK
Yu. Kharin, Belarus
C. Kim, Korea
A. Krishnamurthy, India
U. Krieger, Germany

A. Latkov, Latvia
Q. Li, China
G. Medvedev, Belarus
A. Melikov, Azerbaijan
R. Nobel, The Netherlands
E. Orsingher, Italy
M. Pagano, Italy
G. Saporta, France
K. Samuylov, Russia
Y. Shokin, Russia
J. Sztrik, Hungury
N. Temirgaliev, The Netherlands
H. Tijms, The Netherlands
O. Tikhonenko, Poland

Local Organizing Committee

A. Dudin, Chair
A. Nazarov, Co-chair
R. Yakupov, Co-chair
O. Galazhinskaya
I. Garayshina
B. Gladkikh
E. Glukhova
R. Ivanovskiy
V. Ivnitskii
V. Kochetkov
Yu. Kostyuk

T. Lyubina
A. Moiseev
S. Moiseeva
V. Poddubnyi
V. Rykov
S. Senashov
A. Shkurkin
S. Suschenko
K. Voytikov
O. Zmeev

Table of Contents

A Novel Framework for the Design and Development of Software Routers

Davide Adami[1], Stefano Giordano[2], Michele Pagano[2], and Luis G. Zuliani[3]

[1] CNIT Research Unit - University of Pisa, Pisa, Italy
[2] Dept. of Information Engineering, University of Pisa, Pisa, Italy
[3] Link Protect GmbH, Münchner Straße 92, 85614 Kirchseeon, Germany
m.pagano@iet.unipi.it

Abstract. Flexibility and programmability are key features of open networking platforms to effectively design, develop and assess new Future Internet architectures. This paper introduces a novel framework, based on open source software and PC hardware, that defines all the building blocks necessary to provide a full set of advanced networking capabilities (including QoS support and Traffic Engineering). The framework implementation also takes into account software reuse by facilitating maintenance and customization of the network protocol stack in software routers. The proposed open framework is also available as live distributions, which allow network designers to take advantage of its capabilities with a short learning curve.

Keywords: software routers, open frameworks, QoS, traffic engineering.

1 Introduction

Originally conceived as an experimental packet-switched network, the Internet has become a global communication infrastructure. As scalability and robustness are critical requirements of this evolutionary process, academia and industry are increasingly relying on field trials in order to develop and assess novel network solutions [16]. In this context, PCs started to be used as platforms for the development of router prototypes thanks to the availability of powerful and cheap commodity hardware and the wide spread of open source software. When playing such a role, PCs are called SRs (Software Routers) [11].

Usually, SRs are based on open source, Unix-like OSes, such as GNU/Linux and BSD variants. These systems, in addition to routing and forwarding capabilities, may offer complete solutions to filter and manipulate layer 2 and above information, with a level of flexibility which is comparable only to costly, commercial routers. Moreover, SRs have programmability features, enabling the development of third party extensions and completely new functionalities as well as fine-tuning of existing ones.

During the last years, SRs have started playing an important role even in the market. Leading network vendors are exploring the SOHO (Small Office/Home

A. Dudin et al. (Eds.): ITMM 2014, CCIS 487, pp. 1–10, 2014.

Office) market with embedded SRs, using mixed open and closed source solutions (for the OS and protocol stack, respectively [9]). More recently, SRs are being used for the deployment of NGNs (Next Generation Networks) and also as learning tools (see, for instance, [6]). Despite their increasing popularity, SRs have relevant weaknesses. On the one hand, off-the-shelf SRs are often shipped with closed networking stack that limits (or even prevents) the development of new capabilities. On the other hand, completely open SRs are often developed for specific purposes, like the optimization of an existing routing protocol or the design of a brand new scheduler. These enhancements tend to be developed as ad-hoc, intrusive hacks in the original source code, making its maintenance and reuse a complex and time-consuming task. QoS (Quality of Service) support is a concrete example of this deficiency: there is a large number of open source tools to enforce QoS, but a clear lack of orchestrating procedures to achieve a specific behaviour.

Taking into account the above-mentioned weaknesses, this paper deals with the design and development of a novel open framework for SRs with advanced network functionalities. The goal is twofold: to provide flexible tools for the configuration and management of QoS–aware networks and to allow the evaluation of NGN architectures through fully-operational trials. In more detail, the framework is exclusively based on free and open source software, and delivers out-of-the-box automatic management of virtual circuits and traffic differentiation, by effectively combining the DS (DiffServ) and MPLS (MultiProtocol Label Switching) IETF architectures, still the most advanced QoS and TE (Traffic Engineering) solutions for IP networks.

Indeed, MPLS [18] allows the management of pseudo-virtual circuits over a wide range of layer 2 and 3 protocols, including IP: in an MPLS network the IP header is used exclusively at edge routers to map packets into a FEC (Forwarding Equivalence Class). All packets associated to a given FEC are "tagged" with an identical label; for instance, in an IP over Ethernet network, labels are inserted between the Ethernet and the IP packet headers, in a structure called Shim Header. Inside an MPLS cloud, only labels are used to forward packets, according to label forwarding tables (much simpler than IP routing tables). The path that all packets belonging to the same FEC use to traverse the network is called LSP (Label Switched Path) and the network nodes are denoted as LSRs (Label Switching Routers). TE capabilities of MPLS allow to avoid congestion both in the steady–state and in failure scenarios by establishing LSPs along links with available bandwidth and providing resilience in case of failure by means of built–in mechanisms, such as link protection and fast reroute.

DS [12] maps the incoming packets in classes, and resource reservation is performed on a per-class basis. These service classes, known as PHBs (Per Hop Behaviours), specify how packets must be treated by a router (i.e., how a router must distribute its resources, prioritizing certain classes on detriment of others). Packets at ingress routers are classified in one of the defined PHBs: EF (Expedited Forwarding), AF (Assured Forwarding) and Default PHB, typically used for Best Effort (BE) traffic, by setting the suitable DSCP (DS Code Point) in the IP header TOS field, renamed DS field. While DS assures data delivery with "relative" QoS, there is no TE control in DS networks.

In more detail, two different solutions, DiffServ over MPLS [15] and DS-TE (DiffServ-aware MPLS TE, [13,14]) have been standardized to make DS and MPLS interwork. Thus, DS packets are transported inside LSPs and the network is enabled to perform TE while QoS is guaranteed. More specifically, DiffServ over MPLS allows the creation of TE–LSPs that carry DS marked IP packets from a single or multiple classes (L-LSP and E-LSP, respectively), with assured bandwidth. However, the main problem is that such integration of DS and MPLS is not able to guarantee end-to-end QoS to TE-LSPs on a per-class basis under any operating conditions because QoS is only provided at node level and MPLS is unaware of traffic classes. DS-TE makes MPLS aware of traffic classes, allowing end-to-end resource reservation with traffic class granularity and providing the fault tolerance of MPLS at traffic class level. To achieve such differentiated treatment, in [14] the concept of CT (Class Type) is introduced. A CT is the set of traffic trunks crossing a link that is governed by a specific set of BCs (Bandwidth Constraints). CTs are used for link bandwidth allocation, constraint based routing and admission control.

The rest of the paper is organized as follows. Section 2 describes the general features of the proposed open framework architecture; then, Section 3 and Section 4 detail the Data Plane and Control/Management Plane functionalities, respectively. Finally, conclusions are drawn in Section 5.

2   Open Framework Architecture and Features

Our open framework (see Figure 1) is based on a sharp separation among the functionalities of data, control and management planes. The next subsections detail these functionalities and provide an overview of the key features of DS-MPLS nodes, which will be denoted in the following as DS-LSRs.

2.1   Data Plane

The data plane is responsible for data forwarding in accordance with the rules established by the control and management planes. Its components are briefly described in the following.

1. **Traffic Control:** enforces the QoS of data flows. It is responsible for performing label–based packet switching and includes the following entities:
   - Policy Enforcer: available in edge DS-LSRs, it filters the incoming packets, admitting only those from authorized parties and at rates that are in conformance with Service Level Agreements (SLAs).
   - Classifier: processes incoming packets in a pre-routing phase, preparing them to be forwarded using class-specific resources.
   - Lookup: instead of destination-based routing, it uses multiple routing tables to route packets in LSPs, also taking into account other parameters such as DS classes.

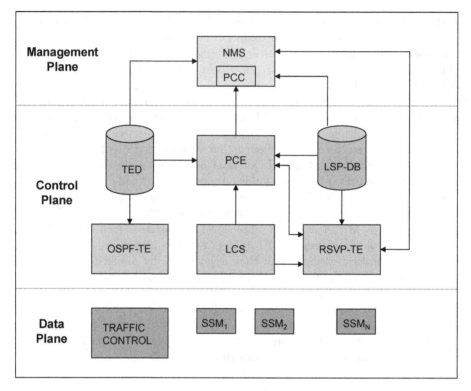

Fig. 1. DS-MPLS Open Framework architecture

- Forwarder: it uses a set of schedulers to forward MPLS packets according to DS classes specification on a per-LSP basis.
2. **SSMs (Service-Specific modules):** are external programs developed to bring intelligence to the network. This way, other than just forwarding packets, SRs can elaborate data, simplifying the creation of upper layer applications and overlays. SSMs can perform basic services, such as cryptography and compression, as also more advanced functions, such as video transcoding.

2.2 Control Plane

The key part of the open framework architecture is the control plane, that has been completely designed and implemented to support both DiffServ over MPLS and DS-TE. Roughly speaking, it collects the network state information, computes the suitable paths for the user traffic and sets up the corresponding LSPs. Its main components are the following:

1. **TED (Traffic Engineering Database):** describes the properties and state of network links and routers. It is defined by using XML.
2. **LSP-DB (LSP Database):** details the characteristics of the active LSPs. As the TED, it is defined by using XML.

3. **PCE (Path Computation Element):** is responsible for processing LSP setup requests (in case of bidirectional circuit requests, the PCE computes a path that is able to accommodate two LSPs in opposite directions). This module may employ different Path Computation Algorithms, working in a centralized or distributed manner [10]. In case of DS-TE architecture, it supports MAM (Maximum Allocation Model) and RDM (Russian Dolls Model) bandwidth allocation models [17]. After the path computation phase, LSPs are established.

4. **LCS (LSR Control System):** configures all routers involved in LSP setup or teardown requests. It uses a secure, centralized approach to contact and configure DS-LSRs. The LCS is formed by two distinct components: the LCA (LSR Control Agent) and the LLEA (LSR Local Enforcement Agent). The LCA is always in the same node as the PCE, no matter if the node is a DS-LSR or a host external to the network. It is responsible for receiving the LSP information from the PCE and translating it into configuration commands for all the routers in the path. The LCA controls all the DS-LSR specific information such as the labels to be configured in the interfaces. Once the set of commands that each router must execute to configure the LSP(s) are defined, the LCA contacts the LLEAs entities in every DS-LSR along the path. Each LLEA is responsible for locally executing the set of commands to configure the LSP(s), and then to report back to the LCA the state of the operation.

5. **RSVP-TE:** is used to perform robust, automatic provisioning of LSPs.

6. **OSPF-TE:** is responsible for maintaining, updating and synchronizing the TED across the DS domain.

2.3 Management Plane

The management plane provides the system interfaces to control and query all other entities. It consists of the following components:

1. **NMS (Network Management System):** provides graphical and command-line user interfaces for network resources administration. NMS users can setup and teardown LSPs, manage FEC-LSP associations, and analyse the operational state of DS-LSRs, links, protocols and virtual circuits.

2. **PCC (Path Computation Client):** generates LSP setup requests and sends them to the PCE.

3 Data Plane Functionalities

The open framework data plane uses both standard GNU/Linux routing and traffic control functionalities, as well as a number of experimental tools to add new advanced capabilities, such as label-switching routing. Usually, these external add-ons require intrusive kernel patching and hacking of userspace tools. In the open framework to build up a full-featured data plane, customized patches have been developed to circumvent version mismatches.

3.1 LSP Establishment

MPLS support in the open framework is granted by the MPLS for Linux project [4]. It consists of several patches in the Linux kernel and in the following userspace tools: iproute2 (traffic control in Linux) [2], iptables (layer 3 packet filtering) [5], and ebtables (layer 2 cell/frame filtering) [3]. A new tool called mpls allows to manage LSPs.

To establish E-LSP and L-LSPs, our open framework takes advantage of an mpls tool feature that allows to set the EXP field of the shim header and the TC index field of the packet buffer descriptor (internal to the DS-LSR) in function of the DSCP value in the IP header. The TC index plays a key role when assigning MPLS packets to schedulers.

To setup an LSP, the following steps are mandatory:

- **ingress and core DS-LSRs:** creation of an entry in the NHLFE (Next Hop Label Forwarding Entry) table, with the corresponding mapping from DSCP to EXP and TC index;
- **core and egress DS-LSRs:** creation of an entry in the ILM (Incoming Label Map) table, to map incoming and outgoing labels;
- **core DS-LSRs:** cross-connection between ILM and NHLFE entries.

3.2 FEC to LSP Binding

Since the standard GNU/Linux networking stack was not designed with MPLS support in mind, routing decisions can only be taken based on the information contained in the IP routing table. The availability of a single routing table is a significant limitation in case TE policies should be applied and, for instance, two packets with the same source and destination IP addresses should be sent towards the destination by means of different LSPs (maybe configured on the same outgoing interface).

On the ingress DS-LSR, the FEC-to-NHLFE table allows to associate each incoming packet belonging to a specific FEC with a set of instructions (NHLFE) that indicate how to forward the packet to the next LSR. Nevertheless, due to limitations of the userspace Linux tools, IP packets can not be directly assigned to FECs at ingress DS-LSRs.

To overcome these issues, the open framework provides a great amount of flexibility regarding FEC definition. Almost every field of layers 2, 3 and 4 packet headers can be used to define a FEC. Moreover, the framework introduces multiple routing tables, whose lookup priorities are higher with respect to the default IP routing table.

In more detail, by using forward marks, each FEC is associated with a specific routing table, which usually contains a single entry related to a specific LSP (in fact, its NHLFE). This solution enables the finest QoS routing granularity, although it is also the most expensive in terms of resources. The open framework also allows the use of multiple NHLFE entries per table.

3.3 Hierarchical Scheduling

The open framework provides QoS support through a modular tree of hierarchical packet schedulers, available on every interface of DS-LSRs. The hierarchical scheduler tree is illustrated in Figure 2.

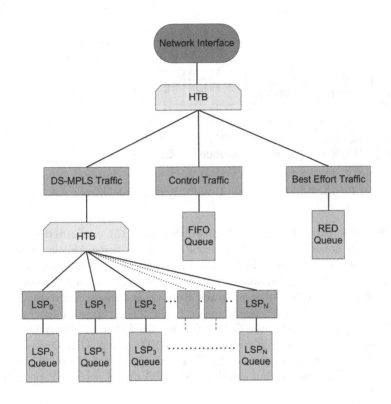

Fig. 2. Hierarchical scheduler tree

The reservable bandwidth for all interfaces is initially divided between three types of traffic: DS-MPLS, control and BE. By reserving a fraction of bandwidth to control traffic, the control plane is isolated from the data plane (out-of-band approach). Also, by reserving a small amount of bandwidth to BE traffic, a minimal service level is guaranteed, avoiding complete starvation of the lowest class. To enable bandwidth separation, the HTB (Hierarchical Token Bucket) [1] packet scheduler is used. While the interface bandwidth share reserved to control traffic is managed by a FIFO queue, the BE traffic is sent to a RED (Random Early Detection) queue. The DS-MPLS traffic bandwidth share is managed by another HTB scheduler, which is responsible for guaranteeing the nominal bandwidth to LSPs.

When establishing a new LSP, a specific hierarchical scheduler subtree is configured according to the type of the LSP. The default scheduler subtrees for E-LSPs and L-LSPs (carrying AF and EF traffic) are depicted in Figure 3.

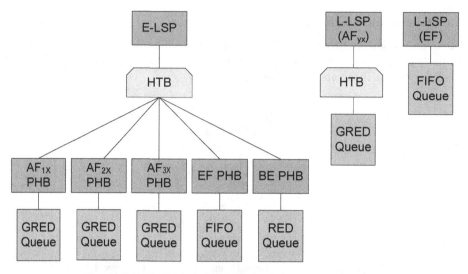

Fig. 3. E-LSP (left) and L-LSP (right) scheduler subtrees

The E-LSP subtree divides the bandwidth among the configured PHBs. According to our default configuration, each one of the three AF classes can use up to 100% of the LSP bandwidth, with a guaranteed bandwidth equal to 25% of the LSP capacity. A guaranteed level of 25% of total bandwidth is offered to the EF PHB, with no borrowing allowed from other classes. There is no bandwidth reservation for the BE traffic inside LSPs. The drop precedences for the AF PHBs are enforced using the GRED (Generalized Random Early Detection) queueing discipline, a multiple virtual queues RED variant. GRED uses the TC index (which is directly related to the EXP shim header field) to discriminate AF packets. EF packets have a small buffer FIFO queue, while BE packets are managed by a RED queue.

L-LSP scheduler subtrees are simpler: for AF and EF traffics, GRED and FIFO queue are used, respectively.

To assign every MPLS packet to the correct subtree according to the label field of its shim header, a specially crafted TC filter rule (`tc` is one of the main `iproute2` tools) is used.

4   Functionalities of Control and Management Planes

The open framework control and management planes provide all the functionalities necessary to automatically setup, monitor and teardown LSPs. The key entity is the PCE, which communicates with the PCC integrated in the NMS. The link metrics usable by routing algorithms are the reservable bandwidth, the maximum delay and a user-defined cost (for instance, monetary cost or power consumption). The TED as well as signalling and routing protocols used by MPLS were extended to support the non-standard metrics.

An LSP setup request must contain the following information: ingress and egress DS-LSRs, bandwidth, and LSP type (Non specified; E-LSP; L-LSP carrying BE, AF_{yx} or EF traffic; LSP carrying TE-Class[0-7] traffic). Optionally, it can also specify the maximum delay, the user-defined cost and the path computation algorithm. Moreover, if a bandwidth allocation model is used (MAM or RDM), it is also necessary to specify the priority and the preemption of the LSP. Finally, when a bidirectional circuit is requested, the PCE computes two LSPs with the same path and properties, but with opposite directions.

An LSP teardown request must specify only the identifier of the LSP to be removed. As already highlighted in section 2.1, due to the centralized architecture of the LSP provisioning process, the PCE relies on the LCS to enforce the configuration of the DS-LSRs involved.

Moreover, to properly flood TE-LSAs describing TE link properties, the OSPF implementation of the Quagga [7] suite of protocols is used.

Finally, the RSVP-TE daemon from the TEQUILA project [8] has been enhanced to become an integral part of the framework.

5   Conclusion

Programmability and virtualisation will be the key features of next generation routers in order to enhance the intelligence of the network and to offer a more effective support for new emerging applications. Since the lack of flexibility is currently the Achilles heel of high-end network devices, partially or completely open SRs are becoming more and more valuable. Taking into account this scenario, the open framework described in the paper is completely based on both open source software and commodity PCs. More specifically, it defines and implements all the building blocks necessary to integrate DS domains over MPLS tunnels with QoS guarantees and provides a full set of advanced networking functionalities (including QoS and TE).

As programmability, flexibility and ease of use are primary features of the open framework, NGNs designers can quickly benefit from virtual circuits with bandwidth guarantees and traffic differentiation, as well as from the flexibility of service-specific modules. Moreover, unlike other similar projects, the open framework does not limit to just shipping software, but provides a cohesive integration of networking tools to enable even greater capabilities.

Finally, the live distribution releases allow to easily use the open framework functionalities.

References

1. HTB Linux kernel implementation (July 2014),
   http://luxik.cdi.cz/~devik/qos/htb
2. iproute2 - Linux Foundation (July 2014), http://www.linuxfoundation.org/
   collaborate/workgroups/networking/iproute2
3. Linux Ethernet bridge firewall tables (July 2014),
   http://ebtables.sourceforge.net

4. MPLS for Linux project (July 2014),
   http://sourceforge.net/projects/mpls-linux/support
5. Netfilter/iptables project (July 2014), http://www.netfilter.org
6. Netkit (July 2014), http://wiki.netkit.org
7. Quagga Software Routing Suite (July 2014), http://www.quagga.net
8. TEQUILA project (July 2014), http://www.ist-tequila.org
9. Vyatta Core (July 2014), http://www.vyatta.org
10. Adami, D., Callegari, C., Giordano, S., Pagano, M.: Distributed and Centralized Path Computation Algorithms: Implementation in NS2 and performance comparison. In: IEEE International Conference on Communications, ICC 2008, pp. 17–21 (May 2008)
11. Bianco, A., Finochietto, J., Mellia, M., Neri, F., Galante, G.: Multistage switching architectures for software routers. IEEE Network 21(4), 15–21 (2007)
12. Blake, S., Black, D., Carlson, M., Davies, E., Wang, Z., Weiss, W.: An Architecture for Differentiated Service. RFC 2475 (Informational) (December 1998), http://www.ietf.org/rfc/rfc2475.txt, updated by RFC 3260
13. Faucheur, F.L.: Protocol Extensions for Support of Diffserv-aware MPLS Traffic Engineering. RFC 4124 (Proposed Standard) (June 2005), http://www.ietf.org/rfc/rfc4124.txt
14. Faucheur, F.L., Lai, W.: Requirements for Support of Differentiated Services-aware MPLS Traffic Engineering. RFC 3564 (Informational) (July 2003), http://www.ietf.org/rfc/rfc3564.txt, updated by RFC 5462
15. Faucheur, F.L., Wu, L., Davie, B., Davari, S., Vaananen, P., Krishnan, R., Cheval, P., Heinanen, J.: Multi-Protocol Label Switching (MPLS) Support of Differentiated Services. RFC 3270 (Proposed Standard) (May 2002), http://www.ietf.org/rfc/rfc3270.txt, updated by RFC 5462
16. Haßlinger, G., Nunzi, G., Meirosu, C., Fan, C., Andersen, F.U.: Traffic engineering supported by Inherent Network Management: analysis of resource efficiency and cost saving potential. International Journal of Network Management 21(1), 45–64 (2011), http://dx.doi.org/10.1002/nem.770
17. Lai, W.: Bandwidth Constraints Models for Differentiated Services (Diffserv)-aware MPLS Traffic Engineering: Performance Evaluation. RFC 4128 (Informational) (June 2005), http://www.ietf.org/rfc/rfc4128.txt
18. Rosen, E., Viswanathan, A., Callon, R.: Multiprotocol Label Switching Architecture. RFC 3031 (Proposed Standard) (January 2001), http://www.ietf.org/rfc/rfc3031.txt

Land Cover Change Analysis
Using Change Detection Methods

Anton Afanasyev[1], Alexander Zamyatin[1,2], and Pedro Cabral[3]

[1] National Research Tomsk Polytechnic University, Tomsk, Russia
[2] National Research Tomsk State University, Tomsk, Russia
[3] Instituto Superior de Estatística e Gestão de Informação,
ISEGI, Universidade Nova de Lisboa, 1070-312 Lisboa, Portugal
{afanasyevaa,zamyatin}@tpu.ru, pcabral@isegi.unl.pt

Abstract. Land cover change detection analysis by remote sensing currently is used for monitoring and accounting. The existing diversity of change detection methods and the absence of the conventional way of change detection methods choice in each case do not allow unsupervised analysis of large amounts of data. A proposed multi-stage approach, involves the use of multiple change detection methods and thresholding functions, as well as a way to assess the scale of changes. This approach allows to select data and deserves expert review from a large amount of incoming data. On the example of Portugal's land cover change analysis project, the paper shows result fragments of change detection which were obtained by different methods and threshold functions. The assessment of changes scale was concluded to see whether further analysis is necessary for landscape fragments.

Keywords: Remote sensing, change detection, Landyn project.

1 Introduction

Amounts of data coming from different remote sensing (RS) satellites and accumulated in the relevant archives are increasing every day and can run into terabytes daily. At the same time, the subject of special interest in such archives is *multi-temporal data*, allowing detection of land cover changes and their dynamics in different territories. However, due to the large amount of data in these archives and complexity of change detection analysis, identification of changes occurring in landscape and defining their scope, speed and the short- and long-term forecasting of these phenomena is extremely difficult and sometimes impossible in practice.

So, ways of solving the problem of land cover change detection on aerospace images (AI) are sought. The results of change detection can be extremely useful in monitoring of emergencies flow and early (preventive) detection of its causes. Searching and selecting appropriate ways and methods to detect the changes in each case have not been yet trivial[1–3]. In addition, the task is complicated by a wide range of potentially available methods of change detection based on

A. Dudin et al. (Eds.): ITMM 2014, CCIS 487, pp. 11–17, 2014.

different mathematical apparatus and the absence of formalized procedures of such search and selection.

There are several basic well-known approaches to the land cover change detection by the RS data (algebraic, transformation, classification)[3]. Each of the approaches alone and its combinations are characterized by its own set of advantages and disadvantages that should be considered under different data distortion, sometimes significantly impede solution of changes detection problem. The sources of such distortions include satellite imagery conditions, atmospheric conditions, illumination of the study area, soil moisture and other factors, which are not always eliminated by data correction[2, 3]. Furthermore, we should take into account different degree of observed changes in different band of multi-channel AI.

Today, the problem change detection usually solved without significant empirical study, and consideration of the aforementioned disturbing factors. However, there are some fairly successful attempt to analyze the land cover change detection methods allowing identifying areas of practical applicability and efficiency of these methods[4].

For the development of features to implement in an extremely large amount of data of quickly land cover changes detection and searching for areas with significant dynamic of change, we propose to discuss the appropriate multi-stage approach and some details of its practical use.

2   Multi-stage Approach

Stage 1. Recharging archive of RS data with new images and formation of multi-temporal series of images for the area of interest -
$$\mathbf{I}^{t1}, \mathbf{I}^{t2}, ..., \mathbf{I}^{tn-1}, \mathbf{I}^{tn}; \mathbf{I}^{'t1}, \mathbf{I}^{'t2}, ..., \mathbf{I}^{'tn-1}, \mathbf{I}^{'tn}, ...; \mathbf{I}^{"t1}, \mathbf{I}^{"t2}, ..., \mathbf{I}^{"tn-1}, \mathbf{I}^{"tn}; ...$$

Stage 2. Change detection analysis in the existing archive of RS data in pairs of images ($\mathbf{I}^{t1} - \mathbf{I}^{tn}$, $\mathbf{I}^{'t1} - \mathbf{I}^{'tn}$, etc.), where $\mathbf{I}^{t1}$ - initial (first) image of the observed period, and $\mathbf{I}^{tn}$ - most actual (last) image of the observed period.

In general, the solution of the change detection problem requires two multi-temporal AI of the same landscape fragment, in the time period of interest. The images that capture the changes are represented in form of two arrays: $\mathbf{I}^{t1} = i^1_{xyz}, x = 1..H, y = 1..W, z = 1..M$ and $\mathbf{I}^{t2} = i^2_{xyz}, x = 1..H, y = 1..W, z = 1..M$, where H and W - the number of elements in rows and columns in initial AIs, M - the number of bands (channels) of the images. Simple form of change detection process can be represented as follows[3]:

- on the basis of multi-temporal data arrays $\mathbf{I}^{t1}$ and $\mathbf{I}^{t2}$ difference image $\mathbf{D} = d_{xy}$, is formed with some change detection method. Each element of $\mathbf{D}$ reflects the degree (probability) of change;
- getting matrix of changes $\mathbf{B} = b_{xy}$, with the thresholding of $\mathbf{D}$; values of matrix $\mathbf{B}$ reflect the presence or absence of significant change at each point; the threshold value may be fixed empirically or calculated using a specific functions.

Details of the most common change detection methods are listed below.

The result of the change detection process usually varies significantly depending on the threshold calculation characteristics, so details of threshold searching should be given. Methods of the threshold value calculation using a different mathematical approaches[6, 7, 12], but the most common are techniques with the matrix $\mathbf{D}$ histogram analysis.

For example, method of Yanni is based on finding histogram peaks[8]. Otsu method determines the threshold value by reducing the variance in the two classes of values - below and above of threshold[7, 15]. In the method of Kittler-Illingworth[6, 15] which assumes that normalized histogram consist of the two components (corresponding to the two classes), each of them has a normal distribution with mean values $\mu_1(\tau)$, $\mu_2(\tau)$ and standard deviations $\sigma_1(\tau)$, $\sigma_2(\tau)$. Threshold τ chosen in such a way to minimize the area of the intersection region of these two distributions in histogram. Also for the values separation in $\mathbf{D}$ into two classes unsupervised classification methods can be used[12].

In the context of the priori information lack about the characteristics of the analyzed image histograms, Kittler-Illingworth and Otsu methods received steady spread in practice, than we choose them as the main methods of threshold searching[7].

Stage 3. Selection multi-temporal series of original images, which scale of change deserves special attention, for further analysis.

To assess the scale of change we identify areas with a high concentration of changes using a "sliding window" of order r. Area covered with "sliding window" considered as high concentration area when concentration of changes d_i above a certain threshold t, where $d_i = n_i/S^{\mathrm{w}}$, n_i - amount of changed pixels in the "window", $S^{\mathrm{w}} = (2 \cdot r + 1)^2$ - the area of "sliding window", and $i = 1..N$, where N - number of regions with a high concentration of change. As an integral criterion of change scale we will use parameter $M^{\mathrm{d}} = \sum_{(i=1)}^{N} d_i \cdot S^{\mathrm{w}}/S$, where S - area of the region under consideration.

Choice of algorithm parameters, such as the order of the window r and the threshold of concentration t deserve individual detailed consideration, so we choose appropriate values empirically $t = 0.5$, $r = 20$.

Stage 4. Expert analysis of selected multi-temporal images, reflecting the trend of interest (fixed and projected) of land cover changes.

As a result of a multi-stage approach, researcher have opportunity to analyze significantly reduced set of multi-temporal data on the final stage, allowing more quickly, carefully and thoroughly assess the existing land cover trends.

3   Change Detection Methods

Image Difference method is used more often than others because of its ease of implementation and performance[2, 3]. Each cell of $\mathbf{D}$ contains absolute value of the respective cells $\mathbf{I}_1$ and $\mathbf{I}_2$ difference, $d_{xyz} = |i^1_{xyz} - i^2_{xyz}|$. The smaller the value of a particular element of the matrix, the less probability that a significant change has occurred in corresponding area.

Ratio method is also relatively wide used method, which differs from the previous one, with using division instead of difference $d_{xyz} = |i^1_{xyz}/i^2_{xyz}|$. In this case, $d_{xyz} \in (0, +\infty)$ and the closer the value d_{xyz} to 1, the less likely in the corresponding region significant change occurred.

Change vector analysis method considers the cell values of matrices $\mathbf{I}_1$ and $\mathbf{I}_2$ with fixed coordinates x and y as the vectors (M-dimensional) coordinates: $V^1_{xy} = \{i^1_{xy1}, i^1_{xy2}, \ldots, i^1_{xyM}\}$ and $V^2_{xy} = \{i^2_{xy1}, i^2_{xy2}, \ldots, i^2_{xyM}\}$. Thus, the values of the difference matrix D cells can be obtained as the Euclidean distance between the corresponding vectors. The advantage of this method is that it allows you to receive a two-dimensional matrix D without additional channels fusion. Also, this method is sometimes used to classify types of changes, since this case, is possible to find M-dimensional vector of changes in addition to the value of changes[3, 9, 10].

The *principal components analysis* method uses statistical Karhunen-Loeve transform to reduce the dimension[11]. As a result, we obtain the transformed matrices $\mathbf{I}_1$ and $\mathbf{I}_2$. In most cases after conversion only one or two principal components leaved[3]. This technique also belongs to the most effective and commonly used despite of low performance.

The method *chi-square* mean transformation $Y = (X - M^x)^T \cdot \Sigma^{-1} \cdot (X - M^x)$ [3, 16], where X-vector of difference $[\mathbf{D}_{xy1}, \mathbf{D}_{xy2}, \ldots, \mathbf{D}_{xyM}]^T$, M^x vector of means, and Σ^{-1}- co-variance matrix for $\mathbf{D}$.

In methods using *regression*, the difference is defined as the difference between the values of cells $\mathbf{I}_2$ and calculated values based on the regression function (values i^1_{xyz} are considered as arguments, and the corresponding i^2_{xyz} as the value of the function)[12, 16]. On the first step form of the regression function $f(x)$ and its coefficients is selected, with some method (e.g., the least squares method). Further, as described above, apply the formula $d_{xyz} = |f(i_{1_{xyz}}) - i_{2_{xyz}}|$.

Unsupervised clustering method, provides preliminary clustering images $\mathbf{I}_1$ and $\mathbf{I}_2$ to get $\mathbf{I}^c_1$ and $\mathbf{I}^c_2$. And then finding the difference $d_{xyz} = |i_{1_{xyz}} - i_{2_{xyz}}|$. The main complexity of this method consist in selection the type and parameters of the clustering algorithm (such algorithms as k-means and ISODATA are used widely)[3, 9].

4   Case Study for Change Detection

Here are some details of the above-described multi-stage approach application for solving applied problems of land cover change detection in Portugal and identify fragments with areas of low and high scale of change, carried out within the framework of an international research project with the support of the Portuguese Landyn Science Foundation[13].

To study used random sampling area (fig. 1) for 1980, 1995 and 2010 (1279 pieces) provided by the General Directorate of planning in Portugal (Direo-Geral do Territrio, DGT), and have a total area of 499,596 hectares (about 6% of the territory). Each element matched area 2×2 km. The samples dataset in ESRI shapefile format in was converted into a raster format (TIFF) with a 100-meter spatial resolution and then imported into the IDRISI Selva software[14].

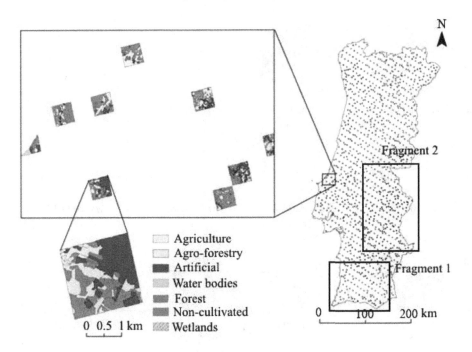

Fig. 1. The study area and test data

Landscape class structure of the study area is represented by two sets of data
- the integrated set of 7-landscaped classes and detailed set of 32 landscape
classes).

To improve the reliability of the change detection analysis to the whole dataset
of Portugal used change detection methods that implement the algebraic (image
difference and change vector analysis) and transformational (principal component and "chi" - square) approaches, also two threshold methods was applied,
that were mentioned above.

Figure 4 shows the fragments of results obtained by various methods of change
detection and threshold identification. It can be seen that the result of the change
detection procedure is strongly dependent on applied methods. Also, the proposed method of change scale assessment allows effectively detect the areas of
high concentration of changes deserving detailed consideration and differentiate
images having such areas of images with minor amounts of high concentration
areas.

After determining the data sets with a high scale changes implemented fourth
stage of approach in which these data are used for the simulation.

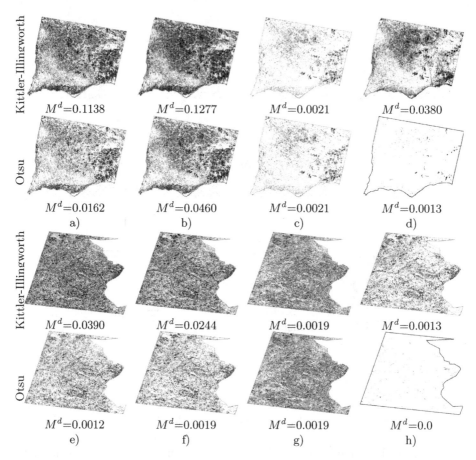

Fig. 2. Binary maps with high (a-d) and low (e-h) scale landscape changes, obtained by different methods of change detection: a) and e) image difference; b) and f) vector analysis; c) and g) the principal component analysis; d) and h)"chi" square, with 2 different thresholding methods: Kittler-Illingworth and Otsu

5 Conclusion

We have proposed multi-stage approach to the analysis of land cover dynamics using change detection methods to the tasks of expediting aerospace monitoring and retrieval of high scale landscape changes.

There are some of the international research project implementation details, which allowed testing proposed multi-stage approach on real data. Testing of multi-stage approach using algebraic and transformational approach change detection methods allowed us to determine areas with significant changes in a given time and to simplify the procedure for selecting data from a cumbersome array of multi temporal date in areas with significant landscape dynamics.

This work was supported by RFBR (grant 14-07-00027a) and the project LANDYN Portuguese Science Foundation (grant PTDC/CS-GEO/101836/ 2008).

References

1. Almutairi, A., Warner, T.A.: Change Detection Accuracy And Image Properties: A Study Using Simulated Data. Remote Sensing 2(6), 1508–1529 (2010)
2. Radke, R.J.: Image Change Detection Algorithms: A Systematic Survey. IEEE Trans. Image Process. 14(3), 294–307 (2005)
3. Lu, D., Mausel, P.: Change Detection Techniques. Remote Sensing 25(20), 2365–2407 (2004)
4. Zamyatin, A.V., Afanasyev, A.A.: The Applicability Analysis of The Approaches to the Identification of Land Cover Changes by Remote Sensing Data. Informational Technologies 4, 38–42 (2014) (in Russian)
5. Cabral, P., Zamyatin, A.: Advanced spatial Metrics Analysis in Cellular Automata Land Cover Change Modeling. DYNA, Sede Medellin 78(170), 42–50 (2011)
6. Kittler, J., Illingworth, J.: Minimum Error Thresholding. Pattern Recognition 19(1), 41–47 (1986)
7. ENVI Tutorial: Image change detection, http://www.exelisvis.com/portals/0/pdfs/envi/ImageChangeTutorial.pdf
8. Horne, E., Yanni, M.K.: New Approach to Dynamic Thresholding. In: EUSIPCO-9: European Conference on Signal Processing, Edinburg, UK, pp. 34–44 (1994)
9. Sohl, T.L.: Change Analysis in the United Arab Emirates: An Investigation of Techniques. Photogrammetric Engineering and Remote Sensing 65(4), 475–484 (1999)
10. Lambin, E.F., Strahlers, A.H.: Change-Vector Analysis in Multitemporal Space: A Tool to Detect Aand Categorize Land-Cover Change Processes Using High Temporal-Resolution Satellite Data. Remote Sensing of Environment 48(2), 231–244 (1994)
11. Shlens, J.: A Tutorial on Principal Component Analysis (2009), http://www.cs.uu.nl/docs/vakken/ddm/texts/Normal/pca.pdf
12. Dianat, R., Kasaei, S.: On Automatic Threshold Selection in Regression Method for Change Detection in Remote Sensing Images. In: The 4th International Symposium on Telecommunications, Tehran, Iran, pp. 1–6 (2008)
13. DGT. LANDYN Alteraes de Uso E Ocupao Do Solo Em Portugal Continental: Caracterizao, ForasMotrizes E CenriosFuturos. RelatrioAnual 2012-2013. Direo-GeraldoTerritrio, Lisboa (2013), http://landyn.isegi.unl.pt/reports/
14. Clark Labs. IDRISI (version Selva) (2013), http://www.clarklabs.org/products/idrisi.cfm
15. Otsu, N.: A Threshold Selection Method from Gray-Level Histograms. Automatica 11(285-296), 23–27 (1975)
16. Ridd, M.K., Liu, J.: A Comparison of Four Algorithms for Change Detection in an Urban Environment. Remote Sensing of Environment 63(2), 95–100 (1998)

Joint Probability Density of the Intervals Length of the Modulated Semi-synchronous Integrated Flow of Events and Its Recurrence Conditions*

Maria Bakholdina and Alexander Gortsev

Department of Operations Research, Faculty of Applied Mathematics and
Cybernetics, National Research Tomsk State University,
Tomsk, Russian Federation
`maria.bakholdina@gmail.com`

Abstract. This paper is focused on the problem of flow parameters estimation of the modulated semi-synchronous integrated flow of events, which is related to the class of doubly stochastic Poisson processes (DSPPs) with a a piecewise constant intensity process, that are typical for telecommunication networks. To solve this problem, first of all, the probabilistic characteristics of the flow should be found. In this paper we propose a technique for obtaining the formulas for calculating the probability density of the length of the interval between the neighboring flow events and the joint probability density of the length of the two neighboring intervals. Also we find the conditions of the flow recurrence.

Keywords: modulated semi-synchronous integrated flow of events, Markovian arrival process (MAP), doubly stochastic Poisson process (DSPP), flow state, flow parameters estimation, probability density, joint probability density.

1 Introduction

Due to the rapid evolution of computing and information technology during the last several decades, a new sphere of queueing theory applications – design and development of computer, telecommunication and other networks – has appeared. The use of mathematical methods developed in a queueing theory allows to find the quality characteristics of network components operation for various problems, such as estimation of probabilistic characteristics of switching and routing nodes, analysis of nodes buffer storage, the local and global flows management and so on.

It is worthwhile to note that the conditions of the real objects and systems operation are such that we can assert that the servers parameters are known and stable as time goes, but we can not tell this about the intensity processes and parameters of the input flows of events that come to the servers. Moreover,

---

* The work is supported by Tomsk State University Competitiveness Improvement Program.

A. Dudin et al. (Eds.): ITMM 2014, CCIS 487, pp. 18–25, 2014.

the intensities of the input flows usually vary within time, and frequently their changes are accidental. As a result, it is necessary to consider the mathematical models of doubly stochastic Poisson processes (DSPPs), which are characterized by having the number of events in any given time interval as being Poisson distributed, conditionally to another positive stochastic process called intensity [1]–[4].

There are two known classes of doubly stochastic flows of events. The first class contains the flows of events, which intensity process is a continuous random process [1], [2]. The second class contains flows, which intensity is a piecewise constant stationary random process with a finite number of states. These flows are typical for telecommunication networks. The flows of the second type were considered for the first time and independently presented by Basharin, Koko-tushkin and Naumov [5] and Neuts [6]. Basharin et al. named these flows as Markov chain (MC) arrival processes; Neuts – as Markov Versatile arrival processes (MVP). Since the early 1990s to date, these flows of events are called as the doubly stochastic flows of events or MAP-flows, or MC-flows [7]–[9].

As has been mentioned above, in the real situations the intensity process of the input flow of events may vary in time in a random way and it is typically unobservable. Also the flow parameters can be unknown. In such situations, the use of adaptive queueing systems, when the unknown parameters or states of the input flow are estimated during the system operation and the service procedure is changed correspondingly, seems to be more rational. That is why, the central problems faced when modeling these processes are: 1) flow states estimation on monitoring the time moments of the events occurrence (the filtering of the underlying and unobservable intensity process) [10]; 2) flow parameters estimation on monitoring the time moments of the events occurrence [11].

This paper is focused on the problem of flow parameters estimation of the modulated semi-synchronous integrated flow of events, which is related to the class of Markovian arrival processes (MAPs). To solve this problem, first of all, the probabilistic characteristics of the flow should be found. So in this paper we propose a technique for obtaining the formulas for calculation the joint probability density of the intervals length of the flow and find the conditions of its recurrence.

The rest of the paper is organized as follows. In Section 2 we present the modulated semi-synchronous integrated flow of events, which provides our framework. In Section 3 we derive the formulas for probability density $p(\tau)$ calculation and in Section 4 – for joint probability density $p(\tau_1, \tau_2)$ calculation. And finally, Section 5 contains the conditions of the flow recurrence.

2   Problem Statement

In this paper we consider the modulated semi-synchronous integrated flow of events (further flow of events), which intensity process is a piecewise constant stationary random process $\lambda(t)$ with two states 1, 2 (first, second correspondingly). In the state 1 $\lambda(t) = \lambda_1$ and in the state 2 $\lambda(t) = \lambda_2$ ($\lambda_1 > \lambda_2$). The duration of the process $\lambda(t)$ staying in the first (second) state is distributed according to the exponential law with parameter $\beta(\alpha)$. If at the time moment t the process is found in the first (second) state, then at the interval $[t, t + \Delta t)$, where Δt (hereinafter) is sufficiently small, with probability $\beta \Delta t + o(\Delta t)$ $(\alpha \Delta t + o(\Delta t))$ the sojourn time of the process $\lambda(t)$ in the first (second) state comes to the end and the process $\lambda(t)$ transits to the second (first) state. During the time interval when $\lambda(t) = \lambda_i$, a Poisson flow of events with intensity λ_i, $i = 1, 2$, arrives. Also at any moment of an event occurrence in state 1 of the process $\lambda(t)$, the process can change its state to state 2 with probability p $(0 \le p \le 1)$ or continue to stay in state 1 with complementary probability $1 - p$. I.e., after an event occurrence the process $\lambda(t)$ can change or not change its state from state 1 to state 2. The transition of the process $\lambda(t)$ from state 2 to state 1 at the moment of an event occurring in the second state is impossible. At the moment when the state changes from the second to the first state, an additional event in state 1 is assumed to be initiated with probability δ $(0 \le \delta \le 1)$. I.e., first the transition from state 2 to state 1 is made and thereafter an additional event is initiated or not. Such flows with additional events initiation are called integrated flows. Under the made assumptions we can assert that $\lambda(t)$ is a Markovian process. So the flow can be characterized by $\{D_0, D_1\}$, in terms of the rate matrices,

$$D_0 = \left\| \begin{matrix} -(\lambda_1 + \beta) & \beta \\ (1 - \delta)\alpha & -(\lambda_2 + \alpha) \end{matrix} \right\|, \quad D_1 = \left\| \begin{matrix} (1 - p)\lambda_1 & p\lambda_1 \\ \delta\alpha & \lambda_2 \end{matrix} \right\|. \tag{1}$$

Intensities of the process $\lambda(t)$ transitions from state to state without the event occurrence fill in the matrix D_0 in (1). Intensities of the process $\lambda(t)$ transitions from state to state with the event occurrence fill in the matrix D_1 in (1). Diagonal elements of the matrix D_0 are intensities of the process $\lambda(t)$ output from its states taken with the opposite signs. Fig. 1 shows the possible variant of the flow formation. Here 1, 2 are the states of the process $\lambda(t)$; additional events, that may occur in the first state at the moment of process $\lambda(t)$ transition from state 2 to state 1, are marked with letter δ; the flow events t_1, t_2, ..., are shown as circles.

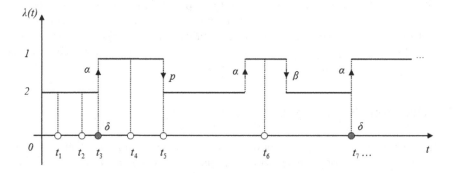

Fig. 1. The formation of the flow

It should be mentioned that the process $\lambda(t)$ is basically unobservable. We register only time moments t_1, t_2, ... of the flow events occurring. The process $\lambda(t)$ is considered in a steady-state conditions. So under the made assumptions we can assert that the sequence of the time moments t_1, t_2, ... corresponds to an embedded Markov chain, i.e. the flow has the Markov property if the evolution of the flow is considered from the time moment t_k, $k = 1, 2, ...$, of the event occurrence.

Denote by $\tau_k = t_{k+1} - t_k$, $k = 1, 2, ...$, the value of interval k length between the neighboring flow events. In a steady-state conditions we may take that the probability density of the interval k length is $p(\tau_k) = p(\tau)$, $\tau \geq 0$, for any k. Thereby we may also take that the time moment t_k is equal to zero, i.e. the moment of the event occurrence is $\tau = 0$. Now let (t_k, t_{k+1}), (t_{k+1}, t_{k+2}) be the neighboring intervals with the corresponding values of interval length $\tau_k = t_{k+1} - t_k$, $\tau_{k+1} = t_{k+2} - t_{k+1}$. Due to the stationary of the flow, the arrangement of the intervals on a time axis is arbitrarily. That is way we may consider the neighboring intervals (t_1, t_2), (t_2, t_3) with the corresponding values of interval length $\tau_1 = t_2 - t_1$, $\tau_2 = t_3 - t_2$; $\tau_1 \geq 0$, $\tau_2 \geq 0$, wherein $\tau_1 = 0$ corresponds to the time moment t_1 and $\tau_2 = 0$ corresponds to the time moment t_2 of the flow events arrival. The respective joint probability density is defined as $p(\tau_1, \tau_2)$, $\tau_1 \geq 0$, $\tau_2 \geq 0$.

3   The Expressions for Probability Density $p(\tau)$

Let us introduce into consideration the probabilities $p_{ij}(\tau)$ that there is no events at the interval $(0, \tau)$ and that at the time moment τ the value of the process $\lambda(t)$ is $\lambda(\tau) = \lambda_j$ in condition that at the time moment $\tau = 0$ the value of the process $\lambda(t)$ is $\lambda(0) = \lambda_i$, $i, j = 1, 2$. Then the probabilities $p_{ij}(\tau)$ satisfy the following systems of differential equations:

$$p'_{11}(\tau) = -(\lambda_1 + \beta)p_{11}(\tau) + \alpha(1 - \delta)p_{12}(\tau) , \qquad (2)$$
$$p'_{12}(\tau) = -(\lambda_2 + \alpha)p_{12}(\tau) + \beta p_{11}(\tau) ;$$

$$p'_{22}(\tau) = -(\lambda_2 + \alpha)p_{22}(\tau) + \beta p_{21}(\tau) , \qquad (3)$$
$$p'_{21}(\tau) = -(\lambda_1 + \beta)p_{21}(\tau) + \beta\alpha(1 - \delta)p_{22}(\tau) ;$$

with the boundary conditions: $p_{11}(0) = 1$, $p_{12}(0) = 0$; $p_{22}(0) = 1$, $p_{21}(0) = 0$. Solving the systems of equations 2, 3, we find the probabilities $p_{ij}(\tau)$, $i, j = 1, 2$:

$$p_{11}(\tau) = \frac{1}{z_2 - z_1}[(\lambda_2 + \alpha - z_1)e^{-z_1\tau} - (\lambda_2 + \alpha - z_2)e^{-z_2\tau})] , \qquad (4)$$

$$p_{12}(\tau) = \frac{\beta}{z_2 - z_1}(e^{-z_1\tau} - e^{-z_2\tau}), \quad p_{21}(\tau) = \frac{\alpha(1 - \delta)}{z_2 - z_1}(e^{-z_1\tau} - e^{-z_2\tau}) , \qquad (5)$$

$$p_{22}(\tau) = \frac{1}{z_2 - z_1}[(\lambda_1 + \beta - z_1)e^{-z_1\tau} - (\lambda_1 + \beta - z_2)e^{-z_2\tau}) , \qquad (6)$$

$$z_1 = \frac{1}{2}[\lambda_1 + \lambda_2 + \alpha + \beta - \sqrt{(\lambda_1 - \lambda_2 - \alpha + \beta)^2 + 4\alpha\beta(1 - \delta)}] , \qquad (7)$$

$$z_2 = \frac{1}{2}[\lambda_1 + \lambda_2 + \alpha + \beta + \sqrt{(\lambda_1 - \lambda_2 - \alpha + \beta)^2 + 4\alpha\beta(1 - \delta)}], \ 0 < z_1 < z_2 . \quad (8)$$

According to the definition of the flow we introduce the probability $p_{11}(\tau) \times \times e^{-\beta\Delta\tau}(1 - e^{-\lambda_1\Delta\tau})(1 - p) = p_{11}(\tau)\lambda_1(1 - p)\Delta\tau + o(\Delta\tau)$ – the joint probability that the process $\lambda(t)$ changes its state from the first state to the first one at the interval $(0, \tau)$ without the event occurring ($\lambda(0) = \lambda_1$, $\lambda(\tau) = \lambda_1$), and at the half-interval $[\tau, \tau + \Delta\tau)$ the duration of the first state does not come to the end, the event of the Poisson flow with intensity λ_1 arrives and the process $\lambda(t)$ remains in the first state. The joint probabilities take the following form for different i and j ($i, j = 1, 2$)

$$p_{11}(\tau)\lambda_1(1 - p)\Delta\tau + o(\Delta\tau) , \quad p_{12}(\tau)\alpha\delta\Delta\tau + o(\Delta\tau) ,$$

$$p_{11}(\tau)\lambda_1 p\Delta\tau + o(\Delta\tau) , \quad p_{12}(\tau)\lambda_2\Delta\tau + o(\Delta\tau) ,$$

$$p_{21}(\tau)\lambda_1(1 - p)\Delta\tau + o(\Delta\tau) , \quad p_{22}(\tau)\alpha\delta\Delta\tau + o(\Delta\tau) ,$$

$$p_{21}(\tau)\lambda_1 p\Delta\tau + o(\Delta\tau) , \quad p_{22}(\tau)\lambda_2\Delta\tau + o(\Delta\tau) .$$

The corresponding probability densities take the form

$$\widetilde{p}_{11}^{(1)}(\tau) = p_{11}(\tau)\lambda_1(1 - p) , \quad \widetilde{p}_{11}^{(2)}(\tau) = p_{12}(\tau)\alpha\delta ,$$

$$\widetilde{p}_{12}^{(1)}(\tau) = p_{11}(\tau)\lambda_1 p , \quad \widetilde{p}_{12}^{(2)}(\tau) = p_{12}(\tau)\lambda_2 ,$$

$$\widetilde{p}_{21}^{(1)}(\tau) = p_{21}(\tau)\lambda_1(1 - p) , \quad \widetilde{p}_{21}^{(2)}(\tau) = p_{22}(\tau)\alpha\delta ,$$

$$\widetilde{p}_{22}^{(1)}(\tau) = p_{21}(\tau)\lambda_1 p , \quad \widetilde{p}_{22}^{(2)}(\tau) = p_{22}(\tau)\lambda_2 .$$

Then the probability densities $\widetilde{p}_{ij}(\tau)$, that the process $\lambda(t)$ changes its state from the state i to the state j without the event occurrence at the interval $(0, \tau)$ and with the event occurrence at the time moment τ, can be written for different i and j ($i, j = 1, 2$) as

$$\widetilde{p}_{11}(\tau) = p_{11}(\tau)\lambda_1(1 - p) + p_{12}(\tau)\alpha\delta , \quad \widetilde{p}_{12}(\tau) = p_{11}(\tau)\lambda_1 p + p_{12}(\tau)\lambda_2 , \quad (9)$$
$$\widetilde{p}_{21}(\tau) = p_{21}(\tau)\lambda_1(1 - p) + p_{22}(\tau)\alpha\delta , \quad \widetilde{p}_{22}(\tau) = p_{21}(\tau)\lambda_1 p + p_{22}(\tau)\lambda_2 .$$

Substituting 4–6 into 9 we find the explicit formulas for probability densities $\widetilde{p}_{ij}(\tau)$, $i, j = 1, 2$.

Let us denote by $\pi_i(0)$ the conditional stationary probability that the process $\lambda(t)$ sojourns in the state i ($i = 1, 2$) at the time moment $\tau = 0$ in condition that at this time moment the flow event has arrived ($\pi_1(0) + \pi_2(0) = 1$). Since the sequence of the time moments of the flow events arrival corresponds to an embedded Markov chain, the following equations take place:

$$\pi_1(0) = \pi_1(0)p_{11} + \pi_2(0)p_{21} , \quad \pi_2(0) = \pi_1(0)p_{12} + \pi_2(0)p_{22} , \quad (10)$$

where p_{ij} is a transitional probability that the process $\lambda(t)$ changes its state from the state i to the state j ($i, j = 1, 2$) during the time from the event arrival at the time moment $\tau = 0$ till the moment of the next flow event arrival. Here the probabilities p_{ij} are determined as

$$p_{ij} = \int_0^\infty \widetilde{p}_{ij}(\tau)\, d\tau \, , \tag{11}$$

where $\widetilde{p}_{ij}(\tau)$ are defined by 9, $p_{ij}(\tau)$ are defined by 4–6. Calculating the corresponding integrals 11 we find

$$p_{11} = \frac{1}{z_1 z_2}[\lambda_1(1-p)(\lambda_2+\alpha) + \alpha\delta\beta] \, , \tag{12}$$

$$p_{12} = \frac{1}{z_1 z_2}[\lambda_1 p(\lambda_2+\alpha) + \lambda_2\beta] \, , \tag{13}$$

$$p_{21} = \frac{1}{z_1 z_2}[\lambda_1\alpha(1-p+p\delta) + \alpha\delta\beta] \, , \tag{14}$$

$$p_{22} = \frac{1}{z_1 z_2}[\lambda_2(\lambda_1+\beta) + \lambda_1 p\alpha(1-\delta)] \, . \tag{15}$$

Substituting 12–15 into 10, we obtain the explicit formulas for $\pi_i(0)$:

$$\pi_1(0) = \alpha\frac{\lambda_1(1-p+p\delta) + \delta\beta}{p\lambda_1\lambda_2 + \lambda_1\alpha + \lambda_2\beta + \alpha\delta(p\lambda_1+\beta)} \, , \tag{16}$$

$$\pi_2(0) = \frac{\lambda_1(1-p+p\delta) + \delta\beta}{p\lambda_1\lambda_2 + \lambda_1\alpha + \lambda_2\beta + \alpha\delta(p\lambda_1+\beta)} \, . \tag{17}$$

And the probability density is defined by the formula

$$p(\tau) = \sum_{i=1}^{2} \pi_i(0) \sum_{j=1}^{2} \widetilde{p}_{ij}(\tau), \ \tau \geq 0 \, . \tag{18}$$

Substituting first 9 into 18 and next 4–6 and 16, 17 into 18, carrying out some transformations, we obtain the explicit expression for probability density $p(\tau)$ calculation:

$$p(\tau) = \gamma z_1 e^{-z_1\tau} + (1-\gamma)z_2 e^{-z_2\tau}, \ \tau \geq 0 \, , \tag{19}$$

$$\gamma = \frac{1}{z_2 - z_1}[z_2 - \lambda_1\pi_1(0) - (\alpha\delta+\lambda_2)\pi_2(0)] \, , \tag{20}$$

where z_1, z_2 are defined by 7, 8; $\pi_1(0)$, $\pi_2(0)$ are defined by 16, 17.

4   The Expressions for Joint Probability Density $p(\tau_1, \tau_2)$

Since the sequence of the time moments of the flow events arrival corresponds to an embedded Markov chain, the following formula for joint probability density $p(\tau_1, \tau_2)$ takes place:

$$p(\tau_1, \tau_2) = \sum_{i=1}^{2} \pi_i(0) \sum_{j=1}^{2} \widetilde{p}_{ij}(\tau_1) \sum_{k=1}^{2} \widetilde{p}_{jk}(\tau_2), \ \tau_1 \geq 0, \ \tau_2 \geq 0 \, . \tag{21}$$

Then substituting first $\widetilde{p}_{ij}(\tau_1)$, $\widetilde{p}_{ij}(\tau_2)$, that are defined by 9, next $p_{ij}(\tau_1)$, $p_{ij}(\tau_2)$, that are defined by 4–6 for $\tau = \tau_1$ and $\tau = \tau_2$, and finally $\pi_1(0)$ and $\pi_2(0)$, that are defined by 16, 17 into 21, carrying out some transformations, we find the formula for $p(\tau_1, \tau_2)$ calculation in the following form:

$$p(\tau_1, \tau_2) = p(\tau_1)p(\tau_2) + \gamma(1 - \gamma)\frac{\lambda_1(\lambda_2 - p\lambda_2 - p\alpha\delta)}{\lambda_1\lambda_2 + \lambda_1\alpha + \lambda_2\beta + \alpha\delta\beta} \times \qquad (22)$$
$$\times (ze_1^{-z_1\tau_1} - z_2e^{-z_2\tau_1})(ze_1^{-z_1\tau_2} - z_2e^{-z_2\tau_2}), \; \tau_1 \geq 0, \; \tau_2 \geq 0 \,,$$

where $p(\tau_1)$, $p(\tau_2)$ are defined by 19 for $\tau = \tau_1$ and $\tau = \tau_2$, γ are defined by 20.

5   The Recurrence Conditions of the Flow

It can be shown by using the equations 20, 16 and 17 that

$$\gamma(1 - \gamma) = \frac{1}{(z_2 - z_1)^2}(\lambda_1 - \lambda_2 - \alpha\delta)\frac{\lambda_1\lambda_2 + \lambda_1\alpha + \lambda_2\beta + \alpha\delta\beta}{p\lambda_1\lambda_2 + \lambda_1\alpha + \lambda_2\beta + \alpha\delta(p\lambda_1 + \beta)} \times \qquad (23)$$
$$\times [\pi_1(0)(p\lambda_1 + \beta) - \alpha\pi_2(0)] \,.$$

It follows from 22, 23 that

1) if $\lambda_1 - \lambda_2 - \alpha\delta = 0$, then the joint probability density 22 becomes factorable: $p(\tau_1, \tau_2) = p(\tau_1)p(\tau_2)$; and it follows from 7, 8 that $z_1 = \lambda_1$, $z_2 = \lambda_2 + \alpha + \beta$; 20 implies $\gamma = 1$. In this case $p(\tau) = \lambda_1 e^{-\lambda_1\tau}$, $\tau \geq 0$.

2) if $\lambda_2 - p\lambda_2 - p\alpha\delta = 0$, $p \neq 1$, then the joint probability density 22 becomes factorable: $p(\tau_1, \tau_2) = p(\tau_1)p(\tau_2)$; and it follows from 16, 17 that $\pi_1(0) = 1 - p$, $\pi_2(0) = p$. 20 implies

$$\gamma = \frac{1}{z_2 - z_1}[z_2 - \lambda_1(1 - p) - \lambda_2] \,,$$

and, consequently,

$$p(\tau) = \gamma z_1 e^{-z_1\tau} + (1 - \gamma)z_2 e^{-z_2\tau}, \; \tau \geq 0 \,.$$

3) if $\pi_1(0)(p\lambda_1 + \beta) - \alpha\pi_2(0) = 0$, then the joint probability density 22 becomes factorable: $p(\tau_1, \tau_2) = p(\tau_1)p(\tau_2)$; it follows from 7 that $z_1 = \lambda_1(1 - p + p\delta) + \delta\beta$; 20 implies $\gamma = 1$. In this case $p(\tau) = z_1 e^{-z_1\tau}$, $\tau \geq 0$.

If one of these conditions is met, the flow of events will be the recurrent flow. For, let $p(\tau_1, ..., \tau_k, \tau_{k+1})$ be the joint probability density of $\tau_1, ..., \tau_k, \tau_{k+1}$, where $\tau_k = t_{k+1} - t_k$, $k = 1, 2, ...$ For $k = 2$ we have $p(\tau_1, \tau_2) = p(\tau_1)p(\tau_2)$. Now we proceed by mathematical induction. Assume that $p(\tau_1, ..., \tau_k) = p(\tau_1)...p(\tau_k)$. Since the sequence of time moments $t_1, t_2, ..., t_k, t_{k+1}$ of the flow events occurrence is an embedded Markov chain, then the flow has the Markov property at the moments of the flow events arrival. Then $p(\tau_1, ..., \tau_k, \tau_{k+1}) = p(\tau_1, ..., \tau_k) \times \times p(\tau_{k+1}|\tau_1, ..., \tau_k) = p(\tau_1, ..., \tau_k)p(\tau_{k+1}|\tau_k)$, where $p(\tau_{k+1}|\tau_k) = p(\tau_k, \tau_{k+1})/p(\tau_k)$. Since for the neighboring intervals (t_k, t_{k+1}) and (t_{k+1}, t_{k+2}), $k = 1, 2, ...$ we have $p(\tau_k, \tau_{k+1}) = p(\tau_k)p(\tau_{k+1})$, then $p(\tau_{k+1}|\tau_k) = p(\tau_{k+1})$. This proves the factorization of the joint probability density $p(\tau_1, ..., \tau_k, \tau_{k+1})$.

6   Conclusion

The obtained results provide the possibility to solve the problem of the flow parameters estimation. In general, the method of moments is used to estimate the unknown flow parameters. For the particular cases of the recurrent flow we can apply the maximum-likelihood technique.

References

1. Cox, D.R.: Some Statistical Methods Connected with Series of Events. J. Royal Statistical Society B 17, 129–164 (1955)
2. Kingman, Y.F.C.: On doubly stochastic Poisson process. Proceedings of Cambridge Phylosophical Society 60(4), 923–930 (1964)
3. Bremaud, P.: Point Processes and Queues: Martingale Dynamics. Springer, New York (1981)
4. Last, G., Brandt, A.: Marked Point Process on the Real Line: The Dynamic Approach. Springer, New York (1995)
5. Basharin, G.P., Kokotushkin, V.A., Naumov, V.A.: Method of equivalent substitutions for calculating fragments of communication networks for a digital computer - 1. Engineering Cybernetics 17(6), 66–73 (1979)
6. Neuts, M.F.: A versatile Markov point process. Journal of Applied Probability 16, 764–779 (1979)
7. Dudin, A.N., Sun, B.: A multiserver MAP/PH/N system with controlled broadcasting by unreliable servers. Automatic Control and Computer Sciences 43(5), 247–256 (2009)
8. Telek, M., Horvath, G.: A minimal representation of Markov arrival processes and a moments matching method. Performance Evaluation 64, 1153–1168 (2007)
9. Okamura, H., Dohi, T., Trivedi, K.S.: Markovian arrival process parameter estimation with group data. IEEE/ACM Transactions on Networking 17, 1326–1339 (2009)
10. Gortsev, A.M., Nezhelskaya, L.A., Solovev, A.A.: Optimal State Estimation in MAP Event Flows with Unextendable Dead Time. Automation and Remote Control 73(8), 1316–1326 (2012)
11. Gortsev, A.M., Nissenbaum, O.V.: Estimation of the dead time period and parameters of an asynchronous alternative flow of events with unextendable dead time period. Russian Physics Journal 48(10), 1039–1054 (2005)

Sets of Bipartite Sets of Events
and Their Application

Irina Baranova

Institute of Mathematics and Computer Science
of Siberian Federal University
79 Svobodny pr., 660041 Krasnoyarsk, Russia
irinabar@yandex.ru
http://www.sfu-kras.ru/

Abstract. One of the significant, fundamental and demanded tasks of modern statistical data analysis and computer science is to develop methods for the analysis of different types of data. The paper considers a situation, when one part of the researching data is numerical and the part is multiple. The notion of the set of bipartite sets of events is offered. This set consists of the sets of events, whose first part corresponds to the random variables, and second part — to the sets.

In this work it is considered all possible types of the set of bipartite sets. The formula of probabilistic distribution for all types of this set is shown. The concepts of the Minkovsky set-operation of the set of bipartite sets of events and its probability are resulted. Also in paper practical problems of application of the given set are considered.

Keywords: Bipartite set of events, probabilistic distribution, set of bipartite sets of events.

1 Introduction

In some fields of science and practical activities different researches result in solving the problems of system analysis. In paper it is considered a situation, when the one part of the events describing the complex system's behavior is numerical and the second part is sets. The main difficulty for analysis of such complex systems lies in the fact that number of all possible events is big and the data describing systems behavior is polytypic. This problem is especially actual for applied fields of science, whose are bound up with analysis of social, economic and natural systems. They are medicine, ecology, biology, actuary, finances, insurance, sociology and others.

In works [1], [2] the bipartite set of random events method was suggested, in which each system's element represents a bipartite set of random events. The first part of this set corresponds to the random variables, and second part – to the sets. The basic idea of this method concludes in reduction of an analysis of system's elements to analysis of corresponding bipartite sets of events.

A. Dudin et al. (Eds.): ITMM 2014, CCIS 487, pp. 26–33, 2014.

In this work the notion of the set of bipartite sets of events is offered. The work considers all possible types of this set, and the formula of probabilistic distribution for these sets is shown. Particular attention is paid for the application of this set.

2   Bipartite Set of Events

In paper it is studied complex system, whose behavior is describing by numeric and set data. Then the results of observation for the behavior of researching object is a set, which consists of random variables and random sets [4], [5]. Consider the probabilistic space $(\Omega, \mathcal{F}, \mathbf{P})$. Let $\mathfrak{X} \subset \mathcal{F}$ be the finite set of events chosen from algebra $\mathcal{F}$ of that space. Let designate $N = |\mathfrak{X}|$.

Definition 1. *Random set of events under a set of the chosen events $\mathfrak{X}$ is decided on probabilistic space as a random element of*

$$K : (\Omega, \mathcal{F}, \mathbf{P}) \to \left(2^{\mathfrak{X}}, 2^{2^{\mathfrak{X}}}\right)$$

on values from measurable space $\left(2^{\mathfrak{X}}, 2^{2^{\mathfrak{X}}}\right)$, where $2^{\mathfrak{X}}$ is power set $\mathfrak{X}$, $2^{2^{\mathfrak{X}}}$ is algebra of all its subsets.

Probabilistic distribution [6] of random set of events K which has been set under a set of the chosen events $\mathfrak{X} \subseteq \mathcal{F}$ can be presented several equivalent distributions of the probabilities generated by a set of events $\mathfrak{X}$ *[3]*. In this work we define it by probability distribution of the I-st sort.

Definition 2. *Probabilistic distribution of the I-st sort is a set from 2^N probabilities of type*

$$\left\{ p(X) = \mathbf{P}(K = X) = \mathbf{P}\left(\left(\bigcap_{x \in X} x \right) \cap \left(\bigcap_{x \in X^c} x^c \right) \right), \quad X \subseteq \mathfrak{X} \right\}.$$

As stated above, our general situation, when the one part of the results of observation for behavior of researching object are numerical and the second part — sets, can be described as the set of the random elements.

Definition 3. *The bipartite set of random elements is set of the random elements, that can be defined in that way:*

$$\{\boldsymbol{\xi}, \boldsymbol{K}\} = \boldsymbol{\xi} \cup \boldsymbol{K} = \{\xi_a, a \in A, K_\beta, \beta \in B\}, \tag{1}$$

where first part of set is the random variables $\boldsymbol{\xi} = \{\xi_a, a \in A\}$, the second part is the random sets of events $\mathbf{K} = \{K_\beta, \beta \in B\}$, A is the indices set of a random variables, and B — indices set of a random sets.

Now we considered the bipartite set of random events, which corresponds the bipartite set of random elements, defined above.

Let $\{\xi_a, a \in A\}$ are random variables with finite set of possible values

$$\mathcal{R}_a = \{r_{a_1}, \ldots, r_{a_{N_a}}\} \subset \mathbb{R}, a \in A.$$

To each random variables can be put in correspondence set of events

$$\xi_a \implies \mathcal{Y}_a = \{\mathcal{Y}_a(r_a), r_a \in \mathcal{R}_a\}.$$

The event $\mathcal{Y}_a(r_a) = \{\xi_a \leq r_a\} = \{\omega : \xi_a(\omega) \leq r_a\}$ is the event from definition of distribution function of random variable and a set with inserting structure of dependences. To first part of set $\{\boldsymbol{\xi}, \boldsymbol{K}\}$ (random variables) can be put in correspondence set of events:

$$\boldsymbol{\xi} \implies \mathcal{Y} = \sum_{a \in A} \mathcal{Y}_a.$$

For each random set of events K_β, $\beta \in B$ can be put in correspondence the finite set of events $\mathfrak{X}_\beta$:

$$K_\beta \iff \mathfrak{X}_\beta.$$

To second part of set $\{\boldsymbol{\xi}, \boldsymbol{K}\}$ can be put in correspondence common set of events $\mathfrak{X}$:

$$\boldsymbol{K} \iff \mathfrak{X} = \sum_{\beta \in B} \mathfrak{X}_\beta, \ \beta \in B.$$

Definition 4. *Bipartite set of events is union of two sets: set of events that determined by random variables, and set of events that determined by random sets of events:*

$$\{\mathcal{Y}, \mathfrak{X}\} = \{\mathcal{Y}_a, \mathfrak{X}_\beta, a \in A, \beta \in B\}. \tag{2}$$

Complete characteristics of bipartite set of events is given by its probabilistic distribution. In paper [3] were found all forms of probabilistic distribution of bipartite set. In paper we present one of them.

Definition 5. *Probabilistic distribution of the I-st sort for the bipartite set of events is a set from probabilities of type $\{\mathcal{Y}, \mathfrak{X}\}$:*

$$\left\{ p(\mathbf{r}, \mathbf{X}) = \mathbf{P}\left(\bigcap_{a \in A} \{ \bigcap_{y \neq r} \{\xi = y\}^c \} \bigcap_{x_\beta \in X} x_\beta \bigcap_{x_\beta \in X^c} x_\beta^c, a \in A, \beta \in B \right) \right\},$$

where $x_\beta \in \mathfrak{X}$, $\beta \in B$, $\mathbf{r} = \{r_a, r_a \in \mathbb{R}\}$, $\mathbf{X} = \{X_\beta, X_\beta \subseteq \mathfrak{X}_\beta\}$.

The comparison between system's elements is difficult, but it is offered the reduction a system's elements to corresponding bipartite sets of events. If we known the probabilistic distributions of the bipartite sets of events then we can compare them by using the Minkovsky set-operations [2].

3 Minkovsky Set-Operations

The operation under events from the each part consisted the bipartite sets of events is called an arbitrary Minkovsky set-operation. For example, the Minkovsky symmetry difference set-operation of two bipartite sets of random events s^1 and s^2:

$$s1(\Delta)s2 = \left\{ \mathcal{Y}_a^1(r_a) \Delta \mathcal{Y}_a^2(r_a), X_\beta^1 \Delta X_\beta^2, X_\beta \subseteq \mathfrak{X}_\beta, r_a \in \mathcal{R}_a, a \in A, \beta \in B \right\}.$$

The probability of the Minkovsky symmetry difference set-operation is

$$\mathbf{P}\left(s^1(\Delta)s^2\right) = \frac{1}{|A|} \sum_{a \in A} \frac{1}{|\mathcal{Y}_a|} \sum_{r_a \in \mathbb{R}} \mathbf{P}\left(\mathcal{Y}_a^1(r_a) \Delta \mathcal{Y}_a^2(r_a)\right) +$$

$$+ \frac{1}{|B|} \sum_{\beta \in B} \frac{1}{|\mathfrak{X}_\beta|} \sum_{X_\beta \subseteq \mathfrak{X}_\beta} \mathbf{P}\left(X_\beta^1 \Delta X_\beta^2\right).$$

It was proved in work [2] that probability of the Minkovsky symmetry difference set–operation of two bipartite sets of random events can be used as a distance between sets.

4 Terrace-Event for Bipartite Set of Event

Let's spend the redenotation of bipartite set of random events presented by formula (2) in that way:

$$Z = \{\mathcal{Y}, \mathfrak{X}\} = \{\mathcal{Y}_a, \mathfrak{X}_\beta, a \in A, \beta \in B\}.$$

Let s is bipartite set of random events, which is a subset of bipartite set of random events Z (i.e. $s \subseteq Z$):

$$s = \{\mathcal{Y}_{s_A}, \mathfrak{X}_{s_B}, s_A \subseteq A, s_B \subseteq B\}.$$

Definition 6. *Terrace–event for bipartite set of events s represents as a set of not intersected events where each event is a subset of appropriate set of events* $\mathcal{Y}_a$ *or* $\mathfrak{X}_\beta$:

$$ter(s) = ter\{\mathcal{Y}_{s_A}, \mathfrak{X}_{s_B}\} = \bigcap_{a \in s_A} ter(\mathcal{Y}_a) \bigcap_{\beta \in s_B} ter(\mathfrak{X}_\beta) =$$

$$= \bigcap_{a \in s_a} \mathcal{Y}_a(r_a) \bigcap_{\beta \in s_B} \left(\bigcap_{x_\beta \in X_\beta} x_\beta \bigcap_{x_\beta \in X_\beta^c} x_\beta^c \right),$$

$s_A \subseteq A$, $s_B \subseteq B$, $r_a \in \mathcal{R}_a$, $X_\beta \subseteq \mathfrak{X}_\beta$, $\mathcal{R}_a = \{r_{a_1}, \ldots, r_{a_{N_a}}\} \subset \mathbb{R}$, $a \in A$.

5   Set of Bipartite Sets of Events

Definition 7. *Set of bipartite sets of events is a set, consisting of the sets of events, whose first part corresponds to the random variables, and second part — to the random sets:*

$$S = \left\{ s^1, \ldots, s^n \right\}. \tag{3}$$

Here are the possible types of the structure of the set S:

1. Set of similar sets of events. All sets are generated by the same bipartite set of random elements, they are described by the same bipartite set Z, hence the terrace–event of set s^i, $i = 1, \ldots, n$ are the same. They differ from each other by the probability of their occurrence:

$$S = \left\{ s^1 = \{\mathcal{Y}_a^1, \mathfrak{X}_\beta^1\}, \ldots, s^n = \{\mathcal{Y}_a^n, \mathfrak{X}_\beta^n\}, \ a \in A, \ \beta \in B \right\}.$$

   This situation is very simple.

2. Set of subsets of the same bipartite set. This is the situation, when bipartite sets of events s^i are the subsets of the same bipartite set Z:

$$S = \left\{ s^1 = \left\{ \mathcal{Y}_{s_A^1}, \mathfrak{X}_{s_B^1}, s_A^1 \subseteq A, s_B^1 \subseteq B \right\}, \ldots, \right.$$

$$\left. s^n = \{ \mathcal{Y}_{s_A^n}, \mathfrak{X}_{s_B^n}, s_A^n \subseteq A, s_B^n \subseteq B \} \right\}.$$

3. Set of subsets of the different bipartite sets. In this case, bipartite sets of events s^i are the subsets of the same bipartite set $Z^i = \{\mathcal{Y}^i, \mathfrak{X}^i\} = \{\mathcal{Y}_a^i, \mathfrak{X}_\beta^i, a \in A^i, \beta \in B^i\}$:

$$S = \left\{ s^1 = \left\{ \mathcal{Y}_{k_A^1}, \mathfrak{X}_{k_B^1}, \ k_A^1 \subseteq A^1, k_B^1 \subseteq B^1 \right\}, \ldots, \right.$$

$$\left. s^n = \{ \mathcal{Y}_{k_A^n}, \mathfrak{X}_{k_B^n}, \ k_A^n \subseteq A^n, k_B^n \subseteq B^n \} \right\}.$$

   This situation is the most common, and set with this type structure is more difficult for studying.

6   Probabilistic Distribution for the Sets of Bipartite Sets of Events

In this part of the paper we present the form of the probability distribution for the set of bipartite sets of events.

6.1 Set of Similar Sets of Events

Let the probabilistic distribution of bipartite set of events $s \subseteq Z$ is given:

$$p(s) = p(\mathcal{Y}_a, \mathfrak{X}_\beta) = p(\mathbf{r}, \mathbf{X}) =$$

$$= \mathbf{P}\Big(\bigcap_{a \in A} \{\mathcal{Y}_a(r_a)\} \bigcap_{x_\beta \in X} x_\beta \bigcap_{x_\beta \in X^c} x_\beta^c, \, a \in A, \, \beta \in B \Big),$$

$x_\beta \in \mathfrak{X}$, $\beta \in B$, $\mathbf{r} = \{r_a, r_a \in \mathbb{R}\}$, $\mathbf{X} = \{X_\beta, X_\beta \subseteq \mathfrak{X}_\beta\}$.

Then probabilistic distribution for the set of bipartite sets of events $S = \left\{ s^1, \ldots, s^n \right\}$ $p(S^t)$, $S^t \subseteq S$ has next form:

$$p(S^t) = \mathbf{P}\left(\bigcap_{s^i \in S^t} s \right) =$$

$$= \mathbf{P}\Big(\bigcap_{s^i \in S^t} \Big\{ \bigcap_{a \in A} \mathcal{Y}_a^i(r_a) \bigcap_{x_\beta \in X} x_\beta \bigcap_{x_\beta \in X^c} x_\beta^c \Big\} \Big), \, S^t \subseteq S.$$

6.2 Set of Subsets of the Same Bipartite Set

As it shown above, in this situation,

$$S = \Big\{ s^1 = \Big\{ \mathcal{Y}_{s_A^1}, \mathfrak{X}_{s_B^1}, \, s_A^1 \subseteq A, s_B^1 \subseteq B \Big\}, \ldots,$$

$$s^n = \{ \mathcal{Y}_{s_A^n}, \mathfrak{X}_{s_B^n}, \, s_A^n \subseteq A, s_B^n \subseteq B \} \Big\}.$$

Let the probabilistic distribution of all bipartite sets of events is given

$$p(s^i), \, s^i \in S.$$

Probabilistic distribution for this type of set of bipartite sets of events S:

$$p(S^t) = \mathbf{P}\left(\bigcap_{s^i \in S^t} s^i \right) == \mathbf{P}\Big(\bigcap_{s^i \in S^t} \Big\{ \bigcap_{a \in s_A^i} ter(\mathcal{Y}_a) \bigcap_{\beta \in s_B^i} ter(\mathfrak{X}_\beta) \Big\} =$$

$$= \mathbf{P}\Big(\bigcap_{a \in s_a^i} \mathcal{Y}_a(r_a) \bigcap_{\beta \in s_B^i} \Big(\bigcap_{x_\beta \in X_\beta} x_\beta \bigcap_{x_\beta \in X_\beta^c} x_\beta^c \Big) \Big), s_A^i \subseteq A, s_B^i \subseteq B, S^t \subseteq S.$$

6.3 Set of Subsets of the Different Bipartite Sets

Now we considered the third type of set, when bipartite sets of events s^i are the subsets of the same bipartite set $Z^i = \{\mathcal{Y}^i, \mathfrak{X}^i\} = \{\mathcal{Y}_a^i, \mathfrak{X}_\beta^i, a \in A^i, \beta \in B^i\}$:

$$S = \Big\{ s^1 = \Big\{ \mathcal{Y}_{k_A^1}, \mathfrak{X}_{k_B^1}, \, k_A^1 \subseteq A^1, k_B^1 \subseteq B^1 \Big\}, \ldots,$$

$$s^n = \left\{ \mathcal{Y}_{k_A^n}, \mathfrak{X}_{k_B^n}, \, k_A^n \subseteq A^n, k_B^n \subseteq B^n \right\} \Big\},$$

here A^j is the indices set for numerical part of the bipartite set $s^j \in S$, B^j — the indices set for plural part. The denotation k_A^j made for index from indices set A^j. Likewise, denotation k_B^j — made for index from indices set B^j. So, set of event $\mathcal{Y}_{k_A^1} \subseteq \mathcal{Y}^i$, and $\mathfrak{X}_{k_B^1} \subseteq \mathfrak{X}_\beta^i$.

Probabilistic distribution for this type of set of bipartite sets of events S is as follows:

$$p\left(S^t\right) = \mathbf{P}\left(\bigcap_{s^i \in S^t} s^i \right) == \mathbf{P}\left(\bigcap_{k_A^1 \in A^1} \mathcal{Y}_{k_A^1} \cdots \bigcap_{k_A^t \in A^t} \mathcal{Y}_{k_A^t} \right.$$

$$\left. \bigcap_{k_B^1 \in B^1} \left(\bigcap_{x_\beta \in X_{k_B^1}} x_\beta \bigcap_{x_\beta \in X_{k_B^1}^c} x_\beta^c \right) \cdots \bigcap_{k_B^t \in B^t} \left(\bigcap_{x_\beta \in X_{k_B^t}} x_\beta \bigcap_{x_\beta \in X_{k_B^t}^c} x_\beta^c \right) \right),$$

$$\mathcal{Y}_{k_A^1} \subseteq \mathcal{Y}^i, \ \mathfrak{X}_{k_B^1} \subseteq \mathfrak{X}_\beta^i, \ S^t \subseteq S.$$

In this formula for simplicity for a subset S^t is determined subset $\left\{ s^1, \ldots, s^t \right\}$ with the indices set for numerical part $A^1, \ldots, A^t$, indices set for plural part $B^1, \ldots, B^t$, and corresponding indices $k_A^1, \ldots, k_A^t$ and $k_B^1, \ldots, k_B^t$. Actually, it considered arbitrary set of bipartite sets $S^t \subseteq S$ with power t (i.e. $|S^t| = t$).

7 Applications of Sets of Bipartite Sets of Events

Now we consider the practical problems of application of the described sets.
Sets of described types can be generated as a result of the following tasks:

- determining the best or the worst elements of the system,
- classification of elements or groups,
- ranking system elements,
- factor analysis,
- cluster analysis,
- data visualization,
- search association rules (finding relationships between the sets of events),
- covariance analysis of polytypic data (identifying links between different data),
- regression analysis of polytypic data (depending on the construction of functions between the data),
- predicting the behavior of the system,
- decision-making in fuzzy data,
- bank (insurance) score.

8 Conclusion

One of the significant, fundamental and demanded tasks of modern statistical data analysis and computer science is to develop methods for the analysis of different types of data. In the paper a situation was considered, when one part of the researching data is numerical and the second part is multiple.

The notion of the set of bipartite sets of events, consisting of the sets of events, whose first part corresponds to the random variables, and second part — to the sets, was offered.

The work considers all possible types of the set of bipartite sets. And it presents the forms of the probability distribution for the described types of set of bipartite sets of events. The concepts of the Minkovsky set-operation of the set of bipartite sets of events and its probability are resulted. Also in the paper practical problems of application of the given set are considered.

Acknowledgments. Author express sincere gratitude to friends and colleagues: O.Yu. Vorobyev, V.V. Bykova, D.V. Semenova, E.E. Goldenok, N.A. Lukyanova and A.A. Novoselov for effective cooperation and support.

References

1. Baranova, I.V.: The bipartite set of events method in eventological analysis of social-economic system/Bulletin of Krasnoyarsk State University, Krasnoyarsk, vol. 1, pp. 142–152 (2006) (in Russian)
2. Vorobyev, O.Y., Baranova, I.V.: The bipartite set method application in eventological analysis of complex system. Siberian Federal University, Krasnoyarsk (2007) (in Russian)
3. Vorobyev, O.Y.: Eventology. Krasnoyarsk, Siberian Federal University (2007) (in Russian)
4. Matheron, G.: Random sets and integral geometry. John Wiley & Sons, New York (1975)
5. Stoyan, D., Stoyan, H.: Fractals, Random Shapes and Point Fields. John Wiley & Sons, New York (1994)
6. Kyburg, H.E.: Probability and Inductive Logic. The Macmillian Company, Collier-Macmillian limited, London (1970)
7. Schweizer, B., Sklar, A.: Probabilistic metric spaces. North Holland, New York (1983)
8. Baranova, I.V.: Regression between the polytypic data, describing behavior of complex system/Proceedings of the Third International Conference "Problems of Cybernetics and Informatics", Baku, Azerbaijan, vol. 2, pp. 244–248 (2010)

Monte Carlo Calculations of Acoustic Wave Propagation in the Turbulent Atmosphere*

Vladimir Belov, Yulia Burkatovskaya, Nikolay Krasnenko, and Luidmila Shamanaeva

Institute of Atmospheric Optics SB RAS,
1, Akademicheskii Ave., Tomsk 634055, Russia
Institute of Cybernetics, National Research Tomsk Polytechnic University,
30 Lenin Prospekt, 634050 Tomsk, Russia
Department of Applied Mathematics and Cybernetics,
National Research Tomsk State University,
36 Lenin Prospekt, 634050 Tomsk, Russia
Institute of Monitoring of Climatic and Ecological Systems SB RAS
10/3, Akademicheskii Ave., Tomsk 634055, Russia
Tomsk State University of Control Systems and Radioelectronics
40 Lenin Prospekt, 634050 Tomsk, Russia
{belov,sima}@iao.ru, tracey@tpu.ru, krasnenko@imces.ru

Abstract. The problem of acoustic wave propagation in the turbulent atmosphere is solved by the Monte Carlo method. A 500-m plane-stratified model of the turbulent atmosphere is considered. Classical and molecular absorption of acoustic radiation and scattering by turbulent temperature and wind velocity fluctuations are taken into account for acoustic radiation frequencies of 1, 2, 3 and 4 kHz. A good agreement of the simulation results with experimentally measured values demonstrates the efficiency of the suggested algorithm.

Keywords: atmosphere, turbulence, acoustic wave propagation, Monte-Carlo calculations.

1 Introduction

Investigations of sound propagation in the atmosphere are necessary for the prediction of its characteristics, finding direction toward a sound source, and quantitative interpretation of the data of acoustic sounding [1]. In the outdoor atmosphere, the sound propagation is influenced by a large number of factors, including the vertical atmospheric stratification, turbulence, viscosity, and effects caused by finite dimensions of sound beams in the transverse direction, that is, by the angular divergence of acoustic beams broadened due to the atmospheric

* This work was supported in part by the Scientific Research executed within the framework of the Special Federal Program "Scientific and Pedagogical Personnel of Innovative Russia" for 2009-2013 (Contracts Nos. 02.740.11.0232 and 14.740.11.0204).

A. Dudin et al. (Eds.): ITMM 2014, CCIS 487, pp. 34–43, 2014.

turbulence [2, 3]. Difficulties of analytical approaches to a solution of the problem of acoustic radiation transfer through the outdoor atmosphere call for the use of numerical methods (for example, see [4, 5]), from which the method of statistical simulation (Monte Carlo) is most promising [6, 7]. This method allows sound scattering on the acoustic refractive index fluctuations caused by wind velocity and temperature inhomogeneities to be taken into account for the most realistic models of the atmosphere.

The equation of acoustic radiation transfer in the turbulent atmosphere in the form of the Neumann series for the acoustic ray intensity was derived in [8]. In [9], the Monte Carlo method was first used to solve the problem of acoustic wave propagation through a vertically stratified turbulent atmosphere. The density of collisions of acoustic particles - phonons - was estimated in terms of the acoustic energy flux density scattered by the atmospheric turbulence derived in [10] in the single scattering approximation.

In the present work, we use the modified Monte Carlo algorithm to solve the problem of acoustic radiation propagation in the atmosphere.

2 Model of the Atmosphere and Geometry of the Numerical Experiment

For a 500-m standard plane-stratified turbulent atmosphere, the total attenuation coefficient was calculated from the formula

$$\sigma_{att}(z_i) = \sigma_{cl} + \sigma_{mol}(z_i) + \sigma_T(z_i) + \sigma_V(z_i), \tag{1}$$

where σ_{cl} and $\sigma_{mol}(z_i)$ are the coefficients of classical and molecular absorption, $\sigma_T(z_i)$ and $\sigma_V(z_i)$ are the coefficients of scattering by turbulent temperature and wind velocity fluctuations, $z_i = z_{i-1} + dz$, $dz = 20$m, $i = 1, \ldots, 26$, and $z_0 = 0$.

The coefficients of classical and molecular absorption σ_{cl} and $\sigma_{mol}(z_i)$, in m^{-1}, were taken from [11, 12].

Analytical expressions for the scattering coefficients were derived in [9, 13] for the von Karman model of the three-dimensional spectra of temperature and wind velocity fluctuations:

$$\begin{aligned}
\sigma_T(z_i) = {}& 0.9\lambda^{-1/3}(z_i)C_T^2(z_i)T^{-2}(z_i)L_0^{-7/3}(z_i) \\
& \times \left\{ 0.07143 \left[B^{7/6}(z_i) - \lambda^{7/3}(z_i) \right] - 0.1A^2(z_i) \left[B^{-5/6}(z_i) - \lambda^{-5/3}(z_i) \right] \right. \\
& \left. - A(z_i) \left[B^{1/6}(z_i) - \lambda^{1/3}(z_i) \right] \right\};
\end{aligned} \tag{2}$$

$$\begin{aligned}
\sigma_V(z_i) = {}& 1.569\varepsilon^{2/3}(z_i)\lambda^{-1/3}(z_i)c^{-2}(z_i)L_0^{-13/3}(z_i) \\
& \times \left\{ 0.1429 \left[B(z_i) + 2A(z_i) \right] \left[B^{7/6}(z_i) - \lambda^{7/3}(z_i) \right] \right. \\
& -0.0769 \left[B^{13/6}(z_i) - \lambda^{13/3}(z_i) \right] - A(z_i) \left[A(z_i) + 2B(z_i) \right] \\
& \left. \times \left[B^{1/6}(z_i) - \lambda^{1/3}(z_i) \right] - 0.2A^2(z_i)B(z_i) \left[B^{-5/6}(z_i) - \lambda^{-5/3}(z_i) \right] \right\},
\end{aligned} \tag{3}$$

where $\lambda(z_i)$ is the wavelength, $c(z_i)$ is the velocity of sound, $L_0(z_i)$ in the outer scale of the atmospheric turbulence, $C_T^2(z_i)$ is the structure function of the wind velocity field $T(z_i)$, $\varepsilon(z_i)$ is the kinetic energy dissipation rate,

$$A(z_i) = 2L_0^2(z_i) + \lambda^2(z_i);$$
$$B(z_i) = 4L_0^2 + \lambda^2(z_i). \tag{4}$$

The normalized scattering phase functions were calculated from the following formulas [9, 13]

$$g_T(z_i, \theta) = 0.1062 L_0^6(z_i) \cos^2 \theta \left[2L_0^2(z_i)(1 - \cos\theta) + \lambda^2(z_i)\right]^{-11/6}$$
$$\times \left\{0.07143 \left[B^{7/6}(z_i) - \lambda^{7/3}(z_i)\right] - 0.1A^2(z_i) \left[B^{-5/6}(z_i) - \lambda^{-5/3}(z_i)\right]\right. \tag{5}$$
$$\left. -A(z_i)\left[B^{1/6}(z_i) - \lambda^{1/3}(z_i)\right]\right\}^{-1};$$

$$g_V(z_i, \theta) = 0.1191 L_0^{13/3}(z_i) \cos^2 \theta (1 + \cos\theta) \left(\frac{A(z_i)}{2L_0^2(z_i)} - \cos\theta\right)^{-11/6}$$
$$\times \left\{0.1429 \left[B(z_i) + 2A(z_i)\right]\left[B^{7/6}(z_i) - \lambda^{7/3}(z_i)\right]\right.$$
$$-0.0763 \left[B^{13/6}(z_i) - \lambda^{13/3}(z_i)\right] - A(z_i)\left[A(z_i) + 2B(z_i)\right] \tag{6}$$
$$\times \left[B^{1/6}(z_i) - \lambda^{1/3}(z_i)\right] - 0.2A^2(z_i)B(z_i)\left[B^{-5/6}(z_i) - \lambda^{-5/3}(z_i)\right]\right\}^{-1}.$$

Calculations were performed for point-sized and finite-aperture (circular aperture with a diameter of 1m.) sound sources with acoustic power of 1W placed at altitude z_s above the Earth's surface for frequencies of 1, 2, 3, and 4kHz typically used in sodars (acoustic radars) [1]. Emitted radiation was continuous in the solid angle subtended by the circular cone of half-angle $\phi = 2.5$, 5, 10, 15, 20, and 25° with respect to the vertical $I|_0(\phi, \varphi) = I_0(\phi)$, that is, independent of the azimuth angle φ. For the finite-aperture sources, calculations were performed for uniform or Gaussian distribution of emitted radiation over the source aperture. According to the data of sodar [14] and lidar measurements [15], the outer scale of turbulence L_0 changes from a few meters to 150 m in the atmospheric boundary layer. In our calculations, it was set equal to 2, 4, 6, 8, 10, 15, 20, 40, 60, and 80m. Acoustic radiation of the source propagated through the plane-parallel layers of the atmosphere with the coefficients of classical and molecular absorption $\sigma_{cl}(i)$ and $\sigma_{mol}(i)$ and scattering on turbulent temperature and wind velocity fluctuations $\sigma_T(i)$ and $\sigma_V(i)$ being constant within these layers, where $i = 1, \ldots, 25$. In calculations of their altitude dependence, the vertical profiles of the atmospheric temperature, pressure, and velocity of sound were taken for the standard model of the atmosphere [16].

Figure 1 shows the vertical profiles of the total attenuation coefficient calculated from Eq. (1) for frequencies $F = 1 - 4$kHz, and Figure 2 shows the vertical profiles of the phonon scattering probability $P_{sc}(i) = [\sigma_T(i) + \sigma_V(i)]/\sigma_{att}(i)$, where σ_T was calculated from Eq. (2) and σ_V was calculated from Eq. (3). It should be noted that at a frequency of 2kHz, the turbulent attenuation becomes comparable with the molecular absorption in the surface layer of the atmosphere for $L_0 \geq 15$m ($_{sc}(i) \geq 0.5$, see Fig. 2.

As demonstrated in [6], at the frequency $F = 1$kHz they are comparable for $L_0 \geq 20$m. In this case, the main contribution to the turbulent attenuation of sound propagating along the vertical direction comes from the dynamic turbulence. The contribution of temperature fluctuations is by 1–2 orders of magnitude

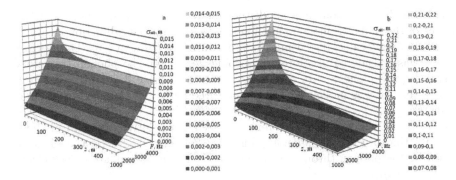

Fig. 1. Vertical profiles of the total attenuation coefficient for $F = 1 - 4$kHz and $L_0 = 10$ (a) and 80m (b)

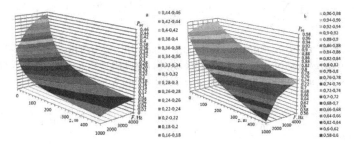

Fig. 2. Vertical profiles of the phonon scattering probability for $F = 1 - 4$kHz and $L_0 = 10$ (a) and 80m (b)

smaller. This was also pointed out in [4]. From Fig. 1 it can also be seen that in the surface layer, the attenuation coefficient increases approximately by an order of magnitude when the outer scale of turbulence L_0 increases from 10 to 80m. In this case, the phonon scattering probability (Fig. 2) increases from 0.35 to 0.95. These data are confirmed by the results presented in [17], where it was concluded that the magnitude of the excess turbulent attenuation fluctuates in wide limits and can be as great as the classical and molecular absorption.

The normalized phase functions of sound scattering on temperature fluctuations calculated by Eq. (5), and the normalized phase functions of sound scattering on wind velocity fluctuations calculated by Eq. (6) for frequencies in the range 1-4 kHz.

3   Computational Algorithm

To construct a computational algorithm, both standard computational procedures borrowed from [18] and procedures developed in [6, 7, 19] with allowance for the specifics of sound interaction with the atmosphere were used. We considered a point-sized source of acoustic radiation placed at an altitude of 35 m above the ground and having an acoustic power of 1 W. A hypothetical receiver was placed above the source at an altitude of 500 m from the ground. The coordinates of the point of phonon emission (x_0, y_0, z_0) and their directional cosines $(\omega_1, \omega_2, \omega_3)$ were calculated using the procedure described in [18]. The Earth's surface was considered absolutely absorbing, and when the phonon trajectory intersected the plane $z = 0$, the phonon was considered absorbed, and a new phonon history was modelled. The phonon free path was modelled by the following scheme.

a) Let c be the cosine of the angle between the positive direction of the z axis and the direction of phonon emission; then $\Delta l = dz/c$ be the distance passed by the phonon through atmospheric layers with attenuation coefficients $\sigma_{att}[1], \ldots, \sigma_{att}[N]$.

b) By subsequent subtraction, we find the number j of the layer such that

$$\frac{z[1] - z_0}{c}\sigma_{att}[1] + \Delta l \sum_{m=2}^{j-1} \sigma_{att}[m] < \ln(rand) \leq \frac{z[1] - z_0}{c}\sigma_{att}[1] + \Delta l \sum_{m=2}^{j} \sigma_{att}[m],$$

where $rand$ is a random number uniformly distributed in the interval [0,1].

c) If there is no number j satisfying condition (14), it is considered that the phonon have been escaped from the medium; otherwise,

$$l_{free} = \frac{z[1] - z_0}{c} + \Delta l(j - 2) - \frac{\ln(rand) + \Delta l \sum_{m=2}^{j-1} \sigma_{att}[m]}{\sigma_{att}[j]}.$$

The point of the next collision was chosen by the well-known formulas [18].

Then the collision type was chosen. The following procedure was used.

d) $p_1 = \sigma_{cl}(j)$, $p_2 = \sigma_{mol}(j)$, $p_3 = \sigma_T(j)$, $p_4 = \sigma_V(j)$.

e) $P_1 = p_1$, $P_2 = p_1 + p_2$, $P_3 = p_1 + p_2 + p_3$, $P_4 = p_1 + p_2 + p_3 + p_4$.

f) $F_1 = P_1/P_4$, $F_2 = P_2/P_4$, $F_3 = P_3/P_4$, $F_4 = P_4/P_4 = 1$.

g) $\alpha = rand$, find the number $k = \min\{l : \alpha < F_l\}$.

h) If $k = 1$, classical absorption was simulated; if $k = 2$, molecular absorption; if $k = 3$, scattering on the temperature fluctuations; otherwise, scattering on the wind velocity fluctuations.

In the case of absorption, the phonon was annihilated, and its statistical weight was added to the element of the array determining the value of the acoustic wave intensity absorbed in the j-th atmospheric layer. In the case of scattering, the scattering angle was determined by the scattering phase function given by Eq. (5) for scattering by temperature fluctuations and by Eq. (6) for scattering by wind velocity fluctuations. The procedure of simulation of the scattering angle was described in detail in [20] Calculations were carried out on a personal computer for 10^6 phonon histories, which provided acceptable calculation errors of 3–10%.

4   Calculation Results and Their Discussion

Figure 3 shows dependencies of the transmitted (I_{tr}, W/m^2) and multiply scattered radiation intensities (I_{msc}, W/m^2) over the detector zones for $F = 1.7$ kHz,

$\phi = 5°$ (a and c) and $15°$ (b and d); $F = 4\text{kHz}$, $\phi = 5°$ (e and g) and $15°$ (f and h), source altitude $z_s = 35\text{m}$, and outer scale of turbulence, in meters, indicated at the upper right of the figure. Results of our calculations demonstrate that the contribution of multiple scattering I_{msc} to the transmitted radiation intensity I_{tr} within the cone of source radiation increases with the outer scale of turbulence from 10.5% (for $L_0 = 10\text{m}$) to 53% (for $L_0 = 20\text{m}$); for $L_0 = 40\text{m}$, the transmitted radiation intensity is completely determined by multiple scattering. In this case, the sharp decrease of I_{tr} and I_{msc} in Fig. 3 is explained by the fact that received radiation is beyond the limits of the cone of source radiation divergence. Within the cone of source radiation divergence, the multiple scattering contribution increases from $4.7 \cdot 10^{-7}$ to $4.3 \cdot 10^{-6}$ W/m^2, that is, by 89% when L_0 increases from 10 to 80 m. This increase in multiple scattering contribution virtually compensates for the decrease in the transmitted radiation intensity with increasing outer scale of turbulence and, as can be seen from Fig.3a, the transmitted radiation intensity for $\leq 50\text{m}$ is virtually independent of the outer scale of turbulence.

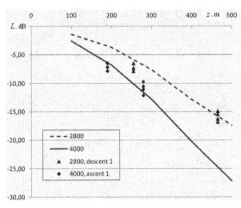

Fig. 4. Total attenuation of acoustic waves propagating along vertical paths versus altitude. Here the solid curves show the results of our Monte Carlo calculations; closed triangles and circles show results of acoustic measurements in [21] with a tethered balloon.

Figure 4 shows the total attenuation of acoustic waves propagating along vertical paths versus distance. Here the solid curves show the results of our Monte Carlo calculations, and closed triangles and circles show results of acoustic measurements performed in [21] with a tethered balloon. Calculations were performed for the vertical profiles of the atmospheric temperature and relative air humidity measured in [21] during first accent and descent of the tethered balloon on April 17, 1973. A good agreement of the results of our Monte Carlo calculations with the experimental data [21] can be seen. This demonstrates the efficiency of the developed Monte Carlo algorithm.

Statistical estimates of the transmitted radiation intensity for finite-aperture sources (with a circular aperture 1 m in diameter and a Gaussian distribution of emitted radiation) with a frequency of 2 kHz demonstrated that in the examined angular source divergence angles, it increased by 66–68% compared to that for the point source. For the uniform distribution of emitted radiation over the source aperture, $I_{tr}(0°, 2.5°)$ remained virtually unchanged. For $F = 3\text{kHz}$, $L_0 = 10\text{m}$, $\phi = 2.5°$, and Gaussian distribution of emitted radiation, $I_{tr}(0°, 2.5°) = 1.6 \cdot 10^{-4}$ W/m^2, that is, it increased by a factor of 2.2 compared to $Itr(0°, 2.5°) = 7.26 \cdot 10^{-5}$ W/m^2 for

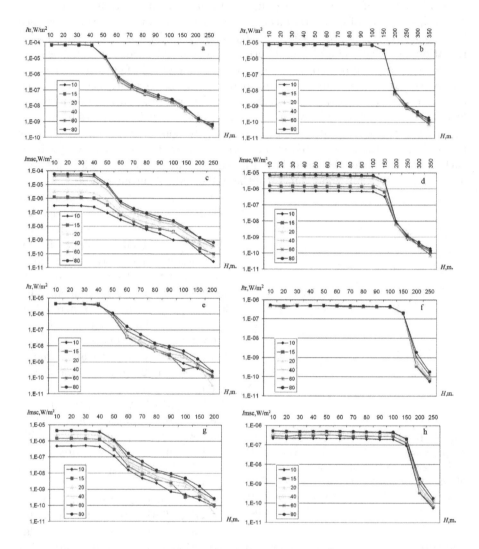

Fig. 3. Distribution of the intensity of transmitted ($I_t r$, W/m^2) and multiply scattered radiation (I_{msc}, W/m^2) over the detector zones for $F = 1.7$kHz, $\phi = 5$ (a and c) and 15° (b and d); $F = 4$kHz, $\phi = 5$ (e and g) and 15° (f and h) the indicated values of the outer scale of atmospheric turbulence

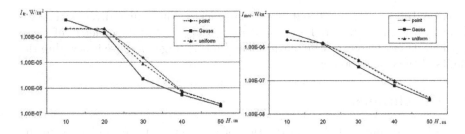

Fig. 5. Effect of the finite circular source aperture ($D = 1$m) with uniform and Gaussian distributions of emitted radiation on the intensity of transmitted (I_{tr}) and multiply scattered acoustic radiation I_{mcs} for $F = 3$kHz, $\phi = 2.5°$, and $L_0 = 10$m

the point source (see Fig. 5). At the same time, it remained virtually unchanged for the uniform distribution of emitted radiation.

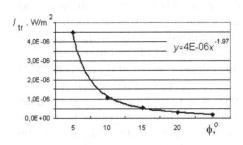

Fig. 6. Dependence of the transmitted radiation intensity on the source divergence angle and its analytical approximation by the power-law dependence (the solid curve) for $F = 4$kHz and $L_0 = 10$m

Analytical approximation of the results of Monte Carlo calculations by power-law, logarithmic, and exponential dependences demonstrated that they are best described by a power-law dependence of the form

$$I_{tr}(0°, \phi) = A\phi^{-B}, \qquad (7)$$

where I_{tr} is in W/m^2 and ϕ is in degrees, with the correlation coefficient close to 1.

Figure 6 shows the dependence $I_{tr}(0°, \phi)$ for the radiation frequency $F = 4$kHz and outer scale of turbulence $L_0 = 10$m. It can be seen that when the source divergence angle increases from 5 to 25°, I_{tr} decreases by 96%, which is essential and confirms the necessity of application of massive protective shields in sodars [1]. Table 1 below gives values of the corresponding constants A and B entering into formula (17) for $L_0 = 10$m and typical sodar frequencies.

Table 1. Values of the coefficients in Eq. (7)

F, kHz	A	B
1	$2.8 \cdot 10^{-3}$	2.00
1.7	$1.7 \cdot 10^{-3}$	2.00
2	$1.4 \cdot 10^{-3}$	2.03
3	$2 \cdot 10^{-5}$	2.02
4	$4 \cdot 10^{-6}$	1.97

From Table 1 it follows that the dependence on the source divergence angle is quadratic in character. It is impossible to obtain the dependence of these coefficients on the outer scale of turbulence in this stage, because it is within the limits of the calculation error.

5 Conclusions

Statistical estimates of the contribution of multiply scattered radiation to the intensity of acoustic radiation transmitted through the lower 500-m atmospheric layer demonstrated that for a frequency of 1.7 kHz, it increases from 15 to 80% with the outer scale of atmospheric turbulence. For a frequency of 4.5 kHz, it increases from 30% to the value comparable with the total transmitted radiation intensity.

The contribution of multiple scattering to the transmitted radiation intensity increased with the outer scale of turbulence from 10.5% (for $L_0 = 10$m) to 53% (for $L_0 = 20$m); for $L_0 = 40$m, the transmitted radiation intensity was completely determined by the contribution of multiple scattering. Statistical estimates demonstrate that the intensity of transmitted radiation within the limits of the cone of source radiation is virtually independent of the outer scale of atmospheric turbulence. The decrease in the transmitted radiation intensity with increase in the source divergence angle is quadratic in character. The transmitted radiation intensity I_{tr} decreases by 96% when the source divergence angle increases from 5 to 25° . This is essential and confirms the necessity of application of massive protective shields in sodars. These quantitative estimates can be used for interpretation of results of acoustic sounding and for prediction of the conditions of acoustic radiation propagation in the atmosphere. The results of Monte Carlo calculations are in good agreement with the available experimental data, which confirms the efficiency of the developed Monte Carlo algorithm.

References

1. Krasnenko, N.P.: Acoustic sounding of the atmospheric boundary layer. Publishing House Vodolei, Tomsk (2001)
2. Matuschek, R., Mellert, V., Kephalopoulos, S.: Model calculations with a fast field programme and comparison with selected procedures to calculate road traffic noise propagation under defined meteorological conditions. Acta Acustika United with Acustica 95, 941–949 (2009)
3. Delany, M.E.: Sound propagation in the atmosphere: A historical review. Acustica 38, 201–223 (1977)
4. Razin, A.V.: The mean field method in the problem of acoustic wave propagation in the turbulent atmosphere. Izv. Vyssh. Uchebn. Zaved. Radiofiz 51, 413–424 (2008)
5. Raspet, R., Lee, S.W., Kuester, E., Chang, D.C., Richards, W.F., Gilbert, R., Bong, N.: A fast-field program for sound propagation in a layered atmosphere above an impedance ground. J. Acoust. Soc. Am. 77, 345–352 (1985)
6. Shamanaeva, L.G., Burkatovskaya, Y.B.: Statistical estimates of the multiple scattering contribution to the acoustic radiation intensity transmitted through the lower 500-meter layer of the atmosphere. Russ. Phys. J. (12), 1297–1306 (2004)

7. Shamanaeva, L.G., Burkatovskaya, Y.B.: Statistical estimates of multiple scattering contribution to the transmitted acoustic radiation intensity. In: Proc. 14th Int. Symp. Adv. Bound. Layer Remote Sens, pp. 14–16. Garmish-Partenkirchen (2006)

8. Ostashev, V.E.: Sound propagation in moving media, Nauka, Moscow (1992)

9. Baikalova, R.A., Krekov, G.M., Shamanaeva, L.G.: Statistical estimates of the multiple scattering contribution in sound propagation through the atmosphere. Opt. Atmos. 1(5), 25–29 (1988)

10. Tatarskii, V.I.: Wave propagation in the turbulent atmosphere. Nauka, Moscow (1967)

11. ANSI Standard S1–26–1995 (R2009). Method for calculation of the absorption of sound by the atmosphere

12. ISO 9613–1:1996–3–05

13. Shamanaeva, L.G.: The dependence of sound extinction on the parameters of thermal turbulence in the atmospheric boundary layer. J. Acoust. Soc. Am. 73(3), 780–784 (1983)

14. Krasnenko, N.P., Shamanaeva, L.G.: Sodar measurements of the structural characteristics of temperature fluctuations and the outer scale of turbulence. Meteorol. Z. 7, 392–397 (1998)

15. Banakh, V.A., Rahm, S., Smalikho, I.N., Falits, A.V.: Measurement of atmospheric turbulence parameters by the coherent pulse wind lidar, scanning vertically. Opt. Atm. Okeana 20, 1115–1120 (2009)

16. Glagolev, Y.A.: Handbook on the physical parameters of the atmosphere, Gidrometeoizdat, Leningrad (1970)

17. DeLoach, R.: On the excess attenuation of sound in the atmosphere. NASA Technical Note D-7832, Washington (1975)

18. Marchuk, G.I., Mikhailov, G.A., Nazaraliev, M.A., Darbinyan, R.A., Kargin, B.A., Elepov, B.S.: Monte Carlo method in atmospheric optics. Nauka, Novosibirsk (1976)

19. Belov, V.V., Burkatovskaya, Y.B., Krasnenko, N.P., Shamanaeva, L.G.: Statistical estimates of the influence of the angular source divergence angle on the characteristics of transmitted acoustic radiation. Russ. Phys. J. (12), 1264–1270 (2009)

20. Krekov, G.M., Shamanaeva, L.G.: Statistical estimates of the spectral brightness of the twilight Earth's atmosphere, pp. 180–186. Atmosphheric Optics, Gidrometeoizdat, Leningrad (1974)

21. Aubry, M., Baudin, F., Weil, A., Rainteau, P.: Measurement of the total attenuation of acoustic waves in the turbulent atmosphere. J. Geophys. Res. 79(36), 5598–5606 (1974)

Parallelization of the Genetic Algorithm in Training of the Neural Network Architecture with Automatic Generation

Lyudmila Bilgaeva and Nikolay Burlov

East-Siberian State University of Technology and Management,
Street Kluchevskaya, 40v, 670013, Ulan-Ude, Russia
bilgaeva@mail.ru,
kohgpat@gmail.com

Abstract. This paper describes genetic algorithm of neural network training with automatic architecture generation, proposed method parallelization of training and modification to this method. And contains A comparative analysis of the original algorithm without the use of parallelization with proposed parallelization algorithm by splitting into groups and exchange of individuals between groups.

Keywords: genetic algorithm, genetic algorithm parallelization, neural network, neural network training, neural network architecture generation.

1 Introduction

Recent time parallelization of neural networks training is one of the trends in the development of neural networks. Due to the fact that the training process is complex, requiring a time consuming task. With the increasing number of inputs and complexity of the neural network architecture of computing the number of operations required for computation and training time will only increase. Parallelization of the neural network training can significantly reduce the amount of time required for training. There is a fairly large number of papers devoted to this subject. Most of them, considered parallelization of classical gradient methods of training neural networks (Backpropagation methhod) [1,3,4,6,7,8,9]. There is also a small number of papers on the parallelization of training neural networks based on genetic algorithms as a mechanism of weights correction with a given architecture of the neural network [2,5]. In this paper, we propose an approach to parallelization of neural networks using a more complex genetic algorithm, belonging to the category of neuroevolution genetic algorithms. The proposed algorithm provides correction weights simultaneously with the automatic construction of a neural network architecture. Additionally offers one of the possible modifications of the proposed method. Computational experiments were carried out for the problem of time series prediction.

A. Dudin et al. (Eds.): ITMM 2014, CCIS 487, pp. 44–49, 2014.

2   Genetic Algorithm of Neural Network Training with Automatic Architecture Generation

Proposed genetic algorithm utilizes direct genetic encoding. Direct encoding allows simplify interpretation of neural network architecture obtained through genetic search. [10] Genetic algorithm based on supervised learning. The choice of supervised learning caused by the usage of the neural networks for the problem of forecasting of time-series in the experiment part. The algorithm starts with the process of generating the initial population of individuals. Each individual of the initial population is a minimum size neural network with pre-established number of input and output neurons, depending on the conditions of the problem. All input and output neurons connects with each other after creation. Minimum size neural network presented in Figure 1 and it's data shown in Table 1.

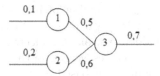

Fig. 1. Minimum size neural network

Table 1. Minimum size neural network data

Neuron, NO	1	2	3
Weights	[0, 1]	[0, 2]	[0,5; 0,6]
Input connections	[network input 1]	[network input 2]	[1, 2]
Output connections	[3]	[3]	[network output 1]

Training starts right after initial population generation. Training process uses training samples. Each individual in population performs computation on every sample in training samples and compares individual result with desired result from the sample. After this each individual calculates it's fitness function value defined as inverse of summarized error on every sample in training samples. Training performs genetic operations, such as reproduction and mutation, over individuals of the population. Modified classic one-point crossover operation used as reproduction operation in algorithm [3]. Crossover point defined as middle value of number of neurons in parent network with bigger number of neurons. After computation of the crossover point, algorithm copies neurons from parent individuals into a new individual. Connections of the neurons also copied. If both neurons, which this connection connects, get copied into a new individual, connection remain unchanged. Otherwise, if one of the neurons is absent, connection changes and instead of non-existing neuron it uses another, randomly chosen neuron. Reproduction operation performs over a given number of individuals in the population with best value of fitness function. Each

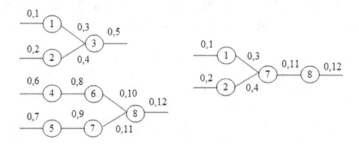

Fig. 2. Reproduction with crossover point equals 2

selected individual selects another individual from the selected group to perform reproduction. Reproduction operation presented in Figure 2.

Mutation performs over a given number of individuals with worst value of fitness function. Mutation consists two phases. On the first phase, mutation randomly adds or remove hidden neurons to selected individuals. On the second phase, mutation randomly changes weights of the connections. After genetic operations algorithm sorts population based on fitness function value and removes individuals with worst fitness function value from the population. Reproduction, mutation and removal of weak individuals performs certain number of times. Later starts comparison of results of best individual on testing samples with provided testing error threshold. If the error of best individual less than a given error threshold, training is considered successful and the algorithm stops. Otherwise, training starts again.

3   Parallelization of Genetic Algorithm

Proposed parallelization method based on splitting entire population of individuals into separate groups. The number of groups is set at the begging of the algorithm. The number of individuals in each group is equal or nearly equal. Each group trains independently. This allows to make several different genetic searches at the same time. After completion of training in each group, algorithm selects best individual from groups. Training with splitting population into groups presented in Figure 3.

The work also addressed method of partitioning the population into groups and splitting the training sample into groups so that each group was trained on its portion of the sample shown in Figure 4. This approach can significantly reduce the time and number of operations required for each training group compared with the previously proposed approach. However, this approach can only be used under certain conditions, namely the absence of a logical connection between sample examples of learning itself. In the context of the problem of time series prediction, this approach can not be used in connection with what has not been considered in detail.

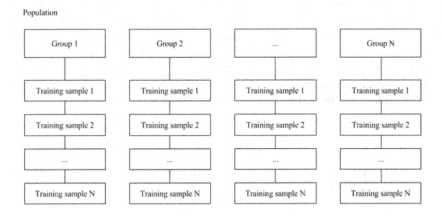

Fig. 3. Training with splitting population into groups

Fig. 4. Training with splitting population and training samples into groups

In addition to the partition of the population into groups, this paper proposes the use of an exchange of individuals between groups. At each learning step, after performing the reproduction, mutation and deletion of weak individuals, each group performs the exchange, in which randomly selected a predetermined number of individuals with the best fixtures and move them to another randomly selected group. This change can increase the diversity of individuals in each group, allowing groups to coordinate their search direction, based on the training results already obtained in the other groups. This modification is also applicable to the approach to the partition of the population into groups and crashed on a training sample group.

4   Results of Experiments

To compare performance of genetic algorithm without parallelization and with parallelization and also with parallelization and exchange of individuals, series of experiments was performed. Each experiment contained neural network training with automatic architecture generation on the problem of forecasting time-series. Experiment time-series had information about numerical parameter over 14 years. Training and testing samples was created by using window method with window size of 4 years. Training samples size was 7 and testing samples size was 3. Neural network gets 4 input values contains information about 4 years and returns value for the next year. Population had 100 individuals. The number of generations equals 30. The number of groups was set to 4. The sampling of results are shown in the Table 2 and it's graph presented in Figure 5.

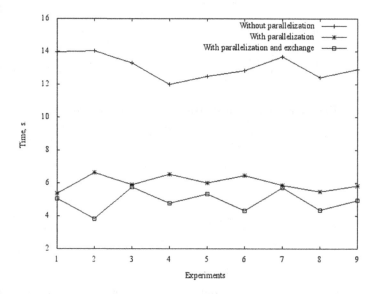

Fig. 5. Time per experiment graph

Table 2. Time per experiment

Method	1	2	3	4	5	6	7	8	9
1	13,98	14,03	13,3	12	12,5	12,84	13,7	12,42	12,9
2	5,37	6,64	5,89	6,53	6	6,46	5,85	5,46	5,84
3	5,06	3,82	5,77	4,77	5,32	4,32	5,71	4,36	4,93

Where Method 1 - without parallelization, Method 2 - parallelization with splitting into groups, Method 3 - parallelization with splitting into groups and exchange.

5   Conclusion

The experimentation results shows the success of the use of proposed parallelization method for presented genetic algorithm based on splitting individuals of population into groups and his modification with the exchange of individuals.

References

1. Sierra-Canto, X., Madera-Raminez, F., Uc-Centina, V.: Parallel training of a back-propaganation neural networks using CUDA. In: 9th International Conference on Machine Learning and Applications, pp. 307–312. IEEE Computer Society, Washington (2010)
2. Kattan, A.R.M., Abdullah, R., Salam, R.A.: Training Feed-Forward Neural Networks Using a Parallel Genetic Algorithm with the Best Must Survive Strategy. In: Conference on Intelligent Systems, Modelling and Simulation, pp. 96–99. IEEE Computer Society, Liverpool (2010)
3. Scanzio, S., Cumani, S., Gemello, R., Manip, F., Laface, P.: Parallel implementation of Artificial Neural Network training for speech recognition. Pattern Recognition Letters 31(11), 1302–1309 (2010), http://Elsevier.com
4. Vesely, K., Burget, L., Grezl, F.: Parallel Training of Neural Networks for Speech Recognition. In: Interspeech 2010 (2010), http://noel.feld.cvut.cz/gacr0811/publ/VES10b.pdf
5. Montana, D.J., Davis, L.: Training Feedforward Neural Networks Using Genetic Algorithms, http://ijcai.org
6. Saratchandran, P., Sundararajan, N., Foo, S.K.: Parallel Implementations of Back-propagation Neural Networks on Transputers: A Study of Training Set Parallelism, River Edge, NJ. World Scientific, Singapore (1996)
7. Sittig, D.F., Orr, J.A.: A parallel implementation of the backward error propagation neural network training algorithm: experiments in event identification. J. Computers and Biomedical Research 25(6), 547–561 (1992)
8. Li, C.H., Yang, L.T., Li, M.: Parallel Training of An Impovered Neural Networks for Text Categorization. International Journal of Parallel Programming 42(3), 505–523 (2013)
9. Guan, S.-W., Li, S.: Parallel Growing and Training of Neural Networks Using Output Parallelism. J. IEEE Transactions on Neural Networks 13(3), 542–550 (2002)
10. Stanley, K., Miikkulainen, R.: Evolving neural networks through augmenting topologies. J. Evolutionary Computation 10, 99–127 (2002)

Stationary Distribution Insensitivity of a Closed Queueing Network with Non-active Customers

Julia Bojarovich and Yuliya Dudovskaya

Francisk Skorina Gomel State University,
Gomel, Belarus
dudovskaya@gmail.com, juls1982@list.ru

Abstract. Stationary functioning of a closed queueing network with temporarily non-active customers is analyzed. Non-active customers are located in the network nodes in queues, being not serviced. For a customer, the opportunity of passing from its ordinary state to the temporarily non-active state (and backwards) is provided. Quantity of work for customer service is a random distributed value. Stationary distribution insensitivity with respect to functional form of distribution of work quantity for customer service is established.

Keywords: closed queueing network, temporarily non-active customers, stationary distribution insensitivity.

1 Introduction

Currently, attention to queueing theory is mainly stimulated by the need to apply results of this theory to important practical problems. During the past years, an important research effort has been devoted to the problem of queueing systems reliability. Herewith, the problem of customer reliability becomes relevant too. Indeed not only queueing system can break down. Customers may also lose their quality indicators. Queueing network with temporarily non-active customers is a model with customers, which are partly unreliable. The necessity of their study was caused by practical considerations, because such networks allow us to consider models with partially unreliable customers. Non-active customers are located in the network systems in queues, being not serviced. For a customer, the opportunity of passing from its ordinary state to the temporarily non-active state (and backwards) is provided. Non-active customers can be interpreted as customers with defect that makes them unfit for service. G. Tsitsiashvili and M. Osipova [1,2] have observed an open exponential queueing network with non-active customers and have established the form of stationary distribution.

The standard assumption in analysis of classical queueing networks [3,4] is that service time is exponentially distributed random value. But real numerous statistical data prove the opposite. Therefore there is an actual problem to develop an analytical apparatus for the study of queueing networks with arbitrary functions of service time distribution. Currently, this problem attracts increasing

A. Dudin et al. (Eds.): ITMM 2014, CCIS 487, pp. 50–58, 2014.

attention of researchers. The first result about stationary distribution insensitivity belongs to B.A. Sevastyanov, who has observed queueing system $M/G/m/0$ and has proved stationary distribution insensitivity [5]. BCMP-theorem (Baskett, Chandy, Muntz, Palacios) [6] is the first result about stationary distribution insensitivity for queueing networks. We have generalized the result [1,2] in the case of random distributed service times [7] – [9]. We have established stationary distribution insensitivity with respect to functional form of service time distribution.

V. A. Ivnitsky [10] has considered quite interesting class of queueing networks: customer service has not "temporal" but so-called "energetical" interpretation. Every service operation is characterized by the random variable of work to be performed. Stationary distribution insensitivity with respect to functional form of distribution of work quantity for customer service has been obtained for different classes of open and closed queueing networks [10].

This paper provides stationary functioning of a closed queueing network with temporarily non-active customers. Quantity of work for customer service is a random distributed value. Stationary distribution insensitivity with respect to functional form of distribution of work quantity for customer service is established.

2 Queueing Network Description

A closed queueing network with the set of systems $J = \{1, 2, \ldots, N\}$ is considered. M customers are circulating in the network. Non-active customers are located in the network systems in queues, being not serviced. There are input Poisson flows of signals with rates ν_i and φ_i, $i \in J$. When arriving at the system $i \in J$ the signal with rate ν_i induces an ordinary customer, if any, to become a non-active. When arriving at the system $i \in J$ the signal with rate φ_i induces an non-active customer, if any, to become an ordinary. Signals do not need service.

Let $n_i(t), n'_i(t)$ are numbers of ordinary and non-active customers in the system $i \in J$ at time t accordingly.

Stochastic process $z(t) = ((n_i(t), n'_i(t)), i \in J)$ is considered. Space of states for process $z(t)$ is $Z = \{((n_1, n'_1), \ldots, (n_N, n'_N)) | n_i, n'_i \geq 0, \sum_{i \in J}(n_i + n'_i) = M, i \in J\}$.

Numbering of ordinary customers in the system queue is made from the "tail" of the queue to the device. Non-active customers in the queue of the system $i \in J$ are numbered as follows: a customer, which has become non-active in the last turn, has number n'_i. When arriving at the system $i \in J$ the signal with rate ν_i induces an ordinary customer with number 1 to become a non-active customer with number $n'_i + 1$. When arriving at the system $i \in J$ the signal with rate φ_i induces a non-active customer with number n'_i to become an ordinary customer with number 1. So, the set of customers numbers in the system $i \in J$ is $(1, \ldots, n'_i, 1, \ldots, n_i)$.

The discipline of service is LCFS-PR. When arriving at the system $i \in J$ a customer receives immediate service and gets number $n_i + 1$. Displaced customer keeps number n_i and becomes the first in the queue to finish its service. Customer service has not "temporal" but so-called "energetical" interpretation. Every service operation is characterized by the random variable of work to be performed. Quantities of work for customer service are independent random distributed values $\eta_i(n_i + n'_i)$ with functions of distribution $B_i(n_i + n'_i, z)$ ($B_i(n_i + n'_i, 0) = 0, i \in J$) and expected values $\tau_i(n_i + n'_i) < \infty$. The speed of customer service is $\alpha_i(n_i + n'_i)$, $i \in J$. Here n_i, n'_i are numbers of ordinary and non-active customers in the system $i \in J$ accordingly. After the service in the system $i \in J$ the customer passes to the system $j \in J$ with the probability $p_{i,j}$ ($\sum_{j=1}^{N} p_{i,j} = 1$). Let $p_{i,i} = 0, i \in J$.

A traffic equations system is:

$$\varepsilon_i = \sum_{j=1}^{N} \varepsilon_j p_{j,i}, \quad i \in J. \tag{1}$$

It has been proved [4], that traffic equations system has the unique non-trivial solution up to constant.

3   Stationary Distribution Insensitivity

The model of closed queueing network with temporarily non-active customers when $B_i(n_i + n'_i, z) = 1 - exp\{-\mu_i z\}$ ($z > 0$, $\mu_i > 0$), $\tau_i(n_i + n'_i) = 1/\mu_i$ and $\alpha_i(n_i + n'_i) = 1$, $i \in J$, has been considered in [7]. The following theorem has been proved.

Theorem 1. *Markov process* $z(t) = ((n_i(t), n'_i(t)), i \in J)$ *is ergodic and has stationary distribution*

$$p((n_1, n'_1), \ldots, (n_N, n'_N)) = \frac{1}{G(M, N)} p_1(n_1, n'_1) \ldots p_N(n_N, n'_N),$$

where $((n_1, n'_1), \ldots, (n_N, n'_N)) \in Z.$

$$p_i(n_i, n'_i) = \left(\frac{\varepsilon_i}{\mu_i}\right)^{n_i} \left(\frac{\varepsilon_i \nu_i}{\mu_i \varphi_i}\right)^{n'_i}, i \in J,$$

ε_i *is the traffic equations system solution.* $G(M, N)$ *– a normalizing constant, which can be found from the condition:*

$$\sum_{((n_1, n'_1), \ldots, (n_N, n'_N)) \in Z} p((n_1, n'_1), \ldots, (n_N, n'_N)) = 1. \tag{2}$$

We consider a closed queueing network, where quantities of work for customer service are independent random distributed values. In this case $z(t)$ is not a Markov process.

Denote by $\psi_{i,k}(t)$ – the remaining quantity of work for service of the customer, which has position k in the system i at time t, $\psi_i(t) = (\psi_{i,1}(t), \ldots, \psi_{i,n_i+n'_i}(t))$, $i \in J$.

$$\frac{d\psi_{i,n_i+n_i'}(t)}{dt} = -\alpha_i(n_i + n_i'), \ i \in J.$$

So we introduce into consideration Markov process $\zeta(t) = (z(t), \psi(t))$, where $\psi(t) = (\psi_1(t), \dots, \psi_N(t))$.

Denote by

$$F(z,x) = F(z, x_{1,1}, \dots, x_{1,n_1+n_1'}; x_{2,1}, \dots, x_{2,n_2+n_2'}; \dots; x_{N,1}, \dots, x_{N,n_N+n_N'}) =$$

$$= \lim_{t\to\infty} P\{z(t) = z, \psi_{i,1}(t) < x_{i,1}, \dots, \psi_{i,n_i+n_i'}(t) < x_{i,n_i+n_i'}, i \in J\}, \ z \in Z,$$

$$x_{k,l} \in \mathbb{R} \ \forall \, k = \overline{1,N}, \ l = \overline{1, n_k + n_k'}.$$

Functions $F(z,x)$ are called stationary functions of probabilities states distribution of the process $\zeta(t)$.

Theorem 2. *Markov process $\zeta(t)$ is ergodic. Stationary functions of probabilities states distribution of the process $\zeta(t)$ are:*

$$F(z,x) = G^{-1}(M,N)p_1(n_1,n_1')p_2(n_2,n_2')\dots p_N(n_N,n_N') \times \qquad (3)$$

$$\times \prod_{i=1}^{N} \prod_{s=1}^{n_i+n_i'} \frac{1}{\tau_i(s)} \int_0^{x_{i,s}} (1 - B_i(s,u))du, \ z \in Z,$$

where

$$p_i(n_i, n_i') = \varepsilon_i^{n_i} \left(\frac{\varepsilon_i \nu_i}{\varphi_i}\right)^{n_i'} \prod_{s=1}^{n_i+n_i'} \frac{\tau_i(s)}{\alpha_i(s)}, \qquad (4)$$

ε_i *is the traffic equations system solution. $G(M,N)$ is a normalizing constant, which can be found from the condition (2).*

Proof. Denote by $e_i \in Z$ – the vector, which coordinates equal 0 with the exception of $(n_i, n_i') = (1,0)$, and denote by $e_i' \in Z$ – the vector, which coordinates equal 0 with the exception of $(n_i, n_i') = (0,1)$, $i \in J$.

We consider the process $\zeta(t)$. In the case of exponentially distributed service times the process $z(t)$ is ergodic by ergodic Markov theorem. The process $\zeta(t)$ is also ergodic, because $\zeta(t)$ is obtained from $z(t)$ by adding of continuous components.

The process $\zeta(t)$ changes its states due to incoming signals. Such changes we will call spontaneous changes.

Suppose that h is a small time interval and consider the probability

$$P\{z(t+h) = z, \psi_{i,1}(t+h) < x_{i,1}, \dots, \psi_{i,n_i+n_i'}(t+h) < x_{i,n_i+n_i'}, i \in J\}.$$

This event may occur in the following ways:

1. From the moment t during time h there were no spontaneous changes and service in any system was not over. The probability of this event is

$$P\{z(t) = z, \psi_{i,1}(t) < x_{i,1}, \ldots, \alpha_i(n_i + n_i')hI_{n_i>0} \leq \psi_{i,n_i+n_i'}(t) <$$

$$< x_{i,n_i+n_i'} + \alpha_i(n_i + n_i')hI_{n_i>0}, \, i \in J\}\times$$

$$\times(1 - \sum_{i=1}^{N}(\nu_i I_{n_i>0} + \varphi_i I_{n_i'>0})h + o(h)).$$

2. During time h a customer has been serviced in the system $j \in J$ and has been routed to the system $i \in J$. There were no spontaneous changes.

$$P\{z(t) = z - e_i + e_j, \psi_{k,1}(t) < x_{k,1}, \ldots, \alpha_k(n_k + n_k')hI_{n_k>0} \leq \psi_{k,n_k+n_k'}(t) <$$

$$< x_{k,n_k+n_k'} + \alpha_k(n_k + n_k')hI_{n_k>0}, \, k \in J, \, k \neq i, \, k \neq j,$$

$$\psi_{j,1}(t) < x_{j,1}, \ldots, \psi_{j,n_j+n_j'}(t) < x_{j,n_j+n_j'}, \psi_{j,n_j+n_j'+1}(t) <$$

$$< \alpha_j(n_j + n_j' + 1)(h - \theta),$$

$$\psi_{i,1}(t) < x_{i,1}, \ldots, \alpha_i(n_i + n_i' - 1)(h - \theta)I_{n_i>1} \leq \psi_{i,n_i+n_i'-1}(t) <$$

$$< x_{i,n_i+n_i'-1} + \alpha_i(n_i + n_i' - 1)(h - \theta)I_{n_i>1}\}\times$$

$$\times B_i(n_i + n_i', x_{i,n_i+n_i'} + \alpha_i(n_i + n_i')\theta)p_{j,i}I_{n_i>0},$$

here $0 < \theta < h$.

3. During time h an informational signal with rate ν_i has arrived at the system $i \in J$. There were no other spontaneous changes. No customer was serviced.

$$P\{z(t) = z + e_i - e_i', \psi_{k,1}(t) < x_{k,1}, \ldots, \alpha_k(n_k + n_k')hI_{n_k>0} \leq \psi_{k,n_k+n_k'}(t) <$$

$$< x_{k,n_k+n_k'} + \alpha_k(n_k + n_k')hI_{n_k>0}, \, k \in J, \, k \neq i,$$

$$\psi_{i,1}(t) < x_{i,1}, \ldots, \alpha_i(n_i + n_i')h \leq \psi_{i,n_i+n_i'}(t) <$$

$$< x_{i,n_i+n_i'} + \alpha_i(n_i + n_i')h\}(\nu_i h + o(h))I_{n_i'>0}.$$

4. During time h an informational signal with rate φ_i has arrived at the system $i \in J$. There were no other spontaneous changes. No customer was serviced.

$$P\{z(t) = z - e_i + e_i', \psi_{k,1}(t) < x_{k,1}, \ldots, \alpha_k(n_k + n_k')hI_{n_k>0} \leq \psi_{k,n_k+n_k'}(t) <$$

$$< x_{k,n_k+n_k'} + \alpha_k(n_k + n_k')hI_{n_k>0}, \, k \in J, \, k \neq i,$$

$$\psi_{i,1}(t) < x_{i,1}, \ldots, \alpha_i(n_i + n_i')hI_{n_i>1} \leq \psi_{i,n_i+n_i'}(t) <$$

$$< x_{i,n_i+n_i'} + \alpha_i(n_i + n_i')hI_{n_i>1}\}(\varphi_i h + o(h))I_{n_i>0}.$$

5. During time h there were more than two changes of queueing network condition. This probability is $o(h)$.

Therefore

$$P\{z(t+h) = z, \psi_{i,1}(t+h) < x_{i,1}, \ldots, \psi_{i,n_i+n'_i}(t+h) < x_{i,n_i+n'_i}, i \in J\} =$$

$$= P\{z(t) = z, \psi_{i,1}(t) < x_{i,1}, \ldots, \alpha_i(n_i + n'_i)hI_{n_i>0} \leq \psi_{i,n_i+n'_i}(t) <$$

$$< x_{i,n_i+n'_i} + \alpha_i(n_i + n'_i)hI_{n_i>0}, i \in J\} \times$$

$$\times (1 - \sum_{i=1}^{N}(\nu_i I_{n_i>0} + \varphi_i I_{n'_i>0})h + o(h)) +$$

$$+ \sum_{i=1}^{N} \sum_{j=1, j \neq i}^{N} P\{z(t) = z - e_i + e_j, \psi_{k,1}(t) < x_{k,1}, \ldots, \alpha_k(n_k + n'_k)hI_{n_k>0} \leq$$

$$\leq \psi_{k,n_k+n'_k}(t) < x_{k,n_k+n'_k} + \alpha_k(n_k + n'_k)hI_{n_k>0}, k \in J, k \neq i, k \neq j,$$

$$\psi_{j,1}(t) < x_{j,1}, \ldots, \psi_{j,n_j+n'_j}(t) < x_{j,n_j+n'_j}, \psi_{j,n_j+n'_j+1}(t) < \alpha_j(n_j + n'_j + 1)(h - \theta),$$

$$\psi_{i,1}(t) < x_{i,1}, \ldots, \alpha_i(n_i + n'_i - 1)(h - \theta)I_{n_i>1} \leq \psi_{i,n_i+n'_i-1}(t) < \quad (5)$$

$$< x_{i,n_i+n'_i-1} + \alpha_i(n_i + n'_i - 1)(h - \theta)I_{n_i>1}\} \times$$

$$\times B_i(n_i + n'_i, x_{i,n_i+n'_i} + \alpha_i(n_i + n'_i)\theta)p_{j,i}I_{n_i>0} +$$

$$+ \sum_{i=1}^{N} P\{z(t) = z + e_i - e'_i, \psi_{k,1}(t) < x_{k,1}, \ldots, \alpha_k(n_k + n'_k)hI_{n_k>0} \leq$$

$$\leq \psi_{k,n_k+n'_k}(t) < x_{k,n_k+n'_k} + \alpha_k(n_k + n'_k)hI_{n_k>0}, k \in J, k \neq i,$$

$$\psi_{i,1}(t) < x_{i,1}, \ldots, \alpha_i(n_i + n'_i)h \leq \psi_{i,n_i+n'_i}(t) <$$

$$< x_{i,n_i+n'_i} + \alpha_i(n_i + n'_i)h\}(\nu_i h + o(h))I_{n'_i>0} +$$

$$+ \sum_{i=1}^{N} P\{z(t) = z - e_i + e'_i, \psi_{k,1}(t) < x_{k,1}, \ldots, \alpha_k(n_k + n'_k)hI_{n_k>0} \leq$$

$$\leq \psi_{k,n_k+n'_k}(t) < x_{k,n_k+n'_k} + \alpha_k(n_k + n'_k)hI_{n_k>0}, k \in J, k \neq i,$$

$$\psi_{i,1}(t) < x_{i,1}, \ldots, \alpha_i(n_i + n'_i)hI_{n_i>1} \leq \psi_{i,n_i+n'_i}(t) <$$

$$< x_{i,n_i+n'_i} + \alpha_i(n_i + n'_i)hI_{n_i>1}\}(\varphi_i h + o(h))I_{n_i>0} + o(h).$$

Every probability from (5) may be expressed in terms of functions

$$F_t(z, x) = P\{z(t) = z, \psi_{i,1}(t) < x_{i,1}, \ldots, \psi_{i,n_i+n'_i}(t) < x_{i,n_i+n'_i}, i \in J\}.$$

Consider the decomposition of $F_t(z, x)$ in a Taylor series, taking into consideration that

$$P\{z(t) = z, \psi_{i,1}(t) < x_{i,1}, \ldots, \alpha_i(n_i + n'_i)h \leq \psi_{i,n_i+n'_i}(t) < x_{i,n_i+n'_i} + \alpha_i(n_i + n'_i)h, i \in J\} =$$

$$= F_t(z, x_{i,1}, \ldots, x_{i,n_i+n'_i} + \alpha_i(n_i + n'_i)h, i \in J) - \sum_{k=1}^{N} F_t(z, x_{i,1}, \ldots, x_{i,n_i+n'_i} + \alpha_i(n_i + n'_i)h,$$

$$i \in J, i \neq k; x_{k,1}, \ldots, x_{k,n_k+n_k'-1}, \alpha_k(n_k + n_k')h) + \ldots$$

$$+ F_t(z, x_{i,1}, \ldots, x_{i,n_i+n_i'-1}, \alpha_i(n_i + n_i')h, \ i \in J).$$

Therefore

$$P\{z(t) = z, \psi_{i,1}(t) < x_{i,1}, \ldots, \alpha_i(n_i + n_i')h \leq \psi_{i,n_i+n_i'}(t) < x_{i,n_i+n_i'} + \alpha_i(n_i + n_i')h, i \in J\} =$$

$$= F_t(z, x_{i,1}, \ldots, x_{i,n_i+n_i'}, \ i \in J) + \sum_{i=1}^{N} \frac{\partial F_t(z, x_{i,1}, \ldots, x_{i,n_i+n_i'}, \ i \in J)}{\partial x_{i,n_i+n_i'}} \alpha_i(n_i + n_i')h -$$

$$- \sum_{i=1}^{N} \frac{\partial F_t(z, x_{l,1}, \ldots, x_{l,n_l+n_l'}, \ l \in J, l \neq i; x_{i,1}, \ldots, x_{i,n_i+n_i'-1}, 0)}{\partial x_{i,n_i+n_i'}} \alpha_i(n_i + n_i')h + o(h).$$

We consider $B_i(n_i + n_i', x_{i,n_i+n_i'} + \theta)$ as a function of the variable θ, use its decomposition in a Taylor series and let t tend to infinity. So we obtain the following equations system:

$$F(z,x) = F(z,x) + h \sum_{i=1}^{N} \alpha_i(n_i + n_i') \left(\frac{\partial F(z,x)}{\partial x_{i,n_i+n_i'}} - \left(\frac{\partial F(z,x)}{\partial x_{i,n_i+n_i'}} \right)_{x_{i,n_i+n_i'}=0} \right) I_{n_i>0} -$$

$$- \left(\sum_{i=1}^{N} (\nu_i I_{n_i>0} + \varphi_i I_{n_i'>0}) h + o(h) \right) F(z,x) + \qquad (6)$$

$$+ h \sum_{j=1}^{N} \sum_{i=1, i \neq j}^{N} \alpha_j(n_j + n_j' + 1) p_{j,i} B_i(n_i + n_i', x_{i,n_i+n_i'}) \times$$

$$\times \left(\frac{\partial F(z + e_j - e_i, x)}{\partial x_{j,n_j+n_j'+1}} \right)_{x_{j,n_j+n_j'+1}=0} I_{n_i>0} +$$

$$+ \sum_{i=1}^{N} F(z + e_i - e_i', x)(\nu_i h + o(h) I_{n_i'>0} +$$

$$+ \sum_{i=1}^{N} F(z - e_i + e_i', x)(\varphi_i h + o(h)) I_{n_i>0} + o(h).$$

Subtracting $F(z,x)$ from both sides of (6), dividing both sides of (6) by h and letting h tend to zero, we obtain the following differential equations system:

$$F(z,x) \sum_{i=1}^{N} (\nu_i I_{n_i>0} + \varphi_i I_{n_i'>0}) =$$

$$= \sum_{i=1}^{N} \alpha_i(n_i + n_i') \left(\frac{\partial F(z,x)}{\partial x_{i,n_i+n_i'}} - \left(\frac{\partial F(z,x)}{\partial x_{i,n_i+n_i'}} \right)_{x_{i,n_i+n_i'}=0} \right) I_{n_i>0} +$$

$$+ \sum_{j=1}^{N} \sum_{i=1, i \neq j}^{N} \alpha_j (n_j + n'_j + 1) p_{j,i} B_i(n_i + n'_i, x_{i,n_i+n'_i}) \times$$

$$\times \left(\frac{\partial F(z + e_j - e_i, x)}{\partial x_{j,n_j+n'_j+1}} \right)_{x_{j,n_j+n'_j+1}=0} I_{n_i>0} + \tag{7}$$

$$+ \sum_{i=1}^{N} F(z + e_i - e'_i, x) \nu_i I_{n'_i>0} + \sum_{i=1}^{N} F(z - e_i + e'_i, x) \varphi_i I_{n_i>0}.$$

Divide (7) into the next local balance equations:

$$F(z, x)\left(\nu_i I_{n_i>0} + \varphi_i I_{n'_i>0} \right) = F(z + e_i - e'_i, x) \nu_i I_{n'_i>0} + F(z - e_i + e'_i, x) \varphi_i I_{n_i>0}, \tag{8}$$

$$\alpha_i (n_i + n'_i)\left(\left(\frac{\partial F(z, x)}{\partial x_{i,n_i+n'_i}} \right)_{x_{i,n_i+n'_i}=0} - \frac{\partial F(z, x)}{\partial x_{i,n_i+n'_i}} \right) I_{n_i>0} = \tag{9}$$

$$= \sum_{j=1, j \neq i}^{N} \alpha_j (n_j + n'_j + 1) p_{j,i} B_i(n_i + n'_i, x_{i,n_i+n'_i}) \times$$

$$\times \left(\frac{\partial F(z + e_j - e_i, x)}{\partial x_{j,n_j+n'_j+1}} \right)_{x_{j,n_j+n'_j+1}=0} I_{n_i>0}, \; i \in J.$$

Substituting $F(z, x)$, determined by means of (3), (4), into local balance equations (8), (9), considering traffic equation system (1), we obtain identity. □

Denote by $\{p(z), z \in Z\}$ – stationary distribution of the process $z(t)$. From the foregoing theorem, considering equality $p(z) = F(z, +\infty)$, we obtain

Corollary 1. *Process $z(t)$ is ergodic and has stationary distribution*

$$p(z) = G^{-1}(M, N) p_1(n_1, n'_1) p_2(n_2, n'_2) \ldots p_N(n_N, n'_N), \; z \in Z,$$

which does not depend on functional form of $B_i(s, x)$, $i \in J$. Probabilities $p_i(n_i, n'_i)$, $i \in J$, may be found by means of (4).

4   Conclusion

We have considered stationary functioning of a closed queueing network with temporarily non-active customers. Expression for stationary distribution has been derived. Finally, stationary distribution insensitivity with respect to functional form of distribution of work quantity for customer service is established. Research results have practical importance and may be used for real networks investigation.

References

1. Tsitsiashvili, G.S., Osipova, M.: Distributions in stochastic network models. Nova Publishers, Inc., US (2008)
2. Tsitsiashvili, G.S., Osipova, M.: Queueing models with different schemes of customers transformations. In: Proceedings of the 19th International Conference, Mathematical Methods for Increasing Efficiency of Information Telecommunication Networks, pp. 128–133 (2007)
3. Jackson, J.R.: Network of Waiting Lines. Oper. Research. 4, 518–521 (1957)
4. Gordon, W.J., Newell, G.F.: Closed queueing networks with exponential servers. Oper. Research. 15, 252–267 (1967)
5. Sevastyanov, B.A.: An ergodic theorem for Markov processes and its application to telephone systems with refusals. Theo. Prohah. Appl. 2, 104–112 (1957)
6. Baskett, F.: Open, Closed and Mixed Networks of Queues with Different Classes of Customers. J. Assoc. Comput. Mach. 22, 248–260 (1975)
7. Boyarovich, Y.S.: The stationary distribution invariance of states in a closed queueing network with temporarily non-active customers. Automation and Remote Control 73, 1616–1623 (2012)
8. Bojarovich, J., Malinkovsky, Y.V.: Stationary distribution invariance of an open queueing network with temporarily non-active customers. Tomsk State University. Journal of Control and Computer Science 20, 62–70 (2012)
9. Bojarovich, J., Malinkovsky, Y.: Stationary Distribution Invariance of an Open Queueing Network with Temporarily Non-active Customers. In: Dudin, A., Klimenok, V., Tsarenkov, G., Dudin, S. (eds.) BWWQT 2013. CCIS, vol. 356, pp. 26–32. Springer, Heidelberg (2013)
10. Ivnitsky, V.A.: Theory of queueing networks. Fizmatlit, Moscow (2004)

On Guaranteed Sequential Change Point Detection for TAR(1)/ARCH(1) Process

Yulia Burkatovskaya, Ekaterina Sergeeva, and Sergei Vorobeychikov

Institute of Cybernetics, Tomsk Polytechnical University,
30 Lenin Prospekt, 634050 Tomsk, Russia
Department of Applied Mathematics and Cybernetics,
Tomsk State University,
36 Lenin Prospekt, 634050 Tomsk, Russia
tracey@tpu.ru, sergeeva_e_e@mail.ru, sev@mail.tsu.ru
http://www.tpu.ru, http://www.tsu.ru

Abstract. The problem of guaranteed parameter estimation and change point detection of threshold autoregressive processes with conditional heteroscedasticity (TAR/ARCH) is considered. The parameters of the process are assumed to be unknown. A sequential procedure with guaranteed quality is proposed. The results of simulation are presented.

Keywords: TAR/ARCH, guaranteed parameter estimation, change point detection.

1 Introduction

The TAR models were first proposed by Tong in [1] and since then they became a standard class of nonlinear time series models. The values of the threshold process are determined not only by values of the process, but by the abrupt changes in its dynamics. The parameters, which regulate moving of the process through regions of state space, are called threshold parameters. Autoregressive heteroscedastic (ARCH) models have proved to be very useful for describing changing in volatility of econometric processes and other time series. The first efforts to combine aforementioned models and to research properties of such models were made in [2], [3], [4].

Most statistical limit theorems require the existence of stationary distribution of the processes, so it is important to obtain necessary and sufficient conditions of ergodicity for a given class of models. Such conditions are especially important for investigation of asymptotic properties of estimators of unknown parameters. For TAR(1) model a necessary and sufficient condition for geometrical ergodicity was considered in [5]. Conditions for existence geometric ergodicity of process generated by AR-ARCH model were proposed in [6], but they are more restrictive than those, that were obtained in [7] since the exact specifications of the AR/ARCH model were not used. A method for determining whether threshold

A. Dudin et al. (Eds.): ITMM 2014, CCIS 487, pp. 59–68, 2014.
© Springer International Publishing Switzerland 2014

AR/ARCH model is ergodic and what moments exist when it is ergodic was proposed in [8].

Estimation of unknown parameters of models with mixed structure, which consists of both linear and nonlinear parts, is very interesting for applications but quite a difficult task. A modified quasi-maximum likelihood estimator for AR(1)/ARCH(1) model based on truncation of the likelihood function was proposed in [7]. Estimators based on least squares method were considered in [9].

The idea of constructing estimators for unknown parameters of models using special stopping rule, in order to guarantee precisely their quality, was first proposed in [10]. Since then, it was widely used in literature for estimating parameters in models with independent and dependent data. In [11] a method for estimation unknown autoregressive parameters of AR/ARCH model with guaranteed accuracy based on weighted least square method was proposed. In [12] sufficient conditions for ergodicity of the process were obtained and guaranteed sequential estimators of the autoregressive parameters of TAR(1)/ARCH(1) model were proposed. The accuracy of the estimation depends on the procedure parameters.

The problem of change point detection arises often in different applications connected with time series analysis, financial mathematics, image processing, etc. Two types of algorithms are used to detect the change point: a posteriori method, when the estimation of the change point is conducted in a sample of a fixed size, and sequential methods, when the decision on change point can be taken after getting a next observation. The properties of the sequential procedures usually can be investigated either for the case of independent observations or when the number of observations tends to infinity [13], [14], [15].

Last decades autoregressive conditional heteroscedasticity processes are widely used in various applications. The problem of change point detection for this type of processes under different assumptions on parameters and noise distributions is considered in [16], [17], [18],[19], [20], etc. The properties of algorithms are studied asymptotically or by simulation. Theoretical investigation of algorithm properties for a fixed sample size is usually impossible.

In [11] we proposed to detect the instant of parameters' change in the AR(p)/ARCH(q) process by making use of guaranteed sequential estimators. The sequence of estimators was constructed and the estimators obtained on different time intervals were compared. In this study such approach is applied to the TAR(1)/ARCH(1) model. The properties (the asymptotic and non-asymptotic) of procedures are investigated. The special construction of the estimators based on the weighted least squares method guarantees prescribed accuracy of the proposed procedure, i.e. prescribed upper boundaries for the false alarm and delay probabilities.

Simulation experiments were conducted and the results showed good performance of the proposed procedure.

2   Problem Statement

We consider TAR(1)/ARCH(1) autoregressive process specified by the equation

$$
\begin{aligned}
x_{k+1} &= \lambda_1 x_k^+ + \lambda_2 x_k^- + \sqrt{\omega + \alpha^2 x_k^2}\,\xi_{k+1}; \\
x_k^+ &= \max\{0, x_k\}; \\
x_k^- &= \min\{0, x_k\},
\end{aligned}
\tag{1}
$$

where $\{\xi_k\}_{k\geq 0}$ is a sequence of independent identically distributed random variables with zero mean and unit variance, $\omega > 0$. The value of the parameter vector $\lambda = [\lambda_1, \lambda_2]$ changes from $\mu^0 = [\mu_1^0, \mu_2^0]$ to $\mu^1 = [\mu_1^1, \mu_2^1]$ at the change point θ. Values of the parameters before and after θ are supposed to be unknown. The difference between μ^0 and μ^1 satisfies the condition

$$
(\mu_i^0 - \mu_i^1)^2 \geq \Delta, \quad i = 1, 2,
\tag{2}
$$

where Δ is the known value defining the minimum difference between the parameters before and after the change point. The problem is to detect the change point θ from observations x_k.

3   Ergodic Region of the Process

In [12] sufficient conditions of the ergodicity according to [21] were obtained

$$
\begin{aligned}
&\lambda_1 < 1, \quad \lambda_2 < 1, \quad \lambda_1\lambda_2 < 1; \\
&\alpha^2 < \frac{\min\{(1-\lambda_1)^2, (1-\lambda_2)^2, (1-\lambda_1)^2(1-\lambda_2)^2, (1-\lambda_1\lambda_2)^2\}}{(2-\lambda_1-\lambda_2)^2\mu^2},
\end{aligned}
\tag{3}
$$

where $\mu = \max\left\{\int_0^{\infty} z f_\xi(z)dz, -\int_{-\infty}^{0} z f_\xi(z)dz\right\}$.

At the Fig.1 one can see an example of ergodic (a) and non-ergodic (b) TAR/ARCH process. Non-ergodic process is characterized by sharp changes.

Note that if in process (1) the parameter $\alpha = 0$ then the noise variance is constant. As a result we have the process TAR(1) with well-known necessary and sufficient conditions of ergodicity [5]: $\lambda_1 < 1, \lambda_2 < 1, \lambda_1\lambda_2 < 1$. Our conditions (3) take the same form in the case $\alpha = 0$.

4   Guaranteed Parameter Estimator

Since the parameters both before and after the change point are unknown, it is logical to use estimators of the unknown parameters in the change point detection procedure. We use the sequential estimators proposed in [12] for a case of symmetric density distribution function. The main advantage of the estimators is their preassigned mean square accuracy depending on the parameter of the estimation procedure. Here we construct modified estimators which allow us to use non-symmetric noise distributions.

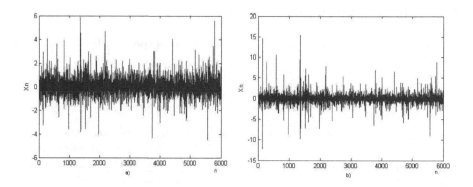

Fig. 1. Ergodic (a) and non-ergodic (b) processes

It should be noted that if parameters ω and α are unknown then process (1) has unknown and unbounded from above noise variance. To obtain a process with bounded noise variance we denote $\max\{1, |x_k|\}$ as m_k and rewrite the process in the form

$$y_{k+1} = \lambda_1 y_{k,1} + \lambda_2 y_{k,2} + \gamma_k \xi_{k+1};$$

$$y_{k+1} = \frac{x_{k+1}}{m_k}, \quad y_{k,1} = \frac{x_k^+}{m_k}, \quad y_{k,2} = \frac{x_k^-}{m_k}, \quad \gamma_k = \frac{\sqrt{w + \alpha x_k^2}}{m_k}. \tag{4}$$

The noise variance of the process $\{y_k\}$ is bounded from above by the unknown value $(\omega + \alpha^2)$. To eliminate the influence of the unknown constant the special factor Γ_N constructed by first N observations is used. If the distribution density function is symmetric, i.e. $f_\xi(x) = f_\xi(-x)$, and satisfy certain conditions [22] then Γ_N can be taken in the following form

$$\Gamma_N = C_N \sum_{k=1}^{N} \left(\frac{x_k}{\min\{1, |x_k|\}} \right)^2, \quad C_N = E \left(\sum_{k=1}^{N} \xi_k^2 \right)^{-1}. \tag{5}$$

It was proved in [12] that

$$E \frac{1}{\Gamma_N} \leq \frac{1}{\omega + \alpha^2}. \tag{6}$$

If $\{\xi_k\}$ have standard normal distribution then the sum $\sum_{k=1}^{N} \xi_k^2$ has χ^2 distribution with N degrees of freedom. In this case one has

$$C_N = \frac{1}{2^{N/2}\Gamma(N/2)} \int\limits_0^{+\infty} x^{N/2-3} e^{-x/2} dx = \frac{\Gamma(N/2)}{4\Gamma(N/2-2)} = \frac{1}{(N-2)(N-4)}.$$

This constant is defined for $N \geq 5$.

If the function $f_\xi(x)$ is not symmetric and $|\lambda_1| < 1$, $|\lambda_2| < 1$ then we propose to construct the factor Γ_N as follows

$$\Gamma_N = C_N \sum_{k=1}^{N} \left(\frac{|x_k| + |x_{k-1}|}{\min\{1, |x_k|\}} \right)^2, \quad C_N = E \left(\sum_{k=1}^{N} \xi_k^2 \right)^{-1}. \tag{7}$$

The factor also satisfies condition (6) because from (1) one has

$$(\omega + \alpha^2)\xi_k^2 \leq (x_{k+1} - \lambda_1 x_k^+ + \lambda_2 x_k^-)^2/(\min\{1, |x_k|\})^2.$$

The proposed estimator of the parameter vector λ is written in the following form

$$\hat{\lambda}_i = \hat{\lambda}_i(H) = \frac{1}{\Gamma_N H} \sum_{k=N+1}^{\tau_i} v_{k,i} y_{k,i} y_{k+1}, \quad i = 1, 2, \tag{8}$$

where the stopping time $\tau_i = \tau_i(H)$ are defined by the following conditions

$$\tau_i = \min \left\{ t > N : \sum_{k=N+1}^{t} y_{k,i}^2 \geq \Gamma_N H \right\}, \quad i = 1, 2. \tag{9}$$

The weights $\{v_{k,i}\}$ possess the value unity at the interval $[N + 1, \tau_i - 1]$. At the moment τ_i they are defined from the equation

$$\sum_{k=N+1}^{\tau_i} v_{k,i} y_{k,i}^2 = \Gamma_N H, \quad i = 1, 2. \tag{10}$$

The properties of the estimator were established and proved in [12].

Theorem 1. *For process (1) satisfying conditions (3) stopping time τ_i (9) is finite with probability one. Estimators (8) are unbiased and the variance of the estimators is bounded from above*

$$E(\hat{\lambda}_i - \lambda_i)^2 \leq \frac{1}{H}, \quad i = 1, 2. \tag{11}$$

Hence, the parameter H defines the accuracy of the estimator.

It is shown in [12] that the random variable $\sqrt{H}\zeta_i = \sqrt{H}(\hat{\lambda}_i - \lambda_i)$ converges by distribution to the random variable Z with the characteristic function $\phi(y) = Ee^{-\eta^2 y^2/2}$, where $\eta^2 = (\omega + \alpha^2)/\Gamma_N$, i.e. $E\eta^2 < 1$. Using this result we can formulate the following theorem which gives an asymptotic upper bound for the probability of large values of the standard deviation for the estimator (8). This result is more precise then one obtained in [12].

Theorem 2. *If process (1) is ergodic, and the compensating factor Γ_N satisfies the following conditions $N \to \infty$, $N/H \to 0$ as $H \to \infty$, then for sufficiently large H*

$$P\left\{ \left(\hat{\lambda}_i - \lambda \right)^2 > x \right\} \leq 2 \left(1 - \Phi\left(\sqrt{xH} \right) \right), \tag{12}$$

where $\Phi(\cdot)$ is the standard normal distribution function

Proof. For ergodic process (1) as $N \to \infty$ for the compensating factor Γ_N (5) the following convergency in probability takes place [21]

$$\eta = (\omega + \alpha^2)/\Gamma_N \to^P E\left((\omega + \alpha^2)/\Gamma_N\right) \leq 1.$$

The condition $N/H \to 0$ guarantees us that the number of observation for the constructing of the compensating factor Γ_n is not too large.

Denote $E\eta$ as β^2 and consider as $H \to \infty$ the probability

$$\mathcal{P}\left\{\left(\hat{\lambda}_i - \lambda\right)^2 > x\right\} \to \int\limits_{y^2 > xH} e^{-\beta^2 y^2/2}\,dy \leq 2\left(1 - \Phi\left(\sqrt{xH}/\beta\right)\right).$$

Since $\beta < 1$ then one obtain (19). The theorem has been proved.

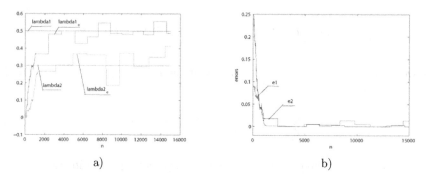

Fig. 2. Parameter estimators (a) and their errors (b)

On Fig. 2 a typical behavior of the proposed parameter estimators and their squared errors with $H = 100$ are shown. One can see that the estimators are rather close to the parameter values.

5   Change Point Detection Procedure

Consider now the change point detection problem for process (1). At the first stage, we define intervals $[\tau_i^{n-1} + 1, \tau_i^n]$, $n \geq 1$. The estimators $\hat{\lambda}_i^n$ of the parameters of process (1) are constructed on each interval. Then the estimators on intervals $[\tau_i^{n-l-1} + 1, \tau_i^{n-l}]$ and $[\tau_i^{n-1} + 1, \tau_i^n]$, where $l > 1$ is an integer, are compared. If the interval $[\tau_i^{n-1} + 1, \tau_i^n]$ does not include the change point θ, then vector λ on this interval is constant. It can be equal to the initial value μ^0 or the final value μ^1. Thus for certain n, if $\tau_i^{n-l} < \theta < \tau_i^{n-1} + 1$, the difference between values of the parameters on intervals $[\tau_i^{n-l-1} + 1, \tau_i^{n-l}]$ and $[\tau_i^{n-1} + 1, \tau_i^n]$ is no less then Δ. This is the key property for the change point detection. Two parameters can change at the moment θ, i.e. λ_1 and λ_2. Note that at the estimator

$\hat{\lambda}_i$ the values $y_{k,i}$ are used, and $y_{k,1}y_{k,2} = 0$. That allows us to construct change point detection procedures separately for every parameter.

We construct a set of sequential estimation plans

$$(\tau_i^n, \hat{\lambda}_i^n) = (\tau_i^n(H), \hat{\lambda}_i^n(H)), \quad n \geq 1, \quad i = 1, 2,$$

where $\{\tau_i^n\}$, $n \geq 0$ is the increasing sequence of the stopping instances ($\tau_0 = N$), and $\hat{\lambda}_i^n$ is the guaranteed parameter estimator on the interval $[\tau_i^{n-1}+1, \tau_i^n]$. The following condition holds true for the estimator

$$E(\hat{\lambda}_i^n(H) - \lambda_i)^2 \leq \frac{1}{H}. \tag{13}$$

Then we choose an integer $l > 1$. We associate the statistic J_i^n with the n-th interval $[\tau_i^n + 1, \tau_i^n]$ for all $n > l$

$$J_i^n = \left(\hat{\lambda}_i^n - \hat{\lambda}_i^{n-l}\right)^2. \tag{14}$$

This statistic is the squared deviation of the estimators with numbers n and $n - l$. Properties of the statistics are given in the following theorem.

Theorem 3. *The expectation of the statistics J_i^n (14) satisfies the following inequality:*

$$E[J_i^n | \tau_i^n < \theta] \leq \frac{2}{H};$$
$$E[J_i^n | \tau_i^{n-l} < \theta \leq \tau_i^{n-1}] \geq \Delta. \tag{15}$$

Proof. Denote the deviation of the estimator $\hat{\lambda}_i^n$ from the true value of the parameter λ_i as ζ_i^n. Let the parameter value remains unchanged until the instant τ_i^n, i.e., $\theta > \tau_i^n$. In this case, $\hat{\lambda}_i^n = \mu_i^0 + \zeta_i^n$, $\lambda_i^{n-l} = \mu_i^0 + \zeta_i^{n-l}$ and statistic (14) can be written in the form

$$J_i^n = \left(\zeta_i^n - \zeta_i^{n-l}\right)^2,$$

where

$$\zeta_i^n = \frac{1}{\Gamma_N H} \sum_{k=\tau_i^{n-1}+1}^{\tau_i^n} v_{k,i} y_{k,i} \gamma_k \xi_{k+1}. \tag{16}$$

The expectation of the statistic is equal to

$$E J_i^n = E(\zeta_i^n)^2 + E(\zeta_i^{n-l})^2 - 2E\zeta_i^{n-l}\zeta_i^n.$$

For every n according to (11) $E(\zeta_i^n)^2 \leq 1/H$. Using conditional expectation properties it can be shown that $E\zeta_i^{n-l}\zeta_i^n = 0$ for $l > 0$. Hence

$$E[J_i^n | \theta > \tau_i^n] \leq \frac{2}{H}. \tag{17}$$

Let the change of the parameter took place on the interval $[\tau_i^{n-l}, \tau_i^{n-1}]$ i.e. $\tau_i^{n-l} < \theta \le \tau_i^{n-1}$. In this case, $\hat{\lambda}_i^= \mu_i^1 + \zeta_i^n$, $\hat{\lambda}_i^{n-l} = \mu_i^0 + \zeta_i^{n-l}$, and statistic (14) is

$$J_i^n = (\mu_i^1 - \mu_i^0 + \zeta_i^n - \zeta_i^{n-l})^2.$$

The expectation of the statistics can de written as

$$EJ_i^n(\mu_i^1 - \mu_i^0)^2 + 2(\mu_i^1 - \mu_i^0)E(\zeta_i^n - \zeta_i^{n-l}) + E(\zeta_i^n - \zeta_i^{n-l})^2.$$

Taking into account that $E\zeta_i^{n-l}\zeta_i^n = 0$, $(\mu_i^1 - \mu_i^0)^2 \ge \Delta$ and (11) one can obtain

$$E\left[J_i^n | \tau_i^{n-l} < \theta \le \tau_i^{n-1}\right] \ge \Delta.$$

Thus (15) is implied. The theorem has been proved.

Hence, the change of the expectation of the statistic J_i^n allows us to construct the following change point detection algorithm. The J_i^n values are compared with a certain threshold δ, where $2/H < \delta < \Delta$. When the value of the statistic exceeds δ then the change point is considered to be detected. If at least one parameter of the vector $\lambda = [\lambda_0, \lambda_1]$ changes then the change point θ can be detected.

The probabilities of false alarm and delay in the change point detection in any observation cycle are important characteristics of any change point detection procedure. Due to the application of the guaranteed parameter estimators in the statistics, we can bound these probabilities from above.

Theorem 4. *The probability of false alarm P_0 and the probability of delay P_1 in any observation cycle $[\tau_i^{n-1} + 1, \tau_i^n]$ are bounded from above*

$$P_0 \le \frac{2}{\delta H}, \quad P_1 \le \frac{2}{(\sqrt{\Delta} - \sqrt{\delta})^2 H}. \tag{18}$$

Proof. First we consider the false alarm probability, i.e. the probability that the statistic J_i^n exceeds the threshold before the change point. Using the Chebyshev inequality, we have

$$P_0 = \mathcal{P}\left\{J_i^n > \delta | \tau_i^n < \theta\right\} = \mathcal{P}\left\{(\zeta_i^n - \zeta_i^{n-l})^2 > \delta\right\} \le \frac{2}{\delta H}.$$

This imply the first inequality from (18).

Then we consider the delay probability, i.e., the probability that the statistic J_i^n does not exceed the threshold after the change point

$$P_1 = \mathcal{P}\left\{J_i < \delta | \tau_i^{n-l} < \theta < \tau_i^{n-1}\right\} = \mathcal{P}\left\{|\mu_i^1 - \mu_i^0 + \zeta_i^n - \zeta_i^{n-l}| < \sqrt{(\delta)}\right\}.$$

Taking into account that $|\mu_i^0 - \mu_i^1| > \sqrt{\Delta}$ and using the absolute value properties and the Chebyshev inequality, one has

$$P_1 \le \mathcal{P}\left\{|\mu_i^0 - \mu_i^1| - |\zeta_i^n - \zeta_i^{n-l}| < \sqrt{\delta}\right\} \le \mathcal{P}\left\{|\zeta_i^n - \zeta_i^{n-l}| > \sqrt{\Delta} - \sqrt{\delta}\right\}$$
$$\le \frac{E(\zeta_i^n - \zeta_i^{n-l})^2}{(\sqrt{\Delta} - \sqrt{\delta})^2} \le \frac{2}{H(\sqrt{\Delta} - \sqrt{\delta})^2}.$$

This imply the second inequality from (18). The theorem has been proved.

Then we consider asymptotic properties of the proposed change point detection procedure for $H \to \infty$ if process (1) is ergodic. At the previous subsection it was shown that $E\zeta_i^n \zeta_i^{n-l} = 0$. So the random variables $\sqrt{H/2}(\zeta_i^n - \zeta_i^{n-l})$ have the same asymptotic distribution as $\sqrt{H}\zeta_i$ and one can obtain the similar result.

Theorem 5. *If process (1) is ergodic, and the compensating factor Γ_N satisfies the following conditions $N \to \infty$, $N/H \to 0$ as $H \to \infty$, then for sufficiently large H*

$$P\left\{ \left(\zeta_i^n - \zeta_i^{n-l}\right)^2 > x \right\} \leq 2\left(1 - \Phi\left(\sqrt{xH/2}\right)\right), \qquad (19)$$

where $\Phi(\cdot)$ is the standard normal distribution function

This result implies the asymptotic inequalities for the probabilities of false alarm and delay.

Theorem 6. *In the conditions of the Theorem 5 the probabilities of false alarm and delay for sufficiently large H*

$$
\begin{aligned}
P_0 &\leq 2\left(1 - \Phi\left(\sqrt{\delta H/2}\right)\right); \\
P_1 &\leq 2\left(1 - \Phi\left(\left(\sqrt{\Delta} - \sqrt{\delta}\right)\sqrt{H/2}\right)\right).
\end{aligned}
\qquad (20)
$$

where $\Phi(\cdot)$ is the standard normal distribution function

Proof. Along the lines of the proof of Theorem 4 one has

$$
\begin{aligned}
P_0 &= P\left\{ J_i^n > \delta \mid \tau_i^n < \theta \right\} = P\left\{ \left(\zeta_i^n - \zeta_i^{n-l}\right)^2 > \delta \right\}; \\
P_1 &= P\left\{ J_i < \delta \mid \tau_i^{n-l} < \theta < \tau_i^{n-1} \right\} \leq P\left\{ \left(\zeta_i^n - \zeta_i^{n-l}\right)^2 > \left(\sqrt{\Delta} - \sqrt{\delta}\right)^2 \right\}.
\end{aligned}
$$

This and the result of Theorem 5 imply inequalities (20). The theorem has been proved.

We can use these estimators instead of (18) for sufficiently large H.

References

1. Tong, H.: On a Threshold Model. In: Chen, C.H. (ed.) Pattern Recognition and Signal Processing, pp. 101–141. Sijhoff & Noordhoff, Amsterdam (1978)
2. Rabemanajara, R., Zakoian, J.M.: Threshold ARCH Models and Asymmetries in Volatility. Journal of Applied Econometrics 8, 31–49 (1993)
3. Zakoian, J.M.: Threshold Heteroskedastic Models. Journal of Econometric Dynamics and Control 18, 931–955 (1994)
4. Liu, J., Li, W.K., Li, C.W.: On a Threshold Autoregression with Conditional Heteroscedastic Variances. Journal of Statistical Planning and Inference 62, 279–300 (1997)
5. Petrucelli, J.D., Woolford, S.W.: A Threshold AR(1) Model. J. Appl. Prob. 21, 270–286 (1984)

6. Cline, D.B.H., Pu, H.: Stability and the Lyapounov Exponent of Threshold AR-ARCH Models. The Annals of Applied Probability 14, 1920–1949 (2004)
7. Lange, T., Rahbek, A., Jensen, S.T.: Estimation and Asymptotic Inference in the AR-ARCH Model. Econometric Reviews 30(I.2), 129–153 (2011)
8. Cline, D.B.H.: Evaluating the Lyapounov Exponent and Existence of Moments for Threshold AR-ARCH Models. Journal of Time Series Analysis 28(2), 241–260 (2006)
9. Hwang, S.Y., Kim, S., Lee, S.D., Basawa, I.V.: Generalized Least Squares Estimation for Explosive AR(1) Processes with Conditionally Geteroscedastic Errors. Statist. Probab. Lett. 77(I.13), 1439–1448 (2007)
10. Vald, A.: Sequential Analisys. Phismatgis, Moscow (1960)
11. Sergeeva, E.E., Vorobejchikov, S.E.: An Efficient Algorithm for Detecting a Change Point of Autoregressive Parameters of AR(p)/ARCH(q) Process. In: Proceedings of the 11th International Conference of Pattern Recognition and Information Processing, pp. 156–159 (2011)
12. Burkatovskaya, Y.B., Vorobeychikov, S.E.: Guaranteed Estimation of Parameters of Threshold Autoregressive Process with Conditional Heteroskedasticity. Tomsk State University Journal of Control and Computer Science 2(23), 32–41 (2013) (in Russian)
13. Page, E.S.: Continuous Inspection Schemes. Biometrica 42(1), 100–115 (1956)
14. Hinkley, D.V.: Inference About Change-Point From Cumulative Sum-Tests. Biometrica 58(3), 509–523 (1971)
15. Lorden, G.: Procedures for Reacting to a Change in Distribution. Annals. Math. Statist. (42), 1897–(1971)
16. Kokoszka, P., Leipus, R.: Change-Point Estimation in ARCH. Bernoulli 6(3), 513–539 (2000)
17. Horvath, L., Liese, F.: Lp Estimators in ARCH Models. Journal of Statistical Planning and Inference 119, 277–309 (2004)
18. Aue, A., Horvath, L., Huskova, M., Kokoszka, P.: Change-Point Monitoring in Linear Models with Conditionally Heteroskedastic Errors. Econometrics J. 79, 373–403 (2006)
19. Cheng, T.-L.: An Efficient Algorithm for Estimating a Change-Point. Statistics and Probability Letters 79, 559–565 (2009)
20. Dupuy, J.F.: Detecting Change in a Hazard Regression Model with Rightcensoring. Journal of Statistical Planning and Inference 139, 1578–1586 (2009)
21. Mein, S., Tweedie, R.: Markov Chains and Stochastic Stability. Springer (1993)
22. Dmitrienko, A.A., Konev, V.V.: On Sequential Classification of Autoregressive Processes with Unknown Noise Variance. Problems of Information Transmission 31(4), 51–62 (1995) (in Russian)

On CLIQUE Problem for Sparse Graphs of Large Dimension

Valentina Bykova and Roman Illarionov

Institute of Mathematics and Computer Science
of Siberian Federal University
79 Svobodny pr., 660041 Krasnoyarsk, Russia
bykvalen@mail.ru, illroman@yandex.ru
http://www.sfu-kras.ru/

Abstract. In this paper the NP-hard Maximum Clique Problem (MCP) is considered. It is supposed that the input graph is sparse. Also, it is believed that the input graph can have a huge number of vertices. A biphasic algorithm for finding the exact solution of the MCP is proposed in the paper. The first phase of the algorithm is the preprocessing of the input graph by decomposing it into atoms. The second phase of the algorithm reduces to an application for each atom classical algorithm Wilf and then to the formation of solutions for the graph as a whole. The level of sparseness of the input graph is given in the form of restrictions on its treewidth. It have been proved that the running time of biphasic algorithm is polynomial in the number of vertices and the exponential of the treewidth of the input graph.

Keywords: Graph algorithms, sparse graphs, decomposition graph, atom graph, preprocessing, biphasic algorithms, treewidth, FPT-algorithms.

1 Introduction

Maximum Clique Problem (MCP) is one of the well-known NP-hard problems of discrete Mathematics and complexity theory. This problem has a wide variety of applications. In modern applications, such as analysis of chemical compounds and genomic database, automating the design of complex technical products and systems, data clustering, searching the maximum cliques have to be carried out in sparse graphs of very large dimension [1], [2], [8]. The input graphs can contain up to a million vertices. Due to this fact, efficient algorithms are demanded to find exact solution of the MCP in a reasonable time in our class of graphs.

This paper proposes a biphasic algorithm for finding the exact solution of MCP, the first phase of which — preprocessing of the input graph by decomposing it into atoms, and the second phase — application for each atom classical algorithm Wilf [14] and then formation of solution for the graph as a whole. Sparsity level of the input graph is given in the form of restrictions on its treewidth and in the form of assumption of the presence of clique minimal separators in the above graph. The paper shows that the running time of the proposed algorithm depends polynomially on the number of vertices and exponentially on the

A. Dudin et al. (Eds.): ITMM 2014, CCIS 487, pp. 69–75, 2014.
© Springer International Publishing Switzerland 2014

treewidth of the input graph, which allows to use it for the processing of large-scale graphs with small treewidth. In this paper the basic concepts and notation of [10] are used. For simplicity, let consider only connected ordinary graphs, i.e. finite, undirected, without loops and multiple edges graphs consisting of one connected component.

2   Formulation of a Problem

Consider a connected ordinary graph $G = (V, E)$ with vertex set V and edge set E, where $n = |V| \geq 2$, $|E| \geq 1$. The set of all vertices of G, adjacent to a vertex $v \in V$, forms in a graph G neighborhood $N(v)$ of this vertex, and the set $N[v] = N(v) \cup \{v\}$ — closed neighborhood. Graph $G' = (V', E')$ is called a subgraph if $V' \subseteq V$, $E' \subseteq E$. If the set of vertices of the subgraph G' is V', and the set of edges E' coincides with the set of all edges of G, both ends of which are owned by V', then $G' = (V', E')$ is called a subgraph which is generated by the set V', and is denoted by $G(V')$.

The set of vertices $V' \subseteq V$ forms a clique graph $G = (V, E)$, if any two vertices belonging to it are adjacent in G, i.e. the subgraph $G(V')$ is complete. Clique is called maximal, if it is not contained in the clique with more of vertices, and maximum, if the number of vertices in it is the greatest among all cliques. Size of the maximum clique of graph G is denoted $\varphi(G)$ and is called as the clique number of a graph G. The set of vertices $V' \subseteq V$ is independent in G, if any two vertices belonging to it are not adjacent, i.e. the subgraph $G(V')$ has no edges. An independent set is called maximal, if it is not a proper subset of another independent set. If the independent set's number of vertices in it is the greatest among all independent sets, it is called maximum. Cardinality of the maximum independent set of vertices of a graph G is called as a number of independence and is denoted by $\alpha_0(G)$.

The notions of clique and independent set are opposites in the sense that every clique (maximal, maximum) of a graph G is an independent set (maximal, maximum) in the complementary graph $\overline{G}$. Therefore, $\varphi(G) = \alpha_0(\overline{G})$.

Recognition version of the MCP traditionally formulated as follows.

INSTANCE: Given a graph $G = (V, E)$ and an integer $0 \leq L \leq n$.
QUESTION: Is there a clique of size at least L in G?

Similarly, let formulate recognition version of the Maximum Independent Set Problem (MISP). Note that the graph G is always $G = \overline{H}$, where $H = \overline{G}$, and the transition to the complement of n-vertex graph feasible in time $O(n^2)$. Thus, problems of MCP and MISP are polynomially reducible to each other and, in this sense, are equivalent. Both of these tasks in recognition formulation are NP-complete and polynomial time algorithms have not been yet found [9]. Further, in this paper the optimization versions of these problems, which is required for a given connected graph to find the maximum clique (maximum independent set) will be considered.

3 Basic Classical Algorithms for Solving

There are many algorithms to find the exact solution of MCP and MISP, the execution of which depends exponentially on the number of vertices and edges of the input graph. The most famous of these are algorithm Bron–Kerbosch and algorithm Wilf. Algorithm Bron–Kerbosch is a recursive procedure that consistently increases candidate clique [3]. Time of the algorithm is

$$O(\text{poly}(n) \cdot 3^{n/3}) = O(\text{poly}(n) \cdot 1,4422^n), \tag{1}$$

where $\text{poly}(n)$ — a polynomial in the number of vertices of the input graph. Since the graph with n vertices can contain up to $3^{n/3}$ maximal cliques [10], then algorithm Bron–Kerbosch is comparable in complexity to the procedure of exhaustive search. There are various modifications of the algorithm [12]. The fastest of them finds an exact solution of MCP for time

$$O(\text{poly}(n) \cdot 2^{0,249n}) = O(\text{poly}(n) \cdot 1,1888^n).$$

Unlike to algorithm Bron–Kerbosch, Wilf recursive algorithm [14] is designed to find the exact solution of the MISP. Its essence is as follows. For any arbitrarily chosen vertex $v \in V$ of $G = (V, E)$ there are two types of independent sets: those, which include the vertex v, and those, which do not include this vertex. Based on this, the original problem can be split into two subtasks (like smaller versions of the original), respectively to the two cases:

- formed independent set contains selected vertex v. In this situation, all the vertices of $N(v)$ can no longer go to the independent set, and its further extensions must be implemented in a graph $G(V \setminus N[v]) = G - N[v]$;
- formed independent set does not contain the vertex v. Further, expansion of this set should continue in the graph $G(V \setminus \{v\}) = G - v$.

A function which returns the cardinality of the maximum independent set of n-vertex graph G is denoted by $maxset(G)$ and by $T(n)$ — its execution time. Then the general scheme of the recursive splitting algorithm for finding solution of MISP is described by the equation

$$maxset(G) = \max\{maxset(G - v), 1 + maxset(G - N[v])\},$$

and evaluation of working time is calculated from the inhomogeneous linear recurrence relation with constant coefficients

$$T(n) = T(n - 1) + T(n - 2) + f(n), \tag{2}$$

where $f(n)$ is a function of polynomial order of growth. Equation (2) is valid, since the graph $G - v$ contains exactly $n - 1$ vertices, and the graph $G - N[v]$ can have up to $n - 2$ vertices (vertex v excluded itself and at least one of the vertices adjacent to it). Application to (2) technology Kullmann–Luckhardt [7] leads to the estimate

$$T(n) = O(\text{poly}(n) \cdot 1,619^n),$$

that is somewhat worse than (1). However, the selection rule of the vertex v, which is used during splitting in $maxset(G)$, can be improved. For example, if a vertex of degree 3 is selected, the recurrence relation is achieved.

$$T(n) = T(n-1) + T(n-4) + f(n)$$

and evaluation

$$T(n) = O(\text{poly}(n) \cdot 1,39^n), \tag{3}$$

which is better than (1).

Next, let propose a modification of the classical algorithm Wilf as a biphasic procedure based on decomposition approach to solving optimization problems on sparse graphs, namely, the preliminary decomposition of the original graph into atoms, retaining all its cliques.

4   Decomposition, Which Retains All Cliques of the Graph

Firstly, it is needed to introduce the necessary concepts. Let $G = (V, E)$, $n = |V|$, $|V| \geq 2$, $|E| \geq 1$, be connected to an ordinary graph. Under the atom graph $G = (V, E)$ is meant its maximum on the inclusion subgraph having no clique minimal separators. It is believed that the set of vertices $S \subseteq V$ separates two non-adjacent vertices x and y of the graph G, if in the subgraph $G(V \setminus S)$ vertices x and y belong to different connected components. The set S is then called (x, y)-separator and minimal (x, y)-separator, if no proper subset is not (x, y)-separator. Separator S is considered to be minimal, if the graph G has at least one pair of vertices x and y that S is minimal (x, y)-separator. Minimal separator S is called as a clique, if S forms a clique in G.

The idea of decomposition of the graph into atoms was proposed by Tarjan [13] in 1985 as a means of implementing the approach of "divide and rule" for solving optimization problems. Tarjan found that atomic decomposition does not destroy the cliques of the graph on the one hand and does not generate new cliques on the other hand. Later in [11] it was proved that the atomic decomposition is unique for each graph, if it is carried out only with help of clique minimal separators. Currently atomic decomposition is especially relevant in modern applications, which are based on graph-theoretic models of large dimension. Therefore, important to create new and improve the known polynomial algorithms for performing such a decomposition.

Decomposition of G into atoms is reduced to its multiple division into parts by one of the found clique minimal separators S, then allocation of connected components of $G(V \setminus S)$ and copy S into these components. This process continues until obtained parts will not contain clique minimal separators. Atomic decomposition algorithm presented in the [6]. This algorithm allows for time $O(n^3)$ for a graph $G = (V, E)$ to find a set of its clique minimal separators $\Delta(G)$ and a set of atoms $\Omega(G)$. It is important to note that a pair of $\Omega(G)$, $\Delta(G)$ — a compact description of any large-scale graphs with preservation of its internal structure. This description can be pre-created and stored in the external memory, and necessary for the processing atoms sequentially loaded into RAM. It is possible to organize parallel and distributed computing by simultaneous processing of several atoms.

Known important properties of atoms [11], [13]:

1. Every atom of $\Omega(G)$ is the induced subgraph of G.
2. Set $\Omega(G)$ always keeps all cliques of the graph G, i.e. every clique in G becomes a clique of one of its atoms, and new cliques does not occur.
3. $1 \leq |\Omega(G)| \leq n$, i.e. the number of atoms in $\Omega(G)$ does not exceed the number of vertices of G.
4. For sparse graph G (for graph with limited by the value of k treewidth) the number of vertices of each atom is bounded above by a positive integer constant $k < n$.

Wilf modified algorithm is based on the specified properties of atoms and proceeds from the assumption that the input graph is sparse.

5   Characterization of Sparse Graphs

There are several definitions of sparse graphs. Connected graph $G = (V, E)$, $n = |V| \geq 2$, $|E| \geq 1$, is called (edge) sparse, if the number of its edges satisfies the condition:

$$|E| \leq an^b, \tag{4}$$

where $a > 0$, $1 \leq b < 2$ — positive real constants and $n = |V|$. It is believed that the smaller the value of b, the more sparse is the graph G. For comparison, in each tree, the number of edges is $n - 1$, i.e. $b = 1$, which corresponds to the lower boundary value of b, and for any complete n-vertex graph always $|E| = n(n - 1)/2$, i.e. $b = 2$, which is equal to the upper boundary value of b.

There is another definition of a sparse graph, which is expressed through a numerical parameter $tw(G)$, called the treewidth of the graph [4]. Boundaries of $tw(G)$: $1 \leq tw(G) \leq n - 1$. So, every n-vertex tree ($n \geq 2$) has a unit treewidth which corresponds to the lower boundary for $tw(G)$, and for complete n-vertex graph inherent treewidth equal to $n-1$, which corresponds to the upper boundary for $tw(G)$. Let k is a given positive integer constant. If $tw(G) \leq k$, then let say that the graph $G = (V, E)$ has a limited (by the value of k) treewidth. It is assumed that the smaller the value of k, the more sparse graph G is. It is known [4] that if $tw(G) \leq k$, then for the number of edges of $G = (V, E)$ takes place the inequality:

$$|E| \leq kn - k(k + 1)/2. \tag{5}$$

When $k = 1$ and $k = n-1$, the relation (5) leads to inequalities (4), corresponding to trees and complete graphs. Consequently, the restriction $tw(G) \leq k$ does not contradict (4) and defines a natural measure of edge sparsity of G. As it is given that always $\varphi(G) - 1 \leq tw(G)$, it is possible to say that the value of $tw(G)$ also limits the size of cliques of the graph G.

6 Description of the Biphasic Algorithm and Analysis of Computational Complexity

Finding the exact solution MCP is applied to a sparse connected graph $G = (V, E)$, $n = |V| \geq 2$, $|E| \geq 1$, such that $tw(G) \leq k$, is to implement two phases.

Phase 1: Building an atomic representation $\Omega(G)$ for a graph G.

Phase 2: Consistently apply the algorithm to all the atoms of an obtained set of atoms $\Omega(G)$. Previously, perform transition to its complement for each atom. Form solution for G based on the results: maximum clique is defined as the largest by inclusion among all maximal cliques found in the atoms. As a result, give the maximum clique of G and its power.

Runtime phase 1 is $O(n^3)$. Taking into account the properties of atoms above, and the estimation (3), the time of solving the problem in relation to a single atom is obtained

$$O(\text{poly}(k) \cdot 1,39^k),$$

and for graph G as a whole

$$O(n \cdot \text{poly}(k) \cdot 1,39^k). \tag{6}$$

From (6) it follows that the running time of biphasic algorithm is polynomially dependent on n and exponentially on k. Consequently, the more sparse is the input graph, the faster the algorithm. Algorithms with estimates of the form (6) are called Fixed Parameter Tractable algorithms (FPT-algorithms) [5]. The existence of such an algorithm for MCP and MISP suggests that these problems are FPT-solvable with respect to treewidth of the graph.

7 Conclusions

In this paper the proposed algorithm which is based on decomposition approach to solving the MCP can be also applied to other algorithms for solving this problem, in particular, to the algorithm Bron–Kerbosch and its other versions. This approach allows the creation of FPT-algorithms for solving NP-hard problems on sparse graphs of large dimension. The main drawback of the proposed approach is: not all graphs have clique minimal separators. However, in modern applications such graphs are rare.

References

1. Boginski, V., Butenko, S., Pardalos, P.M.: Mining market data: A network approach. Computers & Operations Research 33, 3171–3184 (2006)
2. Broder, A., Kumar, R., Maghoul, F., Raghavan, P., Rajagopalan, S., Stata, R., Tomkins, A., Wiener, J.: Graph structure in the Web. J. Computer Networks 33, 309–320 (2000)

3. Bron, C., Kerbosch, J.: Algorithm 457: finding all cliques of an undirected graph. J. Communications of the ACM 16(9), 575–577 (1973)
4. Bykova, V.V.: Computational aspects of treewidth for graph. J. Applied Discrete Mathematics 3(13), 65–79 (2011) (in Russian)
5. Bykova, V.V.: FPT-algorithms on graphs of limited treewidth. J. Applied Discrete Mathematics 2(16), 65–78 (2012) (in Russian)
6. Bykova, V.V.: The Clique Minimal Separator Decomposition of a Hypergraph. J. of Siberian Federal University. Mathematics & Physics 1(5), 36–45 (2012) (in Russian)
7. Bykova, V.V.: On the asymptotic solution of the recurrence relations of special type and the technology Kullmann-Luckhardt. J. Applied Discrete Mathematics 4(22), 56–66 (2013) (in Russian)
8. Gardiner, E., Willett, P., Artymiuk, P., Gardiner, E.: Graph-theoretic techniques for macromolecular docking. J. Chem. Inf. Comput. 40, 273–279 (2000)
9. Garey, M., Johnson, D.: Computers and intractability: A guide to the theory of NP-completeness. W.H. Freeman and Company, New York (1979)
10. Emelichev, V.A., et al.: Lectures on the theory of graphs. Book House LIBROKOM, Moscow (2012) (in Russian)
11. Leimer, H.G.: Optimal decomposition by clique separators. J. Discrete Mathematics 113, 99–123 (1993)
12. Robson, J.M.: Algorithms for maximum independent sets. J. Algorithms 7(3), 425–440 (1986)
13. Tarjan, R.E.: Decomposition by clique separators. J. Discrete Mathematics 55, 221–232 (1985)
14. Wilf, H.S.: Algorithms and complexity. Prentice-Hall, Englewood Cliffs (1986)

Agent Model of Hierarchy Processes in the Ontology with Active Semantics

Soelma Danilova and Alexander Sitnichenko

East Siberia State University of Technology and Management,
40V Klyuchevskaya ul, 670013 Ulan-Ude, Russia
dan-soelma@yansex.ru

Abstract. This article describes the development and research of the hierarchy process modelig in the ontology with active semantics. It based on using the multi-agent technology and multi-threaded programming.

Keywords: ontology, the active semantics, hierarchy processes, multi-agent technology, multi-threaded programming.

1 Introduction

Nowadays, the global information space has applied a technology pervasive computing and multi-agent systems, as well as ontology repository of knowledge to solve problems of information retrieval. So it making the reason to create the ontology with active semantics as one of possible approaches to develop such concept as "Internet with thinking skills".

In ontology with active semantics reflects not only the knowledge domain concepts and relationships between them, but actions (processes) that are relevant to the semantics of these concepts. That is, depending on the task, concepts may play a role of the object or the subject of the action. In this case, there is an opportunity to develop and explore the ontology not only as a source of knowledge, but as a "tool" to solve various problems, especially computing tasks.

Thus, the creation of an ontology with the active semantics is a perspective direction. Its main element is categorical object the "Action". To date, the authors investigated the different implementation models. Proposed in this paper an agent-based model of hierarchy processes uniquely corresponds to the scheme of categorical objects. This enables the implementation of remote computing, which can be the basis of centers of GRID-calculations. They are required to solve many problems, such as automation technology of outsourcing knowledge.

2 Development of a Hierarchy Processes

Since the activity of the ontology depends on its processes, then to ensure the integrity of the knowledge base about the processes of ontology, it is necessary to solve the problem of determining their connectivity, which is necessary to build up the tree interaction processes. To solve this problem, as well as allocation

A. Dudin et al. (Eds.): ITMM 2014, CCIS 487, pp. 76–81, 2014.

of general concepts based on formal descriptions of ontology, it is proposed to use the agent-based approach. View of a tree in the form of ontology processes interacting agents, in our opinion, is the most appropriate way, as this implementation strategy allows not prescribe formal links between processes, thereby reducing the amount of memory required and simplifying the use of ontologies in cloud computing.

One of the most important properties of agents is an ability interoperability. An agent is created for each semantic vertex in the ontology. Each agent has its own purpose, that it must reach when needed. Agents of complex processes designed to solution two tasks: break the task into subtasks and send the request to their decision, the results assembly from the lower-level agents.

Way to communication of agents is one of the major challenges that must be addressed. To implement communication between agents is necessary to use the mechanism of mixed communications, which assumes the division of the whole set of agents in the system into two classes - class of customers and class of contractors, and organize the interaction between them through the bulletin board.

Agents will be represented by the streams that are suitable mechanism for building a system whose functionality is clearly divided into a few computational operations. In addition to the functional separation of the program into a plurality of operations, multi-agent approach allows us to reduce the run time by parallelization on multiprocessor computers.

The author of [1,2] proposed a categorical apparatus of ontology in which causal and propozitional models are the basis for the entry "Action". Its graphical interpretation is shown in Figure 1. All properties of the action displayed by element P, indicating the action parameters, the initial object over which it is necessary to perform an action, or source data to perform the computational process, the final results of the object or action, derived objects or intermediate results of the action, the place of performance of action, etc.

In this work are invited to perform a hierarchy of processes through the use of a causal model. In this model action is activated if the event E, what is happening with the subject of the action S, is interpreted as the cause of an event happening with the object O of action. When implementing process hierarchy tool T can serve as multi-agent system, in which the mapping actions A in agents will be through the definition of D.

All processes in any domain can be classified as simple or complex. [3] If you take a subject area - set theory, the processes - a set-theoretic operations of union, intersection, difference, complement, and symmetric difference. Simple processes are operations of union, intersection and difference, we extend their by logical operations of conjunction, disjunction and checking the validity of the inclusion of one set to another. omplement of a set, symmetric difference and testing the equality of two sets are complex processes.

Initial hierarchy of processes for three complex operations will contain two levels, as shown in Figure 2. When determining new operations the number of levels may increase if the following two conditions: firstly, the operation can

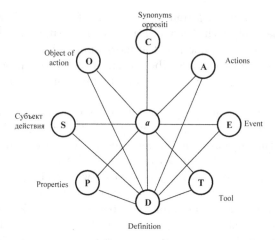

Fig. 1. Graphical interpretation of the entry "Action"

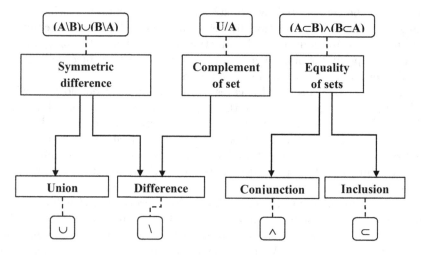

Fig. 2. Initial hierarchy of processes

be expressed through simple operations; secondly, the formula determines the operation which contains the sign of one of the more difficult operations, located at the top level of the hierarchy.

3   Implementation of the Multi-agent Model of Processes Hierarchy

A hierarchy of processes is performed using multi-agent approach, which makes it easy to use ontologies in applications that implement the technology of pervasive computing, since the construction of the agents in the system can be carried out taking into account the peculiarities of the "exit" agents outside the application located on the same computer. Such agents will be able to exchange messages with other agents in the network. Location agents on different computers will enable the computing power of many machines in the network, thereby increasing the capacity of the entire computer system.

To implement the actions in the system, we use a simplified causal model, as it allows you to organize the interaction between multiple agents in the system based on a causal relationship. The relations between the system is implemented on the basis of a mixed communication mechanism that partially combines direct and indirect communications mechanisms. All agents of the system are divided into two classes, "Customers", which serve as complex agents and contractors, which act as simple agents. The system sends data to process "Contractors" on the basis of analysis boards, in which are located requests of agents' "Customers". "After executing computing by agents "Contractor" , data is sent directly to the agent "Customer".

Described above allowed to develop multiagent model of hierarchy of processes, which is presented in Figure 3.

Initial expression, in this case, the set-theoretic formula passes through an I / O module in the operation queue, where the search engine sequentially selects the formula for their processing. Based on their analysis occurs selection of a suitable agent to perform calculations, and sent him to the computational data.

When using multi-agent model of hierarchy of processes in any domain is necessary the decomposition of the received information for calculating or processing. The agent transmits the received information to decomposition module where it enters the separation module. Here, the expression is divided into parts, and the intermediate information is used to correct the division is stored in the buffer memory of decomposition module. After completion of the decomposition processed information is passed back to the agent. Decomposition is not performed if the agent is a base, since the information is already indivisible in this context.

If the agent is a base, its methods evaluate the expression. If the agent is derived, computing parts fit onto the evaluation stack.

After placing the information in the evaluation stack the system sends the request to all basic agents with indicating of calculation parameters. Suitable agents send a response containing the address of the agent and the information about it. After this there is sampling occurs most suitable agent for calculations. After selecting the agent receives computing information along with a return address derivative agent.

After calculating the information by agent, the result is returned to the customer adress. Basic agents return information directly to derivatives agents. When derivative agent receives all the necessary information, it checks whether the calculations are finished. If the calculations is not finished, we return to the stack, if the calculations are finished, the agent issues a message containing the result of the calculation, or returns to its destination.

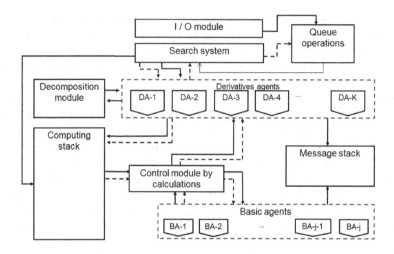

Fig. 3. Multi-agent model of hierarchy of processes

It should be noted that all agents in the system send into messages stack in during process work. Presentation of agents implemented using threads. Each thread defines a class object that represents the process of system. Attributes and methods show properties of actions stored in the ontology. In addition, these components of the model, as a search engine, decomposition module are also agents that perform system functions.

4   Principles of Multi-threaded Programming in the Implementation of the Multi-agent Hierarchy Processes

Consider some of the principles of multi-threaded programming, which are suitable for the implementation of the multi-process hierarchy in the ontology [4]. Functionality of the program clearly and naturally divided into several disparate operations. During the execution of long and complex calculations are not blocked by the GUI. The program can run on a multiprocessor computer.

Of course, the execution of the multi-model assumes that the program is separated into a plurality of heterogeneous operations, which are the processes of ontology. The program also performs some calculations, that should not interfere with the operation of the program. Also be aware that the multithreaded system has several advantages.

Improved application response - any program containing many independent from each other actions can be redesigned so that each action is performed in a separate thread. For example, the user interface multithreaded should not wait for the completion of one task to start the other.

More efficient use of multiprocessing - typically applications that implement concurrency through streams should not consider the number of available processors [5]. Application performance increases uniformly in the presence of additional processors. Computational algorithms and applications with a high degree of parallelism, for example, matrix multiplication can be performed much faster. Improved structure of the program - some programs more efficiently represented as several independent or semi-independent units than a single monolithic program. Multithreaded programs are easier to adapt to changes in user requirements.

Efficient use of system resources - programs that use two or more processes that have access to shared data through shared memory, contain more than one thread of execution. Moreover, each process is complete and the state of the address space of the operating system.

5 Conclusion

To test the proposed agent-based model a prototype software has been developed. Computational experiments were designed to test the correctness and speed of calculation. Now, the experimental results show that the agent model has a high-performance, simplicity of design and also easy way to make changes, compared with the previously developed models.

Thus, this paper shows how to implement an approach to remote computing, which allows to use multi-agent technology to implement the actions prescribed in the ontology. This enables the use of ontologies in the computation processes or information processing cycles. Besides the use of multithreaded programming can significantly reduce processing time in computer networks.

References

1. Naykhanova, L.V.: Technology for creating methods of automatic construction of ontologies using genetic programming and automata, Ulan-Ude (2008)
2. Naykhanova, L.V.: Construction sign conceptual objects and method for constructing terminosistemy. In: Materials II Intern. Conf. on Cognitive Science, pp. 592–593. St. Petersburg (2006)
3. Naykhanova, L.V., Danilova, S.D., Kim, N.B.: On the application-based programming in the construction of the hierarchy of processes in active ontology semantics. Theoretical and applied problems of modern information technologies. Materials of the XXI. Scientific and Engineering Conf., Ulan-Ude, pp. 274–281 (2012)
4. Andrews, G.R.: Fundamentals of multithreaded, parallel and distributed programming. Williams (2003)
5. Linev, A.V.: Parallel programming techniques for new processor architectures, Moscow (2010)

On Estimation of Linear Functional by Utilizing a Prior Guess

Yuri Dmitriev, Peter Tarassenko, and Yuri Ustinov

National Research Tomsk State University,
36 Lenin Ave., Tomsk 634050, Russian Federation
ptara@mail.tsu.ru

Abstract. The problem of statistical estimation of a linear functional of an unknown distribution with prior guess about the value of this functional is considered. A combined estimator of the functional is proposed to be used. The estimator is a linear combination of prior guess and nonparametric estimator. The optimal (in terms of minimal mean square error) weighting factor is subject of estimation itself, that can be done by using of prior guess recursively k times. According to this, k–adaptive combined estimators of the functional are proposed, their limiting distributions are presented. Examples of combined estimators and numerical results are provided.

Keywords: linear functional, prior guess, a priori information, combined estimator, nonparametric estimator.

1 Introduction

A variety of problems in statistical processing of experimental data comprise estimation of a linear functional of an unknown distribution. Linear functionals represent certain numerical (probabilistic) characteristics of the observed random variable. In order to reduce the size of expensive experimental data or to improve the accuracy of estimation for a fixed sample size, the additional (prior, initial) information can be used. The additional information may be used in numerous statistical problems, may exist in various forms and may come from different sources. There are many papers in the literature devoted to the estimation of the probability characteristics with using additional information. Estimators of the mean were proposed in [1]–[4]. Estimators of the variance of finite samples have been considered in [5] and [6]. Estimators of conditional quantile have been developed in [7]. In [8] this problem was considered for dependent data. A new class of M-estimators with auxiliary information has been introduced in [9]. Missing data case represented in [10], censored data case has been considered in [11]. Using a priori information in the processing of tomography data is reported in [12]. Problems of adaptive classification and optimization are considered in [13].

In this paper we consider the case when there exists an assumption on the value of estimated functional. The assumed value we will refer to as a prior

A. Dudin et al. (Eds.): ITMM 2014, CCIS 487, pp. 82–90, 2014.

guess. The term 'prior guess' has been probably first introduced by Ferguson [14] and used later in various contexts. Combined statistical estimators adapting a prior guess and their properties have been considered in [15]–[17]. In this paper we propose k–adaptive combined estimators that use prior guess recursively k times. Asymptotic distributions of the estimators have been obtained, that allow to study the influence of a prior guess to the estimation accuracy. The limit case is considered as $k \to \infty$.

2 Statement of the Problem

Let $X_1, ..., X_n$ be independent observations of size n over a random variable X with unknown distribution function F on R^1. Consider the problem of statistical estimation of a linear functional on a certain class of distributions $\mathcal{F}$.

$$J(F) = M_F[\varphi(X)] = \int_{-\infty}^{\infty} \varphi(x)dF(x), \quad F \in \mathcal{F}, \tag{1}$$

where φ is known real function. Nonparametric estimator of the functional is $\hat{J} = J(F_n)$, where $F_n(x) = n^{-1}\sum_{i=1}^{n} c(x - X_i)$ is empirical distribution function, $c(t) = \{0, t < 0; 1, t \geq 0\}$.

Suppose that there is an assumption that J may be equal to Ψ. The value of Ψ is called a prior guess, it acts as a prior (possible) value of the functional J. The problem is to construct an estimator of functional (1), taking into account nonparametric estimator $\hat{J}$ and prior guess Ψ. Following to [17] and [15], we consider a combined estimator of the form

$$\hat{J}(\lambda) = (1 - \lambda)\hat{J} + \lambda\Psi = \hat{J} - \lambda(\hat{J} - \Psi), \tag{2}$$

where the weighting coefficient λ is selected from the minimum of mean square error (MSE) $S_F(\lambda) = M_F[\hat{J}_\lambda - J]^2$ and its optimal value is given by

$$\lambda_n^* = (1 + n\Delta_F^2/\sigma_F^2)^{-1} = (1 + \bar{b}_n^2(F))^{-1} \tag{3}$$

with minimum of MSE $S_F(\lambda_n^*) = \sigma_F^2(1 - \lambda_n^*)/n$, where $\sigma_F^2 = D_F(\varphi(X))$ is the variance of $\varphi(X)$, $\Delta_F = J(F) - \Psi$ is the value of displacement of the prior guess from the true value $J(F)$, and $\bar{b}_n(F) = \sqrt{n}\Delta_F/\sigma_F$ is the normalized displacement.

The weighting factor λ_n^* varies between $0 < \lambda_n^* \leq 1$, and shows contribution of each estimator to the combined estimator (2). If $\Delta_F = 0$, we have $\lambda_n^* = 1$, and prior guess Ψ should be taken as the estimator of the functional $J(F)$. When $\Delta_F \neq 0$, which usually happens in practice, $\lambda_n^* < 1$, and with the growth of sample size ($n \to \infty$), $\lambda_n^* \to 0$, the influence of a prior guess and the advantage in estimation accuracy decrease.

The asymptotic behavior of optimal estimator $\hat{J}(\lambda_n^*)$ is given by the following theorem. In order to describe asymptotic properties we consider the convergence $\bar{b}_n \to b$, referring to a series of samples from the sequence of random variables X.

Theorem 1. *Let $\sigma_F^2 < \infty$ for each $F \in \mathcal{F}$ and sequence $\bar{b}_n$ converges to b as $n \to \infty$. Then the random sequence $\xi_n^* = \sqrt{n}(\hat{J}(\lambda_n^*) - J)/\sigma_F$ converges in distribution to the random variable $\xi^* = \eta - \lambda^*(\eta + b)$, where $\lambda^* = 1/(1 + b^2)$ and $\eta \in N(0,1)$ is standard normal random variable. The random variable ξ^* has variation $D\xi^* = [b^2/(1 + b^2)]^2$, mean $M\xi^* = -b/(1 + b^2)$, and MSE $S\xi^* = b^2/(1 + b^2)$.*

Proof. The convergency in distribution follows from the central limit theorem and the continuity theorem ([18], Chapter 6). Moments of ξ^* are calculated analytically.

The significant benefit in MSE of the optimal estimator $\hat{J}(\lambda_n^*)$ as compared to regular estimator $\hat{J}$ makes it interesting to use. Unfortunately, the coefficient λ_n^* depends on the unknown distribution function F, and usually, its value is unknown, that complicates the practical use of the formulae (2) and (3). The solution to this situation is building statistical estimators of λ_n^* and adaptive estimators of the functional $J(F)$.

3   Adaptive Estimators

We construct adaptive estimators based on consequent use of a prior guess. The first estimator can be obtained by substitution of F with F_n in formula (3):

$$\hat{\lambda}_1 = \lambda(F_n) = (1 + n\hat{\Delta}^2/\hat{\sigma}^2)^{-1} = (1 + \hat{b}_n^2)^{-1},$$

where $\hat{\Delta} = \hat{J} - \Psi$ is estimator of displacement, $\hat{b}_n = \sqrt{n}\hat{\Delta}/\hat{\sigma}$ is estimator of normalized displacement. Substituting λ with $\hat{\lambda}_1$ in (2), we obtain the first adaptive combined estimator $\hat{J}_1 = \hat{J} - \hat{\lambda}_1(\hat{J} - \Psi)$. Using $\hat{J}_1$ in estimation of displacement Δ_F, we obtain $\hat{\Delta}_1 = \hat{J}_1 - \Psi$ and $\hat{b}_{1,n} = \sqrt{n}\hat{\Delta}_1/\hat{\sigma}$. Then the second estimator will be $\hat{\lambda}_2 = (1 + \hat{b}_{1,n}^2)^{-1}$ and $\hat{J}_2 = \hat{J} - \hat{\lambda}_2(\hat{J} - \Psi)$. After repeating this procedure k times consecutively, we obtain the following expressions for the estimator

$$\hat{J}_k = \hat{J} - \hat{\lambda}_k(\hat{J} - \Psi), \quad \hat{\lambda}_k = \left(1 + \hat{b}_{k-1,n}^2\right)^{-1}, \tag{4}$$

$$\hat{b}_{k,n} = \frac{\sqrt{n}(\hat{J}_k - \Psi)}{\hat{\sigma}} = \hat{b}_n(1 - \hat{\lambda}_k) = \hat{b}_n\left(\frac{\hat{b}_{k-1,n}^2}{1 + \hat{b}_{k-1,n}^2}\right), \quad \hat{b}_{0,n} = \hat{b}_n,$$

Let us refer to $\hat{J}_k$ as k–adaptive estimator. We emphasize here that the prior guess Ψ has been used at each step of estimation of Δ_F.

4   Asymptotic Properties

Consider the asymptotic behavior of $\hat{J}_k$. Let $\xi_{k,n} = \sqrt{n}(\hat{J}_k - J)/\hat{\sigma}$. Denote $\eta_n = \sqrt{n}(\hat{J} - J)/\hat{\sigma}$, $b_n = \sqrt{n}\Delta_F/\hat{\sigma}$ and consider the sequence of functions defined as $q_0(x) = x$, $q_k(x) = xq(q_{k-1}(x))$, $k \in \{1, 2, 3, \ldots\}$, where $q(x) = x^2/(1 + x^2)$.

Using these notations, (4) can be written as $\hat{b}_{k,n} = \hat{b}_n q(\hat{b}_{k-1,n}) = q_k(\hat{b}_n)$ and $\xi_{k,n} = -b_n + q_k(\hat{b}_n)$, $\hat{b}_n = \eta_n + b_n$.

The following lemma gives properties of function family $q_k(x)$ that shown in figure 1.

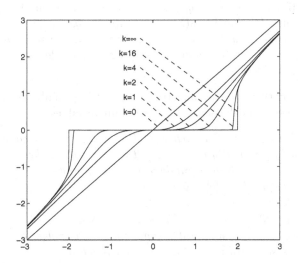

Fig. 1. Sequence of functions $q_k(x)$ for $k \in \{0, 1, 2, 4, 16, \infty\}$

Lemma 1. *The functions $q_k(x)$, $k \in \{0, 1, 2, \ldots\}$ have following properties:*
1. *$q_k(x)$ is odd function by x, i.e. $q_k(-x) = -q_k(x)$;*
2. *$q_k(x)$ is monotonically increasing function by x;*
3. *$q_{k-1}(x) > q_k(x)$ when $x > 0$ and $q_{k-1}(x) < q_k(x)$ when $x < 0$;*
4. *$\lim_{x \to \pm\infty} q_k(x)/x = 1$ for each k;*
5. *$\lim_{k \to \infty} q_k(x) = q_\infty(x)$, where*

$$q_\infty(x) = \begin{cases} (x - \sqrt{x^2 - 4})/2, & x \leq -2, \\ 0, & |x| < 2, \\ (x + \sqrt{x^2 - 4})/2, & x \geq 2. \end{cases}$$

Proof. The function $q_0(x)$ is odd and monotonically increasing, $q(x)$ is even and monotonically increasing when $x > 0$. Based on that, the first and second propositions can be obtained by induction.

The third proposition can be proved using the inequality $q(x) < 1$ that holds for any x. When $x > 0$, we have $q_1(x) = xq(x) < x = q_0(x)$. Suppose that $q_{k-1}(x) < q_{k-2}(x)$ for $x > 0$. Then $q_k(x) = xq(q_{k-1}(x)) < xq(q_{k-2}(x)) = q_{k-1}(x)$. Since $q_k(x)$ is odd function, the third proposition is proved.

The fourth proposition is obvious for $k = 0$ because $q_0(x) = x$. Suppose for some k that $\lim_{x \to \pm\infty} q_{k-1}(x)/x = 1$ and consider $\lim_{x \to \pm\infty} q_k(x)/x = \lim_{x \to \pm\infty} q(q_{k-1}(x)) = \lim_{x \to \pm\infty} q(x) = 1$.

Proving the fifth proposition, denote $p = q_\infty(x)$ for fixed x. The limit in recursive equation $q_k(x) = xq(q_{k-1}(x))$ as $k \to \infty$ leads to the equation $p = xp^2/(1 + p^2)$. Solutions to the equation are $p_0 = 0$, $p_- = (x - \sqrt{x^2 - 4})/2$, $p_+ = (x + \sqrt{x^2 - 4})/2$. The value p_0 is only root for $|x| < 2$. In order to remove extraneous roots for particular x, consider the case of $x > 2$, when $p_- < 1$ and $p_+ > 1$. If we prove that $q_k(x) > 1$ for $x > 2$, then the roots p_- and p_0 are extraneous. Obviously $q_0(x) > 1$, suppose $q_{k-1}(x) > 1$ and consider $q_k(x) = xq(q_{k-1}(x)) > xq(1) = x/2 > 1$. Similarly, the roots p_+ and p_0 are extraneous for $x < -2$. Root values at $x = \pm 2$ we consider as a limit cases for $x = 2 + 0$ and $x = -2 - 0$.

In the following theorem the convergence $b_n \to b$ in probability means utilizing a series of samples from the sequence of random variables X.

Theorem 2. *Let $\sigma_F^2 < \infty$ for each $F \in \mathcal{F}$ and sequence b_n converges to non-random value b in probability as $n \to \infty$. Then for each k the random sequence $\xi_{k,n}$ converges in distribution to random variable ξ_k and*
  1. *$\xi_k = -b + q_k(\eta + b)$ if $|b| < \infty$,*
  2. *$\xi_k = \eta$ if $|b| = \infty$,*
  3. *$P\{\xi_k < x\} = \Phi\left(q_k^{-1}(x + b) - b\right)$, $x \in (-\infty, \infty)$,*
where $\eta \in N(0, 1)$ is the standard normal random variable, $q_k^{-1}(x)$ is inverse function, $\Phi(x)$ is standard normal distribution function.

Proof. The functions $q_k(x)$ are continuous and $\hat{\sigma}^2$ converges to σ_F^2 in probability as $n \to \infty$. Then the first statement of the theorem follows from convergency of η_n to η in distribution by the central limit theorem and the continuity theorem ([18], Chapter 6).

The second statement of the theorem follows from the representation

$$\xi_{k,n} = \eta_n - \frac{\eta_n + b_n}{1 + q_{k-1}^2(\eta_n + b_n)},$$

where the second term converges weakly to zero as $|b_n| \to \infty$ due to the proposition 5 from lemma 1.

Since $q_k(x)$ is monotonically increasing function, then the distribution of the random variable ξ_k is defined by the formula

$$P\{\xi_k < x\} = P\{\eta < q_k^{-1}(x + b) - b\} = \Phi\left(q_k^{-1}(x + b) - b\right) = G_{k,b}(x).$$

In particular, $G_{k,\pm\infty}(x) = \Phi(x)$, $G_{k,0}(x) = \Phi(q_k^{-1}(x))$, and $G_{k,0}(x)$ is symmetric around the point $x = 0$.

5   Examples of k–Adaptive Combined Estimators and Numerical Results

In this section we provide some examples of estimators, their asymptotic properties, and results of numeric calculations.

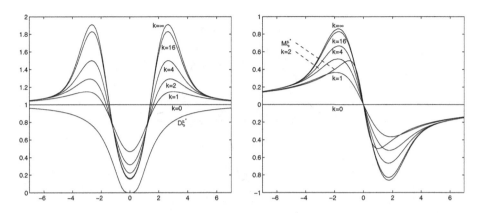

Fig. 2. Dependence of the variation $D\xi^*$ and $D\xi_k$ (left plot), mean $M\xi^*$ and $M\xi_k$ (right plot) on normalized displacement b and $k \in \{0, 1, 2, 4, 16, \infty\}$

Consider the k–adaptive combined estimators $\hat{J}_k$ under $k \in \{1, 2\}$ and corresponding normalized asymptotic random variables ξ_k.

$$\hat{J}_1 = \hat{J} - \left[1 + \hat{b}_n^2\right]^{-1} (\hat{J} - \Psi), \quad \xi_1 = -b + \frac{(\eta + b)^3}{1 + (\eta + b)^2},$$

$$\hat{J}_2 = \hat{J} - \left[1 + \frac{\hat{b}_n^3}{1 + \hat{b}_n^2}\right]^{-1} (\hat{J} - \Psi), \quad \xi_2 = -b + \frac{(\eta + b)^7}{[1 + (\eta + b)^2]^2 + (\eta + b)^6}.$$

If $\Delta_F = 0$ then $b_n = b = 0$ and $\xi_1 = \eta^3/(1 + \eta^2)$, $\xi_2 = \eta^7/[1 + \eta^2]^2 + \eta^6$ has symmetric distribution with mean $M\xi_1 = 0$, $M\xi_2 = 0$ and variation $D\xi_1 = S\xi_1 \approx 0.467$, $D\xi_2 = S\xi_2 \approx 0.316$, which shows potential benefit of using prior guess, it should be compared with variation of regular nonparametric estimator $D\xi_0 = S\xi_0 = 1$. If $\Delta_F \neq 0$ and $b = \pm\infty$ then $\xi_k = \eta$. In that case asymptotic distributions of adaptive estimator $\hat{J}_k$ and nonparametric estimator $\hat{J}$ are identical and equal to normal distribution.

According to lemma 1, the limit estimator (obtained after using the prior guess infinite number of times, $k = \infty$), can be written as

$$\hat{J}_\infty = \begin{cases} \hat{J} - \left[1 + \frac{\left(\hat{b}_n - \sqrt{\hat{b}_n^2 - 4}\right)^2}{4}\right]^{-1} (\hat{J} - \Psi), & \hat{b}_n \leq -2, \\ \Psi, & |\hat{b}_n| < 2, \\ \hat{J} - \left[1 + \frac{\left(\hat{b}_n + \sqrt{\hat{b}_n^2 - 4}\right)^2}{4}\right]^{-1} (\hat{J} - \Psi), & \hat{b}_n \geq 2. \end{cases}$$

Using the change of variables in the moment expression for asymptotic normalized random variable ξ_k (see theorem 2 and lemma 1), we have the expression $M\xi_k^m = \int_{-\infty}^{\infty} x^m d\Phi(q_k^{-1}(x + b) - b) = \int_{-\infty}^{\infty} (q_k(z + b) - b)^m d\Phi(z)$ for numerical calculation of mean $M\xi_k$, dispersion $D\xi_k = M\xi_k^2 - (M\xi_k)^2$, and MSE $S\xi_k = D\xi_k + M\xi_k^2$ for $k \in \{0, 1, 2, \ldots, \infty\}$.

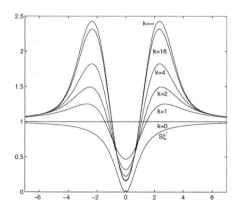

Fig. 3. Dependence of the MSE $S\xi^*$ and $S\xi_k$ on normalized displacement b and $k \in \{0, 1, 2, 4, 16, \infty\}$

According to theorem 1 the optimal estimator $\hat{J}(\lambda_n^*)$ (see (2), (3)) corresponds to normalized asymptotic random variable $\xi^* = \eta - \lambda^*(\eta + b)$, where $\lambda^* = 1/(1 + b^2)$. It has variation $D\xi^* = [b^2/(1 + b^2)]^2$, mean $M\xi^* = -b/(1 + b^2)$, and MSE $S\xi^* = b^2/(1 + b^2)$.

Table 1. Extremal points of $D\xi_k$, $M\xi_k$, $S\xi_k$. Arguments of maximum (rows 3, 6, 9) and points of intersection with level one (rows 4 and 8) are presented with accuracy ± 0.07.

No	k	1	2	4	16	∞
1.	$\min_b D\xi_k = \min_b S\xi_k$, $b = 0$	0.467	0.316	0.221	0.161	0.153
2.	$\max_b D\xi_k$	1.148	1.293	1.501	1.829	1.909
3.	$\arg\max_b D\xi_k$	± 2.94	± 2.8	± 2.66	± 2.66	± 2.66
4.	$b : D\xi_k = 1$	± 1.82	± 1.68	± 1.54	± 1.40	± 1.40
5.	$\max_b |M\xi_k|$	0.366	0.519	0.668	0.829	0.860
6.	$\arg\max_b |M\xi_k|$	± 1.82	± 1.82	± 1.68	± 1.68	± 1.68
7.	$\max_b S\xi_k$	1.251	1.494	1.823	2.315	2.431
8.	$b : S\xi_k = 1$	± 1.40	± 1.26	± 1.12	± 0.98	± 0.98
9.	$\arg\max_b S\xi_k$	± 2.66	± 2.52	± 2.38	± 2.38	± 2.24

Variation $D\xi_k$ and $D\xi^*$, mean $M\xi_k$ and $M\xi^*$ (figure 2), MSE $S\xi_k$ and $S\xi^*$ (figure 3) depend on normalized displacement b and order k. The value $k = 0$ corresponds to regular nonparametric estimator, $\hat{J}_0 = \hat{J}$ and $D\xi_0 = S\xi_0 \equiv 1$, $M\xi_0 \equiv 0$ do not depend on b. Extremal points of that functions are presented at table 1. Potential benefit from using prior guess is shown at the row 1 of the table. This significant advantage at the neighborhood of point $b = 0$ is compensated by loss of efficiency in the neighborhood of $b = \pm 2$.

Acknowledgement. The authors thank Tomsk State University for organizational and financial assistance as the article is prepared within the Tomsk State University Competitiveness Improvement Program.

References

1. Abu-Dayyeh, W.A., Ahmed, M.S., Ahmed, R.A., Muttlak, H.A.: Some estimators of a finite population mean using auxiliary information. Applied Mathematics and Computation 139, 287–298 (2003)
2. Al-Omari, A.I.: Ratio estimation of the population mean using auxiliary information in simple random sampling and median ranked set sampling. Statistics and Probability Letters 82, 1883–1890 (2012)
3. Vishwakarma, G.K., Singh, H.P.: A general procedure for estimating the mean using double sampling for stratification and multi-auxiliary information. Journal of Statistical Planning and Inference 142, 1252–1261 (2012)
4. Haq, A., Shabbir, J.: An improved estimator of finite population mean when using two auxiliary attributes. Applied Mathematics and Computation 241, 14–24 (2014)
5. Arcos, A., Rueda, M., Martinez, M.D., Gonzalez, S., Roman, Y.: Incorporating the auxiliary information available in variance estimation. Applied Mathematics and Computation 160, 387–399 (2005)
6. Yadav, S.K., Kadilar, C.: A two parameter variance estimator using auxiliary information. Applied Mathematics and Computation 226, 117–122 (2014)
7. Qin, Y.S., Wu, Y.: An estimator of a conditional quantile in the presence of auxiliary information. Journal of Statistical Planning and Inference 99, 59–70 (2001)
8. Liang, H.-Y., de Una-Alvarez, J.: Conditional quantile estimation with auxiliary information for left-truncated and dependent data. Journal of Statistical Planning and Inference 141, 3475–3488 (2011)
9. Bravo, F.: Efficient M-estimators with auxiliary information. Journal of Statistical Planning and Inference 140, 3326–3342 (2010)
10. Liu, X., Liu, P., Zhou, Y.: Distribution estimation with auxiliary information for missing data. Journal of Statistical Planning and Inference 141, 711–724 (2011)
11. Baklizi, A.: Preliminary test estimation in the two parameter exponential distribution with time censored data. Applied Mathematics and Computation 163, 639–643 (2005)
12. Kim, J.-H., Tsourlos, P., Yi, M.-J., Karmis, P.: Inversion of ERT data with a priori information using variable weighting factors. Journal of Applied Geophysics 105, 1–9 (2014)
13. Han, F., Ling, Q.-H.: A new approach for function approximation incorporating adaptive particle swarm optimization and a priori information. Applied Mathematics and Computation 205, 792–798 (2008)
14. Ferguson, T.S.: A Bayesian Analysis of Some Nonparametric Problems. The Annals of Statistics 1(2), 209–230 (1973)
15. Dmitriev, Y.G., Tarasenko, P.F.: The use of a priori information in the statistical processing of experimental data. Russian Physics Journal 35(9), 888–893 (1992)
16. Albers, C.J., Schaafsma, W.: Estimating a density by adapting an initial guess. Computational Statistics and Data Analysis 42, 27–36 (2003)

17. Tarima, S.S., Dmitriev, Y.G.: Statistical estimation with possibly incorrect model assumptions. Tomsk State University Journal of Control and Computer Science 8(4), 87–99 (2009)
18. Borovkov, A.A.: Mathematical statistics. Gordon and Breach Science Publishers, Amsterdam (1998)

Modeling EMA and MA Algorithms to Estimate the Bitrate of Data Streams in Packet Switched Networks

Alexander Domnin, Nikolay Konnov, and Victor Mekhanov

Penza State University, Computer Science Department,
St. Krasnaya 40, Penza, Russia
{a,knn,mvb}@pnzgu.ru
http://pnzgu.ru

Abstract. The article considers the problems of colored timed Petri net apparatus application for modeling the procedures of operative estimate of bitrate in packet switching net channels based on algorithms of moving average and exponential moving average.

Keywords: flow rate, packet net, moving average, exponential moving average, quality of service, Petri net, CPN Tools.

1 Introduction

Development of modern telecommunication is to a great extent determined by the widespread implementation of quality of service (QoS) methods assurance, which is understood as the ability of a net to provide specific service to the traffic of each application. The necessary service is characterized by such parameters as carrying capacity (bandwidth), packet delay and its variation (jitter), percent of lost packets [1, 2].

The choice of the most efficient quality of service management methods in corporate networks, development of new QoS methods appear to be a complex problem that may be solved by means of simulation preferably using the instruments of telecommunication net research that are not attached to particular equipment, but based on mathematical models. The report considers the modeling problems of one of QoS mechanisms – traffic control using the apparatus of hierarchical colored timed Petri nets [3]. The given apparatus features the following advantages:

- Petri net is a universal algorithmic system providing description practically of any algorithms;
- colors enable to describe and model algorithms depending on processed data content;
- hierarchical pattern enables to build complex multicomponent models;
- time property enables to model dynamic characteristics of objects.

A. Dudin et al. (Eds.): ITMM 2014, CCIS 487, pp. 91–100, 2014.

CPN Tools is a freely distributable packet. It was chosen as a development system, and in order to design network models the authors used a general approach described in [4–6].

Among other QoS mechanisms (traffic marking and classification, queue management, overload management) of great importance are the mechanisms of traffic management that directly determine the packet delay and loss probability thereof, as well as the efficiency of network equipment resource usage. The following are the standard management methods:

- *traffic profiling*, which is performed at the border network equipment, and which is to limit packet flow rate in accordance with the dedicated bandwidth;
- *shaping* – smoothing of the traffic pulsation to eliminate bursts leading to the loss of frames and buffer thrashing, and unpredictable delay fluctuations that negatively impact the multimedia applications.

At the present time there is a significant development of various adaptive methods of traffic management and rearrangement of nodal equipment (switches and routers) bandwidth in real time [7].

A mandatory component of any traffic management policy is bitrate measurement. There are the following rate measurement algorithms:

- *averaging over adjacent time intervals*. The main disadvantage of the said algorithm is the impossibility to estimate bitrate with high burst;
- *sliding window* provides good approximation of the average bitrate by computing the simple moving average (MA), however, requiring considerable computing resources;
- *leaky bucket* and *token bucket* algorithms are easy to realize and therefore extensively used in profiling and shaping, however, practically do not measure the current bitrate, but only the limit thereof.

In order to measure the bitrate in reference [8] it is suggested to use the more easily realizable algorithm of exponential smoothing (exponential moving average - EMA), applied in short-term forecasting of time series [9, 10].

CPN Tools modeling of traffic rate measurement by algorithms of *averaging complementary intervals*, *leaky bucket* and *token bucket* is considered in [11, 12]. Therefore, further we are to consider the formation principles of colored timed Petri nets for algorithms of bitrate measurement based on EMA and MA algorithms.

2   EMA Algorithm Modeling

In the traditional approach the exponential smoothing of the traffic profile directly in the channel for any moment in time $t_j = jTI$, where TI is the bit length, EMA V_j value equals:

$$V_j = \alpha \cdot X_j + (1 - \alpha) \cdot V_{j-1} \tag{1}$$

where α is smoothing constant $(0 < \alpha < 1)$, X_j - instantaneous data rate at the moment, which may have the value of 0, if at this period of time there is no data transfer in the channel, or the maximum value of physical channel speed.

For purposes of traffic management it is necessary to take into account the data rate not in the random moment of time, but either at the start of frame transfer, or at the end of frame transfer. Besides, the traffic profile represents a time series consisting of zeroes and ones: sequence of ones is determined by the number of bit intervals, during which the frame is transferred, and the sequence of zeroes – by the number of bit intervals, during which there is a pause between frames (fig. 1). Thus, to compute the EMA traffic rate it is possible to use the following recurrence formulas:

$$\begin{cases} VP_i = (1 - \alpha)_i^P \cdot VL_{i-1} \\ VL_i = 1 - (1 - \alpha)_i^L \cdot (1 - VP_i) \end{cases} \tag{2}$$

where VP_i, VL_i – average rates respectively at the start and at the end of transfer of i frame; P_i and L_i - length of the interframe pause and transfer time of i frame respectively.

The average rate in the time interval $t \in (t_{i-1}, t_i)$ may be evaluated as follows:

$$V_i \approx \frac{VP_i \cdot P_i + VL_i \cdot L_i}{P_i + Li} \tag{3}$$

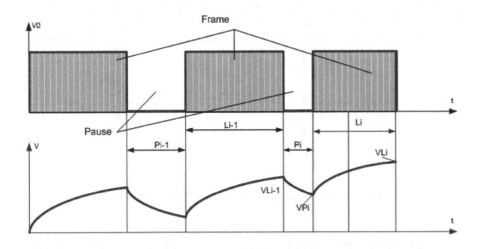

Fig. 1. Timing chart of EMA application

In this case to simulate the researched algorithms we form a frame flow, movement of which is determined by the content of the frames, and due to this fact it is important to effectively use the feature of marker coloration. There were introduced two types markers:

– information markers, the movement of which imitates frame processing in a switch marked by a multiset with frame color, which may occupied either by transfer of the *frame* of *frm* color, reflecting the structure of the transferred frame and consisting of sender and receiver addresses *src* and *dst*, *qos* priority control field, *szfrm* size of transferred data and *delay* total delay to compute the delay of frame movement in the net, or by *avail* color (free), that enables to reveal and process the events related to the presence or absence of the frame in position;
– control markers, the colors of which reflect the condition of frame flow processing in the switch.

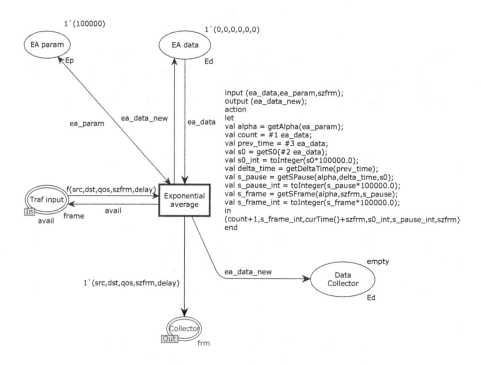

Fig. 2. Petri subnet, modeling the EMA algorithm

Petri net simulating the EMA algorithm is shown in fig. 2 and includes the following positions:

– *Traf_input*, that accepts the markers corresponding to the input frame flow;
– *EA_param* includes the marker with information about the smoothing constant ($\alpha = 1/ea\_param$);
– *EA_data*, that includes a marker with information about the current value of the computed *ea_data* rate;

- *Data_Collector* - the aggregator of markers with information about the computed rate (*count* – i marker number, *s_frame_int*, *curTime()+szfrm*, *s0_int*, *s_pause_int*, *szfrm* – computed values of VP_i, VL_i, VL_{i-1}, P_i and L_i respectively).;
- *Collector* – output position of the model;

In the subnet there is one *Exponential average* transition that realizes traffic bitrate measuring according to the algorithm of exponential smoothing by computing next values of VP_i and VL_i rates using CPN ML language procedures. Herewith, the direct computation by formula (2) realizes user *getSPause* and *getSFrame* functions. It should be noted that the use of the above mentioned functions is connected with conversion of the format of variables with floating point in integers, which may lead to an error, the impact of which should be compensated by introduction of scaling.

3   MA Algorithm Modeling

Using the traditional approach the simple moving average of traffic in a channel for a random moment of time $t_j = jTI$, V_j will be equal to:

$$V_j = \frac{1}{T} \cdot \sum_{i=0}^{T-1} X_{j-i} = V_{j-1} + \frac{X_j - X_{j-T+1}}{T} \tag{4}$$

where T – a sliding window of averaging.

Taking into account the fact that the traffic profile represents a time series consisting of zeroes and ones: rate should be computed at the start of frame transfer and at the end of frame transfer, the window will not move along the time axis uninterruptedly, but "by jerks", as shown in fig. 3 Thus, similarly to EMA, the computation formula for and may be represented as follows:

$$\begin{cases} VP_i = VL_{i-1} - \sum_{\tau 1} L_{\tau 1} \\ VL_i = VP_i + \frac{1}{T} \cdot L_i - \frac{1}{T} \cdot \sum_{\tau 2} L_{\tau 2} \end{cases} \tag{5}$$

where the sums represent the amount of bits transferred into the channel on time intervals $\tau 1 \in (t_i - P_i - T, t_i - T)$ and $\tau 2 \in (t_i - T, t_i + L_i - T)$ respectively, i.e. removed from the window when it moves during i frame arrival and transfer end. The said fact causes certain difficulties in practical realization of the MA algorithm, as it is necessary not just to retain the frame lengths, caught in the window, but also to retain the length of pauses between them and "split" the last frame, if the rear edge of the window matches the time of the transfer thereof.

The variant of Petri net modeling the MA algorithm has been suggested in reference [13], and fig. 4 shows a more compact realization of the subnet including the following positions:

- *Ingress_port*, that accepts the markers describing the frames of the input traffic flow;

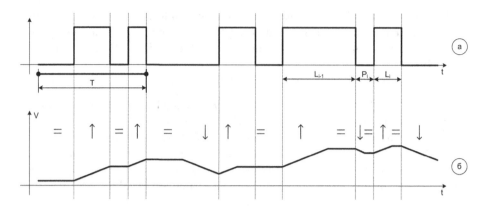

Fig. 3. Time chart of MA application

- *Forefront, Backfront*, that include time markers representing the start (*Forefront*) and the end (*Backfront*) of frame receiving;
- *SW_param*, contains marker with information about the size of averaging window (*fw_param*) in bit intervals;
- *Framefront, Frameback*, that contain markers caught in the current position of averaging window, with time markers representing the start (*Framefront*) and the end (*Frameback*) of the respective frame receiving;
- *Cache* – buffer position used for aggregation of two net branches;
- *SW_size*, contains the volume of data caught in averaging window in the current position thereof;
- *Frame_size_cache*, intended for temporary storage of information about the size of the frame added to averaging window;
- *Data_collector2*, contains markers with information about a sequence number of a marker (*count2*) and volume of data (*fw_data_new*), caught in averaging window at the moment of the end of the respective frame receiving (current value of channel usage may be computed as a ratio of *fw_data_new* to *fw_param*);
- *Packet_counter* – buffer position, calculating marker filling of *Data_collector2* position;
- *Collector* – output port, collecting markers of the processed frames for further processing.

The net operation of MA algorithm modeling is described below.

The next input marker goes to *Ingress_Port* position, after that there takes place *Split_frame* transition actuation. Then *Forefront* position is taken by the marker, containing information about a frame, which will be used for computation of the current position of the window at the moment of frame appearance (the frame will not be registered in the size of the window), and *Backfront* position – by the time marker increased by the size of the frame.

After that the marker in *Backfront* goes through the transition *No_bf_check* or *Check_backfront*, and marker in *Forefront* – through transition *No_ff_check*,

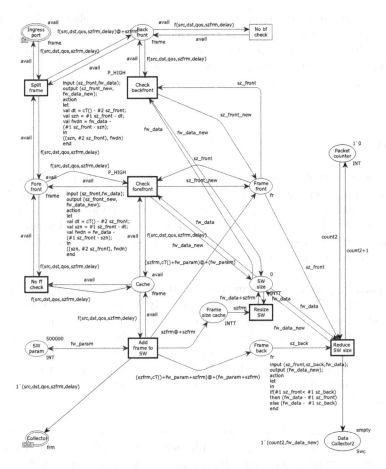

Fig. 4. Petri subnet, modeling the MA algorithm

or *Check_forefront*. Transitions *Check_backfront* and *Check_forefront* have the highest priority and are necessary to process the cases, when at appearance of the next frame (*Check_forefront*) or receiving thereof (*Check_backfront*) the back front of averaging window is in the period of time when the receiving of the already arrived frame took place ("splitting" of the frame according to formula (5)). In such cases the information about the amount of data in "the sliding window" at the current moment should be adjusted. Adjustment values *sz_front_new* and *sz_data_new* are computed by the appropriate procedures taking into account the current simulated time obtained using *cT()* function.

No_bf_check transition or *Check_backfront* transition actuation leads to marker removal from *Back_Front* position, and in case of *No_ff_check* or *Check_forefront* actuation the marker transits to *Cache_buffer* position. Then, there occurs the actuation of *Add_Frame_to_SW* transition that saves information about the processed frame in averaging window (*Frame_front* and *Frame_back*). As the information about the amount of data in the window increases by the size of the appeared frame after receiving thereof, it is necessary to change the values in *SW_size* position after a period of time that equals the size of the frame. The said is realized by *Frame_size_cache* position and *Resize_SW* transition.

Reduce_SW_size transition removes the information about the frames missing the current "sliding window" from *Frame_front*, *Frame_back* and *SW_size* positions.

4   Results

For the example the created models application, below are the results of modeling of bitrate measurement in the Ethernet switch supporting IEEE 802.1 Q/P. Simulated time is represented in TI (one cycle corresponds to 100 ns Ethernet or 10 ns for Fast Ethernet). The input traffic was modeled using the special Petri net [14] and represents a combination of two components, merger of which forms the data flow with pronounced bursts:

- Regular, representing a sequence of 100 double-framed packets with the length of 12176, the period of which linearly increases by 5% from 200000;
- Random, with frames, length distribution of which is of pronounced bimodal nature: 25% of all frames have minimal and maximal length (512 and 12176 respectively), the length of others is equally distributed in the remaining range. The frames follow through time intervals, distributed according to the exponential law (technological interval of 160 cycles is also taken into account).

Fig. 5 shows the realization profile of the input traffic with the length of around 11000000 cycles (the amount of frames of the regular composition is 100, average length of the interframe intervals of the random component of the input flow is 2960 cycles), and realization of rate changes through MA (T =500000) and EMA ($\alpha = 1/250000$) algorithms. The transient process with the length of 500000 bit cycles is not shown. Channel load factor, computed according to the input realization, equals to 0.748.

Both algorithms produce quite similar results and clearly follow the linear trend of implementration concerning the changes in bitrate.

However, the comparison of the Petri nets, presented in figs. 2 and 4, visually shows a clear advantage of EMA algorithm in realization simplicity unlike MA.

Therefore, the presented subnet, modeling MA algorithm, may be implemented not from the point of view of MA feasibility assessment, but as a "sample" model for comparison with new algorithms of bitrate assessment, based on exponential smoothing. Thus, the modeling results show the effectiveness of the operative measurement of the current bitrates in the telecommunication equipment using EMA algorithm that may be used for evaluation of rate not just of the general flow, but of individual components, corresponding to the introduced QoS classes.

Scientific novelty of the research results is the following:

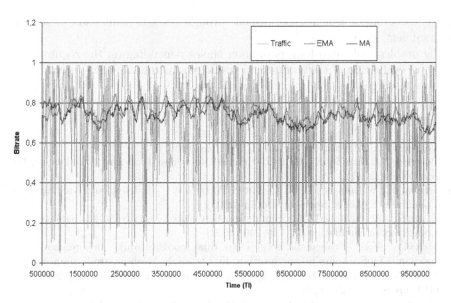

Fig. 5. Results of MA and EMA algorithms modeling

- authors first proposed to use EMA algorithm for traffic bitrate measurements;
- authors developed two models in timed colored Petri nets modeling EMA and MA algorithms, and conducted a comparative analysis of both models;
- research demonstrated the advantage of ease of implementation and lower computational complexity of EMA algorithm in comparison with MA algorithm;

In the engineering realization of EMA algorithm it is possible to decrease the computing complexity via linear approximation of the exponential transfer characteristic, enabling to reduce the number of multiplication in processing of each frame to two.

The obtained values of the current rate are easy to use for solution of routine problems of traffic management:

- in the course of profiling during the frame receiving the value VL is forecasted. This value is compared with the given boundary value, and according to the result of the comparison the frame is either deleted, or marked;
- in the course of shaping of the output traffic in the case, when the forecasted VL is greater than the boundary value, it is necessary to increase the time of the current pause by a calculated value.

The advantage of the suggested method of data rate measurement is the possibility of the efficient usage of measurement results for dynamic redistribution of the bandwidth, load balancing and other tasks of net adaptive management. On the basis of the suggested method it is possible to realize EMA of higher orders, modeling of which is performed through cascade connection of several Petri subnets, shown in fig. 2.

The problems of smoothing parameter choice depending on the required accuracy of evaluation of traffic rate and traffic parameters are the subject of the separate research.

References

1. Vegesna, S.: IP Quality of Service. Cisco Press (2001)
2. Barreiros, M., Lundqvist, P.: QOS-Enabled Networks: Tools and Foundations, 1st edn. Wiley (2011)
3. Jensen, K., Kristensen, L.M.: Coloured Petri Nets – Modeling and Validation of Concurrent Systems. Springer, Berlin (2009)
4. Zaitsev, D.A.: Switched LAN Simulation by Colored Petri Nets. Mathematicsand Computers in Simulation 65, 245–249 (2004)
5. Mehanov, V.B.: Primenenie setej Petri dlja modelirovanija mehanizmov obespechenija QoS v komp'juternyh setjah. In: Materialy Mezhdunarod. Simpoziuma, Novye Informacionnye Tehnologii i Menedzhment Kachestva (NITMQ 2010), Je-GRI, pp. S.209–S.214 (2010)
6. Mehanov, V.B.: Primenenie setej Petri dlja modelirovanija telekommunikacij s podderzhkoj kachestva obsluzhivanija / Trudy XVII Vserossijskoj nauchno-metodicheskoj konferencii "Telematika 2010", Tom 1. SPb, SPbGU ITMO, S.283–S.284 (2010)
7. Muranov, O.S.: Improving the adaptive control of packet telecommunications network traffic. – Manuscript. Candidate Dissertation in Engineering, speciality 05.13.06 – Information Technologies. National Aviation University, Kyiv (2010)
8. Domnin, A.L.: Primenenie metoda jeksponencial'nogo sglazhivanija dlja profilirovanija i shejpinga setevogo trafika / Materialy mezhdunarodnoj nauchnoj konferencii "Informacionnye tehnologii i sistemy (ITS 2013)", S.30–S.32. BGUIR, Minsk (2013)
9. Chou, Y.-L.: Statistical Analysis, 2nd edn. Holt, Rinehart and Winston of Canada Ltd. (1975)
10. Droke, C.: Moving averages. Simplified. Marketplace Books, Columbia (2001)
11. Domnin, A.L., Kizilov, E.A., Pushkarev, V.A.: Modelirovanie mehanizmov QOS v kommutatorah Ethernet cvetnymi setjami Petri / Sb. statej uchastnikov Vserossijskogo konkursa nauchnyh rabot studentov i aspirantov 'Telematika' 2010: telekommunikacii, veb tehnologii, superkomp'juting'. SPb., SPbGU ITMO, S.35–S.38 (2010)
12. Mehanov, V.B., Domnin, A.L.: Modelirovanie algoritmov upravlenija polosoj propuskanija cvetnymi setjami Petri / Trudy IX Mezhdunarod. nauchno-tehnicheskoj konferencii 'Novye informacionnye tehnologii i sistemy', ch. 2, S.77–S.82. Izd-vo PGU, Penza (2010)
13. Konnov, N.N., Domnin, A.S.: Razrabotka modeli algoritma 'skol'zjashhego' okna cvetnymi vremennymi setjami Petri / Estestvennye i matematicheskie nauki v sovremennom mire, vol. 9-10. Sbornik statej po materialam IX-X mezhdunarodnoj nauchno-prakticheskoj konferencii, S.76–S.81. Izd. 'SibAK', Novosibirsk (2013)
14. Nikishin, K.I., Konnov, N.N., Domnin, A.L.: Modelirovanie trafika seti Ethernet cvetnymi setjami Petri / Sb. materialov I Mezhdunar. nauch.-prakt. konf 'Sovremennye problemy komp'juternyh nauk (SPKN 2013)', S.120–S.123. Izd-vo PGU, Penza (2013)

Sensitivity Analysis of Reliability Characteristics to the Shape of the Life and Repair Time Distributions*

Dmitry Efrosinin[1] and Vladimir Rykov[2]

[1] Johannes Kepler University
Altenbergerstrasse 69, 4040 Linz, Austria
[2] State University of Oil & Gas
Leninskiy Prospect 65, 119991 Moscow, Russia
dmitry.efrosinin@jku.at, vladimir_rykov@mail.ru
http://www.jku.at, http://www.gubkin.ru

Abstract. The present paper deals with some simple finite population queueing systems which are used to describe the cold redundancy systems. The systems are assumed to be in a complete failure state as soon as all of the units are failed. For such models the time dependent state probabilities and reliability function are analyzed. It is shown that the reliability function have a weak sensitivity to the shape of the life and repair time distributions and this sensitivity vanishes upon the probability of the complete failure state decreases.

Keywords: Sensitivity analysis, cold redundancy system, finite-population queueing system, reliability function.

1 Introduction

The stability behavior of the systems by varying of their parameters or initial conditions represents a kernel research topic almost in all areas of natural sciences. Particularly, the insensitivity or weak sensitivity of system's characteristic measures to the shape of the input distributions of the corresponding random variables plays an important role by modeling and analyzing of the complex stochastic systems.

One of the first result concerning insensitivity of system's characteristics to the shape of service time distribution has been proposed by B.Sevast'yanov [9], who has proved insensitivity of the Erlang's formulas to the shape of service time distribution for the loss queueing systems. In [4] I.Kovalenko has shown that the necessary and sufficient condition for insensitivity of stationary characteristics to the shape of repair time distribution in case of reliability systems with a Poisson flow of failures and generally distributed repair times of elements. According to

---

* This work was funded by the COMET K2 Center "Austrian Center of Competence in Mechatronics (ACCM)", funded by the Austrian federal government, the federal state Upper Austria, and the scientific partners of the ACCM.

A. Dudin et al. (Eds.): ITMM 2014, CCIS 487, pp. 101–112, 2014.

this condition any element subject to failure must be immediately accepted for the repair. The sufficiency of this condition for recurrent failure flow and general repair time distribution was found in Rykov [6] by means of the theory of multi-dimensional alternating processes. At the same time Koenig et al. [3] has applied the complementary variable method and shown some examples where reliability characteristics strongly depend on the shape of life and repair time distributions.

The problem of insensitivity or weak sensitivity of the reliability characteristics to the shapes of input distributions is not still exhaustively studied. Within the paper we intend to analyze some simple finite source queueing models, which are appropriate to describe cold redundancy systems. Some types of finite population queues were analyzed by Knessl et al.[2], Langaris and Katsaros [5]. In Rykov [7] and Rykov et al. [8] it was shown that while the sensitivity of steady-state probabilities to the form of inter-arrival and service time distributions in case of cold redundancy system are obviously observable, the waiting time distribution exhibits only a weak sensitivity. The loss systems with Poisson arrivals as expected are insensitive to the type of service time distributions and homogeneity of the servers but the same systems with generally distributed inter-arrival time do not possess such a property any more. A missing link in sensitivity analysis of such systems consists in investigation of the probabilistic characteristics on one life cycle of such systems, e.g. reliability function and mean time to failure. This problem will be the topic of the present research.

The following notations are used within the paper. $GI/GI/m/n//n-m$ stands for a closed queueing system with n sources of customers, m servers and $n - m$ places in the buffer, where GI means "General Independent". At the first position of this notation GI specifies the recurrent flow of failures and at the second one – the recurrent service process. These symbols can be changed by M for exponential distributions in a Markov case. In the paper we consider two types of cold redundancy models, namely $M/GI/1/2//1$ and $GI/M/1/2//1$, and compare them with the simple Markov model $M/M/1/2//1$. The mean time to failure, the mean service time as well as the failure and repair intensities are denoted respectively by $\bar{a}, \bar{b}, \alpha(x)$ and $\beta(x)$. Denote also the Laplace-Stiltjes transforms (LST) of the life and repair time by $\tilde{a}(s)$ and $\tilde{b}(s)$.

2   The $M/GI/1/2//1$ Queueing System

2.1   Non-stationary State Probabilities

Consider a two source cold redundancy system $M/GI/1/2//1$ with one repair server. The units have an exponential life time distributions with parameter α and general repair time distribution $B(t)$. Denote by

$$\{Z(t)\}_{t \geq 0} = \{J(t),\, X(t)\}_{t \geq 0}$$

a two-dimensional stochastic process, where the first component stands for the number of failed elements at time t and the second one stands for the elapsed repair time of the unit at time t. The process $\{Z(t)\}_{t \geq 0}$ is obviously Markovian one with state space $E = \{0, (i, x) | i \in \{1, 2\}, x \in \mathbb{R}_+\}$. Define the following state probabilities:

(1) $\pi_0(t) = \mathbb{P}[N(t) = 0]$ – the probability that both of units are operational at time t.

(2) $\pi_i(t; x)dx = \mathbb{P}[N(t) = i; \, x < X(t) \le x + dx]$ – the joint probability that at time t there are i failed units and the unit is being repaired with the elapsed repair time between x and $x + dx$, $i = 1, 2$.

The system of forward partial differential equations for $i = 0, 1, 2$ and $x > 0$:

$$\left[\frac{d}{dt} + \alpha\right]\pi_0(t) = \int_0^\infty \pi_1(t; u)\beta(u)du, \tag{1}$$

$$\left[\frac{\partial}{\partial t} + \frac{\partial}{\partial x} + \alpha + \beta(x)\right]\pi_1(t; x) = 0;$$

$$\left[\frac{\partial}{\partial t} + \frac{\partial}{\partial x} + \beta(x)\right]\pi_2(t; x) = \alpha\pi_1(t; x)$$

with the boundary condition $\pi_1(t; 0) = \alpha\pi_0(t) + \int_0^\infty \pi_2(t; u)\beta(u)du$, $\pi_2(t; 0) = 0$, together with the normalizing equation $\pi_0(t) + \sum_{i=1}^2 \int_0^\infty \pi_i(t; x)dx = 1$ and initial condition $\pi_0(0) = 1, \pi_i(0; x) = 0, i = 1, 2$, for any fixed x.

Theorem 1. *The state probabilities in terms of the LT are given by*

$$\tilde{\pi}_0(s) = \frac{1}{s + \alpha}[1 + \tilde{\pi}_1(s; 0)\tilde{b}(s + \alpha)], \tag{2}$$

$$\tilde{\pi}_1(s; x) = \tilde{\pi}_1(s; 0)e^{-(s+\alpha)x}(1 - B(x)),$$

$$\tilde{\pi}_2(s; x) = \tilde{\pi}_1(s; 0)(e^{-sx} - e^{-(s+\alpha)x})(1 - B(x)),$$

$$\tilde{\pi}_1(s; 0) = \frac{\alpha}{(s + \alpha)(1 - \tilde{b}(s)) + s\tilde{b}(s + \alpha)}.$$

Proof. By taking Laplace transforms of equations from (1), we obtain

$$s\tilde{\pi}_0(s) - 1 = -\alpha\tilde{\pi}_0(s) + \int_0^\infty \tilde{\pi}_1(s; u)\beta(u)du, \tag{3}$$

$$s\tilde{\pi}_1(s; x) + \frac{\partial\tilde{\pi}_1(s; x)}{\partial x} = -(\alpha + \beta(x))\tilde{\pi}_1(s; x),$$

$$s\tilde{\pi}_2(s; x) + \frac{\partial\tilde{\pi}_2(s; x)}{\partial x} = -\beta(x)\tilde{\pi}_2(s; x) + \alpha\tilde{\pi}_1(s; x),$$

$$\tilde{\pi}_1(s; 0) = \alpha\tilde{\pi}_0(s) + \int_0^\infty \tilde{\pi}_2(s; u)\beta(u)du, \quad \tilde{\pi}_2(s; 0) = 0.$$

Solving the second differential equation from (3), we get

$$\tilde{\pi}_1(s; x) = \tilde{\pi}_1(s; 0)e^{-(s+\alpha)x}(1 - B(x)). \tag{4}$$

Combination of the third equation of (3) and (4) leads to

$$\tilde{\pi}_2(s; x) = \tilde{\pi}_1(s; 0)(e^{-sx} - e^{-(s+\alpha)x})(1 - B(x)). \tag{5}$$

Substituting (4) and (5) respectively into the first and fourth equations of (3), after some algebra yields

$$\tilde{\pi}_0(s) = \frac{1}{s+\alpha}\left[1 + \tilde{\pi}_1(s; 0)\tilde{b}(s+\alpha)\right], \tag{6}$$

$$\tilde{\pi}_1(s; 0) = \frac{\alpha}{1 - \tilde{b}(s) + \tilde{b}(s+\alpha)}\tilde{\pi}_0(s),$$

that finishes the proof.

2.2   Reliability Function

Denote by T the random variable of the time to the complete failure of the system, then the reliability function is $R(t) = \mathbb{P}[T > t]$.

Theorem 2. *The Laplace transform of $R(t)$ is of the form*

$$\tilde{R}(s) = \frac{s + \alpha[2 - \tilde{b}(s+\alpha)]}{(s+\alpha)(s+\alpha - \alpha\tilde{b}(s+\alpha))}. \tag{7}$$

Proof. In order to evaluate the reliability of the system we will treat complete failure state, where $N(t) = 2$ for the first time, as absorbing one. We get a new system with the following set of equations:

$$\left[\frac{d}{dt} + \alpha\right]\pi_0(t) = \int_0^\infty \pi_1(t; u)\beta(u)du, \tag{8}$$

$$\left[\frac{\partial}{\partial t} + \frac{\partial}{\partial x} + \alpha + \beta(x)\right]\pi_1(t; x) = 0, \ \pi_1(t; 0) = \alpha\pi_0(t)$$

with the initial condition $\pi_0(0) = 1$. By taking Laplace transforms of these equations, we obtain

$$s\tilde{\pi}_0(s) - 1 = -\alpha\tilde{\pi}_0(s) + \int_0^\infty \beta(u)\tilde{\pi}_1(s; u)du, \tag{9}$$

$$s\tilde{\pi}_1(s; x) + \frac{\partial\tilde{\pi}_1(s; x)}{\partial x} = -(\alpha + \beta(x))\tilde{\pi}_1(s; x), \ \tilde{\pi}_1(s; 0) = \alpha\tilde{\pi}_0(s).$$

From the second and third equations of (9) we obtain

$$\tilde{\pi}_1(s; x) = \tilde{\pi}_1(s; 0)e^{-(s+\alpha)x}(1 - B(x)) = \alpha\tilde{\pi}_0(s)e^{-(s+\alpha)x}(1 - B(x)). \tag{10}$$

Substituting (10) into the first equation of (9) yields

$$(s+\alpha)\pi_0(s) = 1 + \alpha\tilde{\pi}_0(s)\tilde{b}(s+\alpha),$$

this implies $\pi_0(s) = \frac{1}{s+\alpha-\alpha\tilde{b}(s+\alpha)}$. Hence we have

$$\tilde{R}(s) = \tilde{\pi}_0(s) + \int_0^\infty \tilde{\pi}_1(s; x)dx = \tilde{\pi}_0(s)\left[1 + \alpha \int_0^\infty e^{-(s+\alpha)x}(1 - B(x))dx\right]$$

Upon substitution, one gets formula (7).

Corollary 1. *For the mean time to failure* $\mathbb{E}[T]$ *and the variance* $\mathbb{V}[T]$ *we get*

$$\mathbb{E}[T] = \frac{1}{\alpha}\left[1 + \frac{1}{1 - \tilde{b}(\alpha)}\right], \tag{11}$$

$$\mathbb{V}[T] = \frac{2 + (\tilde{b}(\alpha) - 2)\tilde{b}(\alpha) - 2\alpha\tilde{b}'(\alpha)}{[\alpha(1 - \tilde{b}(\alpha))]^2}. \tag{12}$$

Proof. The statement follows from the property of the Laplace transform,

$$\mathbb{E}[T] = \int_0^\infty R(t)dt = \tilde{R}(s)\Big|_{s=0},$$

$$\mathbb{V}[T] = \mathbb{E}[T^2] - \mathbb{E}[T]^2 = -\left[2\tilde{R}'(s) + \tilde{R}^2(s)\right]\Big|_{s=0}.$$

Remark 1. For exponential distribution of the repair time $B(x) = 1 - e^{-\beta x}$,

$$R(t) = \frac{e^{-\frac{1}{2}(2\alpha+\beta)t}\left[\sqrt{\beta(4\alpha + \beta)}\cosh(\frac{1}{2}\sqrt{\beta(4\alpha + \beta)}t) + (2\alpha - \beta)\sinh(\frac{1}{2}\sqrt{\beta(4\alpha + \beta)}t)\right]}{\sqrt{\beta(4\alpha + \beta)}}.$$

and $\mathbb{E}[T] = \frac{1}{\alpha}\left[2 + \frac{\beta}{\alpha}\right]$, $\mathbb{V}[T] = \frac{1}{\alpha^2}\left[\left(2 + \frac{\beta}{\alpha}\right)^2 - 2\right]$.

Remark 2. Denote by N_T – random value of the number of repairs until the system reaches the complete failure state $N(t) = 2$. This value describes also the number of regenerative cycles (e.g. the time between two successive visits of state $N(t) = 1$) without complete failure. Since a complete failure occurs in each regeneration cycle with a probability

$$p_F = \mathbb{P}[A < B] = \int_0^\infty A(u)b(u)du = 1 - \tilde{b}(\alpha),$$

the number of repairs N_T has a geometrical distribution

$$f_{N_T}(n) = \mathbb{P}[N_T = n] = \tilde{b}(\alpha)^n(1 - \tilde{b}(\alpha))$$

with the mean number $\mathbb{E}[N_T] = \frac{1}{p_F} - 1$. This characteristic is a natural discrete-valued counterpart of the time to failure T.

We now turn to the loss system $M/GI/2/2//0$ with two servers. The corresponding LSTs are denoted by $\tilde{b}_i(s)$ and repair intensities are $\beta_i, i = 1, 2$. Consider a Markov process

$$\{Z(t)\}_{t\geq 0} = \{J(t), X_1(t), X_2(t)\}_{t\geq 0}$$

with state space $E = \{0, (1, x_1, 1), (1, x_2, 2), (2, x_1, x_2) | x_1, x_2 \in \mathbb{R}_+\}$. We note that $J(t) = 1$ stands for the case when a certain server 1 or 2 is occupied. By analogy with a previous system define the state probabilities $\pi_0(t), \pi_1(t; x, 1)$, $\pi_1(t; x, 2)$ and $\pi_2(t; x, y)$.

Theorem 3. *The Laplace transform of $R(t)$ is of the form*

$$\tilde{R}(s) = \frac{s + \alpha[2 - p_1\tilde{b}_1(s + \alpha) - p_2\tilde{b}_2(s + \alpha)]}{(s + \alpha)(s + \alpha - \alpha p_1\tilde{b}_1(s + \alpha) - \alpha p_2\tilde{b}_2(s + \alpha))}, \tag{13}$$

where $p_i = \mathbb{P}[arriving\ customer\ to\ the\ empty\ system\ is\ served\ by\ server\ i]$.

Proof. As before the state $J(t) = 2$ stands for the absorption. Hence we get the following system of differential equations,

$$\left[\frac{d}{dt} + \alpha\right]\pi_0(t) = \int_0^\infty \pi_1(t; u, 1)\beta_1(u)du + \int_0^\infty \pi_1(t; u, 2)\beta_2(u)du,$$
$$\left[\frac{\partial}{\partial t} + \frac{\partial}{\partial x} + \alpha + \beta_1(x)\right]\pi_1(t; x, 1) = 0,$$
$$\left[\frac{\partial}{\partial t} + \frac{\partial}{\partial x} + \alpha + \beta_2(x)\right]\pi_1(t; x, 2) = 0;$$
$$\pi_1(t; 0, 1) = \alpha p_1\pi_0(t), \quad \pi_1(t; 0, 2) = \alpha p_2\pi_0(t).$$

By applying the LST to this system we get an expression for the transform $\tilde{\pi}_0(s)$, which has to be substituted to

$$\tilde{R}(s) = \tilde{\pi}_0(s) + \int_0^\infty (\tilde{\pi}_1(s; x, 1) + \tilde{\pi}_1(s; x, 2))dx,$$

in a similar as it was done in Theorem 2.

3   The $GI/M/1/2//1$ Queueing System

3.1   Non-stationary State Probabilities

Consider now analogous system with generally distributed life time of the unit and exponentially distributed repair time. Denote by $\{Z(t)\}_{t\geq 0} = \{J(t), X(t)\}_{t\geq 0}$ stochastic process, where the first component denotes the number of failed units at time t and the second one denotes the elapsed operational time of the working unit. Define the state probabilities:

(1) $\pi_i(t; x)dx = \mathbb{P}[N(t) = i, x < X(t) \leq x + dx]$ – the joint probability that at time t there are i failed units and the operational has elapsed working time between x and $x + dx, i = 0, 1$.

(2) $\pi_2(t) = \mathbb{P}[N(t) = 2]$ – the probability of the "bad" state (complete failure state) of the system at time t.

These probabilities satisfy the system of the forward differential equations,

$$\left[\frac{\partial}{\partial t} + \frac{\partial}{\partial x} + \alpha(x)\right]\pi_0(t; x) = \beta\pi_1(t; x), \tag{14}$$

$$\left[\frac{\partial}{\partial t} + \frac{\partial}{\partial x} + (\alpha(x) + \beta)\right]\pi_1(t; x) = 0,$$

$$\left[\frac{d}{dt} + \beta\right]\pi_2(t) = \int_0^\infty \alpha(u)\pi_1(t; u)du$$

with the boundary conditions $\pi_0(t; 0) = 0$, $\pi_1(t; 0) = \int_0^\infty \pi_0(t; u)\alpha(u)du + \beta\pi_2(t)$, together with normalizing condition $\sum_{i=0}^{1}\int_0^\infty \pi_i(t; x)dx + \pi_2(t) = 1$ For the initial condition $\pi_0(0; x) = \delta(x)$, where $\delta(x)$ is a Dirac-δ-function, the solution of the system (14) is of the form:

Theorem 4. *The state probabilities in terms of the Laplace transform are given by*

$$\tilde{\pi}_0(s; x) = \left[e^{-sx} + \tilde{\pi}_1(s; 0)(e^{-sx} - e^{-(s+\beta)x})\right](1 - A(x)), \tag{15}$$

$$\tilde{\pi}_1(s; x) = \tilde{\pi}_1(s; 0)e^{-(s+\beta)x}(1 - A(x)),$$

$$\tilde{\pi}_2(s) = \frac{1}{s+\beta}\tilde{a}(s+\beta)\tilde{\pi}_1(s; 0),$$

$$\tilde{\pi}_1(s; 0) = \frac{(s+\beta)\tilde{a}(s)}{(s+\beta)(1 - \tilde{a}(s)) + s\tilde{a}(s+\beta)}.$$

Proof. By taking Laplace transform of equations from (14), we obtain

$$s\tilde{\pi}_0(s; x) + \frac{\partial\tilde{\pi}_0(s; x)}{\partial x} - \delta(x) = -\alpha(x)\tilde{\pi}_0(s; x) + \beta\tilde{\pi}_1(s; x), \tag{16}$$

$$s\tilde{\pi}_1(s; x) + \frac{\partial\tilde{\pi}_1(s; x)}{\partial x} = -(\alpha(x) + \beta)\tilde{\pi}_1(s; x),$$

$$s\tilde{\pi}_2(s) = -\beta\tilde{\pi}_2(s) + \int_0^\infty \alpha(u)\tilde{\pi}_1(s; u)du,$$

$$\tilde{\pi}_1(s; 0) = \int_0^\infty \tilde{\pi}_0(s; u)\alpha(u)du + \beta\tilde{\pi}_2(s), \quad \tilde{\pi}_2(s; 0) = 0.$$

From the second equation of (16) it follows,

$$\tilde{\pi}_1(s; x) = \tilde{\pi}_1(s; 0)e^{-(s+\beta)x}(1 - A(x)).$$

Substituting this expression to the first equation of (16) and solving the corresponding differential equation, we get

$$\tilde{\pi}_0(s; x) = \left[e^{-sx} + \tilde{\pi}_1(s; 0)(e^{-sx} - e^{-(s+\beta)x})\right](1 - A(x)).$$

The last two expressions of (15) follows directly via substitution of these results to the remaining equations.

3.2   Reliability Function

Theorem 5. *The Laplace transform of $R(t)$ is of the form*

$$\tilde{R}(s) = \frac{(1 - \tilde{a}(s))(1 + \tilde{a}(s + \beta))}{s(1 - \tilde{a}(s) + \tilde{a}(s + \beta))}. \tag{17}$$

Proof. To derive the function $R(t)$ consider the system of differential equations with absorption in state 2,

$$\left[\frac{\partial}{\partial t} + \frac{\partial}{\partial x} + \alpha(x)\right]\pi_0(t; x) = \beta\pi_1(t; x), \tag{18}$$

$$\left[\frac{\partial}{\partial t} + \frac{\partial}{\partial x} + (\alpha(x) + \beta)\right]\pi_1(t; x) = 0,$$

$$\pi_1(t; 0) = \int_0^\infty \pi_0(t; u)\alpha(u)du$$

with initial state $\pi_0(0; x) = \delta(x)$. By taking Laplace transform, we obtain

$$s\tilde{\pi}_0(s; x) + \frac{\partial \tilde{\pi}_0(t; x)}{\partial x} - \delta(x) = -\alpha(x)\tilde{\pi}_0(s; x) + \beta\tilde{\pi}_1(s; x), \tag{19}$$

$$s\tilde{\pi}_1(s; x) + \frac{\partial \tilde{\pi}_1(s; x)}{\partial x} = -(\alpha(x) + \beta)\tilde{\pi}_1(s; x),$$

$$\tilde{\pi}_1(s; 0) = \int_0^\infty \tilde{\pi}_0(s; u)\alpha(u)du.$$

Expressions for $\tilde{\pi}_0(s; x)$ and $\tilde{\pi}_1(s; x)$ follow from Theorem 4. Substituting to the last equation (19) and expressing through $\tilde{\pi}_1(s; 0)$ leads to

$$\tilde{\pi}_1(s; 0) = \frac{\int_0^\infty a(u)e^{-su}du}{1 - \tilde{a}(s) + \tilde{a}(s + \beta)} = \frac{\tilde{a}(s)}{1 - \tilde{a}(s) + \tilde{a}(s + \beta)}.$$

Hence we have

$$\tilde{R}(s) = \int_0^\infty \sum_{i=0}^1 \tilde{\pi}_i(s; x)dx = \tilde{\pi}_1(s; 0)\frac{1}{s}\left[1 - \tilde{a}(s)\right] + \int_0^\infty (1 - A(x))e^{-sx}dx$$

$$= \frac{1}{s}\left[1 - \tilde{a}(s)\right]\left[\tilde{\pi}_1(s; 0) + 1\right],$$

which implies the required statement.

Corollary 2. *For the mean time to failure $\mathbb{E}[T]$ and the variance $\mathbb{V}[T]$ we get*

$$\mathbb{E}[T] = \bar{a}\left[1 + \frac{1}{\tilde{a}(\beta)}\right], \tag{20}$$

$$\mathbb{V}[T] = \frac{(1 - \tilde{a}^2(\beta))\bar{a}^2 + 2\bar{a}\tilde{a}'(\beta) + (1 + \tilde{a}(\beta))\tilde{a}(\beta)\tilde{a}''(0)}{\tilde{a}^2(\beta)}. \tag{21}$$

Proof. According to the property of Laplace transform $\mathbb{E}[T] = \tilde{R}(s)\Big|_{s=0}$. Since uncertainty $\frac{0}{0}$ occurs upon substitution of $s = 0$, we evaluate a limit using L'Hospital rule,

$$\lim_{s\to 0} \tilde{R}(s) = \lim_{s\to 0} \frac{(1 - \tilde{a}(s))(1 + \tilde{a}(s + \beta))}{s(1 - \tilde{a}(s) + \tilde{a}(s + \beta))} = \lim_{s\to 0} \frac{-\tilde{a}'(s)(1 + \tilde{a}(s + \beta))}{1 - \tilde{a}(s) + \tilde{a}(s + \beta)} = \frac{\bar{a}(1 + \tilde{a}(\beta))}{\tilde{a}(\beta)}.$$

The value $\mathbb{V}[T]$ can be evaluated due to the same arguments.

Remark 3. The probability of the complete failure in one regeneration cycle is given by

$$p_F = \tilde{a}(\beta),$$

and the number N_T has a density function

$$f_{N_T}(n) = \mathbb{P}[N_T = n] = (1 - \tilde{a}(\beta))^n \tilde{a}(\beta).$$

Now we consider the loss system $GI/M/2/2//0$.

Theorem 6. *The Laplace transform of $R(t)$ is of the form*

$$\tilde{R}(s) = \frac{(1 - \tilde{a}(s))(1 + \tilde{a}(s + \beta_1)p_1 + \tilde{a}(s + \beta_2)p_2)}{s(1 - \tilde{a}(s) + \tilde{a}(s + \beta_1)p_1 + \tilde{a}(s + \beta_2)p_2)}. \tag{22}$$

Proof. To derive the function $R(t)$ consider the system of differential equations with absorption in state $J(t) = 2$,

$$\left[\frac{\partial}{\partial t} + \frac{\partial}{\partial x} + \alpha(x)\right]\pi_0(t; x) = \beta_1\pi_1(t; x, 1) + \beta_2\pi_1(t; x, 2), \tag{23}$$

$$\left[\frac{\partial}{\partial t} + \frac{\partial}{\partial x} + (\alpha(x) + \beta_1)\right]\pi_1(t; x, 1) = 0,$$

$$\left[\frac{\partial}{\partial t} + \frac{\partial}{\partial x} + (\alpha(x) + \beta_2)\right]\pi_1(t; x, 2) = 0,$$

$$\pi_1(t; 0, 1) = p_1 \int_0^\infty \pi_0(t; u)\alpha(u)du, \quad \pi_1(t; 0, 2) = p_2 \int_0^\infty \pi_0(t; u)\alpha(u)du$$

with initial state $\pi_0(0; x) = \delta(x)$. The system can be solve by means of the LST. The statement now follows due to the same arguments as in Theorem 5.

4   Numerical Examples

Now we provide sensitivity analysis of the reliability function to the shape of life and repair time distributions. The random value X is assumed to be exponential $\mathcal{E}(1, \lambda)$, Erlang $\mathcal{E}(n, \lambda), n > 1$, uniform $\mathcal{U}(a, b)$ and Weibull $\mathcal{W}(\lambda, b)$ distributed. For evaluation of $\tilde{R}(s)$ we need to specify the corresponding LSTs:

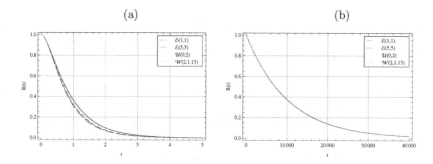

Fig. 1. Reliability function $R(t)$ (a) $\alpha = 2.5$ (b) $\alpha = 0.01$

Table 1. Evaluation results for the $M/GI/1/2//1$ system

Distr.	$\alpha = 2.5$				$\alpha = 0.01$			
	$\mathbb{E}[T]$	$\mathbb{V}[T]$	$\mathbb{E}[N_T]$	p_F	$\mathbb{E}[T]$	$\mathbb{V}[T]$	$\mathbb{E}[N_T]$	p_F
$\mathcal{E}(1,1)$	0.96	0.60	0.41	0.71	$1.02 \cdot 10^4$	$1.04 \cdot 10^8$	100	0.0099
$\mathcal{E}(5,5)$	0.86	0.47	0.15	0.87	$1.02 \cdot 10^4$	$1.03 \cdot 10^8$	100	0.0099
$\mathcal{U}(0,2)$	0.90	0.50	0.25	0.80	$1.02 \cdot 10^4$	$1.03 \cdot 10^8$	100	0.0099
$\mathcal{W}(2,1.13)$	0.87	0.48	0.19	0.84	$1.02 \cdot 10^4$	$1.03 \cdot 10^8$	100	0.0099

Table 2. Evaluation results for the $GI/M/1/2//1$ system

Distr.	$\beta = 2.5$				$\beta = 0.01$			
	$\mathbb{E}[T]$	$\mathbb{V}[T]$	$\mathbb{E}[N_T]$	p_F	$\mathbb{E}[T]$	$\mathbb{V}[T]$	$\mathbb{E}[N_T]$	p_F
$\mathcal{E}(1,1)$	4.50	18.25	2.50	0.29	2.01	2.04	0.01	0.99
$\mathcal{E}(5,5)$	8.59	56.85	6.59	0.13	2.01	0.41	0.01	0.99
$\mathcal{U}(0,2)$	6.03	28.49	4.03	0.19	2.01	0.69	0.01	0.99
$\mathcal{W}(2,1.13)$	7.34	41.63	5.34	0.16	2.01	0.57	0.01	0.99

$$\tilde{f}(s) = \left[\frac{\lambda}{s+\lambda}\right]^n, \ X \sim \mathcal{E}(n,\lambda) \, n \geq 1,$$

$$\tilde{f}(s) = \frac{e^{-as} - e^{-bs}}{s(b-a)}, \ X \sim \mathcal{U}(a,b),$$

$$\tilde{f}(s) = 1 - \frac{sb\sqrt{\pi}}{2} e^{(\frac{sb}{2})^2} erfc\left(\frac{sb}{2}\right), \ X \sim \mathcal{W}(2,b),$$

where $erfc(x) = \frac{2}{\sqrt{\pi}} \int_x^\infty e^{-u^2} du$ – is a complementary error function.

Consider first the system $M/GI/1/2//1$. Figures 1(a,b) illustrates the effect of the shape of the repair time distribution functions on the form of the reliability function $R(t)$. The parameters of distribution functions are chosen in such a way that $\mathbb{E}[B] = 1$. The variances of B are equal, respectively, to $\mathbb{V}[B] = \{1.00; 0.20; 0.33; 0.27\}$. We study two cases, where $\alpha = 2.5$, figure labeled by "a", and the rare event case, where $\alpha = 0.01$, figure labeled by "b". The values of $\mathbb{E}[T]$, $\mathbb{V}[T]$, $\mathbb{E}[N_T]$ and p_F are gathered in Table 1.

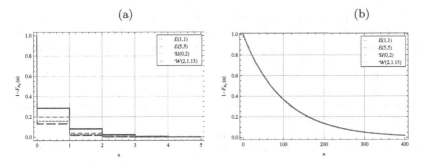

Fig. 2. The function $1 - F_{N_T}(n)$ (a) $\alpha = 2.5$ (b) $\alpha = 0.01$

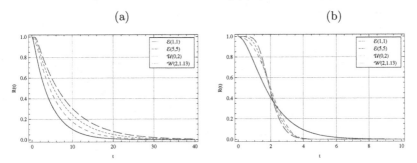

Fig. 3. Reliability function $R(t)$ (a) $\beta = 2.5$ (b) $\beta = 0.01$

The discrete counterpart to the function $R(t)$, i.e. the function $1 - F_{N_T}(n) = 1 - \sum_{k=0}^{n} f_{N_T}(k)$, is analyzed in Figure 2(a,b). The figures show the insensitivity of the shape of distribution functions in case of small p_F. The results for the system $GI/M/1/2//1$ are shown in Figures 3(a,b) is analyzed.

References

1. Gnedenko, B.V., Kovalenko, I.N.: Introduction to the queueing theory. Birkhauser Boston Inc., Cambridge (1966)
2. Knessl, C., Matkowsky, B.J., Schuss, Z., Tier, C.: The two repairman problem: A finite source $M/G/2$ queue. SIAM Journal on Applied Mathematics 47(2), 367–397 (1987)
3. Koenig, D., Rykov, V., Schtoyn, D.: Queueing Theory. Gubkin University Press, Moscow (1979)
4. Kovalenko, I.N.: Investigations and Analysis of Complex Systems Reliability. Naukova Dumka, Kiev (1975)
5. Langaris, C., Katsaros, A.: An $M/G/1$ queue with finite population and gates service discipline. Journal of the Operations Research, Society of Japan 40(1), 133–139 (1997)
6. Rykov, V.V.: Multidimensional alternative processes reliability models. In: Dudin, A., Klimenok, V., Tsarenkov, G., Dudin, S. (eds.) BWWQT 2013. CCIS, vol. 356, pp. 147–156. Springer, Heidelberg (2013)

7. Rykov, V.: To the problem of insensitivity of systems reliability characteristics to the shape of their elements life and repair time distributions. In: Proc. New Informational Technologies in Complex Structures Investigation, p. 98. Publishinh Hous of Tomsk State University, Tomsk (2014)
8. Rykov, V., Efrosinin, D., Vishnevskiy, V.: On sensitivity of reliability models to the shape of life and repair time distributions. submitted to the 2nd Int. Workshop on Statistical Methods in Reliability Assessment of Complex Industrial Multi-state Systems (RAMSS 2014) (2014)
9. Sevastyanov, B.A.: Limit theorem for Markov processes and its application to telephone systems with rejection. Probability Theory and its Applications 2(1) (1957)

Optimal Allocation Problem in the Machine Repairman System with Heterogeneous Servers

Dmitry Efrosinin[1,*], Christopher Spannring[1], and Janos Sztrik[2,**]

[1] Johannes Kepler University Linz,
Altenbergerstrasse 69, 4040 Linz, Austria
[2] University of Debrecen,
Egyetem ter 1, 4032 Debrecen, Hungary
dmitry.efrosinin@jku.at, c.spannring@gmx.at, sztrik.janos@inf.unideb.hu
http://www.jku.at, http://www.irh.inf.unideb.hu

Abstract. A controllable repairman model consists of L machines subject to failures and two repair servers working at different speeds. The problem of optimal allocation of failed machines between the servers is examined. The optimal control policy is calculated versus cost structures. As a result the optimal policy can be of threshold type, hysteretic type or have more complicated form. It is shown that the corresponding Markov process for hysteretic control policy belongs to the class of the Quasi-Birth-and-Death processes (QBD) with three diagonal block infinitesimal matrix. The stationary characteristics in this case are derived in matrix analytic form. Some numerical results are used to illustrate a number of features of the controlled model under study.

Keywords: Machine repairman system, performance analysis, dynamic-programming, optimal allocation, threshold policy, hysteretic policy.

1 Introduction

The machine repairman systems are normally described by means of the closed queueing systems, i.e. the systems with finite population. In such a system the customers of the finite population are the machines which are working at the operation area and during operation time they can fail independently of each other. The failed machines are sent to the repair facility where they can be restored. After the repair the machine becomes as good as a new one and is returned to the operational area. If all repair servers are busy a just failed machine has to wait for the repair at the buffer. In most cases in multi-server case the repair

* This work was funded by the COMET K2 Center "Austrian Center of Competence in Mechatronics (ACCM)", funded by the Austrian federal government, the federal state Upper Austria, and the scientific partners of the ACCM.
** The publication was supported by the TAMOP-4.2.2.C-11/1/KONY-2012-0001 project. The Project has been supported by the European Union, cofinanced by the European Social Fund.

A. Dudin et al. (Eds.): ITMM 2014, CCIS 487, pp. 113–122, 2014.

servers are assumed to be homogeneous, i.e. they repair the machines at equal speeds [2]. Only few papers deal with heterogeneous servers, see e.g. [9].

The problem of optimal allocation between heterogeneous servers was studied exhaustively only for infinite population queues. In [10] it was shown that the optimal allocation policy in heterogeneous system without preemption and switching costs is of threshold type, i.e. the server with larger mean usage cost has to be used if the queue length reaches some prespecified threshold level. The equivalent system with switching cost was analyzed in [5], where the hysteretic allocation policy took place. Due to this policy the usage of the server with higher mean usage cost is performed via the switch-on and switch-off threshold levels. For some other results concerning the hysteretic policy we refer the reader to [1,4,6].

In this paper we combine the finite population queueing system with heterogeneous repair facility and optimal allocation problem which obviously represents a missing subject among the available results. For the fixed threshold level and specified cost structure we have obtained explicitly the corresponding average cost which was minimized. To calculate the policy we use a dynamic-programming approach. Several structural properties of a control policy are established as well.

The rest of the paper is organized as follows: Section 2 describes the mathematical model based on a controllable Markov process. In Section 3 optimization problem is formulated and optimal equations for the dynamic-programming value function are derived. Section 4 deals with explicit evaluation of the mean performance measures. Finally, some numerical examples are presented in Section 5.

2   Mathematical Model

Consider the machine repairman system described in introduction. L machines subject to failure are working in- parallel. The operational time of each machine is exponentially distributed with parameter λ. The machines fail independently of each other. The repair facility consists of two heterogeneous servers with exponential distributed repair times with parameters $\mu_1 > \mu_2 > 0$. The process of the repair is assumed to be without preemption, i.e. the failed machine can not change the server during the repair process. The operational and repair times are assumed to be mutually independent.

Let $Q(t)$ denote the number of failed machines in the buffer and $D_i(t)$ – the state of the ith repair server. The system states at time t are described by a continuous-time Markov process

$$\{X(t)\}_{t\geq 0} = \{Q(t), D_1(t), D_2(t)\}_{t\geq 0}.$$

The controllable model associated with a Markov process $\{X(t)\}_{t\geq 0}$ is a five-tuple

$$\{E, A, \{A(x), x \in E\}, \lambda_{xy}(a), c(x, a)\}.$$

- E is a *state space* of the process $\{X(t)\}_{t \geq 0}$,

$$E = \{x = (q, d_1, d_2); q \in \{0, 1, \ldots, L\}, d_j \in \{0, 1\}, q + \sum_{j=1}^{2} d_j \leq L\}.$$

Further in the paper the notations $q(x), d_j(x), j = 1, 2$, will be used to specify the certain components of the vector state $x = (q, d_1, d_2) \in E$.

- $A = \{0, 1, 2\}$ is an *action space* with elements $a \in A$, where $a = j > 0$ means "to send a failed machine to the server j", $j = 1, 2$, and $a = 0$ means "to send a failed machine to the buffer".
- The *subsets* $A(x) \subseteq A$ *of control actions in state* $x \in E$, where $A(q, 0, 0) \equiv A, A(q, 0, 1) = \{0, 1\}$ and $A(q, 1, 0) = \{0, 2\}$.
- $\lambda_{xy}(a)$ is a *transition intensity* to go from state x to state y under a control action a. It is assumed that the model is *stable* and *conservative*, i.e.

$$\lambda_{xy}(a) \geq 0, \ y \neq x, \ \lambda_{xx}(a) = -\lambda_x(a) = -\sum_{y \neq x} \lambda_{xy}(a), \ \lambda_x(a) < \infty,$$

$$\lambda_{xy}(a) = \begin{cases} \lambda\left[L - q(x) - \sum_{j=1}^{2} d_j(x)\right] & y = x + e_a, \ a \in A(x), \\ \mu_j d_j(x) & y = x - e_j, \ q(x) = 0, \\ \mu_j d_j(x) & y = x - e_j - e_0 + e_a, \ q(x) > 0, \\ & a \in A(x - e_j - e_0). \end{cases}$$

The notation e_j is used for the vector with 1 in the jth position (beginning from 0th) and 0 elsewhere.

- $c(x, a)$ is an *immediate cost* in state x under control action a,

$$c(x, a) = c(x) + c_{01}\lambda\left[L - q(x) - \sum_{j=1}^{2} d_j(x)\right]1_{\{d_2(x)=0, a=2\}} +$$

$$\left[c_{01}\mu_1 d_1(x)1_{\{d_2(x)=0, a=2\}} + c_{10}\mu_2 d_2(x)1_{\{a=0 \lor a=1\}}\right]1_{\{q(x)>0\}},$$

$$c(x) = c_0 q(x) + \sum_{j=1}^{2} c_j d_j(x)$$

where c_0 – *holding cost* per unit of time in the buffer, c_j – *usage cost* of a repair server j per unit of time, c_{01} and c_{10} – fixed costs for switching on and off of the slower repair server. If $c_0 = c_j = 1, j = 1, 2$ and $c_{01} = c_{10} = 0$, then $c(x, a)$ represents the number of failed machines in state x.

We will next explain how the controller chooses its actions. According to the stationary Markov policy $f : E \to A$ whenever at a decision epoch the system state is $x \in E$, the controller choses an action $f(x) = a \in A(x) \subseteq A$ regardless of the past history of the system. We have two types of decision epochs:

- just after a failure of a machine at state x the controller chooses an action $a \in A(x)$, which prescribes to allocate the machine to one of available servers or to the buffer;

– just after a repair completion at server j in state x the controller chooses an action $a \in A(x - e_0 - e_j)$, which prescribes to take another machine from the queue, if it is not empty, and allocate it to one of available repair servers or put it back to the buffer.

3   Optimization Problem for Performance Characteristics

The process $\{X(t)\}_{t \geq 0}$ has a finite state space hence we may guarantee that this process is an irreducible, positive recurrent Markov process defined through its infinitesimal matrix $\Lambda = [\lambda_{xy}(f(x))]$. As it is known [8], for ergodic Markov process with costs the long-run average cost per unit of time (also referred to as *gain*) for the policy f coincides with corresponding assemble average,

$$g^f = \lim_{t \to \infty} \frac{1}{t} V^f(x, t) = \sum_{y \in E} c(y, a) \pi_y^f, \tag{1}$$

where

$$V^f(x, t) = \int_0^t \sum_{y \in E} \mathbb{P}^f[X(u) = y | X(0) = x] c(y, a) du \tag{2}$$

denotes the *total average cost up to time t* when the process starts in state x and π_y^f denotes a stationary probability of the process given policy f. The policy f^* is said to be optimal when for any admissible policy f

$$g^{f^*} = \min_f g^f. \tag{3}$$

We expect that the gain g^{f^*} will be smaller or equal to the gain under other heuristic allocation policies, e.g. Fastest Free Server discipline, which prescribes to use a fastest server among available.

The optimal policy f^* can be evaluated by means of a *Howard iteration algorithm* [3], which constructs a sequence of improved policies until the average cost optimal is reached. The key role in this algorithm is played by the *dynamic programming value function* $v : E \to \mathbb{R}_+$ which indicates a transition effect of an initial state x to the total average cost and satisfies a well-known asymptotic relation,

$$V^f(x, t) = g^f t + v^f(x) + o(1), \ x \in E, \ t \to \infty. \tag{4}$$

The functions V^f, v^f and g^f further in the paper will be denoted by V, v and g without upper index f.

The system will be uniformized as in Puterman [7] with the uniformization constant

$$\lambda L + \mu_1 + \mu_2 = 1,$$

which can be obtained by time scaling. As it is well known, the optimal policy f and the optimal average cost g are solutions of the optimality equation

$$Bv(x) = v(x) + g, \tag{5}$$

where B is the *dynamic programming operator* acting on value function v.

Theorem 1. *The dynamic programming operator B is defined as follows*

$$Bv(x) = c(x) + \left[L - q(x) - \sum_{j=1}^{2} d_j(x)\right]\lambda \min_{a \in A(x)}\{v(x + e_a) + c_{01}1_{\{a=2\}}\} + \quad (6)$$

$$\left[q(x) + \sum_{j=1}^{2} d_j(x)\right]\lambda v(x) + \sum_{j:d_j(x)=1}\mu_j v(x - e_j)1_{\{q(x)=0\}} + \sum_{j:d_j(x)=0}\mu_j v(x) +$$

$$\left[\mu_1 d_1(x) \min_{a \in A(x-e_1-e_0)}\{v(x - e_1 - e_0 + e_a) + c_{01}1_{\{d_2(x)=0,a=2\}}\} + \quad (7)\right.$$

$$\left.\mu_2 d_2(x) \min_{a \in A(x-e_2-e_0)}\{v(x - e_2 - e_0 + e_a) + c_{10}1_{\{a=0\lor a=1\}}\}\right]1_{\{q(x)>0\}}.$$

Proof. The optimality equation is obtained by analyzing the function $V(x, t)$ in some infinitesimal interval $[t, t+dt]$. It leads to the differential equation. Applying further the limit expression

$$\lim_{dt \to 0} \frac{V(x, t + dt) - V(x, t)}{dt} = 0$$

and taking into account Markov property of $\{X(t)\}_{t \geq 0}$ with asymptotic relation (4) ones get (6).

Corollary 1. *From (6) it follows that the optimal policy $f = (f_0, f_1, f_2)$ consists of components which specify the control action just after a new arrival in state x, just after a service completion at server 1 or 2 for nonempty queue,*

$$f_0(x) = \arg\min_{a \in A(x)}\{v(x + e_0), v(x + e_1)1_{\{d_1(x)=0\}}, (v(x + e_2) + c_{01})1_{\{d_2(x)=0\}}\},$$

$$f_1(x) = \arg\min_{a \in A(x-e_1-e_0)}\{v(x - e_1), v(x - e_0), (v(x - e_1 - e_0 + e_2) + c_{01})1_{\{d_2(x)=0\}}\},$$

$$f_2(x) = \arg\min_{a \in A(x-e_2-e_0)}\{v(x - e_2) + c_{10}, v(x - e_2 - e_0 + e_1), v(x - e_0)\}.$$

In case $c_{01} = c_{10} = 0$, $f_j(x) = f_0(x - e_j - e_0), j = 1, 2$.

4   Explicit Evaluation of the Gain Function

As is shown in Section 5 the optimal control policy can be approximated by the hysteretic policy with two threshold levels (U, D), $D \leq U$, i.e. the slower server must be activated when the queue length reaches the upper bound U and deactivated – when the queue length goes below the lower bound D. Under the fixed values U and D the gain g can be evaluated explicitly using the right hand side of (1).

Denote by $\boldsymbol{\pi}$ the row vector of the stationary state probabilities with components $\pi_x = \lim_{t \to \infty} \mathbb{P}[X(t) = x]$. Define the following row-subvectors,

$$\pi_{00} = (\pi_{000}, \pi_{001}), \quad \pi_k = \begin{cases} (\pi_{k10}, \pi_{k11}) & 0 \leq k \leq U - 1, \\ \pi_{k11} & U \leq k \leq L - 2. \end{cases}$$

The corresponding transition diagram is shown in Figure 1.

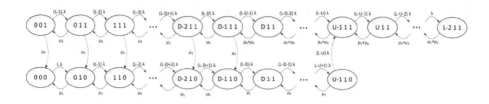

Fig. 1. Transition rate diagram for the hysteretic policy (U, D)

Theorem 2. *The Markov process $\{X(t)\}_{t \geq 0}$ for the thresholds (U, D) is of the QBD-type (Quasi-Birth-Death) with a state space*

$$E = \{x = (q, d_1, d_2); d_j \in \{0, 1\} \text{ if } q \in \{0, 1, \ldots, U-1\}, d_j = 1 \text{ if } U \leq q \leq L-2\}$$

and three-diagonal block infinitesimal matrix $\Lambda = [\lambda_{xy}(U, D)]$ defined as

$$\Lambda = \begin{pmatrix}
-B_0 & A_0^0 & 0 & 0 & 0 & 0 & 0 & 0 & 0 & 0 & 0 & \cdots & 0 \\
C_0 & -B_1^0 & A_0^1 & 0 & 0 & 0 & 0 & 0 & 0 & 0 & 0 & \cdots & 0 \\
0 & \ddots & \ddots & \ddots & & & & & & \ddots & \ddots & \ddots & \vdots \\
0 & \cdots & C_0 & -B_1^{D-1} & A_0^D & 0 & 0 & 0 & 0 & 0 & 0 & \cdots & 0 \\
0 & \cdots & 0 & C_1 & -B_2^D & A_0^{D+1} & 0 & 0 & 0 & 0 & 0 & \cdots & 0 \\
\vdots & \ddots & \ddots & & \ddots & \ddots & \ddots & & & \ddots & \ddots & \ddots & \vdots \\
0 & \cdots & 0 & 0 & C_1 & -B_2^{U-2} & A_0^{U-1} & 0 & 0 & 0 & 0 & \cdots & 0 \\
0 & \cdots & 0 & 0 & 0 & C_1 & -B_3 & a_0^U & 0 & 0 & 0 & \cdots & 0 \\
0 & \cdots & 0 & 0 & 0 & 0 & C_2 & -b_2^U & a_0^{U+1} & 0 & 0 & 0 & \cdots & 0 \\
0 & \cdots & 0 & 0 & 0 & 0 & 0 & c_2 & -b_2^{U+1} & a_0^{U+2} & 0 & \cdots & 0 \\
\vdots & \ddots & \ddots & & \ddots & & & & \ddots & \ddots & \ddots & \ddots & \vdots \\
0 & \cdots & 0 & 0 & 0 & 0 & 0 & 0 & 0 & c_2 & -b_2^{L-4} & a_0^{L-3} & 0 \\
0 & \cdots & 0 & 0 & 0 & 0 & 0 & 0 & 0 & 0 & c_2 & -b_2^{L-3} & a_0^{L-2} \\
0 & \cdots & 0 & 0 & 0 & 0 & 0 & 0 & 0 & 0 & 0 & c_2 & -b_2^{L-2}
\end{pmatrix},$$

where

$$C_0 := \begin{pmatrix} \mu_1 & 0 \\ 0 & \mu_1 \end{pmatrix}, B_0 = \begin{pmatrix} L\lambda & 0 \\ -\mu_2 & (L-1)\lambda + \mu_2 \end{pmatrix}, A_0^k = \begin{pmatrix} (L-k)\lambda & 0 \\ 0 & (L-k-1)\lambda \end{pmatrix}$$

$$B_1^k = \begin{pmatrix} (L-k-1)\lambda + \mu_1 & 0 \\ -\mu_2 & (L-k-2)\lambda + c_2 \end{pmatrix}, C_1 = \begin{pmatrix} \mu_1 & 0 \\ 0 & c_2 \end{pmatrix}, A_0^{D-1} = \begin{pmatrix} (L-D+1)\lambda & 0 \\ 0 & (L-D)\lambda \end{pmatrix},$$

$$B_1^{D-1} = \begin{pmatrix} (L-D)\lambda + \mu_1 & 0 \\ -\mu_2 & (L-D-1)\lambda + c_2 \end{pmatrix}, A_0^k = \begin{pmatrix} (L-k)\lambda & 0 \\ 0 & (L-k-1)\lambda \end{pmatrix},$$

$$B_2^k = \begin{pmatrix} (L-k-1)\lambda + \mu_1 & 0 \\ 0 & (L-k-2)\lambda + c_2 \end{pmatrix}, C_2 = \begin{pmatrix} 0 & c_2 \end{pmatrix}, A_0^{U-1} = \begin{pmatrix} (L-U+1)\lambda & 0 \\ 0 & (L-U)\lambda \end{pmatrix},$$

$$B_3 = \begin{pmatrix} (L-U)\lambda + \mu_1 & -(L-U)\lambda \\ 0 & (L-U-1)\lambda + c_2 \end{pmatrix},$$

$$a_0^k = A_0^k \cdot e_2 = (L-k-1)\lambda, \ b_2^k = B_2^k \cdot e_2 = (L-k-2)\lambda + c_2, \ c_2 = C_2 \cdot e_2 = \mu_1 + \mu_2.$$

Proof. The statement can be proved by simple block identification at the system of balance equations taking into account defined above specifications of the sub-vectors.

Theorem 3. *The stationary state probabilities π_{00} and $\pi_k, 0 \le k \le L - 2$, can be calculated by*

$$\pi_{00} = \pi_{L-2} \prod_{i=0}^{L-2} M_{L-2-i},$$

$$\pi_k = \pi_{L-2} \prod_{i=0}^{L-k-3} M_{L-2-i}, \; 0 \le k \le L - 3,$$

$$\pi_{L-2} = \left[1 + \sum_{k=0}^{U-1} \prod_{i=0}^{L-k-3} M_{L-2-i}e + \sum_{k=U}^{L-3} \prod_{i=0}^{L-k-3} M_{L-2-i} \right]^{-1},$$

where M_k satisfies the recursive relations

$$M_k = \begin{cases} C_0 B_0^{-1}, & k = 0 \\ C_0 \left(B_1^{k-1} - M_{k-1} A_0^{k-1} \right)^{-1}, & 1 \le k \le D - 1 \\ C_1 \left(B_1^{D-1} - M_{D-1} A_0^{D-1} \right)^{-1}, & k = D \\ C_1 \left(B_2^{k-1} - M_{k-1} A_0^{k-1} \right)^{-1}, & D+1 \le k \le U - 1 \\ C_2 \left(B_3 - M_{U-1} A_0^{U-1} \right)^{-1}, & k = U \\ c_2 \left(b_2^{k-1} - a_0^{k-1} M_{k-1} e_1 \right)^{-1}, & k = U + 1 \\ c_2 \left(b_2^{k-1} - a_0^{k-1} M_{k-1} \right)^{-1}, & U+2 \le k \le L - 2. \end{cases}$$

Proof. The main idea consists in deriving the recursive relations for the sub-vectors π_k from the system of balance equations in the form

$$\pi_{00} = \pi_0 M_0, \; \pi_k = \pi_{k+1} M_{k+1},$$

where matrices M_k can be evaluated also by the recursive relations defined in the statement. Note that the inverse matrices which are involved into these formulas are well defined since the matrices are main diagonal dominant and hence non-singular.

Corollary 2. *The main performance measures:*

- *Load factor of the repair server $j = 1, 2$*

$$\bar{U}_1 = \sum_{k=0}^{U-1} \pi_k e + \sum_{k=U}^{L-2} \pi_k, \; \bar{U}_2 = \pi_{00} e_1 + \sum_{k=0}^{U-1} \pi_k e_1 + \sum_{k=U}^{L-2} \pi_k;$$

- *Mean number of busy servers $\bar{C} = \sum_{j=1}^{2} \bar{U}_j$;*
- *Mean number of failed machines in the buffer*

$$\bar{Q} = \sum_{k=0}^{U-1} k \pi_k e + \sum_{k=U}^{L-2} k \pi_k;$$

- *Mean number of failed machines in the system $\bar{N} = \bar{C} + \bar{Q}$;*
- *Mean waiting and sojourn time of the failed machine*

$$\bar{W} = \frac{\bar{Q}}{\lambda(L - \bar{N})}, \quad \bar{T} = \frac{\bar{N}}{\lambda(L - \bar{N})};$$

- $\mathbb{P}[\text{Machine } n \text{ is failed}] = \frac{\lambda\bar{W}}{\lambda\bar{W}+1};$
- *The mean cost per unit of time*

$$g(U, D) = c_0\bar{Q} + \sum_{j=1}^{2} c_j\bar{U}_j + c_{01}\lambda(L - U)\pi_{U-1}e_0 + c_{10}\mu_2 \sum_{k=0}^{D-1} \pi_k e_1.$$

5   Numerical Examples

In this section we discuss some interesting observations about the properties of the optimal control policy $f = (f_0, f_1, f_2)$. To evaluate optimal policies we apply the Howard iteration algorithm [3] and formulas obtained in previous section.

Table 1. Component f_0 of the (a) optimal control policy (b) hysteretic control policy

(a)								(b)							
System State x	Queue Length $q(x)$							System State x	Queue Length $q(x)$						
(d_1, d_2)	0	1	2	3	4	...	15	(d_1, d_2)	0	1	2	3	5	...	15
(0,0)	1	1	1	1	1	...	1	(0,0)	1	1	1	1	1	...	1
(0,1)	0	0	1	1	1	...	1	(0,1)	1	1	1	1	1	...	1
(1,0)	0	0	0	2	2	...	2	(1,0)	0	0	0	2	2	...	2
(1,1)	0	0	0	0	0	...	0	(1,1)	0	0	0	0	0	...	0

Table 2. Component f_1 of the (a) optimal control policy (b) hysteretic control policy

(a)							(b)						
System State x	Queue Length $q(x)$						System State x	Queue Length $q(x)$					
(d_1, d_2)	1	2	3	4	5	... 15	(d_1, d_2)	1	2	3	4	5	... 15
(1,0)	1	1	1	1	1	... 1	(1,0)	1	1	1	1	1	... 1
(1,1)	0	0	1	1	1	... 1	(1,1)	1	1	1	1	1	... 1

Table 3. Component f_2 of the (a) optimal control policy (b) hysteretic control policy

(a)							(b)						
System State x	Queue Length $q(x)$						System State x	Queue Length $q(x)$					
(d_1, d_2)	1	2	3	4	5	... 15	(d_1, d_2)	1	2	3	4	5	... 15
(0,1)	2	2	2	2	2	... 2	(0,1)	2	2	2	2	2	... 2
(1,1)	2	2	2	2	2	... 2	(1,1)	2	2	2	2	2	... 2

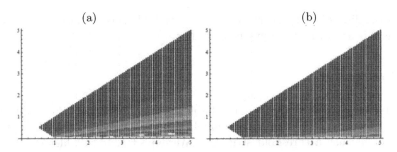

(a)   (b)

Fig. 2. The regions of optimality for threshold U in (a) infinite population (b) finite population $L = 10$ model

If the switching costs c_{01} and c_{10} are set to be equal to 0, than, as expected, the optimal control policy f is of threshold type, i.e. is defined through a single threshold level, $U = D > 0$, like at a similar system with infinite population. Due to this policy the second server with a higher cost per service period $\frac{c_2}{\mu_2} > \frac{c_1}{\mu_1}$ must be used in state $x = (q, 1, 0)$ whenever the number of failed machines in the buffer exceeds the threshold level U, $q(x) > U$.

If the switching costs are differs from 0 for the most values of system parameters λ, μ_1 and μ_2 the optimal policy belongs to the hysteretic policy with two threshold levels, (U, D), discussed at the beginning of the previous section. This policy is optimal in infinite population case as well. But surprisingly this policy for the closed system is not optimal everywhere, i.e. for some values of system parameters, e.g. if c_{01} is very large comparing to other costs, one more threshold level appears for the activation of the first server in state $x = (q, 0, 1)$.

Tables 1–3 illustrate the components $f_j, j = 0, 1, 2,$ of the optimal control policy (OCP) and the hysteretic control policy (HCP), $(4, 1)$, for the following values,

$$L = 17, \lambda = 1, \mu_1 = 5, \mu_2 = 1, c_0 = c_1 = c_2 = 1, c_{01} = 50, c_{10} = 5.$$

The gain function for these policies are equal to

$$g_{\text{OCP}} = 11.0125, \quad g_{\text{HCP}} = 11.0524.$$

As we can see, the difference is not sufficient. In other numerical examples the observable difference in performance was not more than 0.5%, so the hysteretic policy can be treated as a quasi-optimal one.

Another observation concerns the optimal threshold policy with threshold level U if $c_0 = c_1 = c_2 = 1$ and $c_{01} = c_{10} = 0$. The areas of optimality for threshold level depending on ratios $r_1 = \frac{\mu_1}{\lambda}$ and $r_2 = \frac{\mu_2}{\lambda}$ for infinite and finite population models are shown in Figure 2. The larger upper region stands for the case $U = 1$, below is shown the region for $U = 2$ and so on. We observe that the slope in the finite case is flatter. In finite population case the faster server must be more than four time faster as the slower one to get non-trivial solution $U > 1$, otherwise the optimal threshold policy will coincide with the fastest free server discipline $U = 1$, where the fastest available server must be used each time there is a waiting machine in the buffer.

6   Conclusion

In this paper we have studied a controllable machine repairman model with heterogeneous repair servers. For the model without switching costs the optimal control policy has a threshold structure. We expect that this fact can be rigorously proved using event-based dynamic programming approach to prove the monotonicity properties of the value function in the same way as it was done for infinite population models. In general case with switching costs the optimal control policy can be more complicated as a known hysteretic policy but the difference in performance between the policies is negligible.

References

1. Artalejo, J.R., Economou, A.: Markovian controllable queueing systems with hysteretic policies: Busy period and waiting time analysis. Method. and Comp. in App. Prob. 7, 353–378 (2005)
2. Falin, G.I.: A multiserver retrial queue with a finite number of sources of primary calls. Math. and Comp. Mod. 20, 33–49 (1999)
3. Howard, R.: Dynamic Programming and Markov Processes. Wiley Series, New York (1960)
4. Mitrani, I.: Managing performance and power consumption in a server farm. Ann. Oper. Res. 202, 121–134 (2013)
5. Nobel, R., Tijms, H.C.: Optimal control of a queueing system with heterogeneous servers and set-up costs. IEEE Transactions on Automation Control 45(4), 780–784 (2000)
6. Le Ny, L.M., Tuffin, B.: A simple analysis of heterogeneous multi-server threshold queues with hysteresis. In: Proceedings of Applied Telecommunication Symposium (ATS) (2002)
7. Puterman, M.L.: Markov Decision Process: Discrete Stochastic Dynamic Programming. John Wiley&Sons, New-York (1994)
8. Tijms, H.C.: Stochastic Models. An Algorithmic Approach. John Wiley&Sons, New-York (1994)
9. Roszik, J., Sztrik, J.: Performance analysis of finite-source retrial queues with non-reliable heterogeneous servers. J. of Math. Sciences 146, 6033–6038 (2007)
10. Rykov, V., Efrosinin, D.: On the slow server problem. Autom. and Rem. Control 70(12), 2013–2023 (2009)

Quasi-geometric and Gamma Approximation for Retrial Queueing Systems*

Ekaterina Fedorova

National Research Tomsk State University, Tomsk, Russia
moiskate@mail.ru

Abstract. In the paper, we propose methods of quasi-geometric and gamma approximation of the probability distribution of the calls number in the orbit for retrial queueing systems. The description and analysis of the application area of each method for retrial queueing system $M|GI|1$ are given. In addition, the results of both approximations are compared and a table of decision making on the choice of the approximation method are composed. Numerical examples of using the approximation methods for retrial queueing system $MMPP|M|1$ are presented.

Keywords: retrial queueing systems, number of calls in the orbit, gamma-approximation, quasi-geometric approximation.

Introduction

In queueing theory, there are two classes of queueing systems: systems with queue and loss systems. In life real systems, there are situations when queue cannot be explicitly identified, but also we cannot say that calls are lost if they come when the service device is unavailable. Usually, primary call does not refuse to be serviced and performs repeated calls to get the desired service in random time intervals. Examples of these situations are telecommunication systems. Thus a new class of queueing systems has been appeared: the systems with a source of repeated calls or Retrial queueing systems.

The first papers about systems with repeated calls were published in the middle of 20th century. The most of them were devoted to practical problems and influence of repeated attempts on telephone traffic, communication systems etc. [1–6]. The most comprehensive description and detailed comparison of classical queueing systems and retrial queues are contained in books by Artalejo J.R., Gomez-Corral A., Falin G.I. and Templeton J.G.C. [7–9].

The majority of studies of retrial queueing systems are perfomed numerically or via computer simulation [10, 11]. Analytical results are obtained only in cases of simple input and service processes (e.g. stationary Poisson input process or the exponential distribution of service law) [8]. Retrial queueing systems with BMAP input flow [12] are investigated by Dudin A.N. and. Klimenok V.I [14,

---

* This work is performed under the state order No. 1.511.2014/K of the Ministry of Education and Science of the Russian Federation.

15] in whose works matrix methods are mainly used. Also matrix methods for retrial queues analysis are used by Neuts M.F., Artalejo J.R., Gomez-Corral A. [16], Diamond J.E., Alfa A.S. [17] and etc. The most extensive review of retrial queues studies via matrix methods is presented in A. Gomez-Corrals work [18]. Asymptotic and approximate methods were developed by Falin G.I. [19], Artalejo J.R. [20], Anisimov V.V. [21] and others [22–24].

In a number of our previous papers devoted to the study of various single-server retrial queueing system [25, 26], we proposed the asymptotic analysis method for retrial queueing systems under a heavy load condition. During the investigation, we showed that asymptotic characteristic functions of the probability distribution of the number of calls in the orbit in systems with different input process and services laws ($M|M|1$, $M|GI|1$, $MMPP|M|1$, $MMPP|GI|1$) have form of characteristic function of gamma-distribution. However, we have demonstrated that the proposed method has a fairly narrow range of applicability: for a load rate $\rho > 0.95$ Kolmogorov distance between exact and asymptotic distributions has values $\Delta \leq 0.05$. In addition, we have obtained [25, 26] second-order asymptotic characteristic functions of the probability distribution of the number of calls in the orbit which increased the range of the method applicability to a load value equal to 0.8.

In this paper we propose methods of quasi-geometric and gamma approximation of the probability distribution of the calls number in the orbit for retrial queueing systems to increase the range of the applicability. The description of these methods is given for retrial queueing system $M|GI|1$, but they can be applied for all types of retrial queues where the mean and variance of distribution (or their estimates) can be obtained (e.g. the $MMPP|M|1$ retrial queues).

The paper consists of five sections. In Section 1, there is the description of the mathematical model of retrial queue $M|GI|1$. In Section 2, the method of quasi-geometric approximation is described and its numerical analysis is carried out. In the next section, we depict the method of gamma approximation and give some numerical results. In Section 4, we compare results of quasi-geometric and gamma approximations and make conclusions about what type of approximation is to be used for different sets of parameters. In the last section, the numerical results for quasi-geometric and gamma approximations for retrial queue $MMPP|M|1$ are presented.

1 Mathematical Model and the Process under Study

Let us consider retrial queueing system of $M|GI|1$ type. Structure of the system is presented in Figure 1.

The input process is Poisson Arrival Process with rate λ. The service time of each call has general distribution function $B(x)$. If a call arrives when a service device is free, the call occupies the device for the service. If the device is busy, the call goes to the orbit where it is staying during a random time. A duration of that time has an exponential distribution with parameter σ. After this random time, the call from the orbit makes an attempt to reach the device. If the device is free, the call occupies it, otherwise the call instantly returns to the orbit.

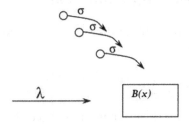

Fig. 1. Retrial queueing system $M|GI|1$

Falin G.I. [8] investigated retrial queueing system $M|GI|1$ by the method of elapsed service time. Obtained formula for generating function of probability distribution of the number of calls in the orbit has the following form:

$$g(z) = \frac{1-z}{k(z)-z} \exp\left\{\frac{\lambda}{\sigma} \int_1^z \frac{1-k(x)}{k(x)-x}\right\} \tag{1}$$

where $k(z) = \beta(\lambda - \lambda z) = \int_0^\infty e^{-(\lambda-\lambda z)x} dB(x)$ is Laplace-Stieltjes transform of the service time distribution function.

Also using the method of introducing the additional variable of the remaining service time, it can be derived that the characteristic function of the number of calls in the orbit has form:

$$H(u) = (1-\rho)\left[\frac{1-e^{ju}}{1-e^{ju}B^*(u)}\right] \cdot \exp\left\{-j\frac{\lambda}{\sigma}\int_1^u \frac{1-B^*(x)}{1-e^{-jx}B^*(x)}\right\} \tag{2}$$

where $B^*(u) = \int_0^\infty e^{-(\lambda-\lambda e^{ju})x} B(x)dx$ is Fourier transform of the service time distribution function.

It is easy to show that result of investigation of retrial queueing system $M|GI|1$ by the method of remaining service time (5) and the result (5) obtained by Falin G.I. are equivalent.

Applying the inverse Fourier transform to the characteristic function $H(u)$, the exact distribution P_i of the number of call in the orbit can be found.

However, for more complex retrial queues (where the incoming flow is not Poisson process) analytical results for calculating the probability distribution are not known in the scientific literature. In this regard, in this paper we propose an approximate method of quasi-geometric and gamma approximations of the probability distribution of the number of calls in the orbit by means of obtaining first and second moments. Firstly we investigate retrial queueing system $M|GI|1$ by proposed methods for the purpose of comparing approximate and exact distributions. In last section some numerical examples of applying quasi-geometric and gamma approximations for retrial queueing system $MMPP|M|1$ are demonstrated.

2   Quasi-geometric Approximation

Definition. A discrete probability distribution Kg_i for $i \geq 0$, for which the following equality holds

$$Kg_i = \begin{cases} (1 - p_0)(1 - \delta)\delta^{i-1}, & \text{for } i \geq 1, \\ p_0, & \text{for } i = 0, \end{cases}$$

is called [27] a quasi-geometric distribution of defect 1.

It is easy to show that mean and variance of the distribution Kg_i are equal to $\dfrac{1 - p_0}{1 - \delta}$, and $2\delta\dfrac{1 - p_0}{(1 - \delta)^2}$ respectively.

Let $i(t)$ denote the number of calls in the orbit and P_i be the probability that there are i calls in the orbit at the moment t where $i \geq 0$. The mean and the variance of the distribution P_i are denoted as $E\{i(t)\}$ and $\text{var}\{i(t)\}$ respectively.

The method of quasi-geometric approximation of the probability distribution of the number of calls in the orbit consists in the approximation of the probability distribution P_i by the quasi-geometric distribution Kg_i which parameters p_0 and δ are calculated through equating means and variances of distributions P_i and Kg_i.

So parameters p_0 and δ of the distribution of quasi-geometric approximation are defined through first and second order moments as follows

$$\delta = \frac{\text{var}\{i(t)\}}{2E\{i(t)\} + \text{var}\{i(t)\}} \quad \text{and} \quad p_0 = 1 - (1 - \delta)E\{i(t)\}. \tag{3}$$

Let us note that the value of p_0, defined by the formula (3) may be negative. In this case we assume $p_0 = 0$. Then quasi-geometric distribution will be shifted.

For the construction of the proposed approximation, it is enough to know the first and second order moments, whereas the distribution function may be unknown. Thus, the proposed method can be applied for more complex systems, in which we can find means and variance of the distribution of the number of calls in the orbit.

In the considered retrial queueing system $M|GI|1$ values of the first and second order moments can be found from the form of the characteristic function (5) using the known formula for calculation of k-th order initial moments of distribution:

$$E\{i^k(t)\} = (-j)^k \frac{d^k H(u)}{du^k}. \tag{4}$$

Let us compare the probability distribution of the number of calls in the orbit P_i and its quasi-geometric approximation Kg_i for different values of the system parameters. We show results on the approximation of the figures.

As distribution law of service time we choose gamma distribution with parameters $\alpha_s = \beta_s = 0.1$, thus the mean of service time distribution is equal $b = 1$.

Let the rate of input process be equal to $\lambda = \rho/b$ where variable ρ is the system load. We will analyze the results of approximation depending on the parameters ρ and σ.

In Figure 2, we show comparison of distributions (P_i is exact distribution and Kg_i is its quasi-geometric approximation) for $\rho = 0.5$ and $\sigma = 10$.

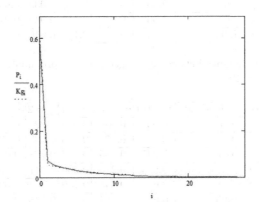

Fig. 2. Comparison of quasi-geometric and exact distributions for $\rho = 0.5$ and $\sigma = 10$

In the following tables (1, 2, 3) we show the Kolmogorov distance [28] between approximate and exact distributions:

$$\Delta = \max_{0 \leq i \leq N} \left| \sum_{n=0}^{i} Kg_n - \sum_{n=0}^{i} P_n \right|$$

for different values of the system parameters.

Table 1. Kolmogorov distance between quasi-geometric approximation and exact distribution for $\alpha_s = \beta_s = 0.1$

Values of the load rate	$\sigma = 0.1$	$\sigma = 0.5$	$\sigma = 1$	$\sigma = 2$	$\sigma = 10$
$\rho = 0.1$	0.032	0.014	0.011	0.009	0.008
$\rho = 0.3$	0.095	0.054	0.050	0.028	0.022
$\rho = 0.5$	0.014	0.068	0.029	0.040	0.017
$\rho = 0.8$	0.205	0.017	0.022	0.027	0.026
$\rho = 0.9$	0.286	0.046	0.023	0.009	0.016
$\rho = 0.95$	0.337	0.192	0.058	0.064	0.057

We assume the criterion of applicability of methods is the condition $\Delta \leq 0.05$ (where Δ is Kolmogorov distance between the approximate and exact distributions). Then, from the above tables it can be concluded that the quasi-geometric

Table 2. Kolmogorov distance between quasi-geometric approximation and exact distribution for $\alpha_s = \beta_s = 1$

Values of the load rate	$\sigma = 0.1$	$\sigma = 0.5$	$\sigma = 1$	$\sigma = 2$	$\sigma = 10$
$\rho = 0.1$	0.034	0.010	0.006	0.005	0.003
$\rho = 0.3$	0.107	0.068	0.048	0.036	0.026
$\rho = 0.5$	0.198	0.096	0.089	0.074	0.056
$\rho = 0.8$	0.504	0.138	0.035	0.032	0.057
$\rho = 0.9$	0.566	0.245	0.119	0.049	0.016
$\rho = 0.95$	0.595	0.288	0.174	0.040	0.012

Table 3. Kolmogorov distance between quasi-geometric approximation and exact distribution for $\alpha_s = \beta_s = 10$

Values of the load rate	$\sigma = 0.1$	$\sigma = 0.5$	$\sigma = 1$	$\sigma = 2$	$\sigma = 10$
$\rho = 0.1$	0.034	0.009	0.005	0.004	0.002
$\rho = 0.3$	0.112	0.070	0.046	0.033	0.021
$\rho = 0.5$	0.247	0.105	0.097	0.078	0.054
$\rho = 0.8$	0.574	0.215	0.060	0.047	0.060
$\rho = 0.9$	0.629	0.333	0.189	0.058	0.036
$\rho = 0.95$	0.657	0.384	0.267	0.138	0.016

Table 4. The range of the applicability of quasi-geometric approximation method

Values of the retrial rate σ	Values of the load rate ρ
$\sigma = 0.1$	$\rho < 0.1$
$\sigma = 0.5$	$\rho < 0.2$
$\sigma = 1$	$\rho \leq 0.3$
$\sigma = 2$	$\rho < 0.5$ or $0.7 < \rho < 0.9$
$\sigma = 10$	Any ρ

approximation of the probability distribution of the number of calls in the orbit provides the best results when the rate σ increases. Note that in the case $\sigma = 10$ Kolmogorov distance between the exact distribution and its quasi-geometric approximation does not exceed 0.06. In the Table 4, the range of the method applicability is demonstrated depending on the system parameters values.

3   Gamma Approximation

In this section, we offer to approximate the probability distribution of the number of calls in the orbit P_i by discrete analogue (defined below) of the gamma distribution G_i with a shape parameter α and an inverse scale (rate) parameter β.

The method of gamma approximation consist in approximation the probability distribution P_i by the discrete analogue of the gamma distribution G_i which parameters is calculated through equating means and variances of distributions

P_i and G_i. So parameters α and β are the following

$$\alpha = \frac{\mathrm{E}\{i(t)\}}{\mathrm{var}\{i(t)\}}, \quad \beta = \frac{\mathrm{E}^2\{i(t)\}}{\mathrm{var}\{i(t)\}}$$

where $\mathrm{E}\{i(t)\}$ is mean and $\mathrm{var}\{i(t)\}$ is variance of the distribution P_i of the number of calls in the orbit, which are calculated by known formula (4).

Several discrete analog alternatives of gamma distribution can be offered. In particular, they are the following:

1. $G_1(i) = c_1 f(i)$ where c_1 is a normalizing constant, and $f(i)$ is the density of the gamma distribution at point i.
2. $G_2(i) = F(i+1) - F(i)$ where $F(i)$ is function of the gamma distribution at point i.
3. $G_3(i) = \begin{cases} c_3 \cdot F(0.5), & \text{if } i = 0, \\ F(i+0.5) - F(i-0.5), & \text{if } i \geq 1 \end{cases}$
   where c_3 is a normalizing constant.

We use the second way for calculation of probability distribution G_i.

Let us compare the probability distribution of the number of calls in the orbit P_i and its gamma approximation G_i for different values of the system parameters.

As distribution law of service time we choose gamma distribution with parameters $\alpha_s = \beta_s = 0.1$, thus the mean of service time distribution is equal $b = 1$.

Let the rate of input process equals $\lambda = \rho/b$ where variable ρ is the system load. We will analyze the results of approximation depending on the parameters ρ and σ.

In Figure 3 we show comparison of distributions (P_i is exact distribution and G_i is its gamma approximation) for $\rho = 0.9$ and $\sigma = 1$.

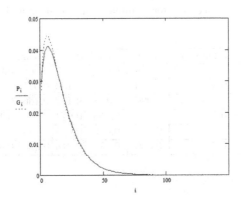

Fig. 3. Comparison of gamma approximation and exact distributions for $\rho = 0.9$ and $\sigma = 1$

In the following tables (5, 6, 7) we show the Kolmogorov distance [28] between approximate and exact distributions:

$$\Delta = \max_{0 \leq i \leq N} \left| \sum_{n=0}^{i} G_n - \sum_{n=0}^{i} P_n \right|$$

for different values of the system parameters.

Table 5. Kolmogorov distance between gamma approximation and exact distribution for $\alpha_s = \beta_s = 0.1$

Values of the load rate	$\sigma = 0.1$	$\sigma = 0.5$	$\sigma = 1$	$\sigma = 2$	$\sigma = 10$
$\rho = 0.1$	0.067	0.028	0.023	0.021	0.019
$\rho = 0.3$	0.122	0.076	0.058	0.049	0.039
$\rho = 0.5$	0.031	0.038	0.037	0.038	0.038
$\rho = 0.8$	0.009	0.032	0.061	0.075	0.121
$\rho = 0.9$	0.013	0.034	0.041	0.055	0.101
$\rho = 0.95$	0.023	0.062	0.039	0.054	0.081

Table 6. Kolmogorov distance between gamma approximation and exact distribution for $\alpha_s = \beta_s = 1$

Values of the load rate	$\sigma = 0.1$	$\sigma = 0.5$	$\sigma = 1$	$\sigma = 2$	$\sigma = 10$
$\rho = 0.1$	0.077	0.022	0.015	0.011	0.008
$\rho = 0.3$	0.199	0.148	0.107	0.081	0.059
$\rho = 0.5$	0.082	0.162	0.162	0.142	0.112
$\rho = 0.8$	0.021	0.042	0.052	0.060	0.066
$\rho = 0.9$	0.009	0.019	0.024	0.029	0.035
$\rho = 0.95$	0.005	0.010	0.012	0.014	0.030

Table 7. Kolmogorov distance between gamma approximation and exact distribution for $\alpha_s = \beta_s = 10$

Values of the load rate	$\sigma = 0.1$	$\sigma = 0.5$	$\sigma = 1$	$\sigma = 2$	$\sigma = 10$
$\rho = 0.1$	0.078	0.020	0.012	0.008	0.005
$\rho = 0.3$	0.213	0.155	0.106	0.075	0.048
$\rho = 0.5$	0.096	0.198	0.197	0.166	0.119
$\rho = 0.8$	0.027	0.056	0.074	0.087	0.105
$\rho = 0.9$	0.013	0.026	0.035	0.043	0.057
$\rho = 0.95$	0.006	0.018	0.018	0.021	0.028

Table 8. The range of the applicability of quasi-geometric approximation method

Values of the retrial rate σ	Values of the load rate ρ
$\sigma = 0.1$	$\rho \geq 0.8$
$\sigma = 0.5$	$\rho \leq 0.1$ or $\rho \geq 0.8$
$\sigma = 1$	$\rho \leq 0.1$ or $\rho > 0.8$
$\sigma = 2$	$\rho < 0.2$ or $\rho \geq 0.9$
$\sigma = 10$	$\rho \leq 0.2$ or $\rho \geq 0.95$

As in the previous section, we assume that the criterion of applicability of methods is the condition $\Delta \leq 0.05$. Then the following table 8 for the range of the method applicability depending on the system parameters values can be composed

4   Comparison of Quasi-geometric Ang Gamma Approximations

In following figures 4 and 5 we demonstrate the comparison of results of proposed methods: quasi-geometric ang gamma approximations of probability distribution of the number of calls in the orbit.

In Figure 4 there are presented gamma and quasi-geometric approximations and exact distribution for $\alpha_s = \beta_s = 0.1$, $\rho = 0.8$ and $\sigma = 0.1$.

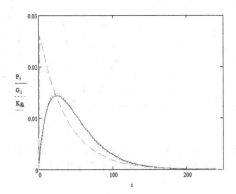

Fig. 4. Comparison of quasi-geometric and gamma approximations for $\rho = 0.8$ and $\sigma = 0.1$

From the figure, it is obvious that quasi-geometric approximation gives unsatisfactory results and it is necessary to apply gamma approximation in this case.

In Figure 5 there are presented gamma and quasi-geometric approximations and exact distribution for $\alpha_s = \beta_s = 10$, $\rho = 0.9$ and $\sigma = 10$.

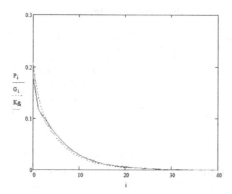

Fig. 5. Comparison of quasi-geometric and gamma approximations for $\rho = 0.9$ and $\sigma = 10$

From the figure, it is obvious that exact distribution is similar to geometric ones, so the quasi-geometric approximation is applied in this case.

By comparing Table 4 and Table 8, an overall table of decision making about type of using approximation is composed (see Table 9).

Table 9. Decision making about the type of approximation in use

Values of the load rate	$\sigma = 0.1$	$\sigma = 0.5$	$\sigma = 1$	$\sigma = 2$	$\sigma = 10$
$\rho = 0.1$	Kg_i	Kg_i	Kg_i	Kg_i	Kg_i
$\rho = 0.3$			Kg_i	Kg_i	Kg_i
$\rho = 0.5$					Kg_i
$\rho = 0.8$	G_i	G_i	G_i	Kg_i	Kg_i
$\rho = 0.9$	G_i	G_i	G_i	G_i	Kg_i
$\rho = 0.95$	G_i	G_i	G_i	G_i	G_i

5 Quasi-geometric and Gamma Approximation for Retrial Queueing System $MMPP|M|1$

In this section, we demonstrate some numerical examples of applying proposed approximation methods for retrial queuing system $MMPP|M|1$. Structure of such system is presented in Figure 6.

The input process is Markov Modulated Poisson Process. The underlying process $n(t)$ is Markov chain with continuous time and finite set of states $n = 1, 2, \ldots, W$. MMPP is a particular case of Markovian Arrival Process (MAP) and it is defined by matrix $\mathbf{D_0}$ and $\mathbf{D_1}$ [13]. Elements of the matrix $\mathbf{D_0}$ describe represent transitions with and elements of $\mathbf{D_1}$ observable transitions.

We denote the generator of the underlying process $n(t)$ by matrix $\mathbf{Q} = \mathbf{D_0} + \mathbf{D_1}$. And the matrix $\mathbf{Q}$ has elements q_{mv} where $m, v = 1, 2, \ldots, W$.

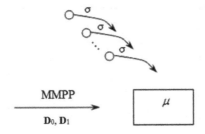

Fig. 6. Retrial queueing systems $MMPP|M|1$

Matrix $\mathbf{D_1}$ is diagonal one with diagonal elements $\rho\lambda_n$ where $n = 1, 2, \ldots, W$ and ρ is some parameter defined below. We introduce a matrix $\mathbf{\Lambda} = \text{diag}\{\lambda_n\}$. Then the following equality holds: $\mathbf{D_1} = \rho\mathbf{\Lambda}$.

The vector-row $\boldsymbol{\theta}$ is the probability distribution of underlying process of MMPP $n(t)$ which is defined as the unique solution of the system:

$$\begin{cases} \boldsymbol{\theta}\mathbf{Q} = \mathbf{0}, \\ \boldsymbol{\theta}\mathbf{e} = 1 \end{cases}$$

where $\mathbf{e}$ is unit column-vector, $\mathbf{0}$ is zero row-vector.

The service time of each call is distributed by exponential law with parameter μ. If a call arrives when a service device is free, the call occupies the device for the service. If the device is busy, the call goes to the orbit where it is staying during a random time. A duration of that time has an exponential distribution with parameter σ. After this random time, the call from the orbit makes an attempt to reach the device. If the device is free, the call occupies it, otherwise the call instantly returns to the orbit.

It is known that the rate of MMPP is defined as $\lambda = \boldsymbol{\theta} \cdot \rho\mathbf{\Lambda} \cdot \mathbf{e}$.

Let the system parameters be such that the following equation holds: $\boldsymbol{\theta} \cdot \mathbf{\Lambda} \cdot \mathbf{e} = \mu$. So the parameter $\rho = \dfrac{\lambda}{\boldsymbol{\theta} \cdot \mathbf{\Lambda} \cdot \mathbf{e}} = \dfrac{\lambda}{\mu}$ and it is called the load in the system.

In numerical example let the system parameters be the following:

$$\mu = 1, \ \sigma = 1,$$

$$\mathbf{\Lambda} = \begin{pmatrix} 0.488 & 0 & 0 \\ 0 & 0.976 & 0 \\ 0 & 0 & 1.463 \end{pmatrix}, \ \mathbf{Q} = \begin{pmatrix} -0.5 & 0.2 & 0.3 \\ 0.1 & -0.3 & 0.2 \\ 0.3 & 0.2 & -0.5 \end{pmatrix}.$$

In Figure 7 and Figure 8, we show comparison of distributions (P_i is exact distribution, which is obtained numerically, Kg_i and G_i are the quasi-geometric and gamma approximations correspondingly) for different values of ρ.

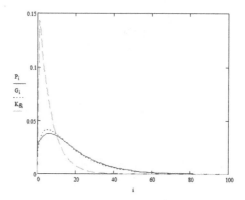

Fig. 7. Comparison of quasi-geometric and gamma approximations for $\rho = 0.9$ in retrial queue $MMPP|M|1$

From the Figure 7 it is obvious that quasi-geometric approximation gives unsatisfactory results and it is necessary to apply gamma approximation in this case. The same result we have for the retrial queue $M|GI|1$ (Table 9). For these values of system parameters the Kolmogorov distance between the exact distribution and gamma approximation is equal to 0.023.

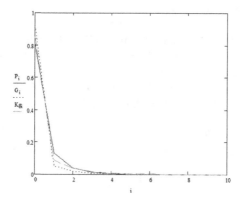

Fig. 8. Comparison of quasi-geometric and gamma approximations for $\rho = 0.3$ in retrial queue $MMPP|M|1$

From the Figure 8 we can conclude that quasi-geometric approximation gives the better results then gamma approximation. The same result we also have for the retrial queue $M|GI|1$ (in Table 9). In this case the Kolmogorov distance between the exact distribution and quasi-geometric approximation is equals to 0.039.

Thus the proposed method of quasi-geometric and gamma approximations can be applied for systems with not only Poisson arrival process but also for systems with more complex models of input process (MMPP, MAP, etc).

Conclusion

In the paper, methods of quasi-geometric and gamma approximation of the probability distribution of the calls number in the orbit for retrial queueing systems are proposed. Based on a study of the $M|GI|1$ system, methods are described and conclusions about their applicability are made by means of the numerical comparison of approximate and exact distributions. The table for decision making about the type of approximation in use is composed. Thus we conclude that these approximations can be applied for almost all values of system parameters ρ and σ. The main advantage of the proposed approximate methods is that for their application we need to know only the mean and variance (or their estimates). In addition, numerical examples of using of approximation methods for retrial queueing systems $MMPP|M|1$ were presented in the paper.

References

1. Wilkinson, R.I.: Theories for toll traffic engineering in the USA. The Bell System Technical Journal 35(2), 421–507 (1956)
2. Cohen, J.W.: Basic problems of telephone trafic and the influence of repeated calls. Philips Telecommunication Review 18(2), 49–100 (1957)
3. Elldin, A., Lind, G.: Elementary Telephone Trafic Theory. Ericsson Public Telecommunications (1971)
4. Jonin, G.L., Sedol, J.J.: Telephone systems with repeated calls. In: Proceedings of the 6th International Teletrafic Congress (ITC-6), Munich, pp. 435/1–433/5 (1970)
5. Shneps-Shneppe, M.A.: The efect of repeated calls on communication system. In: Proceedings of the 6th International Teletrafic Congress(ITC-6), Munich, pp. 433/1–433/5 (1970)
6. Gosztony, G.: Repeated call attempts and their efect on trafic engineering. Budavox Telecommunication Review 2, 16–26 (1976)
7. Artalejo, J.R., Gomez-Corral, A.: Retrial Queueing Systems. A Computational Approach. Springer (2008)
8. Falin, G.I., Templeton, J.G.C.: Retrial queues. Chapman & Hall, London (1997)
9. Artalejo, J.R., Falin, G.I.: Standard and retrial queueing systems: A comparative analysis. Revista Matematica Complutense 15, 101–129 (2002)
10. Neuts, M.F., Rao, B.M.: Numerical investigation of a multiserver retrial model. Queueing Systems 7(2), 169–189 (2002)
11. Ridder, A.: Fast simulation of retrial queues. In: Third Workshop on Rare Event Simulation and Related Combinatorial Optimization Problems, Pisa, pp. 1–5 (2000)
12. Neuts, M.F.: Versatile Markovian Point Process. Journal of Applied Probability 16(4), 764–779 (1979)
13. Lucantoni, D.M.: New results on the single server queue with a batch Markovian arrival process. Stochastic Models 7, 1–46 (1991)

14. Dudin, A.N., Klimenok, V.I.: Queueing System BMAP/G/1 with repeated calls. Mathematical and Computer Modelling 30(3-4), 115–128 (1999)
15. Kim, C.S., Klimenok, V.I., Birukov, A.V., Dudin, A.N.: Optimal multi-threshold control by the $BMAP|SM|1$ retrial system. Annals of Operations Research 141, 193–210 (2006)
16. Artalejo, J.R., Gomez-Corral, A., Neuts, M.F.: Analysis of multiserver queues with constant retrial rate. European Journal of Operational Research 135, 569–581 (2001)
17. Diamond, J.E., Alfa, A.S.: Matrix analytical methods for M/PH/1 retrial queues. Stochastic Models 11, 447–470 (1995)
18. Gomez-Corral, A.: A bibliographical guide to the analysis of retrial queues through matrix analytic techniques. Annals of Operations Research 141, 163–191 (2006)
19. Falin, G.I.: Asymptotic properties of probability distribution of the number of request in system M/G/1/1 with repeated calls. VINITI, 5418–5483 (1983) (in Russian)
20. Artalejo, J.R.: Information theoretic approximations for retrial queueing systems. In: Transactions of the 11th Prague Conference on Information Theory, Statistical Decision Functions and Random Processes, pp. 263–270. Kluwer Academic Publishers, Dordrecht (1992)
21. Anisimov, V.V.: Asymptotic analysis of highly reliable retrial systems with finite capacity. In: Queues, Flows, Systems, Networks: Proceedings of the International Conference Modern Mathematical Methods of Investigating the Telecommunication Networks, Minsk, pp. 7–12 (1999)
22. Yang, T., Posner, M.J.M., Templeton, J.G.C., Li, H.: An approximation method for the M/G/1 retrial queue with general retrial times. European Journal of Operational Research 76, 552–562 (1994)
23. Diamond, J.E., Alfa, A.S.: Approximation method for M/PH/1 retrial queues with phase type inter-retrial times. European Journal of Operational Research 113, 620–631 (1999)
24. Pourbabai, B.: Asymptotic analysis of G/G/K queueing-loss system with retrials and heterogeneous servers. International Journal of Systems Sciences 19, 1047–1052 (1988)
25. Nazarov, A.A., Moiseeva, E.A.: Investigation of retrial queueing system $M|GI|1$ by asymptotic analysis method under heavy load. In: Queues: Flows, Systems, Networks: Proceedings of the International Conference Modern Probabilistic Methods for Analysis, Design, and Optimization of Information and Telecommunication Networks, vol. 22, pp. 106–113. BSU, Minsk (2013) (in Russian)
26. Nazarov, A.A., Moiseeva, E.A.: The research of retrial queueing system $MMPP|M|1$ by the method of asymptotic analysis under heavy load. In: The Bulletin of Tomsk Polytechnic University. Mathematics, Physics and Mechanics, vol. 322(2), pp. 19–23. TPU, Tomsk (2013) (in Russian)
27. Nazarov, A.A., Lyubina, T.V.: The non-Markov dynamic RQ-system with the incoming MMP-flow of requests. Automation and Remote Control 74(7), 1132–1143 (2013)
28. Kovalenko, I.N., Filippova, A.A.: Probability Theory and Mathematical Statistics. A Textbook. Vyschaya shkola, Moscow (1982) (in Russian)

Information Approach to Signal-to-Noise Ratio Estimation of the Speech Signal

Vasiliy Gai

Nizhny Novgorod State Technical Universty n.a. R.E. Alekseev,
Nizhny Novgorod, Russia
vasiliy.gai@gmail.com
http://www.nntu.nnov.ru/

Abstract. The article describes the method of signal-to-noise ratio estimation for speech signals. The proposed method is based on the theory of active perception. Within the scope of work assumes that the speech signal includes a desired signal (system formation) and noise. The conversions, which were described in the theory of active perception, allow allocating the desired signal and solving the problem of signal to noise ratio estimation. The work includes experimental data confirming workability of the proposed method.

Keywords: signal to noise ratio (SNR) estimation, speech signal, theory of active perception (TAP).

1   Introduction

Speech processing system (speaker identification, speech recognition), working under noise conditions, must possess stability to various distortion of input signal. Therefore, to adjust the signal processing algorithm by noise level, such systems must possess the ability of level rating of signal's distortion (signal to noise ratio, SNR), and that kind of rating must be done only by a distorted signal.

Let us consider identity of existent methods of signal to noise ratio estimation:

1. by SNR estimation analyzable signal is divided into segments length by 20-30 ms., the overlap between segments is 50 percent, then the spectrum of each segment is calculated and by the spectra noise estimation [1] are done;
2. methods of SNR estimation are developed taking into account that analyzable signal contain speech and pauses [2].

The analysis of works permits to sort out the following classes of methods of SNR estimation:

1. methods based on voice activity determination are to signal segmentation active speech and pauses, to signal and noise power estimate and to computing SNR [2];

A. Dudin et al. (Eds.): ITMM 2014, CCIS 487, pp. 137–144, 2014.
© Springer International Publishing Switzerland 2014

2. methods using controlled recursive averaging [3], in such methods estimation of noise is performed by averaging the previous values of the spectral power using a smoothing parameter, which depends on probability of the presence of the signal in different frequency band: it shows that the presence of speech in some segment in a certain frequency band may be determined by relation to local energy of noisy speech to its minimum in that segment. If the value of this ratio is less than the threshold, we can conclude that there is no speech signal in segment;

3. methods of noise assessment based on the minimum statistic, methods of this class are based on two assumptions [4]. The first consists in independence of noise and speech, the second - the power spectrum of the noisy speech signal is similar to the noise power spectrum. Therefore, noise dispersion estimate is to calculate the minimum of the spectral concentration of noisy speech signal in fixed length segment. Disadvantages of the method are that it is necessary to select the length of the segment. In this case the wrong choice can considerably affect the assessment's result. Another disadvantage of the method is entering of a delay in the noise estimation parameters, as the length of the segment (1.6 - 2.8 s.) is chosen to ensure including part of speech and pauses;

4. methods of noise parameters assessment based on the histogram using observation that the most frequently occurring value of the energy in some frequency band corresponds to the noise level in this frequency band, i.e. the noise level corresponds to the maximum energy histogram [5].

In this work the solution of SNR estimation is based on using a systematic approach to signal processing described in the theory of active perception [6].

Let us consider the basic propositions of the theory of active perception. From the point of view of an observer, sound signal contains a desired signal (information message) and a hindrance. Desired signal is the information, which is required for the observer to make decisions under the task, and the noise - all the other information. In this work, desired signal is considered as a system formation. In this case, it must contain the structural elements and connections.

Theory of Active Perception (TAP) contains description of operations that allow to allocate the structural elements of the signal and links between them. For detection system elements in TAP integral conversion is used, and to identify links between elements - differentiation. The result of the identification of the differential structure is the signal's spectral description.

Conversion integration and differentiation together form a composition which is called U-transform: $U = d \circ \int$.

Transformations of integration and differentiation for one-dimensional signals realized with the help of four-base dimensional filter-coatings (F_0, F_1, F_2, F_3, see fig.1).

Let $f(t)$ – analyzable sound signal, observed on a finite time interval. Result of applying the U-transform to the signal f – multilevel (roughly exact) spectral representation $D = d_{ij}$, $i = \overline{1, K}$, $j = \overline{1, M_i}$, where K – is number of dissection level, M_i – number of signal's segments on the i-th dissection level,

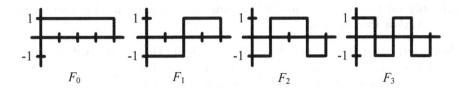

Fig. 1. Basis functions

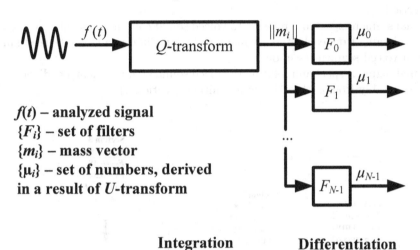

Integration   Differentiation

Fig. 2. U-transform circuit

d_{ij} – spectrum, which is included N spectrum factor (number of using filters), $d_{ij}\{k\}$ – k-th spectrum factor ($k = \overline{1, L}$), f_{ij} – signal's segment f, by which is calculated the spectrum d_{ij} (see fig.2). In calculating the spectral representation of the signal segments are not overlapped. Example U-transform computation is given in [7].

Considered methods of SNR estimation based on use of the Fourier transform. Comparing the Fourier transform with U-transform (major transformation of TAP) can be noted following [6]:

1. Fourier coefficients, except for the certain their lack – complexity, there are integral characteristics that do not contain information about the structural properties of the signal;

2. filters are used in the U-transform, which is endowed with differentiating properties which allows to highlight structural elements of the signal and links between them.

1.1 Signal to Noise Ratio Estimation on the Basis of the Theory of Active Perception

Conducted researches have established properties of the spectra (within the U-transform) relating to the desired signal and pause. The desired signal can be represented as a set of voiced and unvoiced segments (see fig.3):

1. elements of signal spectrum, which is related to pause, close to zero values (section 1).
2. signal segment spectrum relating to voiced sound contains elements diverged considerably from each other by magnitude, but differ in a lesser degree than for a voiced segment (section 2);
3. signal segment spectrum relating to voiced sound contains elements diverged considerably from each other by magnitude (section 3).

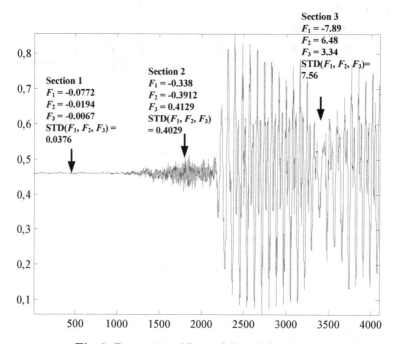

Fig. 3. Properties of Desired Signal And Pause

Let us consider the table (see tab. 1) in which gives values of the spectral coefficients for different sections of the signal and the standard deviation (STD). It may be noted that the result of exposure of noise on the signal:

Table 1. Influence of Noise on Speech

Signal Section	Uniform Noise (SNR = 8 dB)			Normal Noise (SNR = 4 dB)		
	1	2	3	1	2	3
Experiment N. 1 (segment size = 8 ms)						
STD(F_1, F_2, F_3)	0.4747	1.3124	13.0893	1.0637	1.2250	7.0924
F_1	0.6516	-1.1670	-3.3792	0.8600	0.5248	1.5579
F_2	0.8198	-0.0529	17.7059	1.6997	-1.9050	10.2402
F_3	-0.0735	1.4483	-6.2735	-0.4129	-0.4175	-3.8153
Experiment N. 2 (segment size = 8 ms)						
STD(F_1, F_2, F_3)	0.7981	0.5843	12.7010	1.3569	1.3808	5.7403
F_1	-0.7303	1.0742	-4.3398	0.7722	3.3808	0.2616
F_2	0.1213	0.9121	15.6942	-1.9404	1.2583	7.9577
F_3	-1.4737	1.9954	-7.8480	-0.5106	0.7893	-3.2680

1. on the set of spectra factors (under certain noise level) it becomes impossible to distinguish the unvoiced sections of the speech signal from the pause sections (this is also corresponds to the subjective perception of the person of noisy speech signal);
2. values of spectra factors relating to voiced segments decrease with increasing noise.

It is known that an increase in the noise level rhythmic pattern of word or phrase is a parameter which is destroyed in the last turn [10]. Therefore, when SNR estimation we can use the signal segments for which the standard deviation of spectra factors max (desired signal) and the lowest (noise).

Let a finite time interval $[0; T]$ is taken sound signal $f(t)$, which is a function of speech signal $s(t, \lambda)$ and signal-independent noise $n(t)$:

$$f(t) = F(s(t, \lambda), n(t)), 0 \leq t \leq T, \tag{1}$$

where $\lambda = \lambda_1, \ldots, \lambda_m$ is vector of parameters of speech signal. It is assumed that direct observation is only the received signal $f(t)$ available. Let $s_a(t, \lambda)$ is active speech without pauses, then signal to noise ratio estimation (ξ) can be written as follows:

$$\xi = \frac{E\{s_a^2(t, \lambda)\}}{E\{n^2(t)\}}, \tag{2}$$

where E – calculating expectation's operator.

The proposed method of SNR estimation (at the i-th level of decomposition) includes the following operations:

1. the definition of the standard deviation (SD) of spectra factors relating to the signal: $\hat{\sigma} = \max[(\sigma(\{d_{ij}(t)\}))]$, $j = \overline{1, 2^{i-1}}$, $t = \overline{1, N}$;
2. definition STD of spectra factors relating to noise: $\hat{\sigma}_n = \min[(\sigma(\{d_{ij}(t)\}))]$, $j = \overline{1, 2^{i-1}}$, $t = \overline{1, N}$;
3. calculation of SNR estimation $\hat{\xi}$: $\hat{\xi} = 10 \log_{10} \frac{(\hat{\sigma}_s - \hat{\sigma}_n)^2}{\hat{\sigma}_n^2}$.

The proposed method has two parameters:

1. signal segment length (in milliseconds) to compute the spectrum;
2. number of filters which is used in the calculation of the spectrum.

2   Computing Experiment

Let us consider results of SNR estimation based on proposed and existing methods. When testing was used database of records votes 100 speakers (audio sampling frequency – 16 kHz, depth coding – 8 bits). Computing experiment involved deformation records and signal to noise ratio estimation.

Table 2 - Table 3 shows the results of estimation of accuracy of calculation the SNR for existing algorithms.

Table 2. Results of SNR Estimation Method Proposed in [8]

Noise / SNR (in dB)	-2	4	10	16	22	28	34	40	Estimation error
Noise 1	-1.7	2.9	8.9	15.6	22.5	29.4	36.6	38.7	1.08
Noise 2	-1.35	3.11	9.11	15.9	22.9	30.2	37.3	38.6	1.29
Noise 3	-1.16	3.52	9.86	16.5	23.5	30.5	37.6	39.0	1.32
Noise 4	0.03	5.26	11.87	18.8	26.0	33.1	38.4	39.7	2.72

Table 3. Results of SNR Estimation Techniques

SNR (in dB) / Source	-10	-5	0	5	10	15	20	25	30	35	Error
[9], stationary noise	-	-6.1	-3.4	1.9	7.92	13.1	18.1	23.5	29.2	32.2	2.06
[10], white noise	-8.6	-3.7	1.2	6.01	11.02	16.02	21.57	26.85	32.96	-	1.48
[10], white noise	-0.2	-0.95	1.9	3.5	6.6	10	14.1	17.9	21.9	-	5.19
[2], stationary noise	-2	1.3	0.1	4.8	8.13	-	-	-	-	-	1.73

Table. 4 - Table. 7 shows the average test results of the proposed method (for two types of voices: male and female).

Table. 8 shows a calculation error signal / noise ratio under varying conditions. The table shows that the smallest error is achieved using 64 filters, with the duration of the analyzed signal has low effect on the accuracy of SNR estimation.

Analyzing the given table can be noted that the proposed method by accuracy of SNR estimation is not inferior to the existing methods, and in some cases - shows the best results. The advantages of the proposed method is also easy to implement and a wide range of SNR estimation: from -5 to +35 dB.

Table 4. Results of Accuracy of Method (Women's Voices, 8 Seconds)

	Algorithm Parameters / SNR (in dB)	-5	0	5	10	15	20	25	30	35
Uniform Noise	8 ms., 128 filters	-7.11	-2.85	4.01	10.15	15.93	21.07	26.30	31.50	36.11
Normal Noise		-5.82	-6.68	-1.55	4.52	10.07	15.57	20.55	25.49	31.03
Uniform Noise	4 ms., 64 filters	-2.87	0.20	5.29	11.28	17.03	22.24	27.22	32.76	37.65
Normal Noise		-1.68	0.38	2.49	6.76	11.12	16.68	21.93	26.62	31.97
Uniform Noise	2 ms., 32 filters	0.01	2.83	8.26	13.95	19.32	24.63	29.06	34.18	39.35
Normal Noise		5.03	5.79	8.48	9.36	14.00	20.24	24.41	29.99	34.93

Table 5. Results of Accuracy of Method, the Men's Voices, 4 Seconds

	Algorithm Parameters / SNR (in dB)	-5	0	5	10	15	20	25	30	35
Uniform Noise	8 ms., 128 filters	-9.59	-8.54	-2.07	4.17	10.47	16.31	21.72	27.17	31.98
Normal Noise		-7.36	-4.90	-6.13	-1.15	3.92	10.32	15.38	21.10	26.28
Uniform Noise	4 ms., 64 filters	-5.81	-3.20	0.28	5.37	12.23	18.22	22.89	28.89	33.41
Normal Noise		-1.87	-1.69	-0.26	2.04	7.11	11.69	17.34	22.39	28.36
Uniform Noise	2 ms., 32 filters	-0.45	0.48	3.60	9.68	14.86	20.34	25.51	30.47	35.80
Normal Noise		4.19	5.02	3.25	6.03	10.42	15.28	20.39	26.17	31.25

Table 6. Results of Accuracy of Method, Women's Voices, 8 Seconds

	Algorithm Parameters / SNR (in dB)	-5	0	5	10	15	20	25	30	35
Uniform Noise	8 ms., 128 filters	-7.34	-3.77	2.95	8.53	13.85	20.01	25.28	30.73	36.00
Normal Noise		-5.54	-6.38	-1.95	2.46	8.88	14.60	19.70	24.70	30.76
Uniform Noise	4 ms., 64 filters	-3.95	-0.42	5.18	10.77	15.30	21.21	26.58	31.31	37.02
Normal Noise		-0.34	-1.43	0.86	5.20	10.46	15.66	21.09	26.02	31.71
Uniform Noise	2 ms., 32 filters	1.41	4.86	9.04	15.37	19.14	24.40	31.39	34.15	39.96
Normal Noise		4.41	3.28	5.52	8.56	14.89	20.09	25.19	30.35	35.41

Table 7. Results of Accuracy of Method, Women's Voices, 4 Seconds

	Algorithm Parameters / SNR (in dB)	-5	0	5	10	15	20	25	30	35
Uniform Noise	8 ms., 128 filters	-8.35	-4.27	2.71	8.60	14.35	20.13	25.60	30.40	36.15
Normal Noise		-6.99	-6.57	-3.42	1.97	8.26	13.79	19.83	24.78	30.45
Uniform Noise	4 ms., 64 filters	-5.10	0.50	3.91	10.02	15.12	21.66	26.74	31.19	37.63
Normal Noise		-3.92	-1.11	0.01	4.79	9.88	15.81	20.96	27.10	31.46
Uniform Noise	2 ms., 32 filters	0.47	5.24	8.16	14.55	18.07	24.73	29.56	34.52	40.17
Normal Noise		3.39	5.00	6.01	8.75	13.99	18.99	24.98	29.97	35.75

Table 8. Calculation Error SNR (in dB)

	Men			Women		
Noise Type / filter amount	128	64	32	128	64	32
Uniform noise, 4 sec.	4.82	2.52	1.00	1.58	1.01	4.50
Normal noise, 4 sec.	8.62	6.24	4.60	5.88	3.58	2.05
Uniform noise, 8 sec.	1.33	1.76	4.07	1.42	0.98	4.97
Normal noise, 8 sec.	4.65	2.90	2.43	5.31	3.90	1.76

3   Conclusion

Method of signal/noise ratio estimation in the observed speech signal was devised. The proposed method is based on the theory of active perception. Using this theory we revealed the structural elements of the signal and links between them, and thus detected desired signal and hindrance.

Research of the algorithm on testing and real signals confirmed its efficiency and ability to be used in speech signal processing, which requires adjustment to the quality of the analyzed signal.

References

1. Rangachari, S.: A noise-estimation algorithm for highly non-stationary environments. J. Speech Communication 48, 220–231 (2006)
2. Vondrasek, M., Pollak, P.: Methods for Speech SNR Estimation: Evaluation Tool and Analysis of VAD Dependency. J. Radioengineering 14, 6–11 (2005)
3. Cohen, I.: Noise estimation by minima controlled recursive averaging for robust speech enhancement. J. IEEE Signal Processing Letters 9, 12–15 (2002)
4. Martin, R.: Noise power spectral density estimation based on optimal smoothing and minimum statistics. J. IEEE Transactions on Speech and Audio Processing 9, 504–512 (2001)
5. Hirsch, H.-G., Ehrlicher, C.: Noise estimation techniques for robust speech recognition. In: International Conference on Acoustics, Speech, and Signal Processing, pp. 153–156 (1995)
6. Utrobin, V.A.: Physical interpretation of the elements of image algebra. J. Advances in Physical Sciences 47, 1017–1032 (2004)
7. Gai, V.E.: Metod ocenki chastoty osnovnogo tona v uslovijah pomeh 4, 65–71 (2013) (in Russian)
8. Stolbov, M.B.: Algoritm ocenki otnoshenija signal/shum rechevyh signalov. J. Nauchno-tehnicheskij Vestnik Informacionnyh Tehnologij, Mehaniki i Optiki 82, 67–72 (2012)
9. Gerkmann, T.: Unbiased MMSE-Based Noise Power Estimation With Low Complexity and Low Tracking Delay. J. IEEE Transactions on Audio, Speech, and Language Processing. 20, 1383–1393 (2012)
10. Kim, C., Stern, R.M.: Robust Signal-to-Noise Ratio Estimation Based on Waveform Amplitude Distribution Analysis. In: InterSpeech 2008, Brisbane, Australia, pp. 2598–2601 (2008)

Joint Probability Density Function of Modulated Synchronous Flow Interval Duration*

Aleksandr Gortsev and Mariya Sirotina

National Research Tomsk State University, Tomsk, Russia
gam@fpmk.tsu.ru, mashuliagol@mail.ru

Abstract. An explicit form of a probability density function of interval duration between two adjacent events of modulated synchronous doubly stochastic flow is derived. Also an explicit form of a joint probability density function for modulated synchronous flow interval duration is obtained. This flow is one of the mathematical models of information flows, which take place in digital networks with integral service. The flow is considered in stationary mode when there are no transition processes. A recurrent conditions for modulated synchronous flow are obtained using the formula for joint probability density function.

Keywords: modulated synchronous doubly stochastic flow, probability density function of interval duration, joint probability density function of interval duration, flow recurrent conditions.

1 Introduction

Mathematical models of queueing theory are widely used to describe real physical, technological and other processes and systems. In connection with rapid development of computer equipment and information technologies an important sphere of queueing theory applications appeared. This sphere was called as design and creation of data-processing networks, computer communication networks, satellite networks and telecommunication networks [1].

In practice, an intensity of input flow varies along with time. Moreover, these variations are often of a random nature. This leads to consideration of a doubly stochastic flow of events [2,3,4,5,6]. An example of such flow is a modulated synchronous doubly stochastic flow [7,8].

2 Problem Statement

Let us consider the modulated synchronous doubly stochastic flow of events, whose rate is a piecewise constant random process $\lambda(t)$ with two states: $\lambda_1, \lambda_2(\lambda_1 > \lambda_2)$. The sojourn time of the process $\lambda(t)$ in state λ_i has exponential

---

* The work is supported by Tomsk State University Competitiveness Improvement Program.

A. Dudin et al. (Eds.): ITMM 2014, CCIS 487, pp. 145–152, 2014.

probability distribution function with the parameter $\alpha_i, i = 1, 2$. If at the moment t the process $\lambda(t)$ sojourns in the state λ_i than in the small half-interval $[t, t + \Delta t)$, with probability $\alpha_i \Delta t + o(\Delta t)$ process finishes its stay in the state λ_i and moves to the state λ_j with probability is one $(i, j = 1, 2, i \neq j)$. During the time random interval when $\lambda(t) = \lambda_i$ Poisson flow with rate $\lambda_i, i = 1, 2$ arrives. A state transition of the process $\lambda(t)$ may also occur at the moment of Poisson flow event arrival. Moreover, the passing from the state λ_1 to the state λ_2 is realized only at the moment of event occurrence with probability $p \, (0 < p \leq 1)$. With the complementary probability $1 - p$ the process remains at the state λ_1. The passing from the state λ_2 to the state λ_1 is also realized only at the moment of event occurrence with probability $q \, (0 < q \leq 1)$. With the complementary probability $1 - q$ the process remains at the state λ_2. In the described conditions, $\lambda(t)$ is the Markovian process.

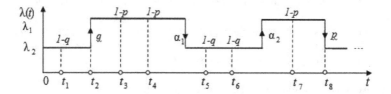

Fig. 1. Forming the modulated synchronous flow

An example of this situation is shown on the fig. 1, where λ_1, λ_2 are the states of process $\lambda(t)$, $t_1, t_2, \ldots$ are the moments of the flow events occurrence.

Block matrixes of infinitesimal coefficients are of the form:

$$D_1 = \begin{vmatrix} (1-p)\lambda_1 & p\lambda_1 \\ q\lambda_2 & (1-q)\lambda_2 \end{vmatrix}, D_0 = \begin{vmatrix} -(\lambda_1 + \alpha_1) & \alpha_1 \\ \alpha_2 & -(\lambda_2 + \alpha_2) \end{vmatrix}.$$

The elements of the matrix D_1 are intensities of the process $\lambda(t)$ passing from the state to the state with an event occurrence. Off-diagonal elements of matrix D_0 are intensities of the process $\lambda(t)$ passing from the state to the state without an event occurrence. Diagonal elements of matrix D_0 are the intensities of process $\lambda(t)$ leaving its states, which are taken with the opposite sign. We should note that if $\alpha_i = 0, i = 1, 2$ there is a usual synchronous flow of events [9].

3   Probability Density Function of Modulated Synchronous Flow Interval Duration

A stationary mode of the flows is considered. A sequence of time moments of event occurence $t_1, t_2, \ldots t_k, \ldots$ is an imbedded Markov chain. So the flow has a markovian chain characteristic when its evolution is considered from the moment t_k.

Let τ is a value of the random variable of interval duration between the moments of two adjacent flow events. Then a probability density function of interval duration between the moments of modulated synchronous flow adjacent event occurrence is written in the form of:

$$p(\tau) = \sum_{i=1}^{2} \pi_i(0) \sum_{j=1}^{2} \widetilde{p}_{ij}(\tau),\tag{1}$$

where $\pi_i(0)$ is a conditional stationary probability that the process $\lambda(t)$ sojourns in the state λ_i at the moment $\tau = 0$ when the flow event occured in the moment $\tau = 0$, $i = 1, 2$ ($\pi_1(0) + \pi_2(0) = 1$); $\widetilde{p}_{ij}(\tau)$ is a probability density function of the interval between two adjacent flow events and probability of $\lambda(\tau) = \lambda_j$ when $\lambda(0) = \lambda_i$ ($i, j = 1, 2$).

Let us introduce $p_{ij}(\tau)$, $i, j = 1, 2$ is a transition probability that there are no flow events on the interval $(0; \tau)$ and $\lambda(\tau) = \lambda_j$ at the moment τ when $\lambda(0) = \lambda_i$ ($i, j = 1, 2$). Then for introduced probabilities $p_{ij}(\tau)$ we have the folowing system of differencial equations:

$$p'_{11}(\tau) = -(\lambda_1 + \alpha_1)p_{11}(\tau) + \alpha_2 p_{12}(\tau),$$
$$p'_{12}(\tau) = -(\lambda_2 + \alpha_2)p_{12}(\tau) + \alpha_1 p_{11}(\tau),$$
$$p'_{22}(\tau) = -(\lambda_2 + \alpha_2)p_{22}(\tau) + \alpha_1 p_{21}(\tau),$$
$$p'_{21}(\tau) = -(\lambda_1 + \alpha_1)p_{21}(\tau) + \alpha_2 p_{22}(\tau),$$
$$p_{11}(0) = 1,\; p_{12}(0) = 0,\; p_{22}(0) = 1,\; p_{21}(0) = 0.$$

Solving obtained system of differential equtions we find out that

$$p_{11}(\tau) = \lambda_1(1-p)\frac{\lambda_2+\alpha_2}{z_1 z_2} + \lambda_2 q\frac{\alpha_1}{z_1 z_2},\tag{2}$$
$$p_{12}(\tau) = \lambda_1 p\frac{\lambda_2+\alpha_2}{z_1 z_2} + \lambda_2(1-q)\frac{\alpha_1}{z_1 z_2},$$
$$p_{22}(\tau) = \lambda_2(1-q)\frac{\lambda_1+\alpha_1}{z_1 z_2} + \lambda_1 p\frac{\alpha_2}{z_1 z_2},$$
$$p_{21}(\tau) = \lambda_2 q\frac{\lambda_1+\alpha_1}{z_1 z_2} + \lambda_1(1-p)\frac{\alpha_2}{z_1 z_2},$$
$$z_1 z_2 = \lambda_1\lambda_2 + \lambda_1\alpha_2 + \lambda_2\alpha_1.$$

Let us derive a formulas for probability densities $\widetilde{p}_{ij}(\tau)$, $i, j = 1, 2$. For this purpose we should consider the interval $(0, \tau + \Delta\tau)$ and how the process $\lambda(t)$ behaves oneself on this interval. The interval $(0, \tau + \Delta\tau)$ consists of two parts: the interval $(0, \tau)$, where $\tau = 0$ is a moment of event occurence (for this interval the probabilities $p_{ij}(\tau)$, $i, j = 1, 2$ are already defined in formula 2), and the small enough half-interval $[\tau, \tau + \Delta\tau)$.

Then $\widetilde{p}_{ij}(\tau)\Delta\tau + o(\Delta\tau)$, $i, j = 1, 2$ is a joint probability that there are no events on the interval $(0, \tau)$, the process $\lambda(t)$ moves from the state λ_i to the state λ_k on the interval $(0, \tau)$, then during the half-interval $[\tau, \tau + \Delta\tau)$ the process $\lambda(t)$ doesn't leave the state λ_k, the Poisson flow event occures with the intensity λ_k and the process $\lambda(t)$ moves from the state λ_k to the state λ_j on this half-interval $[\tau, \tau + \Delta\tau)$.

As an example, we consider a formula derivation for a joint probability $\widetilde{p}_{11}(\tau)\Delta\tau + o(\Delta\tau)$. To carry out this probability let us consider two cases:

1. $\widetilde{p}_{11}^{(1)}(\tau)\Delta\tau + o(\Delta\tau)$ is a joint probability that there are no events on the interval $(0, \tau)$, the process $\lambda(t)$ remains at the state λ_1 on the interval $(0, \tau)$, then during the half-interval $[\tau, \tau + \Delta\tau)$ the process $\lambda(t)$ doesn't leave the state λ_1, then the Poisson flow event occures with the intensity λ_1 and the process $\lambda(t)$ remains at the state λ_1 on this half-interval $[\tau, \tau + \Delta\tau)$. This joint probability is of the form:

$$\widetilde{p}_{11}^{(1)}(\tau)\Delta\tau + o(\Delta\tau) =$$
$$= p_{11}(\tau)e^{-\alpha_1\Delta\tau}(1 - e^{-\lambda_1\Delta\tau})(1 - p) =$$
$$= p_{11}(\tau)\lambda_1\Delta\tau(1 - p) + o(\Delta\tau).$$

Moving $\Delta\tau \to 0$ we get the following expression for probability density $\widetilde{p}_{11}^{(1)}(\tau)$:

$$\widetilde{p}_{11}^{(1)}(\tau) = p_{11}(\tau)\lambda_1(1 - p).$$

2. $\widetilde{p}_{11}^{(2)}(\tau)\Delta\tau + o(\Delta\tau)$ is a joint probability that there are no events on the interval $(0, \tau)$, the process $\lambda(t)$ moves from the state λ_1 to the state λ_2 on the interval $(0, \tau)$, then during the half-interval $[\tau, \tau + \Delta\tau)$ the process $\lambda(t)$ doesn't leave the state λ_2, then the Poisson flow event occures with the intensity λ_2 and the process $\lambda(t)$ moves from the state λ_2 to the state λ_1. This joint probability is of the form:

$$\widetilde{p}_{11}^{(2)}(\tau)\Delta\tau + o(\Delta\tau) =$$
$$= p_{11}(\tau)e^{-\alpha_2\Delta\tau}(1 - e^{-\lambda_2\Delta\tau})q =$$
$$= p_{11}(\tau)\lambda_2\Delta\tau q + o(\Delta\tau).$$

Moving $\Delta\tau \to 0$ we get the following expression for probability density $\widetilde{p}_{11}^{(2)}(\tau)$:

$$\widetilde{p}_{11}^{(2)}(\tau) = p_{11}(\tau)\lambda_2 q.$$

Then, using the equation $\widetilde{p}_{11}(\tau) = \widetilde{p}_{11}^{(1)}(\tau) + \widetilde{p}_{11}^{(2)}(\tau)$ the formula for probability density $\widetilde{p}_{11}(\tau)$ is written in the form of:

$$\widetilde{p}_{11}(\tau) = p_{11}(\tau)\lambda_1(1 - p) + p_{11}(\tau)\lambda_2 q.$$

On the base of the derivation described above another probability densities $\widetilde{p}_{12}(\tau)$, $\widetilde{p}_{21}(\tau)$, $\widetilde{p}_{22}(\tau)$ are obtained.

Therefore the probability densities $\widetilde{p}_{ij}(\tau)$, $i, j = 1, 2$ are of the form:

$$\begin{aligned}
\widetilde{p}_{11}(\tau) &= p_{11}(\tau)\lambda_1(1 - p) + p_{12}(\tau)\lambda_2 q, &\qquad(3)\\
\widetilde{p}_{12}(\tau) &= p_{11}(\tau)\lambda_1 p + p_{12}(\tau)\lambda_2(1 - q),\\
\widetilde{p}_{22}(\tau) &= p_{22}(\tau)\lambda_2(1 - q) + p_{21}(\tau)\lambda_1 p,\\
\widetilde{p}_{21}(\tau) &= p_{22}(\tau)\lambda_2 q + p_{21}(\tau)\lambda_1(1 - p),
\end{aligned}$$

where probabilities $p_{ij}(\tau)$, $i, j = 1, 2$ are defined in 2.

To find out probability $\pi_i(0)$, $i = 1, 2$ let us introduce probability p_{ij}, $i, j = 1, 2$ that during the time interval between the moment $\tau = 0$ (the moment of event occurence) and the moment of the next event occurence the process $\lambda(t)$ moves from the state λ_i to the state λ_j, $i, j = 1, 2$. The probability p_{ij} doesn't depend on time, it is a stationary probability of the process $\lambda(t)$ transition from the state λ_i to the state λ_j between the moments of two adjacent flow events.

Then we get the following system for $\pi_i(0)$, $i = 1, 2$:

$$\pi_1(0) = \pi_1(0)p_{11} + \pi_2(0)p_{21},$$
$$\pi_2(0) = \pi_2(0)p_{22} + \pi_1(0)p_{12},$$
$$\pi_1(0) + \pi_2(0) = 0.$$

Solving the system defined above the expressions for $\pi_i(0)$, $i = 1, 2$ are written in the form of:

$$\pi_1(0) = \frac{p_{21}}{p_{12} + p_{21}}, \quad \pi_2(0) = \frac{p_{12}}{p_{12} + p_{21}}. \tag{4}$$

Because τ is an undefined time moment, then the transition probability p_{ij} that the process $\lambda(t)$ moves from the state λ_i to the state λ_j during a time interval between the moment $\tau = 0$ and the moment of the next flow event occurence, should be written as follows:

$$p_{ij} = \int\limits_0^\infty p_{ij}(\tau)\Delta\tau.$$

Inserting the formulas 2 into integral for p_{ij}, $i, j = 1, 2$ the folowing expressions are derived:

$$p_{11} = \frac{\lambda_1(1-p)(\lambda_2+\alpha_2)+\lambda_2 q\alpha_1}{z_1 z_2}, \tag{5}$$
$$p_{12} = \frac{\lambda_1 p(\lambda_2+\alpha_2)+\lambda_2(1-q)\alpha_1}{z_1 z_2},$$
$$p_{21} = \frac{\lambda_2 q(\lambda_1+\alpha_1)+\lambda_1(1-p)\alpha_2}{z_1 z_2},$$
$$p_{12} = \frac{\lambda_2(1-q)(\lambda_1+\alpha_1)+\lambda_1 p\alpha_2}{z_1 z_2},$$

where $z_1 z_2$ is defined in 2.

Therefore, inserting formulas 5 into 4 the following expressions for probability $\pi_i(0)$, $i = 1, 2$ are obtained:

$$\pi_1(0) = \frac{q\lambda_2(\lambda_1+\alpha_1)+(1-p)\lambda_1\alpha_2}{(p+q)\lambda_1\lambda_2+\lambda_2\alpha_1+\lambda_1\alpha_2}, \tag{6}$$
$$\pi_2(0) = \frac{p\lambda_1(\lambda_2+\alpha_2)+(1-q)\lambda_2\alpha_1}{(p+q)\lambda_1\lambda_2+\lambda_2\alpha_1+\lambda_1\alpha_2}.$$

Inserting 2 into 3, then inserting 3 and 6 into 1 we derive the explicit form of a probability density function $p(\tau)$:

$$p(\tau) = \gamma z_1 e^{-z_1\tau} + (1 - \gamma)z_2 e^{-z_2\tau}, \tag{7}$$

where $\gamma = \frac{1}{z_2-z_1}(z_2 - \pi_1(0)\lambda_1 - \pi_2(0)\lambda_2)$, $1 - \gamma = \frac{1}{z_2-z_1}(-z_1 + \pi_1(0)\lambda_1 + \pi_2(0)\lambda_2)$, $z_{1,2} = \frac{1}{2}(\lambda_1 + \alpha_1 + \lambda_2 + \alpha_2) \mp \sqrt{(\lambda_1 + \alpha_1 - \lambda_2 - \alpha_2)^2 + 4\alpha_1\alpha_2}$.

4   Joint Probability Density Function of Modulated Synchronous Flow Interval Duration

Let τ_1, τ_2 are a values of the random variable of duration of two complementary intervals between the moments of two adjacent flow events. Then a joint probability density function $p(\tau_1, \tau_2)$ is written in the form of:

$$p(\tau_1, \tau_2) = \sum_{i=1}^{2} \pi_i(0) \sum_{j=1}^{2} \widetilde{p}_{ij}(\tau_1) \sum_{k=1}^{2} \widetilde{p}_{jk}(\tau_2).$$

Besides, according to the formula 1 $p(\tau_i) = \sum_{i=1}^{2} \pi_i(0) \sum_{j=1}^{2} \widetilde{p}_{ij}(\tau_i)$, $i = 1, 2$.
Then a difference $p(\tau_1, \tau_2) - p(\tau_1)p(\tau_2)$ is of the form:

$$p(\tau_1, \tau_2) - p(\tau_1)p(\tau_2) =$$
$$= \pi_1(0) \sum_{j=1}^{2} \widetilde{p}_{1j}(\tau_1) \sum_{k=1}^{2} \widetilde{p}_{jk}(\tau_2) + \pi_2(0) \sum_{j=1}^{2} \widetilde{p}_{2j}(\tau_1) \sum_{k=1}^{2} \widetilde{p}_{jk}(\tau_2) -$$
$$-(\pi_1(0) \sum_{j=1}^{2} \widetilde{p}_{1j}(\tau_1) + \pi_2(0) \sum_{j=1}^{2} \widetilde{p}_{2j}(\tau_1)) \times$$
$$\times (\pi_1(0) \sum_{j=1}^{2} \widetilde{p}_{1j}(\tau_2) + \pi_2(0) \sum_{j=1}^{2} \widetilde{p}_{2j}(\tau_2)).$$

Making a difficult enough manupulations the difference $p(\tau_1, \tau_2) - p(\tau_1)p(\tau_2)$ is written as follows:

$$p(\tau_1, \tau_2) - p(\tau_1)p(\tau_2) =$$
$$= \sum_{k=1}^{2} (\widetilde{p}_{1k}(\tau_2) - \widetilde{p}_{2k}(\tau_2)) \times \tag{8}$$
$$\times \left[\pi_1(0)\pi_2(0)(\widetilde{p}_{11}(\tau_1) - \widetilde{p}_{22}(\tau_1)) - \pi_1^2(0)\widetilde{p}_{12}(\tau_1) + \pi_2^2(0)\widetilde{p}_{21}(\tau_1) \right]$$

The first multiplier from 8 is written out as follows:

$$\sum_{k=1}^{2} (\widetilde{p}_{1k}(\tau_2) - \widetilde{p}_{2k}(\tau_2)) = (\lambda_2 - \lambda_1)(z_1 e^{-z_1 \tau_2} - z_2 e^{-z_2 \tau_2})/(z_2 - z_1), \tag{9}$$

where z_1, z_2 are defined in 7.
The second multiplier from 8 is written out as follows:

$$\pi_1(0)\pi_2(0)((\widetilde{p}_{11}(\tau_1) - (\widetilde{p}_{22}(\tau_1)) - \pi_1^2(0)\widetilde{p}_{12}(\tau_1) + \pi_2^2(0)\widetilde{p}_{21}(\tau_1) =$$
$$= -\lambda_1(p - \pi_2(0))(\pi_1(0)p_{11}(\tau_1) + \pi_2(0)p_{21}(\tau_1)) + \tag{10}$$
$$+\lambda_2(q - \pi_1(0))(\pi_1(0)p_{12}(\tau_1) + \pi_2(0)p_{22}(\tau_1)).$$

Inserting $p_{ij}(\tau)$, $i, j = 1, 2$ from 2 into 10 we get:

$$\pi_1(0)\pi_2(0)((\widetilde{p}_{11}(\tau_1) - (\widetilde{p}_{22}(\tau_1)) - \pi_1^2(0)\widetilde{p}_{12}(\tau_1) + \pi_2^2(0)\widetilde{p}_{21}(\tau_1) =$$
$$= (z_1 e^{-z_1 \tau_1} - z_2 e^{-z_2 \tau_1}) \left[\lambda_1(p - \pi_2(0))\pi_1(0) - \lambda_2(q - \pi_1(0))\pi_2(0) \right] /(z_2 - z_1). \tag{11}$$

Inserting 9 and 11 into 8 and making a difficult enough manupulations we obtain:

$$p(\tau_1, \tau_2) = p(\tau_1)p(\tau_2) + \gamma(1 - \gamma)\frac{\lambda_1 \lambda_2 (1 - p - q)}{z_1 z_2} \times$$
$$\times (z_1 e^{-z_1 \tau_1} - z_2 e^{-z_2 \tau_1})(z_1 e^{-z_1 \tau_2} - z_2 e^{-z_2 \tau_2}), \tag{12}$$

where z_1, z_2, γ, $1 - \gamma$ are defined in 7; $\pi_1(0)$, $\pi_2(0)$ are defined in 6; $\gamma(1 - \gamma) = \frac{(\lambda_1 - \lambda_2)((\lambda_1 p + \alpha_1)\pi_1(0) - (\lambda_2 q + \alpha_2)\pi_2(0))}{(p + q)\lambda_1 \lambda_2 + \lambda_2 \alpha_1 + \lambda_1 \alpha_2}$.

From the formula 12, it follows that modulated synchronous flow is a correlated flow in the general case.

5   Recurrence Conditions for Modulated Synchronous Flow

From the formula 12, it follows that for the joint probability density function $p(\tau_1, \tau_2)$ there are two conditions of modulated synchronous flow recurrence:

1. The flow is recurrent if $p + q = 1$. In this case the probability density function $p(\tau)$ is of the form:

$$p(\tau) = \gamma z_1 e^{-z_1 \tau} + (1 - \gamma) z_2 e^{-z_2 \tau},$$
$$\gamma = \frac{1}{z_2 - z_1}(z_2 - \lambda_1 + p(\lambda_1 - \lambda_2)),$$
$$1 - \gamma = \frac{1}{z_2 - z_1}(-z_1 + \lambda_1 - p(\lambda_1 - \lambda_2)),$$

where z_1, z_2 are defined in 7.

2. The flow is recurrent if $(\lambda_1 p + \alpha_1)\pi_1(0) = (\lambda_2 q + \alpha_2)\pi_2(0)$. In this case the probability density function $p(\tau)$ is of the form:

$$p(\tau) = \gamma z_1 e^{-z_1 \tau} + (1 - \gamma) z_2 e^{-z_2 \tau},$$
$$\gamma = \frac{1}{z_2 - z_1}(z_2 - \pi_1(0)(\lambda_1 + \lambda_2 \tfrac{\lambda_1 p + \alpha_1}{\lambda_2 q + \alpha_2})),$$
$$1 - \gamma = \frac{1}{z_2 - z_1}(-z_1 + \pi(0)(\lambda_1 + \lambda_2 \tfrac{\lambda_1 p + \alpha_1}{\lambda_2 q + \alpha_2})),$$

where z_1, z_2 are defined in 7, $\pi_1(0)$, $\pi_2(0)$ are defined in 6.

6   Conclusion and Future Research

During this research the explicit form of the probability density function $p(\tau)$ was derived as well as the explicit form of the joint probability density function $p(\tau_1, \tau_2)$. The formulas obtained allow us to carry out an estimation of flow parameters using the maximum likelihood method or method of matching moments.

References

1. Dudin, A.N., Klimenuk, V.N.: Queue systems with correlated flows, p. 175. Belorussian State University, Minsk (2000)
2. Kingman, J.F.C.: On doubly stochastic Poisson process. Proceedings of Cambridge Phylosophical Society 60(4), 923–930 (1964)
3. Basharin, G.P., Kokotushkin, V.A., Naumov, V.A.: About the method of renewals of subnetwork computation. AN USSR, Techn. Kibernetics 6, 92–99 (1979)
4. Neuts, M.F.: A versatile Markov point process. Journal of Applied Probability 16, 764–779 (1979)
5. Lucantoni, D.M.: New results on the single server queue with a batch markovian arrival process. Communication in Statistics Stochastic Models 7, 1–46 (1991)
6. Card, H.C.: Doubly stochastic Poisson processes in artifical neural learning. IEEE Transactions on Neural Networks 9(1), 229–231 (1998)

7. Gortsev, A.M., Golofastova, M.N.: Optimal state estimation of modulated synchronous doubly stochastic flow of events. Control, Computation and Informstics, Tomsk State University Journal 2(23), 42–53 (2013)
8. Sirotina, M.N.: Optimal state estimation of modulated synchronous doubly stochastic flow of events in conditions of fixed dead time. Control, Computation and Informstics, Tomsk State University Journal 1(26), 63–72 (2014)
9. Bushlanov, I.V., Gortsev, A.M., Nezhel'skaya, L.A.: Estimating parameters of the synchronous twofold-stochastic flow of events. Automation and Remote Control 69(9), 1517–1533 (2008)

Using of Event-Driven Molecular Dynamics Method at the Computer Simulation of Atomic Structures of Amorphous Metals

Vladimir Jordan and Timofei Belov

Altai State University, Barnaul, Russia
jordan@phys.asu.ru,
signofone@yandex.ru

Abstract. The article highlights some lesser known issues related to the study of the atomic structure of liquid and amorphous metals, semiconductors, alloys and with the processes occurring in them at the atomic level at the thermal and stress effects. Features of implementation of event-driven molecular dynamics algorithm in sequential and parallel variants to study the atomic structure of amorphous metals are presented. The stages and the basic problems of computer simulation of atomic structure of metals, as well as an analysis of the results of computational experiments to study the atomic structure of amorphous aluminum at different speeds its superfast cooling liquid melt are discussed.

Keywords: computer simulation, event-driven molecular dynamics method, atomic structure, amorphous metal, calculable block, unit cell, interaction potential function.

1   Introduction

Great contribution to science of metals was made by the discovery of amorphous metals with unusual properties: they are very strong and at the same time have plastic properties. They have soft magnetic material properties, corrosion resistance, which leads to broad prospects of their application in industry. The basic method of producing amorphous metal is extremely fast cooling of the liquid melt. In this way there were obtained amorphous alloys named as metallic glass (cooling rate is very high and is in the range 10^{10}–10^{13} K/s). Microstructure of amorphous metals is a metastable state of structure with local atomic order and this structure is not a polycrystalline structure. There are several models to explain the nature of such a structural organization in amorphous metals: crystal, a dislocation model, a model of random close packing of hard spheres and a cluster model.

Currently there are not fully resolved issues related to the study of the atomic structure of liquid and amorphous metals, semiconductors, alloys and with the processes occurring in them at the atomic level at the thermal and stress effects. These issues, for example, include correctness and the limits of applicability of

A. Dudin et al. (Eds.): ITMM 2014, CCIS 487, pp. 153–161, 2014.

different structural models of amorphous metals. For example, in various amorphous structures for the cluster model, you need to figure out the dimensions, stability and conditions of cluster formation of an ordered structure (of conjugate tetrahedra or Frank-Kasper polyhedra). Also of interest is the question concerning the structure in the interface of the clusters, as well as the mechanisms of their transformation at the thermal and stress effects. Insufficiently explored is the process of crystallization of amorphous metals. Limited possibilities of direct experimental methods are due to the specific properties of the studied medium, in particular, the lack of long-range order in the arrangement of atoms, for example, when trying to create a structural model of liquid, including molten metal. In liquids the kinetic energy of atoms is comparable to their potential energy, so in relation to such medium it is impossible to find an effective small parameter.

To solve these issues the most effective computer simulation method is a method of molecular dynamics, in particular, its kind – event-driven molecular dynamics method [1], which allows with sufficient accuracy in the model to take into consideration and control the parameters of the phenomenon under investigation, to study the dynamics of the processes flowing on the atomic level using a variety of realistic visualizers of structure.

2   Event-Driven Molecular Dynamics Method and Its Parallelization

In classical method of molecular dynamics system is represented as a set of particles with interaction, which is described by the interaction potential function, and the evolution of the system is simulated by numerical integration of the equations of motion using a time step [2].

For a closed system the force acting on the i-th atom:

$$F_i = -\sum_{j \neq i}^{N} \frac{d\phi_{ij}\left(|r_i - r_j|\right)}{d\left(r_i - r_j\right)} \ , \tag{1}$$

where $\phi_{ij}\left(|r_i - r_j|\right)$ – the interaction potential function between i-th and j-th atoms; r_i, r_j – radius-vectors of atoms. Equations of motion are

$$\frac{dr_i}{dt} = v_i \ , \qquad m_i \frac{dv_i}{dt} = F_i \ , \tag{2}$$

where m_i and v_i - mass and velocity vector of i-th atom; t – temporary variable.

Positions and velocities of all N atoms of calculable system characterized by $2 \cdot J \cdot N$ coordinates (J – dimension of the calculable system): the coordinates $x_{j,i}(t)$ describe a position in space (j - number of coordinate axis), values $v_{j,i}(t) = \dot{x}_{j,i}(t)$ describe a j-component of velocity vector of i-th atom. To solve (2) can apply a numerical method for the integration of differential equations, for example, the Runge-Kutta method of 4th order accuracy or commonly known half-step Euler method:

$$\begin{aligned} v_{j,i}\left(t + \Delta t/2\right) &= v_{j,i}\left(t - \Delta t/2\right) + \Delta t \cdot F_{j,i}\left(t\right)/m_i \ , \\ x_{j,i}\left(t + \Delta t\right) &= x_{j,i}\left(t\right) + \Delta t \cdot v_{j,i}\left(t + \Delta t/2\right) \ , \end{aligned} \tag{3}$$

where Δt - integration step (time step), which is chosen by criterion: the fluctuations of total energy of the system must not exceed the fluctuations of the potential energy. Step Δt should be less than $1/4$ of the smallest period of atomic vibrations (about 10^{-13}–10^{-15} s). Otherwise the vibrations of the atoms become aperiodic (increasing energy of the system). Temperature of calculable system of atoms is given by means of set of initial velocities of atoms the according to the Maxwell distribution (the velocities have identical modulus but have random directions). The total kinetic energy must correspond to the desired temperature, and the total momentum must equal zero [2]:

$$|v_i| = \sqrt{2}v_{\text{sq}} = \sqrt{2JkT/m_i} , \qquad \sum_{i=1}^{N} m_i v_i = 0 , \qquad T = 2E/(J \cdot N \cdot k) , \qquad (4)$$

where k - Boltzmann constant, T - temperature of calculable system at each iteration of the numerical experiment, v_{sq} – root-mean-square (rms) velocity of the atom, E – kinetic energy of calculable atom system.

Integration of the equations of motion, which requires a sufficiently high accuracy by using non-smooth potentials, compel to significantly reduce the global time step. Since this affects on the performance of computing, into modeling algorithm requires to include effective analysis of the need the estimation of various atom interactions. Idea of the algorithm based on the sequential control of events [1], ranked by time, is that in the time of the next event is processed only one pair of interacting atoms for which this event (collision between them) immediately comes. After all processing of all remaining pairwise interactions between atoms is delayed until the following time points, but in each successive time again handled the interaction of only one relevant to this event a pair of atoms [1]. Using of this event-driven dynamics algorithm allowed effectively take into account non-smooth potentials and accurately satisfy the modeling constraints [3]. I.e., between the events in accordance with the known analytical formulas the changes in position of the atoms are taken into account as motions of atoms under the influence of force, which is constant on the time step. So is achieved the considerable savings of calculations volume and acceleration of calculations. An additional effect of accelerating the calculations can be achieved through parallelization because in the sequence of events satisfied the condition of their independence. The following is an informal description of the event-driven molecular dynamics algorithm in serial form (non-parallel algorithm):

1. Calculation of interaction forces between the atoms
2. Initialization of current time t in the interval $(0, T)$: $t = 0$
3. Definition and addition to the queue of events for all atoms
4. Determining the next i-th event with the minimum time t_i
5. If t_i is greater than T (end time of simulation interval), then:
   (a) Setting the current time t to the end of the interval: $t = T$
   (b) Go to step 11
6. Setting the current time t to the time of event: $t = t_i$

7. Changing of states of participants of the event in accordance with the changed time
8. Processing of event, directly corresponding to the moment t_i
9. Defining and adding to the queue new events with the participants from processed event
10. Go to step 4
11. Changing of states of all atoms in accordance with the current time t
12. Go to step 1

In the described algorithmic diagram the end of the action interval constant interaction force can be considered as another kind of regular event, in which take part all atoms. It allows you to simulate the presence in the system of different potentials, integrating them with various steps.

Taking between the sufficiently remote atoms the interaction strength equal to zero, a high computational efficiency is achieved by calculating the interaction forces only for atoms in the same block or in adjacent blocks (by means of the coordinate space decomposition on the parallelepiped elements). I.e., using the limitedness of the interaction radius between the atoms, we can achieve almost complete parallelization of processes occurring in contiguous blocks and significant acceleration of calculations. Take into account the interaction of the atom with the other atoms in the same cell and cell entourage of 26 cells (cell block with size 3x3x3 contains 27 cells). Due to the symmetry of the commonly used potentials the amount of computation is reduced 2-fold. Then for each cell having at the average n atoms, the number of verifiable interactions will be defined as $n(n-1)/2 + 13n^2$.

For event-driven molecular dynamics algorithm the global queue of events makes it difficult to parallelize, but keeping of the local time for each atom creates the possibility of a controlled parallel processing of spatially distant events. Namely, processing can last as long as the time elapsed in the absolute frame of reference, is not suitable to the nearest event in the bordering cells. After processing this event the parallel processes exchange messages [4]. If at the time of detection of the new event on the border two cells one process has already moved to the processing of later events, appear the probability of error (the lost interaction), which corrects by the recovery procedure, using a reverse pass in turn saved processed events. However, despite the cost of calculations in the case of detecting inconsistencies-errors, the possibility of acceleration of parallelization is largely preserved, since the number of events in the boundary layer is considerably lower than in cells. Since event management algorithm is applied only within a certain time step, the parallelization of periodic operations (such as the calculation of the interaction forces) can be performed with relatively high efficiency.

3   Basic Problems of Computer Simulation

Molecular dynamic experiment comprises the following stages. Primarily built the initial structure of calculable system (calculable block): are given the block size (10^3–10^6 atoms) and the initial coordinates of the atoms. At this stage the temperature is usually defined by setting appropriate initial velocities by (4) or atomic displacements. The next main stage is to calculate the trajectories of each atom in the calculable block. For this are calculated the displacements of atoms from (3), using the short time step (10^{-13}–10^{-15} s – integration step). To determine the velocity and displacement of atoms, it is necessary to know the force of interaction between them. Forces are defined by (1) using the model selected for the potential functions of the interatomic interaction. When calculating the interactions of atoms at the borders of the block it is necessary to define different, depending on the task, the boundary conditions (periodic, rigid, flexible, highly viscous or free). In the computer simulation process is performed monitoring of different characteristics (temperature, free volume, the coefficient of diffusion, pressure, etc.), as well as when it is required to explore the dynamics of atomic structure, at certain intervals of time the structures of calculable block are memorized. Study using different sub-modules of a computer program, responsible for structure visualization, charting, calculating the required parameters can be attributed to the third stage of the molecular dynamics simulation.

At the quest for a more reliable study using molecular dynamics simulations have to deal with problems that can be reduced to four basic: adequacy of interatomic interaction potential, the count rate (the speed of the experiment performance on the computer), the calculation errors, the variety of visualizers to display the structure of calculable block [5].

The correct description of the interatomic interaction is achieved by selecting an adequate potential function of interaction between atoms. There are theoretical methods based on the approximate solution of the quantum-mechanical problem of calculating the energy of the crystal: the empirical and semi-empirical methods, in which the potential is given as a function (Born-Mayer, Lennard-Jones, Morse, Mi-Gruneisen various power functions, etc.) with a set of parameters, which is chosen according to the reference values for given material. There are also a combination of first-principles (ab initio) and of potential (density functional method) approaches, or many-body potentials of Finnis-Sinclair type and the potentials obtained by embedded atom method (EAM). Accuracy of potential usually associated with its complexity and with execution speed of the computer experiment. Therefore, the question of finding an optimal potential function remains opened, the select of potential type in each case depends on the method of computer simulation and task. In applying the semiempirical potentials using their fitting by a three types of empirical parameters (structural parameter – lattice period, energy parameters – energy of sublimation and the energy of defect formation and etc., power parameters – elastic modules) can be investigated with high accuracy the structural and power changes occurring in the material. Should be used cautiously the energy parameters, take into account in the computer experiment only their relative changes in the study of structural-phase transitions, especially in the case of nanocrystals and low-dimensional systems.

Another major problem of molecular dynamics simulation is the count rate on the computer. High count rate allows to execute the longer experiments, increase the size of the calculable block, use a more complex and realistic interatomic potentials. Count rate indirectly affects the accuracy of computer experiments. There are several ways to increase the count rate:

1. Decrease of size of calculable block (effect – decreased reality of model, increases error at calculation of various structural-energy characteristics)
2. increase of time step of integration (effect – significantly increases the calculation error of trajectories of atoms)
3. Simplification of the potential function, tabulation of potential function, the introduction of so-called "cutoff radius" of potential (effect – strength of the interaction decreases rapidly with the distance between the atoms, so the interaction with remote atoms must be null)
4. Optimization of code – getting rid of extra functions in the main loop of a computer program (looping through all the atoms in the calculation of their interactions and displacements)
5. In the case of pair potentials calculation of force between two atoms occurs only once (the forces between atoms equal in magnitude and opposite in direction)
6. Partition of calculable block on the cells with assigning them numbers – in the calculation of interaction forces it is sufficient to consider only the atoms in neighboring cells and in compliance with "cutoff radius"
7. Acceleration of processing by means multiprocessor parallel calculations

The third problem of molecular dynamics simulation – errors that arise in calculating of the trajectories of motion of the atoms during the experiment – calculation errors, relating to the numerical method used to solve the equations of motion, or to the fact that the variables are rounded in the computer program. Errors, relating to the digitization step of the grid schemes in the solving process of the motion equations, can be reduced by using numerical methods of higher order than the Euler method, for example, the Runge-Kutta method. Furthermore, errors can be reduced by reducing the integration time step, but this leads to an increase in the duration of computer simulations, amount of experiments, or size of calculable block.

Another problem is the realistic visualization for the study of the atomic structure dynamics of the calculable block – you need a diverse set of visualizers for display of structures. There are three main types of visualization:

1. Using charts and graphs that reflect structurally dependent parameters (radial distribution diagrams, phase composition diagrams, crystallographic orientation diagrams, the intensity distribution of diffusion, etc.)
2. Two-dimensional sections of three-dimensional models, the visualizers of two-dimensional structure (areas of tensile and compression, the potential energy distribution, the phase distribution, the pattern of high-packed atomic series, trajectories and displacements of atoms in two-dimensional models, etc.)

3. Three-dimensional visualizations (atomic displacements, visualization of defect regions, fill of areas of particular phase composition, etc.)

Molecular dynamics method allows us to consider the block comprising about a billion atoms. Thus, almost any nanostructures can be modeled with a high degree of accuracy on multiprocessor computers.

4   Analysis of the Results of Computer Simulation of the Atomic Structure of Amorphous Aluminum by Event-Driven Molecular Dynamics Method

To study the structure of some metals, obtained by ultrafast cooling of the melt, in computer simulations it is necessary to change the speed of quenching. Process of simulation, in this case, comprises two stages:

1. Melting of the initial crystalline block
2. Quenching of the melt at a given rate

For example, to simulate the liquid Al structure, the temperature of initial block in crystalline state was set to $4400\,\mathrm{K}$ and sustainment was performed for $10^{-11}\,\mathrm{s}$ (no more than 0.5% of all atoms evaporated - compensation wasn't been made for these atoms). Phase transition from solid to liquid state was determined by radial distribution functions, angular distribution functions, an abrupt change in the specific volume (expansion of metal) or decrease in the temperature of simulation block. The initial temperature and sustainment time were selected to minimize the time of complete destruction of the crystal structure and to prevent the evaporation of particles from simulation block, as the initial temperature is higher than the condensation temperature of aluminum ($2792\,\mathrm{K}$).

Simulation of amorphous aluminum by ultrafast cooling of liquid Al, obtained in the first stage, was performed in the second stage by lowering the temperature with linear decrease in velocities of atoms (it is closer to the real picture). Atoms' initial directions of velocity were assigned randomly, so the results of repeated computer experiments differed from each other. Simulation block comprised a different number of atoms (up to 20000) and represents a portion of a thin film. The interaction of atoms at a distance r was described by a paired Morse potential

$$U\left(r\right) = D \cdot \left(\exp\left(-2\alpha\left(r - \sigma\right)\right) - 2\exp\left(-\alpha\left(r - \sigma\right)\right)\right) \tag{5}$$

with parameters taken from [6]: $\alpha = 1.1646\,\left(\text{Å}\right)^{-1}$, $\sigma = 3.253\,\text{Å}$, $D = 0.2703\,\mathrm{eV}$. Along two axes boundary conditions were periodic, along the third – free. The initial temperature was $4400\,\mathrm{K}$. Such a high temperature is chosen to minimize the time of producing the melt. The time step was $0.01\,\mathrm{ps} = 10^{-14}\,\mathrm{s}$.

Judging from experience of calculations performed by different authors, at the start it is sufficient to place the atoms as a regular lattice of Al (fcc crystal), and choose the initial impulses of the same modulo (subject to compliance with a given temperature) with a random distribution of their directions so that the total momentum of the simulation block was zero. Due to instability of the trajectories of individual atoms (according to Lyapunov) initially some atomic structure containing individual atoms and clusters of atoms, as well as pores, are formed randomly in the melt. The first coordination sphere of those atoms corresponds to one or another well-known ordered structure. The structural units of the melt inside the simulation block occur chaotically. The melting in this model was conducted for 10 picoseconds.

Numerical calculations show that the structural chaos in the liquid state is largely inherited in the solid amorphous state in the case of rapid cooling, as the melt viscosity becomes so large that structural units of the melt come to a stop and retain only internal degrees of freedom. The crystallization process becomes impossible, since in a very short time, the atoms do not have time to move to a distance that would allow them to form a crystal lattice. This metastable state is not the lowest energy state, and the energy of the formed amorphous metal is determined by the arisen atomic structure. These atomic structures are largely random and varied depending on the initial conditions and the cooling rate.

Content analysis of fcc, hcp phases and Frank-Kasper polyhedra was conducted. For each atom of the simulation block the positions of nearest neighbors was evaluated, and then comparison with the reference samples was performed (results in Table 1).

Table 1. Percentages of the fcc, hcp and Frank-Kasper unit cells for three different cooling rates and simulation block of 20000 atoms. Percentage – the ratio of the number of atoms that are the "center" of cells to the total number of atoms in the system.

Cooling rate	Unit cells	Fcc / all unit cells	Hcp / all unit cells	F-K / all unit cells
10^{13} K/s	$48 \pm 15\%$	$59 \pm 24\%$	$21 \pm 15\%$	$37 \pm 32\%$
10^{14} K/s	$42 \pm 2\%$	$43 \pm 3\%$	$25 \pm 3\%$	$31 \pm 3\%$
10^{15} K/s	$37 \pm 2\%$	$46 \pm 3\%$	$26 \pm 3\%$	$27 \pm 3\%$

As expected from the distribution of the number of atoms in the first co-ordination sphere, in all cases, 16-vertex Frank-Kasper figures were not found. Among the figures of Frank-Kasper there were found more icosahedra than any others. Based on the data given in the Table 1, for cooling rate of 10^{14} and 10^{15} K/s discrepancy in the results is of only a few percent, while for the rate of 10^{13} K/s, the scatter of results is very significant (defined statistical regularity is possible in simulation blocks hardened at high speeds). Large scatter in the data for the cells obtained at the slowest cooling may be due to the random nature of solidification centers, and the fcc, hcp and Frank-Kasper elementary cells of these systems are randomly scattered in volume and do not form a conjugate ordered structures (amorphous-nanocrystalline structure is formed).

As is well known, the choice of the interaction potential of atoms plays an important role in the application of molecular dynamics method [7]. If with this results change qualitatively, then the collection of atoms considered under conditions of cooling to be regarded as structurally unstable system (according to

Pontryagin), that is more likely in the neighborhood of the critical cooling rate. Complete theoretical analysis of the structural stability of systems of molecular dynamics is currently appear to be difficult to implement. However, the qualitative picture of the structural stability can be obtained by numerical analysis with the variation of empirical constants in the interparticle interaction potential [8].

5   Conclusions

Thus, it is necessary to use the concepts and methods of the modern theory of nonlinear dynamical systems in the study of atomic structure of liquids and amorphous metals. In addition, our calculations using molecular dynamics suggest that a particular atomic structure of amorphous metal is fundamentally unpredictable and irreproducible at the level of the model, as well as in a real experiment.

References

1. Miller, S., Luding, S.: Event-driven molecular dynamics in parallel. J. Comput. Phys. 193, 306–316 (2004)
2. Adcock, S.A., McCammon, J.A.: Molecular dynamics: Survey of methods for simulating the activity of proteins. Chem. Rev. 106(5), 1589–1615 (2006), http://pubs.acs.org/doi/abs/10.1021/cr040426m, PMID: 16683746
3. Lubachevsky, B.D.: How to Simulate Billiards and Similar Systems. J. Comput. Phys. 94, 255–283 (1991)
4. Plimpton, S.: Fast parallel algorithms for short-range molecular dynamics. J. Comput. Phys. 117(1), 1–19 (1995), http://dx.doi.org/10.1006/jcph.1995.1039
5. Jordan, V.I., Belov, T.A.: The use of the method of the dynamics of "mesocells" for 3D modeling of packaging structures of spheroidal particles in multi-component mixtures. In: Information Technologies and Mathematical Modelling (ITMM 2012): Materials of XI All-Russian Scientific-practical Conference with International Participation (ITMM, Anzhero-Sudzhensk, Russia, November 2012), vol. 1, pp. 45–50. Practica, Kemerovo (2012)
6. Girifalco, L.A., Weizer, V.G.: Application of the morse potential function to cubic metals. Phys. Rev. 114, 687–690 (1959), http://link.aps.org/doi/10.1103/PhysRev.114.687
7. Roth, J., Beck, P., Brommer, P., Chatzopoulos, A., Gahler, F., Hocker, S., Schmauder, S., Trebin, H.R.: Molecular dynamics simulations with long-range interactions. In: Nagel, W.E., Kroner, D.H., Resch, M.M. (eds.) High Performance Computing in Science and Engineering 2013, pp. 141–154. Springer International Publishing (2013), http://dx.doi.org/10.1007/978-3-319-02165-2_11
8. Stukowski, A.: Computational analysis methods in atomistic modeling of crystals. JOM 66(3), 399–407 (2014), http://dx.doi.org/10.1007/s11837-013-0827-5

Theoretical Aspects of Mathematical Modeling of SHS-Process with Consideration of the Diffusion Kinetics and Interphase Transformations in "Mesocells" of Heterogeneous Powder Mixture

Vladimir Jordan and Stanislav Kotenev

Altai State University, Barnaul, Russia
{w_jordan,avalon_real}@mail.ru

Abstract. This article reviews the important aspects of creation of an adequate mathematical model of the propagation of the flame front at the SH-synthesis in the "mesocells" of heterogeneous powder mixture. The model takes into account the phase formation processes in accordance with the phase diagram of the system components and in assumption of staging of interphase chemical transformations (formation and decay) according to the scheme of metal-chemical reactions. The chemical reaction rate is determined by taking into account the function of the source of the exothermic heat generation in chemical reaction and the solutions of problems of the diffusion kinetics and balance ratios on the moving interfacial borders in reactionary "mesocell".

Keywords: mathematical modeling, self-propagation high-temperature synthesis (SHS), diffusion kinetics, interphase transformations, mesocell, heterogeneous powder mixture.

1   Introduction

Basic researches allow to study the different aspects, related to the update of the physicochemical and mathematical models for the various structural and phase transformations in SH-synthesis of new materials. The results of such studies play an important role in solving the problem of optimization of technological conditions for SH-synthesis of materials that meet the set of required functional and performance properties. For example, the intermetallic compounds, i.e. chemical compounds of metals with each other, are formed in many systems. The complexity of their study is associated with the staging and with complex mechanism of formation and the structure of distribution of intermetallic phases in the volume of the resulting material. Composition of the final reaction products of SHS and therefore, the mechanism of component interaction are dependent on many parameters: the initial temperature of the mixture and porosity, the degree of dilution, the heat loss, dispersion of the reagents and their ratio, scale

A. Dudin et al. (Eds.): ITMM 2014, CCIS 487, pp. 162–167, 2014.
© Springer International Publishing Switzerland 2014

parameter and etc. Metal-chemical analysis typically considers two main types of interaction of metal elements: the formation of solid solutions of metals and formation of intermetallic compounds, based on the diagram of the equilibrium states of physicochemical system, depending on the chemical composition and environmental conditions – temperature and pressure (external pressure has no significant effect on the condensed state system where no gaseous phase). Mechanism of the "reaction diffusion" in solid-phase interaction in bimetallic compact specimens is fairly well studied [1]. Kinetics of the "reaction diffusion" of gasless combustion in such systems (including its stationary) sufficiently obey certain laws (linear, parabolic, cubic, exponential), which are expressed by relevant "kinetic" functions. But not so good is the case in the study of solid-phase chemical reactions in heterogeneous powder systems [1], particle size distribution and pore structure significantly affects on the kinetics of formation of the "diffusion layers" on the borders of contacting particles. The thickness of the diffusion layer interacting particles increases with time due to continuing of diffusion process through the diffusion layer of the metal component that supports solid phase chemical reaction (topochemical reaction in diffusion mode, [1]). The variety of defects in the particle packing structure of SHS-sample does not provide isothermal interaction and stationary diffusion kinetics at the microlevel of powder system of particles. Furthermore, in a heterogeneous mixture of powders in result of reaction across the reaction space occurs significant heat generation with nonuniform distribution on the contact surfaces of the particles, leading to self-heating and ignition of heated layers with further propagation of the combustion wave. Therefore to describe the diffusion kinetics in heterogeneous powder mixtures (significantly different from the homogeneous kinetics), considering nonisothermality and unsteady diffusion interaction, on the micro-and mesolevel instead of using the kinetic functions more appropriate is approach, which should be based on the solution of the diffusion equations of the form, that have identity with a heat equation [1].

For the simulation of chemical reactions of various compositions of the powder mixture in long samples of small cross section can be used as in [2] – [7] approach. In these studies, the initial heterogeneous composition is modeled as periodic cellular structure, which is assumed homogeneous in the case of a thermal process (temperature quickly aligned in the cell) and heterogeneous in the case of a chemical process. Heat transfer process is modeled on a macro level, taking into account the local dynamics of heat generation in each elementary reaction cell (mesocell of the particulate mixture). At high-temperature phase changes, in which occur the processes of growth and decay of the new phases, the particles often have a shape close to spherical [7], i.e. diffusion flow is spherically symmetric. Therefore, we understand under mesocells the averaged elements of the heterogeneous structure in the form of powder mixture of face-centered cubic (fcc) lattices, the centers of which are spherical particles of one kind and surrounded by a layer of particles of another kind. We can assume that the final mixture of particles as "mesoparticle" is shaped as an equivalent sphere with a certain radius RS. Thus, the sample volume by means of the planes of fcc symmetry is divided into discrete mesocells. The mutual influence of neighboring mesocells in which there is growth and decay of new phases in spherical mesoparticles, reflected in the fact that the total diffusion flux through the boundary of mesocells or through the plane of fcc symmetry equal zero. More precisely, through the surface of the equivalent sphere of radius Rs the total diffusion flux equal zero.

2   Mathematical Formulation of the Problem

For example, considering a mixture, in which the radii of the aluminum particles substantially smaller than the radii of nickel particles, and the number of particles of aluminum significantly more number of nickel particles, we can assume that the nickel particles are the centers of mesocells and local sources of production of intermetallic phases, which are formed in layers around nickel particles [4].

Assuming staging of chemical transformations (formation and decay) of intermediate phases and scheme metal-chemical reactions corresponding to diagram of phase states, reaction between aluminum and nickel begins upon reaching the melting point of aluminum with instantaneous formation of the product layer consisting of one or more layers of all sorts of equally intermetallic phases simultaneously [4], forming a thin diffusion layer. Considering only regions of single-phase intermediate products in the solid state and eliminating regions of two-phase transient products, the distribution of nickel concentration in the diffusion layer present itself as the function with discontinuity points at the interfaces. Two-phase regions are excluded by the assumption of the smallness of their linear dimensions [4].

Consequently, the mathematical formulation of the problem, taking into account the above stated and model representations, according to [4], [5], consists of heat equation, diffusion equation and balance sheet ratios on the moving interphase boundaries. Heat equation for long sample can be written as a one-dimensional equation.

$$c\rho\frac{\partial T}{\partial t} = \lambda\frac{\partial^2 T}{\partial x^2} + Q \cdot \varPhi(T(x,t)) \ . \tag{1}$$

The initial and boundary conditions for temperature:

$$T(x,0) = T_0 \ , \ \ T(0,t) = T_\alpha \ , \ \ \frac{\partial T(\infty,t)}{\partial t} = 0 \ . \tag{2}$$

In equation (1) introduced the following notation: t – temporary variable, x – coordinate, c – specific heat of the mixture, ρ – the mixture density, T – temperature, λ – thermal conductivity coefficient of the mixture, Q – heat effect of reaction, $\varPhi(T(x,t))$ - the chemical reaction rate, T_0 - initial value of temperature, T_α - the ignition temperature of the sample on the border $x = 0$.

To estimate the rate of a chemical reaction $\varPhi(T(x,t))$, it is necessary at each point previously to solve the problem of chemical reaction in the reaction cell.

Namely:

$$\frac{\partial C_A^{(i)}}{\partial t} = D_A^{(i)}(T)\frac{1}{r^2}\frac{\partial}{\partial r}\left(r^2\frac{\partial C_A^{(i)}}{\partial r}\right) - C_A^{(i)}C_B^{(i)}k_0 exp\left(-\frac{E^{(i)}}{RT}\right) , \qquad (3)$$

$$\frac{\partial C_B^{(i)}}{\partial t} = D_B^{(i)}(T)\frac{1}{r^2}\frac{\partial}{\partial r}\left(r^2\frac{\partial C_B^{(i)}}{\partial r}\right) - C_A^{(i)}C_B^{(i)}k_0 exp\left(-\frac{E^{(i)}}{RT}\right) , \qquad (4)$$

$$D_A^{(i)}(T) = D_{A,0}^{(i)}exp\left(-\frac{E_{d,A}^{(i)}}{RT}\right) , D_B^{(i)}(T) = D_{B,0}^{(i)}exp\left(-\frac{E_{d,B}^{(i)}}{RT}\right) , \qquad (5)$$

$$r_i(t) \le r \le r_{i+1}(t) . \qquad (6)$$

In formulas (3) – (6), the index i denotes the order number of the phase layer ($i = 1 \div 6$) in a reaction cell: 1 - Ni, 2 - $\varepsilon(Ni_3Al)$, 3 - $\delta(NiAl)$, 4 - $\gamma(Ni_2Al_3)$, 5 - $\beta(NiAl_3)$, 6 - Al. Indices A and B relate, respectively, to the particles of nickel and aluminum. Value $E^{(i)}$ - the activation energy of a chemical reaction for the i-th phase in the Arrhenius law (k_0 - pre-exponential factor in the required scale, R - universal gas constant); $E_{d,A}^{(i)}$ and $E_{d,B}^{(i)}$ - values of the activation energy of diffusion through layer of i-th phase for atoms, respectively, Ni and Al; $D_{A,0}^{(i)}$ and $D_{B,0}^{(i)}$ - corresponding diffusion constants of Ni and Al for the diffusion coefficients $D_A^{(i)}(T)$ and $D_B^{(i)}(T)$. Values $C_A^{(i)}$ and $C_B^{(i)}$ represent the mass concentrations in i-th phase layer for atoms, respectively, Ni and Al. Radii $r_{i+1}(t)$ and $r_i(t)$ represent, respectively, the initial and final radii of spherical surfaces bounding the i-th phase layer.

To solve the equations (3) and (4) can use the balance sheet ratios on the moving interphase boundaries [4]:

$$r = r_1(t): \quad \left(1 - C_A^{(2)}(T)\right)\frac{\partial r_1}{\partial t} = D_A^{(2)}(T)\frac{\partial C_A^{(2)}(T)}{\partial r}\bigg|_{r=r_1+0} , \qquad (7)$$

$$r = r_2(t): \quad \left(C_A^{(2)}(T) - C_A^{(3)}(T)\right)\frac{\partial r_2}{\partial t} =$$
$$-D_A^{(2)}(T)\frac{\partial C_A^{(2)}(T)}{\partial r}\bigg|_{r=r_2-0} +D_A^{(3)}(T)\frac{\partial C_A^{(3)}(T)}{\partial r}\bigg|_{r=r_2+0} , \qquad (8)$$

$$r = r_3(t): \quad \left(C_A^{(3)}(T) - C_A^{(4)}(T)\right)\frac{\partial r_3}{\partial t} =$$
$$-D_A^{(3)}(T)\frac{\partial C_A^{(3)}(T)}{\partial r}\bigg|_{r=r_3-0} +D_A^{(4)}(T)\frac{\partial C_A^{(4)}(T)}{\partial r}\bigg|_{r=r_3+0} , \qquad (9)$$

$$r = r_4(t): \quad \left(C_A^{(4)}(T) - C_A^{(5)}(T)\right)\frac{\partial r_4}{\partial t} =$$
$$-D_A^{(4)}(T)\frac{\partial C_A^{(4)}(T)}{\partial r}\bigg|_{r=r_4-0} +D_A^{(5)}(T)\frac{\partial C_A^{(5)}(T)}{\partial r}\bigg|_{r=r_4+0} , \qquad (10)$$

$$r = r_5(t): \quad C_A^{(5)}(T)\frac{\partial r_5}{\partial t} = -D_A^{(5)}(T)\frac{\partial C_A^{(5)}(T)}{\partial r}\bigg|_{r=r_5-0} . \qquad (11)$$

A similar system of balance sheet ratios on the moving interphase boundaries for the concentration of aluminum $C_B^{(i)}$ in the phase interlayers too can be determined.

We can assume that over time the rate of phase formation is proportional to change of its mass. Since the nickel particles are local centers of formation of intermetallic compounds, then the rate of production or consumption of the i-th phase can be written:

$$J_i = m_{Ni}\rho_i \frac{\partial}{\partial t}\left(y_i^3 - y_{i-1}^3\right), \quad y_i = \frac{r_i}{R_s}, \quad i = 2 \div 5 . \tag{12}$$

It should be noted that at the temperature between the melting points ($T_{Al} \leq T \leq T_i$) the reaction rate is positive ($J_i > 0$ – layer increases), and when $T > T_i$ – is negative ($J_i < 0$ - layer splits).

3   Approximate Task Solution

In conditions where the temperature is slowly increased over time or when it is constant (the so-called diffusion annealing mixture), assumed to be valid a state of local thermodynamic equilibrium temperature. The solution of equations (3) – (11) in the unsteady problem definition with so many matching conditions at the interfaces, moving with the times, it is extremely difficult. Therefore, to obtain approximate solution use the method [4], which assumes that the distribution of concentrations within each sub-layer is not very different from the stationary: $C_A^{(i)} = A_i/r + B_i$. Constants A_i and B_i determined from the condition that the function $C_A^{(i)}(T)$ equal to equilibrium concentration at the interfaces. Substitution of the stationary solutions $C_A^{(i)} = A_i/r + B_i$ in formulas (7) – (11) give a system of ordinary differential equations which is solved numerically, for example, by method of the Runge-Kutta of 4-th order accuracy [4]. I.e., the solution of this system is the trajectories of interfaces (the functions) which allow to translate the concentrations $C_A^{(i)}$ and $C_B^{(i)}$.

The chemical reaction rate $\Phi(T(x,t))$ is defined as the average speed of a chemical reaction in volume of the cell:

$$\Phi(T(x,t)) = \sum_{i=2}^{5} \int_{r_{i-1}}^{r_i} C_A^{(i)} C_B^{(i)} k_0 exp\left(-\frac{E^{(i)}}{RT(x,t)}\right) r^2 dr . \tag{13}$$

The average speed of a chemical reaction depends on the temperature, used in equation (1). Practical solution to the problem (1) – (13) is best done in the traditional (for combustion theory) dimensionless variables and parameters [5], taking into account the initial and boundary conditions using the effective difference schemes.

4   Conclusions

Thus, the work developed and proved important aspects of creation of an adequate mathematical model of a propagation of the flame front in the SH-synthesis on the level of mesocells of heterogeneous powder mixture. The model takes into account the phase formation processes in accordance with the phase diagram of the system components and in assumption of staging of interphase chemical transformations (formation and decay) according to the scheme of metal-chemical reactions. The chemical reaction rate is determined by taking into account the function of the source of the exothermic heat generation in chemical reaction, the solutions of problems the diffusion kinetics and balance ratios on the moving interfacial borders in reactionary "mesocell". This model will more accurately study the different modes of propagation of the combustion front and more accurately define the boundaries of the transition from one regime to another.

References

1. Itin, V.I., Nayborodenko, Y.S.: High-temperature synthesis of intermetallic compounds, pp. 84–125. Publisher of Tomsk. University, Tomsk (1989) (in Russian)
2. Lapshin, O.V., Ovcharenko, V.E.: A mathematical model of high-temperature synthesis of nickel aluminide Ni_3Al by thermal shock of a powder mixture of pure elements. J. Combustion, Explosion, and Shock Waves 32(3), 299–305 (1996)
3. Markov, A.A., Filimonov, I., Martirosyan, K.S.: Simulation of front motion in a reacting condensed two phase mixture. J. Comput. Phys. 231, 6714–6724 (2012)
4. Kovalev, O.B., Neronov, V.A.: Metallochemical Analysis of the Reaction in a Mixture of Nickel and Aluminum Powders. J. Combustion, Explosion and Shock Waves 40(2), 172–179 (2004)
5. Shults, D.S., Krainov, A.Y.: Numerical simulation of gasless combustion taking into account the heterogeneity of the structure and the temperature dependence of diffusion. J. Combustion, Explosion, and Shock Waves 48(5), 620–624 (2012)
6. Kovalev, O.B., Belyaev, V.V.: Mathematical modeling of metallochemical reactions in a two-species reacting disperse mixture. J. Combustion, Explosion, and Shock Waves 49(5), 563–574 (2013)
7. Lyubov, B.Y.: A diffusion processes in a non-uniform solid mediums, pp. 15–55. Nauka, Moscow (1981) (in Russian)

Model of Foundation-Base System under Vibration Load

Mikhail Kapustin, Alla Pavlova, Sergei Rubtsov, and Ilya Telyatnikov

Kuban State University,
Stavropolskaya st. 149, 350040 Krasnodar, Russia
pavlova@math.kubsu.ru

Abstract. In the article integral representations of the solutions describing the displacements in the medium and on the surface are obtained, allowing the investigation of patterns of the forming displacement field, excited by the surface load and vertically oriented inclusions in an elastic medium. Computational experiments that allow making conclusions about the impact of vertically oriented inclusions on the interference pattern of the total wave field are completed.

Keywords: elastic medium, steady oscillations, system of vertical inclusions, surface load, wave field.

Modern methods of studying the stability of structures and buildings, the calculation of their durability include a variety of approaches [1–6, etc.]. Problems of the evaluation criteria formation *concerning* condition of surface structures, in particular – the base-foundation complex are closely associated with problems of the development of rational construction schemes of foundation. Data on the spatial and temporal stress distribution in the base and on the nature of the wave energy outflow from the load zone may be useful in engineering practice. The paper describes a research method of results of vibration influence on plate-pile foundations.

The problem of steady-state (with frequency ω) of an elastic layer under the influence of surface and vertically oriented internal loads is considered as a model of the foundation-base system. Set of deepening vertical sources forms a cylindrical surface: $r = r_0$ $(r_0 > a)$, $-h_0 \leq z \leq 0$. The displacement of points of the medium are described in a cylindrical coordinate system by vector of displacement amplitudes $\mathbf{u} = \{u_r, u_z\}$, which corresponds to Lame equations. Load's distribution in depth is modeled by components of the body force – X_r, X_z. Vertical axisymmetric load is applied on the surface of an elastic medium $(z = 0)$ in a circular area $(r \leq a)$. It's described by the function $\mathrm{Re}\left[p(r)e^{-i\omega t}\right]$ $(p$ – a given function of amplitude, r – radius vector of the point of the plane, ω – frequency, t – time). The problem is solved in the class of generalized functions. Load distributed along the length of the inclusions is modeled by Dirac delta function $X_r = \mathrm{Re}\left[f_r(z)\delta(r - r_0)e^{-i\omega t}\right]$, $X_z = \mathrm{Re}\left[f_z(z)\delta(r - r_0)e^{-i\omega t}\right]$. Since we are considering a steady process, all task functions are represented as

A. Dudin et al. (Eds.): ITMM 2014, CCIS 487, pp. 168–173, 2014.

$\psi_1(r, z, t) = \psi(r, z) e^{-i\omega t}$. Below the time factor is omitted. Statement of the problem can be written as follows:

$$(\lambda + 2\mu) \left[\frac{1}{r} \frac{\partial}{\partial} \left(r \frac{\partial u_r}{\partial r} \right) - \frac{u_r}{r^2} \right] + (\lambda + \mu) \frac{\partial}{\partial r} \left[\frac{\partial u_z}{\partial z} \right] + \mu \frac{\partial^2 u_r}{\partial z^2} + \omega^2 \rho u_r$$
$$= f_r(z) \, \delta(r - r_0),$$

$$(\lambda + 2\mu) \frac{\partial^2 \tilde{u}_z}{\partial z^2} + \frac{\mu}{r} \frac{\partial}{\partial r} \left(r \frac{\partial u_z}{\partial r} \right) + (\lambda + \mu) \frac{\partial}{\partial z} \left[\frac{1}{r} \frac{\partial}{\partial r} (r u_r) \right] + \omega^2 \rho u_z$$
$$= f_z(z) \, \delta(r - r_0),$$

where λ, μ – Lame parameters of the elastic medium, ρ – density.

Boundary conditions on the surface of the medium ($z = 0$) have the form

$$\mu \left[\frac{\partial u_z}{\partial r} + \frac{\partial u_r}{\partial z} \right] = 0, \quad (\lambda + 2\mu) \frac{\partial u_z}{\partial z} + \frac{\lambda}{r} \frac{\partial}{\partial r} (r u_r) = \begin{cases} p(r), & r \leq a, \\ 0, & r > a. \end{cases}$$

For an elastic layer of thickness $h \, (0 < r < +\infty; \, -h \leq z \leq 0)$, on the lower boundary in conditions of rigid linkage with non-deformable base

$$u_r(r, -h) = u_z(r, -h) = 0.$$

As the radiation conditions at infinity, we use the limiting absorption principle.

Using Bessel transformation the task is reduced to a system of ordinary differential equations [7, 8]. Solution of the latter is constructed as a superposition of general and private solutions. By applying inverse transform to the components of the obtained vector-valued function, we obtain the solution of the original problem in integral form

$$u_r(r, z) = \int_0^\infty (D_1(\alpha)P(\alpha, z) + D_2(\alpha)M(\alpha, z) + K_1(\alpha, z)) \, \alpha J_1(\alpha r) \, d\alpha,$$

$$u_z(r, z) = \int_0^\infty (D_1(\alpha)R(\alpha, z) + D_2(\alpha)S(\alpha, z) + K_2(\alpha, z)) \, \alpha J_0(\alpha r) \, d\alpha,$$

where

$$D_1(\alpha) = \left(\frac{p(\alpha)}{\rho c_1^2} - 2C_{21}^2 \left[s \left(g_1^+(-h_0) + g_1^-(-h_0) \right) \right. \right.$$
$$\left. \left. - \alpha \sigma_2 \left(g_2^+(-h_0) - g_2^-(-h_0) \right) \right] \right),$$

$$D_2(\alpha) = 2\rho c_2^2 \left[\alpha \sigma_1 \left(g_1^+(-h_0) - g_1^-(-h_0) \right) - s \left(g_2^+(-h_0) + g_2^-(-h_0) \right) \right],$$

$$P\left(\alpha, z\right) = \left[\alpha e^{\sigma_1 z} \Delta_{11} + \alpha e^{-\sigma_1 z} \Delta_{12} - \sigma_2 e^{\sigma_2 z} \Delta_{13} + \sigma_2 e^{-\sigma_2 z} \Delta_{14}\right] \Delta^{-1}\left(\alpha\right),$$

$$R\left(\alpha, z\right) = \left[\alpha e^{-\sigma_2 z} \Delta_{14} - \sigma_1 e^{\sigma_1 z} \Delta_{11} + \sigma_1 e^{-\sigma_1 z} \Delta_{12} + \alpha e^{\sigma_2 z} \Delta_{13}\right] \Delta^{-1}\left(\alpha\right),$$

$$M\left(\alpha, z\right) = \left[\alpha e^{\sigma_1 z} \Delta_{21} + \alpha e^{-\sigma_1 z} \Delta_{22} - \sigma_2 e^{\sigma_2 z} \Delta_{23} + \sigma_2 e^{-\sigma_2 z} \Delta_{24}\right] \Delta^{-1}\left(\alpha\right),$$

$$S\left(\alpha, z\right) = \left[-\sigma_1 e^{\sigma_1 z} \Delta_{21} + \sigma_1 e^{-\sigma_1 z} \Delta_{22} + \alpha e^{\sigma_2 z} \Delta_{23} + \alpha e^{-\sigma_2 z} \Delta_{24}\right] \Delta^{-1}\left(\alpha\right),$$

$$\Delta = 16 C_{21}^2 \rho c_2^2 \left[\sigma_1 \sigma_2 \left(s^2 + \alpha^4\right) \operatorname{ch}\left(\sigma_1 h\right) \operatorname{ch}\left(\sigma_2 h\right)\right.$$
$$\left. - \alpha^2 \left(s^2 + \sigma_1^2 \sigma_2^2\right) \operatorname{sh}\left(\sigma_1 h\right) \operatorname{sh}\left(\sigma_2 h\right) - 2 s \alpha^2 \sigma_1 \sigma_2\right],$$

$$\Delta_{11} = 4\rho c_2^2 \left[\alpha^2 \sigma_1 \sigma_2 - s e^{\sigma_1 h} \chi_2^-\right], \quad \Delta_{12} = 4\rho c_2^2 \left[\alpha^2 \sigma_1 \sigma_2 - s e^{-\sigma_1 h} \chi_2^+\right],$$

$$\Delta_{13} = -4\rho c_2^2 \alpha \sigma_1 \left[e^{\sigma_2 h} \varphi_1^- + s\right], \quad \Delta_{14} = 4\rho c_2^2 \alpha \sigma_1 \left[s - e^{-\sigma_2 h} \varphi_1^+\right],$$

$$\Delta_{21} = -4 C_{21}^2 \alpha \sigma_2 \left[e^{\sigma_1 h} \varphi_2^- + s\right], \quad \Delta_{22} = 4 C_{21}^2 \alpha \sigma_2 \left[s - e^{-\sigma_1 h} \varphi_2^+\right],$$

$$\Delta_{23} = 4 C_{21}^2 \left[\alpha^2 \sigma_1 \sigma_2 - s e^{\sigma_2 h} \chi_1^-\right], \quad \Delta_{24} = 4 C_{21}^2 \left[\alpha^2 \sigma_1 \sigma_2 - s e^{-\sigma_2 h} \chi_1^+\right],$$

$$\chi_j^\pm = \sigma_1 \sigma_2 \operatorname{ch}\left(\sigma_j h\right) \pm \alpha^2 \operatorname{sh}\left(\sigma_j h\right), \quad \varphi_j^\pm = \sigma_1 \sigma_2 \operatorname{sh}\left(\sigma_j h\right) \pm \alpha^2 \operatorname{ch}\left(\sigma_j h\right), \quad j = 1, 2;$$

$$K_1\left(\alpha, z\right) = \alpha \left[g_1^+\left(z\right) - g_1^+\left(-h_0\right)\right] e^{\sigma_1 z} + \alpha \left[g_1^-\left(z\right) - g_1^-\left(-h_0\right)\right] e^{-\sigma_1 z}$$
$$- \sigma_2 \left[g_2^+\left(z\right) - g_2^+\left(-h_0\right)\right] e^{\sigma_2 z} + \sigma_2 \left[g_2^-\left(z\right) - g_2^-\left(-h_0\right)\right] e^{-\sigma_2 z},$$

$$K_2\left(\alpha, z\right) = -\sigma_1 \left[g_1^+\left(z\right) - g_1^+\left(-h_0\right)\right] e^{\sigma_1 z} + \sigma_1 \left[g_1^-\left(z\right) - g_1^-\left(-h_0\right)\right] e^{-\sigma_1 z}$$
$$+ \alpha \left[g_2^+\left(z\right) - g_2^+\left(-h_0\right)\right] e^{\sigma_2 z} + \alpha \left[g_2^-\left(z\right) - g_2^-\left(-h_0\right)\right] e^{-\sigma_2 z},$$

$$K_1\left(\alpha, z\right) = \alpha \left[g_1^+\left(z\right) - g_1^+\left(-h_0\right)\right] e^{\sigma_1 z} + \alpha \left[g_1^-\left(z\right) - g_1^-\left(-h_0\right)\right] e^{-\sigma_1 z}$$
$$- \sigma_2 \left[g_2^+\left(z\right) - g_2^+\left(-h_0\right)\right] e^{\sigma_2 z} + \sigma_2 \left[g_2^-\left(z\right) - g_2^-\left(-h_0\right)\right] e^{-\sigma_2 z},$$

$$K_2\left(\alpha, z\right) = -\sigma_1 \left[g_1^+\left(z\right) - g_1^+\left(-h_0\right)\right] e^{\sigma_1 z} + \sigma_1 \left[g_1^-\left(z\right) - g_1^-\left(-h_0\right)\right] e^{-\sigma_1 z}$$
$$+ \alpha \left[g_2^+\left(z\right) - g_2^+\left(-h_0\right)\right] e^{\sigma_2 z} + \alpha \left[g_2^-\left(z\right) - g_2^-\left(-h_0\right)\right] e^{-\sigma_2 z},$$

$$g_1^\pm\left(z\right) = \frac{r_0}{2\rho\omega^2 \sigma_1} \left(\pm \alpha J_1\left(\alpha r_0\right) \phi_{r1}^\mp\left(z\right) + \sigma_1 J_0\left(\alpha r_0\right) \phi_{z1}^\mp\left(z\right)\right),$$

$$g_2^\pm\left(z\right) = \frac{r_0}{2\rho\omega^2 \sigma_2} \left(\sigma_2 J_1\left(\alpha r_0\right) \phi_{r2}^\mp\left(z\right) \pm \alpha J_0\left(\alpha r_0\right) \phi_{z2}^\mp\left(z\right)\right);$$

$$\phi_{rk}^\mp\left(z\right) = \int_0^z f_r\left(\zeta\right) e^{\mp\sigma_k \zeta} d\zeta, \quad \phi_{zk}^\mp\left(z\right) = \int_0^z f_z\left(\zeta\right) e^{\mp\sigma_k \zeta} d\zeta \quad (k = 1, 2);$$

$$p\left(\alpha\right) = \int_0^\infty p\left(r\right) r J_0\left(\alpha r\right) dr, \quad \Delta\left(\alpha\right) = 2 C_{21}^2 \left(\alpha^2 \sigma_1 \sigma_2 - s^2\right), \quad s = \alpha^2 - 0.5\kappa_2^2,$$

$$\kappa_j^2 = (\omega/c_j)^2, \quad C_{21}^2 = (c_2/c_1)^2,$$

$$c_1 = \sqrt{(\lambda + 2\mu)/\rho}, \quad c_2 = \sqrt{\mu/\rho}, \quad \sigma_j = \sqrt{\alpha^2 - \kappa_j^2}, \quad (j = 1, 2).$$

Where κ_j are respectively the wave numbers of longitudinal and transverse waves; $J_n(\alpha r)$ $(n = 0, 1)$ – Bessel functions of the first order.

Using integral representations for displacement amplitudes on the surface of the elastic layer $(z = 0)$ at $r \to \infty$, using the procedure of closing the contour, the analytical expressions for the amplitudes of Rayleigh wave are obtained at $r \to \infty$

$$u_\beta(r, 0) = i\sqrt{\frac{2\pi\zeta}{r}} \frac{Q_{\beta 1}(\zeta) - Q_{\beta 2}(\zeta)}{2\rho c_2^2 \bar{\Delta}(\zeta)} e^{i(\zeta r - \frac{3\pi}{4})} + O\left(r^{-3/2}\right), \quad \beta = r, z,$$

where

$$Q_{r1}(\alpha) = p(\alpha)\alpha\Big(\sigma_1\sigma_2(\alpha^2 + s)(1 - \mathrm{ch}(\sigma_1 h)\,\mathrm{ch}(\sigma_2 h))$$
$$+ (\alpha^2 s + \sigma_1^2\sigma_2^2)\,\mathrm{sh}(\sigma_1 h)\,\mathrm{sh}(\sigma_2 h)\Big),$$

$$Q_{r2}(\alpha) = \rho c_2^2\kappa_2^2\sigma_2\Big(g_1^+(-h_0)\alpha\sigma_1\left(s - e^{-\sigma_1 h}\varphi_2^+\right) + g_1^-(-h_0)\alpha\sigma_1\left(s + e^{\sigma_1 h}\varphi_2^-\right)$$
$$+ g_2^+(-h_0)\left(\alpha^2\sigma_1\sigma_2 - se^{-\sigma_2 h}\chi_1^+\right) - g_2^-(-h_0)\left(\alpha^2\sigma_1\sigma_2 - se^{\sigma_2 h}\chi_1^-\right)\Big),$$

$$Q_{z1}(\alpha) = p(\alpha)\kappa_2^2\left(\alpha^2\,\mathrm{ch}(\sigma_1 h)\,\mathrm{sh}(\sigma_2 h) - \sigma_1\sigma_2\,\mathrm{sh}(\sigma_1 h)\,\mathrm{ch}(\sigma_2 h)\right),$$

$$Q_{z2}(\alpha) = 2\rho c_2^2\kappa_2^2\sigma_1\Big(g_1^+(-h_0)\left(\alpha^2\sigma_1\sigma_2 - se^{-\sigma_1 h}\chi_2^+\right)$$
$$- g_1^-(-h_0)\left(\alpha^2\sigma_1\sigma_2 - se^{\sigma_1 h}\chi_2^-\right) + g_2^+(-h_0)\alpha\sigma_2\left(s - e^{-\sigma_2 h}\varphi_1^+\right)$$
$$+ g_2^-(-h_0)\alpha\sigma_2\left(s + e^{\sigma_2 h}\varphi_1^-\right)\Big),$$

$$\bar{\Delta}(\alpha) = \alpha\Big(\left((\sigma_1^2 + \sigma_2^2)\sigma_1^{-1}\sigma_2^{-1}(s^2 + \alpha^4) + 4\sigma_1\sigma_2(s + \alpha^2)\right)\mathrm{ch}(\sigma_1 h)\,\mathrm{ch}(\sigma_2 h)$$
$$+ h\left(\sigma_1(s^2 + \alpha^4) - \alpha^2\sigma_1^{-1}\left(s^2 + \sigma_1^2\sigma_2^2\right)\right)\mathrm{ch}(\sigma_1 h)\,\mathrm{sh}(\sigma_2 h)$$
$$+ h\left(\sigma_2(s^2 + \alpha^4) - \alpha^2\sigma_2^{-1}\left(s^2 + \sigma_1^2\sigma_2^2\right)\right)\mathrm{sh}(\sigma_1 h)\,\mathrm{ch}(\sigma_2 h) - 4\sigma_1\sigma_2(s + \alpha^2)$$
$$- 2\left(s^2 + \sigma_1^2\sigma_2^2 + 2s\alpha^2 + \alpha^2\left(\sigma_1^2 + \sigma_2^2\right)\right)\mathrm{sh}(\sigma_1 h)\,\mathrm{sh}(\sigma_2 h) - 2s\alpha^2\sigma_1^{-1}\sigma_2^{-1}\left(\sigma_1^2 + \sigma_2^2\right)\Big).$$

Below are the results for vertical loads. Surface forces $\iint_\Omega p(r, \phi)d\Omega$ and loads on vertical inclusions $\int_{-h_0}^{0} f(z)dz$ were considered to have opposite directions and equal magnitudes. Linear function used to describe the distribution of the load on the inclusions $f(z) = kz + b$, where $k = b(1 - \varepsilon_1)h_0^{-1}$, $b = 2h_0^{-1}(1 + \varepsilon_1)^{-1}$, $z \in [0, -h_0]$, and the function describing the stress on the rigid inclusion, corresponding to the solution of integral equation of the contact problem, in form

$$f(z) = \varepsilon_2(h_0 + z) + e^{z-h_0} + \frac{1}{\sqrt{-z}} + \frac{1}{\sqrt{z + h_0}}, \quad z \in (0, -h_0).$$

Performed calculations allow to draw conclusions about the effect of vertically oriented inclusions on the interference pattern of general wave field. The obtained results indicate that the effect of vertically oriented inclusions on general wave field depends on the length and frequency of their oscillation. The graph shows u_z – the coefficient with $r^{-1/2}$ for the amplitude of the Rayleigh wave vertical component. The following values were taken as characteristics of the elastic layer: $\rho = 1.4 \cdot 10^3$ kg/m^3, $_1 = 0.2 \cdot 10^3$ m/s, $_2 = 0.12 \cdot 10^3$ m/s. Reduced frequency determined by formula $\omega = 2\pi\nu l_0/c_0$, where ν – frequency (Hz), $l_0 = 1$ m, $c_0 = 10^3$ m/s.

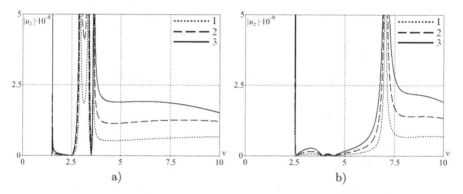

Fig. 1. $h = 20$ m; $1 - a = 2$ m; $2 - a = 3$ m; $3 - a = 4$ m: (a) – mode 1, (b) – mode 2

Increasing the size of the surface load will cause an increase in the amplitude of the Rayleigh wave vertical component (Fig. 1). Increasing the location radius of the vertical inclusions with increasing frequency complicates the interference pattern of the wave field also increasing the number of "locking" frequencies and changes their values (Fig. 2).

Variation of the load distribution function along the generatrix inclusions also complicates the interference pattern of the wave field, this increases the number of "Locking" frequencies and change their values.

The obtained results allow to investigate the features of the displacement field generated by using vertically oriented inclusions and surface load. The given approach can be used in case of the base with parallel planar defects [9, 10].

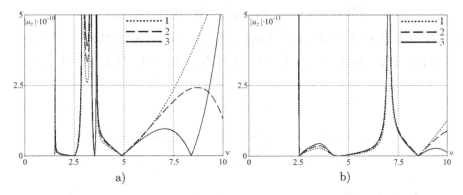

Fig. 2. $h = 20$ m; $h_0 = 10$ m, $\varepsilon_2 = 1.0$, $1 - r_0 = 3$ m; $2 - r_0 = 4$ m; $3 - r_0 = 5$ m: (a) – mode 1, (b) – mode 2

Acknowledgement. This work was supported by grants RFFR 13-01-00132, 13-01-96503.

References

1. Sargsyan, A.E., Gerashchenko, V.S., Shaposhnikov, N.N.: The computational model of pile foundations for the effects of their interaction with the soil environment. Herald MGRS 4, 69–71 (2012) (in Russian)
2. Liang, F.Y., Chen, L.Z., Shi, X.G.: Numerical Analysis of Composite Piled Raft with Cushion Subjected to Vertical Load. Computers and Geotechnics 30, 443–453 (2003)
3. Muir Wood, D., Hu, W., Nash, D.F.T.: Group effects in stone column foundations: model tests. Geotechnique 50(6), 689–698 (2000)
4. Small, J.C., Zhang, H.H.: Piled Raft Foundation Subjected to General Loading. Developments in Theoretical Geomechanics, pp. 57–72. Balkema, Rotterdam (2000)
5. Wolf John, P., Deeks Andrew, J.: Foundation vibration analysis: a strength-of-materials approach, p. 224. Elsevier, Amsterdam (2004)
6. Xie, J., Vaziri, H.H.: A numerical model for analysis of flexible beam-columns on elastic foundations. Computers and Geotechnics 13(1), 51–62 (1992)
7. Pryakhina, O.D., Smirnova, A.V., Evdokimov, A.A., Kapustin, M.S.: Fluctuations in the presence of half-rigid inclusions. Doklady RAS 389(1), 193–197 (2003) (in Russian)
8. Evdokimov, A.A., Kapustin, M.S.: Contact stresses in the problem of the action of a stamp on an elastic layer with a vertically oriented inclusions. Ecological Bulletin of Research Centers of the Black Sea Economic Cooperation (BSEC) 1, 60–66 (2005) (in Russian)
9. Babeshko, V.A., Williams, R., Pavlova, A.V., Ratner, S.V.: Solution to the problem on vibration of an elastic solid with inner cavities. Doklady Physics 47, 141–144 (2002)
10. Pavlova, A.V., Rubtsov, S.E.: Differential factorization method for block-layered media with defects. Bulletin of the University of Nizhni Novgorod Lobachevskii 4(5), 2410–2412 (2011) (in Russian)

Electromagnetic Scattering
by a Three-Dimensional Magnetodielectric Body
in the Presence of Closely Adjacent Thin Wires

Yuri Keller

National Reseach Tomsk State University, Tomsk, Russia
kua1102@rambler.ru

Abstract. This paper presents a version of auxiliary sources method generalized for the case of the presence of thin wires in the vicinity of a magnetodielectric body. A mathematical formulation of this version and a brief description of the created computer code are given. Some numerical results concerning the influence of thin wires on the bistatic cross-section of a dielectric body and the influence of a dielectric body on current distribution are presented.

Keywords: numerical method, electromagnetic scattering, magnetodielectric body, thin wire, scattering cross-section.

1 Introduction

The study of electromagnetic-wave scattering by a three-dimensional magnetodielectric body in the presence of closely adjacent thin wires is of considerable interest. This is explained by the need for solving important applied problems of electromagnetic compatibility and radar detectability, designing multi-element antenna systems, etc. If the distance between a magnetodielectric body and thin wires is less or comparable with the wavelength, the correct formulation of such problems requires solving boundary-value problems of scattering taking into account the electromagnetic interaction between the scatteres. The latter problems can be solved numerically by using, for example, finite-element methods [1]-[3] or integral-equation methods[4]-[6], but the algorithms yielded by them require considerable computer resources. In the recent years, the method of auxiliary sources[7]-[8] has been applied to solving the problems of electromagnetic-wave scattering by structures composed of a finite number of three-dimensional perfectly conducting bodies [9] and structures composed of a finite number of perfectly conducting bodies and thin wires [10]. In this paper, we solve the problem of electromagnetic-wave scattering by structures composed of a magnetodielectric body and thin wires using a version of the auxiliary sources method.

2 Mathematical Formulation

The geometry of the problem is shown in Fig. 1. We consider the stationary problem of diffraction of an electromagnetic field $\{\mathbf{E}_0, \mathbf{H}_0\}$ by a structure comprising

A. Dudin et al. (Eds.): ITMM 2014, CCIS 487, pp. 174–180, 2014.

magnetodielectric body D_i with permittivity ε_i and magnetic permeability μ_i bounded by the surface S and U thin conductors (wires) bounded by the surfaces S'_u $(u = 1, 2, \ldots, U)$ and located arbitrarily with respect to the body D_i. A time dependence of the form $\exp\{-i\omega t\}$ is assumed. The body and wires are disjoint. By a thin wires we mean a perfect conductor of a circular cross-section whose diameter is finite but is much less than the wavelength. The structure considered is located in a homogeneous unbounded medium D_e with dielectric permittivity ε_e and magnetic permeability μ_e. We take a Cartesian coordinate system whose origin is located inside the magnetodielectric body. Let us find the scattered field $\{\mathbf{E}_e, \mathbf{H}_e\}$ in the domain D_e.

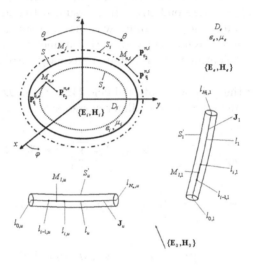

Fig. 1. Geometry of the problem

The mathematical formulation of the problem is as follows:

$$\left.\begin{array}{l}\nabla \times \mathbf{E}_e = i\omega\mu_e\mathbf{H}_e \\ \nabla \times \mathbf{H}_e = -i\omega\varepsilon_e\mathbf{E}_e\end{array}\right|_{D_e}, \quad \left.\begin{array}{l}\nabla \times \mathbf{E}_i = i\omega\mu_i\mathbf{H}_i \\ \nabla \times \mathbf{H}_i = -i\omega\varepsilon_i\mathbf{E}_i\end{array}\right|_{D_i}, \tag{1}$$

$$\left.\begin{array}{l}\mathbf{n} \times (\mathbf{E}_i - \mathbf{E}_e) = \mathbf{n} \times \mathbf{E}_0 \\ \mathbf{n} \times (\mathbf{H}_i - \mathbf{H}_e) = \mathbf{n} \times \mathbf{H}_0\end{array}\right|_S, \quad \mathbf{n}_u \times \mathbf{E}_e = -\mathbf{n}_u \times \mathbf{E}_0\Big|_{S'_u}, u = \overline{1, U}, \tag{2}$$

$$\{\sqrt{\varepsilon_e}\mathbf{E}_e; \sqrt{\mu_e}\mathbf{H}_e\} \times \mathbf{R}/R + \{\sqrt{\mu_e}\mathbf{H}_e; -\sqrt{\varepsilon_e}\mathbf{E}_e\} = O(R^{-1}), R \longrightarrow \infty. \tag{3}$$

Here, $\mathbf{E}_e$, $\mathbf{H}_e$ and $\mathbf{E}_i$, $\mathbf{H}_i$ are the fields in domains D_e and D_i, $\mathbf{n}$ and $\mathbf{n}_u(u = 1, 2, \ldots, U)$, respectively, are the unit vectors normal to the surface S of the body D_i and to the surfaces S'_u of the thin wires, $R = (x^2 + y^2 + z^2)^{1/2}$ and $\mathbf{a} \times \mathbf{b}$ is vector product.

The gist of the suggested method is the following. We introduce two auxiliary surfaces S_i and S_e which are homothetic, with the center at the point O, to the surface S of magnetodielectric body. The surface $S_e = K_e S$ is located inside of the body D_i and is characterized by homothety coefficient K_e smaller than unity; the surface $S_i = K_i S$ lies outside of the body D_i and is characterized by the similarity coefficient K_i greater than unity. If $K_e = K_i = 1$ the auxiliary surfaces S_e and S_i coincide with S.

Let us choose a finite set of points $\{M_{n,e}\}_{n=1}^{N_e}$ on the auxiliary surface S_e and place of each point $M_{n,e}$ a pair of independent auxiliary elementary electric dipoles with momenta $\mathbf{p}_{\tau_1}^{n,e} = p_{\tau_1}^{n,e}\mathbf{e}_{\tau_1}^{n,e}$ and $\mathbf{p}_{\tau_2}^{n,e} = p_{\tau_2}^{n,e}\mathbf{e}_{\tau_2}^{n,e}$. Analogously, on the auxiliary surface S_i we choose a set of points $\{M_{n,i}\}_{n=1}^{N_i}$ and place at each point $\{M_{n,i}$ a pair of independent auxiliary elementary electric dipoles with momenta $\mathbf{p}_{\tau_1}^{n,i} = p_{\tau_1}^{n,i}\mathbf{e}_{\tau_1}^{n,i}$ and $\mathbf{p}_{\tau_2}^{n,i} = p_{\tau_2}^{n,i}\mathbf{e}_{\tau_2}^{n,i}$. Unit vectors $\mathbf{e}_{\tau_1}^{n,e}$ and $\mathbf{e}_{\tau_2}^{n,e}$ lie in the plane tangential to S_e at the point $M_{n,e}$ and unit vectors $\mathbf{e}_{\tau_1}^{n,i}$, $\mathbf{e}_{\tau_2}^{n,i}$ lie in the plane tangential to S_i at the point $M_{n,i}$. It is assumed that dipoles placed on S_e radiate into the homogeneous medium with parameters ε_e, μ_e and dipoles placed on S_i radiate into the homogeneous with parameters ε_i, μ_i.

We also introduce a continuously distributed auxiliary current $\mathbf{J}_u$ on the axis of each thin wire.

Now we represent the unknown scattered field $\{\mathbf{E}_e, \mathbf{H}_e\}$ in D_e as a sum of the auxiliary dipoles located on S_e and auxiliary currents:

$$\mathbf{E}_e(M) = \frac{i\omega}{k_e^2}\{\sum_{n=1}^{N_e}\nabla\times(\nabla\times\mathbf{\Pi}_{n,e}) + \sum_{u=1}^{U}\nabla\times(\nabla\times\mathbf{\Pi}_u)\},$$

$$\mathbf{H}_e(M) = \frac{1}{\mu_e}\{\sum_{n=1}^{N_e}\nabla\times\mathbf{\Pi}_{n,e} + \sum_{u=1}^{U}\nabla\times\mathbf{\Pi}_u\}, \mathbf{\Pi}_{n,e} = \Psi_e(M, M_{n,e})\mathbf{p}_\tau^{n,e}, \quad (4)$$

$$\mathbf{p}_\tau^{n,e} = p_{\tau_1}^{n,e}\mathbf{e}_{\tau_1}^{n,e} + p_{\tau_2}^{n,e}\mathbf{e}_{\tau_2}^{n,e}, \mathbf{\Pi}_u = \int_{l_u}\Psi_e(M, M_{l,u})\mathbf{J}_u dl, M \in D_e,$$

and express field $\mathbf{E}_i$, $\mathbf{H}_i$ in D_i as a sum of fields of auxiliary dipoles placed on the auxiliary surface S_i

$$\mathbf{E}_i(M) = \frac{i\omega}{k_i^2}\{\sum_{n=1}^{N_i}\nabla\times(\nabla\times\mathbf{\Pi}_{n,i})\}, \mathbf{H}_i(M) = \frac{1}{\mu_i}\sum_{n=1}^{N_i}\nabla\times\mathbf{\Pi}_{n,i}, \quad (5)$$

$$\mathbf{\Pi}_{n,i} = \Psi_i(M, M_{n,i})\mathbf{p}_\tau^{n,i}, \mathbf{p}_\tau^{n,i} = p_{\tau_1}^{n,i}\mathbf{e}_{\tau_1}^{n,i} + p_{\tau_2}^{n,i}\mathbf{e}_{\tau_2}^{n,i}, M \in D_i.$$

Here, $k_e = \omega\sqrt{\varepsilon_e\mu_e}$ and $k_i = \omega\sqrt{\varepsilon_i\mu_i}$, $\Psi_e(M, M_{n,e}) = \exp(ik_e R_{MM_{n,e}})/4\pi R_{MM_{n,e}}$, $\Psi_e(M, M_{l,u}) = \exp(ik_e R_{MM_{l,u}})/4\pi R_{MM_{l,u}}$, $\Psi_i(M, M_{n,i}) = \exp(ik_i R_{MM_{n,i}})/4\pi R_{MM_{n,i}}$; $R_{MM_{n,e}}$ and $R_{MM_{l,u}}$, respectively, are the distances from the points $M_{n,e}$ inside the body D_i and the points $M_{l,u}$ on the wire axes to the observation point M in the region D_e, $R_{MM_{n,i}}$ is the distance from points $M_{n,i}$ on S_i to the point M in D_i; N_e and N_i are the numbers of dipoles placed on S_e and S_i, respectively; $p_{\tau_1}^{n,e}$, $p_{\tau_2}^{n,e}$ ($n = 1, 2, \ldots, N_e$) and $p_{\tau_1}^{n,i}$, $p_{\tau_2}^{n,i}$ ($n = 1, 2, \ldots, N_i$) are unknown dipole moments; $\mathbf{J}_u$ ($u = 1, 2, \ldots, U$) are unknown auxiliary currents. The integration is performed along axes of the wires l_u.

Fields (4)-(5) satisfy Maxwell's equations (1) and radiation conditions (3), we should properly select the dipole moments $p_{T_1}^{n,e}$, $p_{T_2}^{n,e}$ ($n = 1, 2, \ldots, N_e$) and $p_{T_1}^{n,i}$, $p_{T_2}^{n,i}$ ($n = 1, 2, \ldots, N_i$), and the axial-current distributions $\mathbf{J}_u$ ($u = 1, 2, \ldots, U$).

Let us use the piecewise-constant approximation for the axial currents. We divide the line l_u of each current $\mathbf{J}_u$ in $N_u$small intervals in which the current can be considered constant. Then the formula for $\mathbf{\Pi}_u$ in equation (4) can be represented in the following approximate form:

$$\mathbf{\Pi}_u = \sum_{i=1}^{N_u} J_{u,i} \mathbf{e}_{u,i} \int_{l_{i-1,u}}^{l_{i,u}} \Psi_e(M, M_{l,u}) dl, \tag{6}$$

where $J_{u,i}$ is the current in the i-th interval of the wire with number u and $\mathbf{e}_{u,i}$ is the unit vector directed along the tangent to the central point of the considered interval. Within the framework of such an approach, the problem of determination of the unknown axial-current distributions is reduced to the problem of finding $\sum\limits_{u=1}^{U} N_u$ current elements.

To find the dipole moments and the current elements, we use boundary conditions (2) which are satisfied according to the following method. Let M_j, where ($j = 1, 2, \ldots, L$) and M'_j, where ($j = 1, 2, \ldots, L_u$), be the collocation points on the surfaces S and S'_u, respectively. Note that within the framework of the conventional approach to an analysis of thin conductors, the azimuthal component of the surface current is neglected compared with the longitudinal component. Then the unknown quantities $p_{T_1}^{n,e}$, $p_{T_2}^{n,e}$ ($n = 1, 2, \ldots, N_e$) and $p_{T_1}^{n,i}$, $p_{T_2}^{n,i}$ ($n = 1, 2, \ldots, N_i$) and $J_{u,i}$ ($u = 1, 2, \ldots, U, i = 1, 2, \ldots, N_u$) can be found from the following system of linear algebraic equations:

$$\mathbf{n}^j \times (\mathbf{E}_i^j - \mathbf{E}_e^j) = \mathbf{n}^j \times \mathbf{E}_0^j, \mathbf{n}^j \times (\mathbf{H}_i^j - \mathbf{H}_e^j) = \mathbf{n}^j \times \mathbf{H}_0^j, j = 1, 2, \ldots, L, \tag{7}$$

$$E_{e,u,l}^j = -E_{0,u,l}^j, u = 1, 2, \ldots, U, j = 1, 2, \ldots, L_u,$$

where $\mathbf{n}_j$ is the unit normal vector to the point M_j on the surface of dielectric body; $\mathbf{E}_e^j$, $\mathbf{H}_e^j$ and $\mathbf{E}_i^j$, $\mathbf{H}_i^j$ are the vectors of scattered fields (4) and (5), respectively, at the collocation point M_j; $\mathbf{E}_0^j$ and $\mathbf{H}_0^j$ are the vectors of exciding field at M_j; $E_{e,u,l}$ and $E_{0,u,l}$ are components of scattered and incident fields directed along the axis of u-th wire at collocation points on it's surface.

The solution of system (7) is found by minimizing the functional

$$\Phi = \sum_{j=1}^{L} \{|\mathbf{n}^j \times (\mathbf{E}_i^j - \mathbf{E}_e^j) - \mathbf{n}^j \times \mathbf{E}_0^j|^2 + \frac{\mu_e}{\varepsilon_e} |\mathbf{n}^j \times (\mathbf{H}_i^j - \mathbf{H}_e^j) - \mathbf{n}^j \times \mathbf{H}_0^j|^2\} + \tag{8}$$

$$+ \sum_{u=1}^{U} \sum_{j=1}^{L_u} |E_{e,u,l}^j + E_{0,u,l}^j|^2.$$

After solving the minimization problem by conjugate gradients method, we determine from eq. (4) the required parameters of the scattered field.

The accuracy of the solution is controlled by calculating the relative norm of the discrepancy of boundary conditions on the surfaces of the dielectric body and thin wires determined by the expression

$$\Delta = (\Phi' / \Phi_0)^{1/2}, \Phi_0 = \sum_{j=1}^{L'} \{ |\mathbf{n}^j \times \mathbf{E}_0^j|^2 + \frac{\mu_e}{\varepsilon_e} |\mathbf{n}^j \times \mathbf{H}_0^j|^2 \} + \sum_{u=1}^{U} \sum_{j=1}^{L'_u} |E_{0,u,l}^j|^2, \quad (9)$$

where Φ' is value of functional (8) provided that a grid of intermediate points is chosen instead of collocation points j, L' is the number of intermediate points on the surface of a dielectric body, and L'_u is the number of intermediate points on the surface of a wire with number u.

3   Numerical Results

Based on the method described in section 2, we developed a code for calculating the scattered-field components and controlling the accuracy of the obtained solution. Using this code we obtained the characteristics of scattering by certain structures and the distributions of currents along thin wires in the presence of nearly three-dimensional magnetodielectric body. Some of the obtained results are presented below.

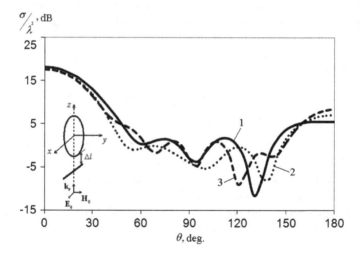

Fig. 2. The effect of thin wires with length $l = 1,6\lambda$ and radius $r_0 = 0,02\lambda$ on bistatic cross-section σ for the ellipsoid with semiaxes $a = b = 0,64\lambda; c = 0,8\lambda$ and parameters $\varepsilon_i/\varepsilon_e = 4, \mu_i/\mu_e = 1$. Curve 1 corresponds to the case where thin wire is absent, *curve 2* - to the case where the distance Δl between the ellipsoid and the wire is equal to $0,05\lambda$ and *curve 3* - to the case where the distance Δl between the ellipsoid and the wire is equal to $0,5\lambda$.

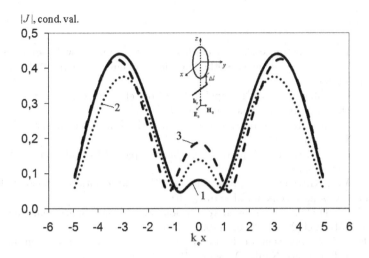

Fig. 3. The current distribution along the wire for the same distances between ellipsoid and the wire are shown. *Curve 1* corresponds to the case where the ellipsoid is absent, *curve 2* - to $\Delta l = 0,05\lambda$ and *curve 3* - to $\Delta l = 0,5\lambda$.

Figure 2 illustrated the effect of thin wires with length $l = 1,6\lambda$ and radius $r_0 = 0,02\lambda$ on bistatic cross-section σ for the ellipsoid with semiaxes $a = b = 0,64\lambda; c = 0,8\lambda$ and parameters $\varepsilon_i/\varepsilon_e = 4, \mu_i/\mu_e = 1$. The semiaxes a, b and c were directed along the x, y and z-axes of the Cartesian system of coordinates, and the wire was directed along the x-axes. A linearly polarizated plane wave exciting the ellipsoid and the wire was propagated along z-axes, and the vector $\mathbf{E}_0$ was directed along the x-axes. The angle θ (see Fig. 1) is plotted on the abscissa in Fig. 2, and the quantity σ/λ^2 , in dB is plotted on the ordinate. Curve 1 corresponds to the case where thin wire is absent, curve 2 - to the case where the distance Δl between the ellipsoid and the wire is equal to $0,05\lambda$ and curve 3 - to the case where the distance Δl between the ellipsoid and the wire is equal to $0,5\lambda$. The results are given in the E–plane (the plane in which vectors $\mathbf{E}_0$ and $\mathbf{k}_e$ lie). In Figure 3 the current distribution along the wire for the same distances between ellipsoid and the wire are shown. Curve 1 corresponds to the case where the ellipsoid is absent, curve 2 - to $\Delta l = 0,05\lambda$ and curve 3 - to $\Delta l = 0,5\lambda$. For results presented in figures 2 and 3 the parameters of the method were chosen: $K_e = 0,6, K_i = 4, N_e = N_i = 484, N_u = 64, L_u = 256$.

The conclusions can be drawn from the results presented in Fig.2 and Fig.3. The presence of the thin wire changes the bistatic cross-section of the magnetodielectric body. The maximum effect of the thin wire is observed in the directions $100 < \theta < 160$. Bistatic cross-section of the structure depends on distance between the magnetodielectric body and the thin wire.

References

1. Hughes, T.J.R.: The finite element method. Prentice Hall, New Jersey (1987)
2. Jin, J.-M.: The finite element method in electromagnetics. John Wiley & Sons, New York (1993)
3. Volakis, J.L., Chatterjee, A., Kempel, L.C.: Finite element method for electromagnetics. Wiley-IEEE Press, New York (1998)
4. Colton, D., Kress, R.: Integral Equation Methods in Scattering Theory. John Wiley & Sons, New York (1983)
5. Khizhnyak, N.A.: Integral Equations of Macroscopical Electrodynamics. Naukova dumka, Kyev (1986)
6. Il'insky, A.S., Kravtsov, V.V., Sveshnikov, A.G.: Mathematical models of electrodynamics. Vysshaya schkola, Moscow (1991)
7. Kupradze, V.D.: On the approximate solution of problems in mathematical physics. Success Math. Sci. 22(2), 59–107 (1967)
8. Alexidze, M.A.: Solution of Boundary Problems by Decomposition on Nonorthogonal Functions. Nauka, Moscow (1978)
9. Dmitrenko, A.G., Kolchin, V.A.: Numerical methods in electromagnetic scattering theory. Journal of Quantitative Spectroscopy & Radiative Transfer 79-80, 775–824 (2003)
10. Dmitrenko, A.G., Keller, Y.A.: Computer modelling of electromagnetic scattering by structures comprising several thin wires. Tomsk State University Journal 290, 150–157 (2006)

Model Predictive Control
for Discrete-Time Linear Systems
with Time Delays and Unknown Input*

Marina Kiseleva and Valery Smagin

Department of Applied Mathematics and Cybernetics,
National Research Tomsk State University, 36 Lenin str., 634050 Tomsk, Russia
kiselevamy@gmail.com, vsm@mail.tsu.ru

Abstract. The paper deals with Model Predictive Control synthesis
based on the system output tracking with control and state delays. Input and state constraints are taken into account when solving the MPC
problem for systems with unknown input. A prediction is carried out on
the base of object states estimates that is obtained by the Kalman filter.
The criteria function is assumed to be convex quadratic. The proposed
algorithm allows to get around the state space extension.

Keywords: Model Predictive Control, discrete systems, state delay,
input delay, unknown input, Kalman filter.

1 Introduction

One of the modern formalized approaches to the system control synthesis based
on mathematical methods of optimizadtion is Dynamic object control theory
with predictive models - Model Predictive Control (MPC).

This approach began to develop in the early 1960s. It was destined for the
process control in petrochemical and energy industries for which the application
of traditional synthesis methods was extremely complicated according to mathematical models complication. For the last years, field of MPC application has
been considerably extended covering technologic fields for object with time delay
[1-6], inventory control [7-8]; and portfolio control and optimization [9].

The paper is devoted to Model Predictive Control synthesis based on the
system output tracking allowing for input and state delays. It has been suggested
to make a synthesis of predictive control using estimates of unknown input that
can be evaluated on the base of modified LSM [10-12].

A new algorithm proposed in the paper allows to include control and state
delays into the model getting around the state space extension. This reduces the
dimension of the block matrices used in the algorithm significantly.

---

* Supported by Tomsk State University Competitiveness Improvement Program, and
the project "Goszadanie Minobrnauki Rossii".

A. Dudin et al. (Eds.): ITMM 2014, CCIS 487, pp. 181–188, 2014.

2   Problem Statement

Suppose the object can be described by the following system of linear-difference equations:

$$x_{t+1} = Ax_t + \sum_{i=1}^{s} A_i x_{t-i} + Bu_{t-h} + Ir_t + w_t,$$
$$x_k = \bar{x}_k, (k = \overline{-s,0}), u_i = \bar{u}_i, (i = \overline{-h,-1}), \tag{1}$$
$$\psi_t = Hx_t + v_t, \tag{2}$$
$$y_t = Gx_t. \tag{3}$$

Here $x_t \in R^n$ is the object state ($x_k = \bar{x}_k, k = -s, \ldots, -1, 0, \bar{x}_k$ is considered to be given), $u_k \in R^m$ is the control input ($u_k = \bar{u}_k, k = -h, \ldots, -1, \bar{u}_k$ is given), $r_t \in R^q$ is the unknown input, $y_t \in R^p$ is the output (to be controlled), $\psi_t \in R^l$ is the observation (measured output), s, h are the state and conrol input delay values respectively. Further, the state noise w_t and measurement noise v_t are assumed to be Gaussian distributed with zero mean and covariances W and V respectively, i.e. $M\{w_t w_k^T\} = W\delta_{t,k}$, $M\{v_t v_k^T\} = V\delta_{t,k}$, where $\delta_{t,k}$ is the Kronecker delta.

In the simple case, when r_t is a zero-mean random vector with the known variance, the optimal filtering problem for the model (1)-(3) comes to the Kalman filtering algorithm. If the input r_t is a deterministic component and its evolution in time is governed by the known linear system, the optimal estimates of r_t and x_t can be obtained using the extended state Kalman filter. In this paper we consider the case when prior knowledge about the time evolution of r_t is not available. Vector r_t is supposed to be completely unknown.

The model under consideration is used to make predictions about the plant behavior over the prediction horizon denoted by N using information (measurements of inputs and outputs) up to and including the current time t. The plant is supposed to operate under the constrained conditions:

$$a_1 \leq S_1 x_t \leq a_2, \tag{4}$$
$$\phi_1(x_{t-h}) \leq S_2 u_{t-h} \leq \phi_2(x_{t-h}). \tag{5}$$

Here S_1 and S_2 are structural matrices that are composed of zeros and units, identifying constrained components of vectors x_t and u_t ; a_1, a_2, $\phi_1(x_t)$, $\phi_2(x_t)$ are given constant vectors and vector-functions. The problem is to determine an acting strategy on the base of the observation ψ_t according to which the output vector of the system y_t will be close to the reference taking into account constraints on the state and control input.

3   Prediction

With the Gaussian assumptions on the state and the measurement noise it is possible to make optimal (in the minimum variance sense) predictions of state and output using a Kalman filter, see e.g. [13].

Let $\hat{x}_{i|j}$ and $\hat{y}_{i|j}$ to be estimates of the state and the output at time i giving information up to and including time j where $j \leq i$. Then

$$\hat{x}_{t+1|t} = A\hat{x}_{t|t-1} + \sum_{i=1}^{s} A_i \hat{x}_{t-i|t-i-1} + Bu_{t-h} + I\hat{r}_t + K_t(\psi_t - H\hat{x}_{t|t-1}),$$

$$\hat{x}_{t|t-1} = \bar{x}_k, k = \overline{-s, 0},$$

$$\hat{y}_{t+1|t} = G\hat{x}_{t+1|t},$$

$$K_t = AP_t H^{\mathrm{T}}(HP_t H^{\mathrm{T}} + V)^{-1},$$

$$P_{t+1} = W + AP_t A^{\mathrm{T}} + AP_t H^{\mathrm{T}}(HP_t H^{\mathrm{T}} + V)^{-1} HP_t A^{\mathrm{T}}, P_0 = P_{x_0}, \qquad (6)$$

where P_{x_0} is the given initial value of the variance matrix. Equation (6) for P_t is known as the discrete-time Riccati-equation.

Evaluate estimates of the unknown input using LSM [10] in order to develop a pedictive model. In this case there is no need to know a behavioral model of the unknown input. Let evaluate state predictions as a result of solving a new optimal control problem where by "control" we will mean the unknown input $\hat{r}_t$. The following quadratic function is proposed to use as an optimal criterion:

$$J(\hat{r}_{t-1}) = \sum_{i=1}^{t} \{\|\psi_t - H\hat{x}_{i|i-1}\|_{C_R}^2 + \|\hat{r}_{i-1}\|_{D_R}^2\}, \qquad (7)$$

where C_R and D_R are symmetric positive definite matrices.

Optimization of the criterion (7) up to the current time t comes to the criterion minimization in each time $i = \overline{1, t}$.

$$J(\hat{r}_{t-1}) = \min_{\hat{r}_0} \min_{\hat{r}_1} \ldots \min_{\hat{r}_{t-1}} \sum_{i=1}^{t} \{\|\psi_t - H\hat{x}_{i|i-1}\|_{C_R}^2 + \|\hat{r}_{i-1}\|_{D_R}^2\}.$$

An optimal estimate of the unknown input at the first step $(t = 1)$:

$$J(\hat{r}_0) = \min_{\hat{r}_0} \{\|\psi_1 - H\hat{x}_{1|0}\|_{C_R}^2 + \|\hat{r}_0\|_{D_R}^2\}.$$

Taking into account $\hat{x}_{1|0} = Ax_0 + Bu_0 + I\hat{r}_0$, we get the following:

$$J(\hat{r}_0) = \min_{\hat{r}_0} \{\|\psi_1 - HAx_0 - HBu_0 - HI\hat{r}_0\|_{C_R}^2 + \|\hat{r}_0\|_{D_R}^2\}. \qquad (8)$$

After some manipulations, we have:

$$J(\hat{r}_0) = \min_{\hat{r}_0} \{\hat{r}_0^{\mathrm{T}}(I^{\mathrm{T}} H^{\mathrm{T}} C_R HI + D_R)\hat{r}_0 -$$

$$-2\hat{r}_0^{\mathrm{T}} I^{\mathrm{T}} H^{\mathrm{T}} C_R(\psi_1 - HAx_0 - HBu_0) + \alpha_0\},$$

where α_0 - variable independent of $\hat{r}_0$.

An optimal estimate of the unknown input at the 1st instant can be found from the following condition

$$\frac{\partial J(\hat{r}_0)}{\partial \hat{r}_0} = 2(I^{\mathrm{T}} H^{\mathrm{T}} C_R HI + D_R)\hat{r}_0 - 2I^{\mathrm{T}} H^{\mathrm{T}} C_R(\psi_1 - HAx_0 - HBu_0) = 0,$$

and have the following expression:

$$\hat{r}_0 = S_R(\psi_1 - HAx_0 - HBu_0),$$

where $S_R = (I^T H^T C_R H I + D_R)^{-1} I^T H^T C_R$. We can get criterion's value at the instant $t = 1$ using the obtained expression for $\hat{r}_0$ in (8), ,

$$J(\hat{r}_0) = (\psi_1 - HAx_0 - HBu_0)^T M_R(\psi_1 - HAx_0 - HBu_0),$$

where $M_R = C_R - 2C_R H I S_R + S_R^T (I^T H^T C_R H I + D_R) S_R$.

At the instant $t = 2$ an optimal estimate of the unknown input is found by optimizing the following function:

$$J(\hat{r}_1) = \min_{\hat{r}_0} \min_{\hat{r}_1} \{ \|\psi_2 - H\hat{x}_{2|1}\|^2_{C_R} + \|\hat{r}_1\|^2_{D_R} + \|\psi_1 - H\hat{x}_{1|0}\|^2_{C_R} + \|\hat{r}_0\|^2_{D_R} \}.$$

Expression for $J(\hat{r}_1)$ can be rearranged in the following way using the Bellman's optimality principle,

$$J(\hat{r}_1) = \min_{\hat{r}_1} \{ \|\psi_2 - H\hat{x}_{2|1}\|^2_{C_R} + \|\hat{r}_1\|^2_{D_R} + J(\hat{r}_0) \} =$$
$$= \min_{\hat{r}_1} \{ \|\psi_2 - HA\hat{x}_{1|0} - HBu_1 - HI\hat{r}_1\|^2_{C_R} +$$
$$+ \|\hat{r}_1\|^2_{D_R} + \|\psi_1 - HAx_0 - HBu_0\|^2_{M_R} \} =$$
$$= \min_{\hat{r}_1} \{ \hat{r}_1^T (I^T H^T C_R H I + D_R)\hat{r}_1 - 2\hat{r}_1^T I^T H^T C_R(\psi_2 - HA\hat{x}_{1|0} - HBu_1) + \alpha_1 \},$$

where α_0 - variable independent of $\hat{r}_1$. Differentiate with respect to $\hat{r}_1$ like in the first step and get the following:

$$\hat{r}_1 = S_R(\psi_2 - HA\hat{x}_{1|0} - HBu_1),$$
$$J(\hat{r}_1) = (\psi_2 - HA\hat{x}_{1|0} - HBu_1)^T M_R(\psi_2 - HA\hat{x}_{1|0} - HBu_1).$$

Applying the Bellman's optimality principle for the next steps and using a method of mathematical induction, we get $\hat{r}_t$:

$$\hat{r}_t = S_R(\psi_t + 1 - HA\hat{x}_{t|t-1} - HBu_t). \tag{9}$$

So, taking into account expressions for unknow input estimates, state and output prediction can be performed in accordance with the following formulas

$$\hat{x}_{t+i|t} = A^{i-1}\hat{x}_{t+1|t} + \sum_{k=1}^{i-1} A^{i-k-1} Bu_{t+k-h} + \sum_{k=1}^{i-1} A^{i-k-1} I\hat{r}_{t+k},$$
$$\hat{y}_{t+i|t} = G\hat{x}_{t+i|t}, i = \overline{1, N}, \tag{10}$$

where $u_{t+k|t}$ - the control input used for prediction, $\hat{r}_{t+k}$ - predicted unknown input estimates that can be obtained on the base of time series forecasting methods [14].

MPC usually requires estimates of the state and/or output over the entire prediction horizon N from time $t + 1$ until time $t + N$, and can only make these predictions based on information up to and including the current time

t. Equations (6) can be used to obtain $\hat{x}_{t+1|t}$, $\hat{y}_{t+1|t}$. Optimal state/output estimates from instant $t+2$ to $t+N$ can be obtained as follows

$$\hat{x}_{t+i+1|t} = A\hat{x}_{t+i|t} + \sum_{j=1}^{s} A_j \hat{x}_{t+i-j|t-j-1} + Bu_{t-h+i|t} + I\hat{r}_{t+i}, \qquad (11)$$

$$\hat{y}_{t+i|t} = G\hat{x}_{t+i|t}, i = \overline{1, N}. \qquad (12)$$

In the above the notation $u_{t-h+i|t}$ is used to distinguish the actual input at the instant $t+i$, namely u_{t-h+i}, from that used for prediction purposes, namely $u_{t-h+i|t}$.

Equation (11) can be expanded in terms of the initial state $\hat{x}_{t+1|t}$ and future control actions $u_{t-h+i|t}$ as follows

$$\hat{x}_{t+i|t} = A^{i-1}\hat{x}_{t+1|t} + \sum_{k=1}^{i-1} A^{i-k-1} \sum_{j=1}^{s} A_j \hat{x}_{t+k-j|t-j-1} +$$
$$+ \sum_{k=1}^{i-1} A^{i-k-1} Bu_{t-h+k|t} + \sum_{k=1}^{i-1} A^{i-k-1} I\hat{r}_{t+k}, i = \overline{1, N}. \qquad (13)$$

Now in terms of predicting the output, equation (12) can be expanded in terms of the above expression for $\hat{x}_{t+i|t}$, which results in series of equations that provide optimal output predictions. The key point to note is that each output prediction is a function of the initial state $\hat{x}_{t+1|t}$ and future inputs $u_{th+i|t}$ only:

$$\hat{y}_{t+i|t} = GA^{i-1}\hat{x}_{t+1|t} + G\sum_{k=1}^{i-1} A^{i-k-1} \sum_{j=1}^{s} A_j \hat{x}_{t+k-j|t-j-1} +$$
$$+ G\sum_{k=1}^{i-1} A^{i-k-1} Bu_{t-h+k|t} + G\sum_{k=1}^{i-1} A^{i-k-1} I\hat{r}_{t+k}, i = \overline{1, N}. \qquad (14)$$

These series of prediction equations can be stated in an equivalent manner using matrix vector notation. Denote

$$\hat{X}_t = \begin{bmatrix} \hat{x}_{t+1|t} \\ \vdots \\ \hat{x}_{t+N|t} \end{bmatrix}, \hat{X}_i^0 = \begin{bmatrix} \hat{x}_{t+1-i|t-i} \\ \vdots \\ \hat{x}_{t+N-i|t-i} \end{bmatrix}, i = \overline{1, s}, \hat{Y}_t = \begin{bmatrix} \hat{y}_{t+1|t} \\ \vdots \\ \hat{y}_{t+N|t} \end{bmatrix}, \hat{R}_t = \begin{bmatrix} \hat{r}_{t+1} \\ \vdots \\ \hat{r}_{t+N} \end{bmatrix},$$

$$U_{t-h} = \begin{bmatrix} u_{t-h+1|t} \\ \vdots \\ u_{t-h+N|t} \end{bmatrix}, \Psi = \begin{bmatrix} E_n \\ A \\ A^2 \\ \vdots \\ A^{N-1} \end{bmatrix}, \Lambda = \begin{bmatrix} G \\ GA \\ GA^2 \\ \vdots \\ GA^{N-1} \end{bmatrix},$$

$$\Psi_i^0 = \begin{bmatrix} 0 & 0 & 0 & \dots & 0 \\ A_i & 0 & 0 & \dots & 0 \\ AA_i & A_i & 0 & \dots & 0 \\ \vdots & \vdots & \vdots & \ddots & \vdots \\ A^{N-2}A_i & A^{N-3}A_i & \dots & A_i & 0 \end{bmatrix}, \Lambda_i^0 = \begin{bmatrix} 0 & 0 & 0 & \dots & 0 \\ GA_i & 0 & 0 & \dots & 0 \\ GAA_i & GA_i & 0 & \dots & 0 \\ \vdots & \vdots & \vdots & \ddots & \vdots \\ GA^{N-2}A_i & GA^{N-3}A_i & \dots & GA_i & 0 \end{bmatrix},$$

$$P = \begin{bmatrix} 0 & 0 & 0 & \dots & 0 \\ B & 0 & 0 & \dots & 0 \\ AB & B & 0 & \dots & 0 \\ \vdots & \vdots & \vdots & \ddots & \vdots \\ A^{N-2}B & A^{N-3}B & \dots & B & 0 \end{bmatrix}, \Phi = \begin{bmatrix} 0 & 0 & 0 & \dots & 0 \\ GB & 0 & 0 & \dots & 0 \\ GAB & GB & 0 & \dots & 0 \\ \vdots & \vdots & \vdots & \ddots & \vdots \\ GA^{N-2}B & GA^{N-3}B & \dots & GB & 0 \end{bmatrix},$$

$$S = \begin{bmatrix} 0 & 0 & 0 \ldots 0 \\ I & 0 & 0 \ldots 0 \\ AI & I & 0 \ldots 0 \\ \vdots & \vdots & \vdots \ddots \vdots \\ A^{N-2}I & A^{N-3}I & \ldots & I & 0 \end{bmatrix}, Q = \begin{bmatrix} 0 & 0 & 0 \ldots 0 \\ GI & 0 & 0 \ldots 0 \\ GAI & GI & 0 \ldots 0 \\ \vdots & \vdots & \vdots \ddots \vdots \\ GA^{N-2}I & GA^{N-3}I & \ldots & GI & 0 \end{bmatrix}. \quad (15)$$

Here E_n is the n-by-n identity matrix.

The predictive model (13)-(14) in matrix notation is as follows

$$\hat{X}_t = \Psi \hat{x}_{t+1|t} + \sum_{i=1}^{s} \Psi_i^0 \hat{X}_i^0 + PU_{t-h} + S\hat{R}_t,$$

$$\hat{Y}_t = \Lambda \hat{x}_{t+1|t} + \sum_{i=1}^{s} \Lambda_i^0 \hat{X}_i^0 + \Phi U_{t-h} + Q\hat{R}_t. \quad (16)$$

4   Model Predictive Control Synthesis

It is proposed to use the following criterion in order to solve the posed problem

$$J(t) = \frac{1}{2} \sum_{k=1}^{N} \{\|\hat{y}_{t+k|t} - \bar{y}_t\|_C^2 + \|u_{t-h+k|t} - u_{t-h+k-1|t}\|_D^2\}, \quad (17)$$

where weighing matrices C and D are assumed to be symmetric and positive definite.

In case when the reference trajectory $\bar{y}_{t+k}$ is unknown for $k \geq 0$ it is reasonable to assume that $\bar{y}_{t+k} = \hat{y}_t$, i.e. the same reference point is held throughout the entire prediction horizon.

The summation terms in (17) can be expanded to offer a quadratic objective function in terms of $\bar{x}_{t+1|t}$ and U_{t-h}. Let

$$\bar{Y}_t = \begin{bmatrix} \bar{y}_{t+1} \\ \vdots \\ \bar{y}_{t+N} \end{bmatrix}.$$

Then using (16) we can get the following expression

$$\frac{1}{2}\sum_{k=1}^{N}\|\hat{y}_{t+k|t} - \bar{y}_t\|_C^2 = \frac{1}{2}\|\hat{Y}_t - \bar{Y}_t\|_{\bar{C}}^2 = \frac{1}{2}U_{t-h}^T \Phi^T \bar{C} \Phi U_{t-h} +$$
$$+ U_{t-h}^T[\Phi^T \bar{C} \Lambda \hat{x}_{t+1|t} + \Phi^T \bar{C} \sum_{i=1}^{s} \Lambda_i^0 \hat{X}_i^0 - \Phi^T \bar{C} \bar{Y}_t] + c_1, \quad (18)$$

where c_1 is a constant term that does not depend either on U_{t-h} or $\hat{x}_{t+1|t}$; and $\bar{C}$ is given by

$$\bar{C} = \begin{bmatrix} C & 0 & \ldots & 0 \\ 0 & C & \ldots & 0 \\ \vdots & \vdots & \ddots & \vdots \\ 0 & 0 & \ldots & C \end{bmatrix}.$$

In a similar manner rearrange the second term of sum in (17)

$$\frac{1}{2}\sum_{k=1}^{N}\|u_{t-h+k|t} - u_{t-h+k-1|t}\|_D^2 = \frac{1}{2}U_{t-h}^T\bar{D}U_{t-h} - u_{t-h+1|t}^T Du_{t-h} + c_2, \quad (19)$$

where c_2 is a constant term that does not depend on u_{t-h+k} $(k = \overline{1,N})$; and $\bar{D}$ is given by

$$\bar{D} = \begin{bmatrix} 2D & -D & 0 & \cdots & 0 \\ -D & 2D & -D & \cdots & 0 \\ \vdots & \ddots & \ddots & \ddots & \vdots \\ 0 & \cdots & -D & 2D & -D \\ 0 & 0 & \cdots & -D & 2D \end{bmatrix}.$$

Combining the above, the criteria function can be expressed as

$$J(t) = \frac{1}{2}U_{t-h}^T FU_{t-h} + U_{t-h}^T f + c_3. \quad (20)$$

Here c_3 is the combination of previous constant terms c_1 and c_2 and may be safely ignored. The terms F and f are given by

$$F = \Phi^T\bar{C}\Phi + \bar{D}, f = \Gamma\left[\begin{array}{c} \hat{x}_{t+1|t} \\ \sum_{i=1}^{s}\Lambda_i^0\hat{X}_i^0 \\ \hat{Y}_t \end{array}\right] - \begin{bmatrix} Du_{t-h} \\ 0 \\ \vdots \\ 0 \end{bmatrix}, \Gamma = \begin{bmatrix} \Phi^T\bar{C} & \Lambda\bar{C}Q & -\Phi^T\bar{C} \end{bmatrix}.$$

In the absence of constraints an analytical solution of the posed problem can be obtained from the condition $\frac{dJ}{dU_{t-h}} = 0$ using vector derivative formulas, see e.g. [15]:

$$\frac{\partial J}{\partial U_{t-h}} = \frac{\partial J}{\partial U_{t-h}}\left[\frac{1}{2}U_{t-h}^T FU_{t-h} + U_{t-h}^T f + c_3\right] =$$

$$= \frac{1}{2}\frac{\partial(trFU_{t-h}U_{t-h}^T)}{\partial U_{t-h}} + \frac{\partial(U_{t-h}^T f)}{\partial U_{t-h}} = \frac{1}{2}[F^T U_{t-h} + FU_{t-h}] + f = 0. \quad (21)$$

As the matrix F is symmetric, the equation (21) can be expressed as follows

$$FU_{th} + f = 0.$$

So, the criteria function can be rearranged as

$$U_{t-h}^* = -(\Phi^T\bar{C}\Phi + \bar{D})^{-1}(\Phi^T\bar{C}\Lambda\hat{x}_{t+1|t} + \Phi^T\bar{C}Q\hat{R}_t - \Phi^T\bar{C}\hat{Y}_t) - \begin{bmatrix} Du_{t-h} \\ 0 \\ \vdots \\ 0 \end{bmatrix},$$

and the optimal predictive control has the form:

$$u_{t-h+1|t}^* = \begin{bmatrix} E_n & 0 \ldots 0 \end{bmatrix} U_{t-h}^*.$$

Optimization of the model with constraints (4), (5) can be performed numerically using Matlab function quadprog.

5 Conclusion

The Model Predictive Control of the system allowing to state and input delays with unknown input is solved, guaranteeing constraints satisfaction and feasibility. The problem of the MPC synthesis is solved without the extension of the state space. The extrapolator is offered to use in order to obtain predicted values of the system output.

References

1. Maciejowski, J.M.: Predictive control with constraints. Prentice Hall (2002)
2. Camacho, E.F., Bordons, C.: Model predictive control. Springer, London (2004)
3. Jeong, S.C., Park, P.: Constrained MPC algorithm for uncertain time-varying systems with state-delay. IEEE Transactions on Automatic Control 50(2), 257–263 (2005)
4. Qin, S.J., Badgwell, T.A.: A survey of industrial model predictive control technology. Control Engineering Practice 11, 733–764 (2003)
5. Kubacik, M., Bobal, V.: Predictive Control of Time-delay Processes. In: Proc. 26th European Conference on Modelling and Simulation, pp. 1–6 (2012)
6. Kiseleva, M.Y., Smagin, V.I.: Model Predictive Control of Discrete Systems with State and Input Delays. Tomsk State University Journal of Control and Computer Science 1(14), 5–12 (2011)
7. Conte, P., Pennesi, P.: Inventory control by model predictive control methods. In: Proc. 16th IFAC World Congress, Prague, pp. 1–6 (2005)
8. Aggelogiannaki, E., Doganis, P., Sarimveis, H.: An Adaptive Model Predictive Control configuration for Production-Inventory Systems. Int. Jour. of Production Economics 114(13), 165–178 (2008)
9. Dombrovskii, V., Obyedko, U.: Predictive control of systems with Markovian jumps under constraints and its application to the investment portfolio optimization. Automation and Remote Control 72(5), 989–1003 (2011)
10. Janczak, D., Grishin, Y.: State estimation of linear dynamic system with unknown input and uncertain observation using dynamic programming. Control and Cibernetics 35(4), 851–862 (2006)
11. Gillijns, S., Moor, B.: Unbiased minimum-variance input and state estimation for linear discrete-time systems. Automatica 43, 111–116 (2007)
12. Hsieh, C.-S.: On the optimality of two-stage Kalman filtering for systems with unknown input. Asian Journal of Control 12(4), 510–523 (2010)
13. Brammer, K., Siffling, G.: Kalman-Bucy Filters. Artech House, Inc., Norwood (1989)
14. Box, G.E.P., Jenkins, G.M.: Time Series Analysis: Forecasting and Control. Holden-Day, San Francisco (1976)
15. Athans, M.: The matrix minimum principle. Information and Control 11(5/6), 592–606 (1968)

Diffusion Appoximation in Inventory Management with Examples of Application[*]

Anna Kitaeva, Valentina Subbotina, and Oleg Zmeev

National Research Tomsk Polytechnic University, Lenin Avenue 30,
National Research Tomsk State University, Lenin Avenue 36, 634050 Tomsk, Russia
kit1157@yandex.ru, valsubbotina@mail.ru, ozmeev@gmail.com

Abstract. Single-product inventory management model with both random and controllable demand and continuous input product flow with fixed uncontrolled rate under finite storage capacity is considered. We consider the stock level process as asymptotically diffusion process and obtain its stationary distribution. An application of the approximation to on/off inventory control is considered and simulation results are given.

Keywords: Stochastic demand, diffusion approximation, inventory management, on/off control.

1 The Problem Statement

Diffusion methods have been applied in a variety of domains, see Janssen, Manca, and Manca [1]; as to application to inventory models, see, for example, Bather [2], Harrison [3], and Puterman [4]. Nowadays a set of stochastic models are available to solve the inventory control problem under various conditions encountered in practice, for example, see Chopra and Meindl [5], and Beyer, Cheng, Sethi, and Taksar [6].

We consider the following stochastic inventory model. Let the product flow be continuous with fixed rate ν_0, the demands be a Poisson process with constant intensity λ, the values of purchases be i.i.d. random variables having a distribution $F(\cdot)$ with finite the first and second moments equals respectively a_1 and a_2. Under certain conditions (for example, a threat of overflow) the product is delivered to outlets and the output flow is assumed to be continuous with a rate $\nu_1(Q)$.

Let $Q(t)$ denotes the level of inventory at time t. We consider the diffusion approximation of Marcovian process $Q(\cdot)$.

The paper consists of three parts: the first part is devoted to the approximation, in the second part we solve the optimization problem using the approximation, and in the third part the results of simulation are given.

---

[*] This work is performed under the state order No. 1.511.2014/K of the Ministry of Education and Science of the Russian Federation.

A. Dudin et al. (Eds.): ITMM 2014, CCIS 487, pp. 189–196, 2014.

2   The Diffusion Appoximation

Let the density exists $P(Q,t) = \dfrac{Pr\,[Q \leq Q(t) < Q + dQ]}{dQ}$, and the rate of the products movement due to non-random factors $\nu_0 - \nu_1(Q) = \nu(Q)$.

Theorem 1. *Let $P(Q,t)$ be a differentiable function of t, $\nu(Q)P(Q,t)$ be a differentiable function of Q, and $\int_0^\infty P(Q,t)dF(u) < \infty$.*

Then the Kolmogorov forward equation holds

$$\frac{\partial P(Q,t)}{\partial t} = -\frac{\partial \{\nu(Q)P(Q,t)\}}{\partial Q} - \lambda P(Q,t) + \lambda \int_0^\infty P(Q+u,t)dF(u). \quad (1)$$

Proof. Derive the Kolmogorov backward equation for a functional of Markov process $Q(t)$

$$\varphi(Q,t) = E\Big\{H\left(Q(\tau)\right) | Q(t) = Q\Big\},$$

and write the adjoint equation, which is the Kolmogorov forward equation for density function $P(Q,t)$, see Barucha-Reid [7].

Consider

$$\varphi(Q,t-\Delta t) = E\Big\{H\Big(Q(\tau)\Big) | Q(t-\Delta t) = Q\Big\} =$$

$$= (1 - \lambda\Delta t)E\Big\{H\left(Q(\tau)\right) | Q(t) = Q + \nu(Q)\Delta t\Big\} +$$

$$+\lambda\Delta t\int_0^Q E\Big\{H\left(Q(\tau)\right) | Q(t) = Q - u\Big\}dF(u) + o(\Delta t) =$$

$$= (1 - \lambda\Delta t)\varphi\left(Q + \nu(Q)\Delta t, t\right) + \lambda\Delta t\int_0^Q \varphi\left(Q - u, t\right)dF(u) + o(\Delta t) =$$

$$= \varphi\left(Q + \nu(Q)\Delta t, t\right) - \lambda\Delta t\varphi(Q,t) + \lambda\Delta t\int_0^Q \varphi\left(Q - u, t\right)dF(u) + o(\Delta t) =$$

$$= \varphi(Q,t) + \nu(Q)\Delta t\frac{\partial\varphi(Q,t)}{\partial Q} - \lambda\Delta t\varphi(Q,t) + \lambda\Delta t\int_0^Q \varphi\left(Q - u, t\right)dF(u) + o(\Delta t),$$

which implies

$$-\frac{\partial\varphi(Q,t)}{\partial Q} = \nu(Q)\frac{\partial\varphi(Q,t)}{\partial Q} - \lambda\varphi(Q,t) + \lambda\int_0^Q \varphi\left(Q - u, t\right)dF(u).$$

So the adjoint equation is

$$\frac{\partial P(Q,t)}{\partial t} = -\frac{\partial \{\nu(Q)P(Q,t)\}}{\partial Q} - \lambda P(Q,t) + \lambda \int\limits_0^\infty P\left(Q+u,t\right)dF(u).$$

The theorem is proved.

Suppose the values of $Q(\cdot)$ are large enough. The idea is to consider some infinitesimal parameter ε so that the process $\varepsilon^2 Q(\cdot)$ is not degenerate.

Denote

$$\nu_1(Q) = v\left(\varepsilon^2 Q\right), t\varepsilon^2 = \tau, Q\varepsilon^2 = x(\tau) + \varepsilon y, P(Q,t) = \varepsilon \Pi(y,\tau,\varepsilon), \qquad (2)$$

where $x(\cdot)$ is a differentiable function, and substitute (2) in (1).

Let $\lim\limits_{\varepsilon \to 0} \Pi(y,\tau,\varepsilon) = \Pi(y,\tau)$ exists.

Let $\nu_1(\cdot)$ be a differentiable function, $\Pi(y,\tau)$ be a differentiable function with respect to τ and twice differentiable with respect to y.

We obtain the equation

$$\varepsilon^2 \frac{\partial \Pi(y,\tau,\varepsilon)}{\partial \tau} - \varepsilon x'(\tau)\frac{\partial \Pi(y,\tau,\varepsilon)}{\partial y} = -\varepsilon \frac{\partial}{\partial y}\left\{\nu_1\left(x(\tau)+\varepsilon y\right)\Pi(y,\tau,\varepsilon)\right\}- \\ -\lambda \Pi(y,\tau,\varepsilon) + \lambda \int\limits_0^\infty \Pi(y+\varepsilon u,\tau,\varepsilon)dF(u). \qquad (3)$$

Rewrite (3)

$$\varepsilon^2 \frac{\partial \Pi(y,\tau,\varepsilon)}{\partial \tau} - \varepsilon x'(\tau)\frac{\partial \Pi(y,\tau,\varepsilon)}{\partial y} =$$

$$= -\varepsilon \frac{\partial}{\partial y}\left\{\left[\nu\left(x(\tau)\right) + \varepsilon y \nu_1'\left(x(\tau)\right)\right]\Pi(y,\tau,\varepsilon)\right\} - \lambda \Pi(y,\tau,\varepsilon)+$$

$$+\lambda \int\limits_0^\infty \left[\Pi(y,\tau,\varepsilon) + \varepsilon u \frac{\partial \Pi(y,\tau,\varepsilon)}{\partial y} + \frac{\varepsilon^2 u^2}{2}\frac{\partial^2 \Pi(y,\tau,\varepsilon)}{\partial y^2}\right]dF(u) + o(\varepsilon^2).$$

It follows

$$\varepsilon^2 \frac{\partial \Pi(y,\tau,\varepsilon)}{\partial \tau} = \varepsilon\left[x'(\tau) - \nu_1(x(\tau)) + \lambda a_1\right]\frac{\partial \Pi(y,\tau,\varepsilon)}{\partial y} - \\ -\varepsilon^2 \nu_1'(x(\tau))\frac{\partial\{y\Pi(y,\tau,\varepsilon)\}}{\partial y} + \varepsilon^2 \frac{\lambda a_2}{2}\frac{\partial^2 \Pi(y,\tau,\varepsilon)}{\partial y^2} + o(\varepsilon^2). \qquad (4)$$

Let function $x(\cdot)$ be a solution of the equation

$$\frac{dx(\tau)}{d\tau} = \nu_1(x(\tau)) - \lambda a_1. \qquad (5)$$

Then it follows that function $\Pi(\cdot,\cdot)$ satisfies the Fokker-Planck equation

$$\frac{\partial \Pi(y,\tau)}{\partial \tau} = -\nu_1'(x(\tau))\frac{\partial\{y\Pi(y,\tau)\}}{\partial y} + \frac{\lambda a_2}{2}\frac{\partial^2 \Pi(y,\tau)}{\partial y^2}.$$

Consequently process $y(\tau,\varepsilon) = \dfrac{\varepsilon^2 Q(t) - x(\tau)}{\varepsilon}$ converges in distribution to the OrnsteinUhlenbeck process $y(\cdot)$ as $\varepsilon \to 0$ satisfying the following stochastic differential equation

$$dy(\tau) = \nu_1'(x(\tau))y d\tau + \sqrt{\lambda a_2}dw(\tau), \tag{6}$$

where $w(\cdot)$ is a standard Brownian motion.

Let $\nu_1(\cdot)$ be a twice differentiable function. From (5) and (6) we get that the process

$$z(\tau) = x(\tau) + \varepsilon y(\tau), \tag{7}$$

satisfies

$$dz(\tau) = (\nu_1(z) - \lambda a_1)d\tau + \varepsilon\sqrt{\lambda a_2}dw(\tau) + \frac{\varepsilon^2}{2}R_2 d\tau, \tag{8}$$

where $R_2 = -y^2\nu''(\varepsilon y\theta), 0 \le \theta \le 1$.

Indeed it is clear

$$dz(\tau) = x'(\tau) + \varepsilon dy(\tau) = \left(\nu_1(x(\tau)) - \lambda a_1\right)d\tau + \varepsilon\left[\nu_1'(x(\tau))y d\tau + \sqrt{\lambda a_2}dw(\tau)\right] =$$

$$= \left(\nu_1(x(\tau)) + \varepsilon y\nu_1'(x(\tau)) - \lambda a_1\right)d\tau + \varepsilon\sqrt{\lambda a_2}dw(\tau). \tag{9}$$

By Taylor expansion with Lagrange remainder we get

$$\nu_1(z) = \nu_1(x + \varepsilon y) = \nu_1(x) + \varepsilon y\nu_1'(x) + \frac{\varepsilon^2}{2}y^2\nu_1''(x + \theta\varepsilon y).$$

So, the equation holds

$$\nu_1(x) + \varepsilon y\nu_1' = \nu_1(z) - \frac{\varepsilon^2}{2}y^2\nu_1''(x + \theta\varepsilon y).$$

From (9) we get

$$dz(\tau) = \left(\nu_1(z(\tau)) - \lambda a_1\right)d\tau + \varepsilon\sqrt{\lambda a_2}dw(\tau) - \frac{\varepsilon^2}{2}y^2\nu_1''(x + \theta\varepsilon y)d\tau.$$

We obtain (8) by taking into account Lagrange remainders properties. We use (2) and (7) to get asymptotic equation

$$\varepsilon^2 Q(t) = x(\tau) + \varepsilon y(\tau) = z(\tau).$$

From (9) we get

$$\varepsilon^2 dQ(t) = \left(\nu_1(\varepsilon^2 Q) - \lambda a_1\right)d\tau + \varepsilon\sqrt{\lambda a_2}dw(\tau) - \frac{\varepsilon^2}{2}R_2 d\tau,$$

which implies

$$dQ(t) = \left(\nu_1(Q(t)) - \lambda a_1\right)\frac{d\tau}{\varepsilon^2} + \sqrt{\lambda a_2}\frac{dw(\tau)}{\varepsilon} - \frac{1}{2}R_2 d\tau$$

taking into account $\nu(Q) = \nu_1(\varepsilon^2 Q)$.
   Since $t\varepsilon^2 = \tau$, we have $dw(\tau)/\varepsilon = dw(t)$ and

$$dQ(t) = \left(\nu_1(Q(t)) - \lambda a_1\right)d\tau + \sqrt{\lambda a_2}dw(\tau) - \frac{1}{2}R_2 d\tau.$$

So approximately

$$dQ(t) = \left(\nu(Q) - \lambda a_1\right)d\tau + \sqrt{\lambda a_2}dw(\tau).$$

Because of the boundedness of $Q(\cdot)$ the stationary distribution exists

$$p(s) = C \cdot exp\left(\frac{2}{a_2\lambda}\int(\nu(s) - a_1\lambda)ds\right), \tag{10}$$

where C is the normalization constant.

3   On/Off Control

The storage capacity let be bounded by Q_{max}. To illustrate the application of the approximation consider the following control of the inventory level: if $Q(\cdot)$ is above a base-stock level $Q_{max}-Q_0$ we begin to deliver the product to outlets with a rate $\nu_1 > \nu_0 - a_1\lambda$ to prevent the stocks overflow, otherwise $\nu(Q) = \nu_0 > a_1\lambda$. The condition $\nu_0 > a_1\lambda$ means that if the inventory level is below the base-stock level, then the stock level is replenished in the mean, i.e., the resources are accumulated.
   We use (10) to get the probability density function

$$p(x) = C \cdot exp\left(\frac{2}{a_2\lambda}(\nu_0 - a_1\lambda)\left(x - (Q_{max} - Q_0)\right)\right), \text{if } x < Q_{max} - Q_0,$$

$$p(x) = C \cdot exp\left(-\frac{2}{a_2\lambda}(a_1\lambda - \nu_0 + \nu_1)\left(x - (Q_{max} - Q_0)\right)\right), \text{if } x > Q_{max} - Q_0,$$

where

$$C = \frac{2(a_1\lambda - \nu_0 + \nu_1)(\nu_0 - a_1\lambda)}{a_2\lambda\nu_1}.$$

The probability of the overflow is

$$\alpha = P(Q(t) > Q_{max}) =$$

$$= C \int_{Q_{max}}^{\infty} exp\left(-\frac{2}{a_2\lambda}(a_1\lambda - \nu_0 + \nu_1)\left(x - (Q_{max} - Q_0)\right)\right) dx =$$

$$= \frac{\nu_0 - a_1\lambda}{\nu_1} exp\left(-\frac{2}{a_2\lambda}(a_1\lambda - \nu_0 + \nu_1)Q_0\right). \tag{11}$$

The probability of the stock-out (the warehouse is empty) is

$$\gamma = P\{Q < 0\} = C \int_{-\infty}^{0} exp\left(\frac{2}{a_2\lambda}(\nu_0 - a_1\lambda)\left(x - (Q_{max} - Q_0)\right)\right) dx =$$

$$= \frac{a_1\lambda - \nu_0 + \nu_1}{\nu_1} exp\left(-\frac{2(\nu_0 - a_1\lambda)}{a_2\lambda}(Q_{max} - Q_0)\right). \tag{12}$$

Parameters of control Q_0 and ν_1 can be found from (11) and (12) if α and γ are fixed.

In Kitaeva [8] the on/off control when the rate of the output flow is proportional to the difference $Q - (Q_{max} - Q_0)$, that is

$$\nu(Q) = \begin{cases} \nu_0, & \text{if } Q < Q_{max} - Q_0, \\ \nu_0 - \beta\left(x - (Q_{max} - Q_0)\right), & \text{if } Q > Q_{max} - Q_0, \end{cases}$$

where $\nu_0 > a_1\lambda, \beta > 0$, is considered. In this case it follows from (10)

$$p(x) = C \cdot exp\left(2d\left(x - (Q_{max} - Q_0)\right)\right), \text{if } x < Q_{max} - Q_0,$$

$$p(x) = C \cdot exp\left(2d\left(x - (Q_{max} - Q_0)\right)\right) - \beta\frac{\left(x - (Q_{max} - Q_0)\right)^2}{a_2\lambda}, \text{if } x > Q_{max} - Q_0,$$

$$C^{-1} = \frac{1 - 2b\Phi(b)exp(b^2)}{2d}, \qquad d = \frac{\nu_0 - a_1\lambda}{a_2\lambda} > 0, \qquad b = -d\sqrt{\frac{a_2\lambda}{\beta}} < 0,$$

$$\Phi(b) = \int_{b}^{\infty} exp(-t^2)dt.$$

The system of equations analogous to (11) and (12) has the form

$$\alpha = \frac{2b\Phi\left(b - \frac{d}{b}Q_0\right)exp(b^2)}{2b\Phi(b)exp(b^2) - 1}, \gamma = \frac{exp\left(-2d(Q_{max} - Q_0)\right)}{1 - 2b\Phi(b)exp(b^2)}. \tag{13}$$

We need to solve (13) with respect to b and Q_0.

In [8] another approach have been used: the variances of the stock level process and the rate of delivering the product to outlets given the probability of the base-stock level exceeding are minimized for continuous and discontinuous linear on/off control respectively.

4   Simulation

Consider the model with constant rate of the output flow.

Ten replications are performed with the model.

The storage capacity is bounded by 3 conventional units. Random demand has beta distribution with shape parameters $\alpha = 0.429$ and $\beta = 0.286$. It follows that $a_1 = 0.6$ and $a_2 = 0.5$. The intensity λ equals 1. The rate of the input product flow ν_0 equals also 1.

For three sets of the probabilities (α, γ) we numerically solve equations (11) and (12) and get the values of the control parameters Q_0 and ν_1, that are using in simulation.

The mean of the stock level process derived from the diffusion approximation is

$$\overline{Q} = Q_{max} - Q_0 + \frac{a_2\lambda}{2\nu_1}\left(\frac{\nu_0 - a_1\lambda}{a_1\lambda + \nu_1 - \nu_0} - \frac{a_1\lambda + \nu_1 - \nu_0}{\nu_0 - a_1\lambda}\right). \qquad (14)$$

We estimate the mean $\widehat{\overline{Q}}$ and compare the estimates with the theoretical results.

The numerical results are shown in Table 1. In the last column of the table the corresponding relative mean absolute error $(RMAE)$ is given.

In Figure 1 three realizations of $Q(t)$ are shown for $\alpha = 0.15, \gamma = 0.10$.

The results of the numerical simulation are consistent with the approximate analytical results, so the diffusion approximation is suitable for the system under consideration.

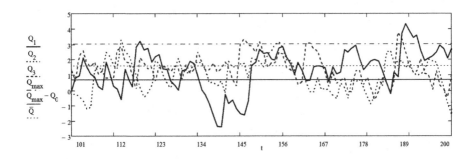

Fig. 1. Stock level process

Table 1. The numerical results

	Q_0	ν_1	$\overline{Q}$	$\widehat{\overline{Q}}$	$RMAE$
$\alpha = 0.01, \gamma = 0.10$	1.933	0.892	0.950	0.956	0.096
$\alpha = 0.10, \gamma = 0.05$	1.705	0.663	1.620	1.690	0.059
$\alpha = 0.15, \gamma = 0.10$	2.326	0.566	1.555	1.582	0.118

References

1. Janssen, J., Manca, O., Manca, R.: Applied Diffusion Processes from Engineering to Finance. John Wiley & Sons, London (2013)
2. Bather, J.A.: A Continuous Time Inventory Model. Journal of Applied Probability 3, 538–549 (1966)
3. Harrison, J.M.: Brownian Motion and Stochastic Flow Systems. John Wiley and Sons, New York (1985)
4. Puterman, M.: A diffusion process model for a storage system. In: Geisler, M.A. (ed.) Studies in the Management Sciences, Logistics, I, pp. 143–159. North-Holland Press, Amsterdam (1975)
5. Chopra, S., Meindl, P.: Supply chain management. Prentice Hall, London (2001)
6. Beyer, D., Cheng, F., Sethi, S.P., Taksar, M.: Markovian demand inventory models. Springer, New York (2010)
7. Barucha-Reid, A.T.: Elements of the Theory of Markov Processes and Their Applications. McGraw-Hill, New York (1960)
8. Kitaeva, A.V.: Stabilization of Inventory System Performance: On/Off Control. In: The 19th World Congress of the International Federation of Automatic Control, Cape Town, South Africa, August 24-29 (2014)

Organization of Onboard Digital Computer System with Reconfiguration

Ekaterina Kniga, Anatoly Shukalov, and Pavel Paramonov

Saint Petersburg National Research University of Information Technologies,
Mechanics and Optics (University ITMO),
Kronverksky pr.49, 197101 Saint-Petersburg, Russian Federation
ekovinskaya@gmail.com

Abstract. The problem of reconfiguration of onboard digital computer systems in case of aircraft constructive functional modules failure. Reliability of arrangements for different types of redundancy. An original algorithm of an onboard computer system in normal conditions and in reconfiguring hardware. Examples of options to assign task to available (serviceable) computing resources.

Keywords: Testing, computing systems.

1   Introduction

Modern onboard digital computing systems (ODCS) of perspective aircraft are [1-12] complex integrated computing devices consisting of different purpose constructive functional modules (CFM).

ODCS of an aircraft perform functional tasks of determining the parameters of flight and navigation modes:

- takeoff, horizontal flight, landing maneuvers;
- monitoring the technical condition of avionics;
- coordination of onboard avionics subsystems;
- collection, storage, processing and delivery to the pilot of data received from the information-measuring system and control field of the cockpit, etc.

To meet the set parameters of fault-tolerance ODCS redundancy is implemented in the computer system:instrumental, functional, informative, etc. In practice, the instrumental redundancy is most often realized (by reservation of CFM), thereby mass and dimensions of avionics increase. Implementing of reservation on ODCS level generally means that in case of one of computing systems CPM failure, whole ODCS is considered as faulty.

Performing ODCS reconfiguration on board of an aircraft in a situation of failure allows to move from the computing system redundancy scheme at ODCS level generally to the computing system redundancy scheme at CFM level, which significantly reduces the mass and dimensions of the avionics hardware.

A. Dudin et al. (Eds.): ITMM 2014, CCIS 487, pp. 197–204, 2014.

2   ODCS Functional Diagram

ODCS consists of a set of CPM different in their function: computing module (CM), graphical module (GM), mass storage module (MSM), switch module (SM), input-output module (IOM) and voltage module (VM). Presented module nomenclature is sufficient to build onboard computers for various purposes:

- onboard digital computer system (ODCS);
- onboard digital mapping system (ODMS);
- onboard interface station (OIS);
- onboard graphics station (OGS).

Functional circuits examples are shown in fig. 1. Further, the structure and operation of computing systems are shown for ODCS, reconfiguration of ODMS, OIS, OGS are applied in a similar manner.

MSM in ODCS is the master module, which performs the arbitrator function. Other CFM act as slave units.

The system functions as described further. When power is applied CFM performs the initialization of each of its components field-programmable gate array, microcontrollers, microprocessors, onboard SpaceWire link switches, etc. After initialization, each CFM receives functional software from MSM memory and puts it into a cell of its internal random access memory (RAM). Further CFM operation as a part of ODCS is determined by functional software algorithm. Next, the MSM initiates functional software starting algorithm for each CFM.

During ODCS operation MSM analyzes test result data of built-in control of each module and in case of detection of failure it initiates the ODCS reconfiguration procedure excluding failed unit from onboard exchange with its blackout and including a serviceable reserved unit from cold or hot standby.

Equivalent reservation circuit of ODCS on ODSC level as a whole is shown in fig. 2, on CFM level in fig. 3.

3   ODCS Work Algorithm

ODCS operation algorithm is presented in fig.4. After the power is supplied the initial test is performed on the four CFM of both subsystems. The initial test includes:

- RAM test of CFM (performed by writing, reading and comparing read and written words);
- ROM test (performed by calculating checksum for memory as a whole or for ROM sector and comparing the obtained values with the previously recorded value);
- Input-output test (performed by software controlled CFM transceivers loop control switching and transmission of test data words through formed channels).

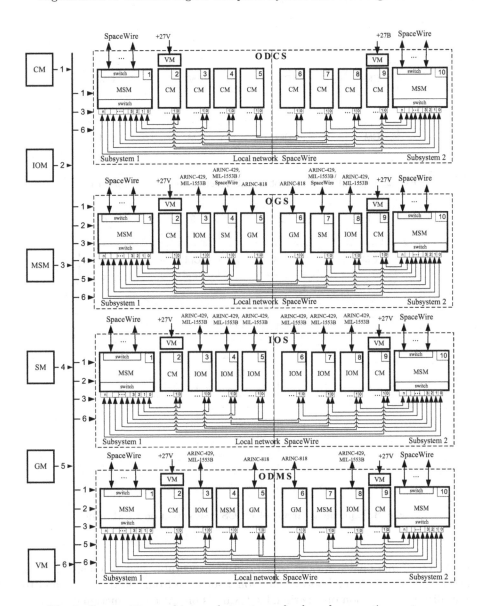

Fig. 1. Composition and internal structure of onboard computing systems

The test result of each CFM is passed into MSM, where it is stored in a dedicated memory sector. If the test result reports the CFM failure, MSM initiates ODCS reconfiguration procedure within available (serviceable, available for use) hardware resources.

If there is not any serviceable resources id ODCS, integral signal computing system serviceability is off and ODCS considered faulty. If completion of all test components is successful, each CFM set serviceability signal on. Then, MSM

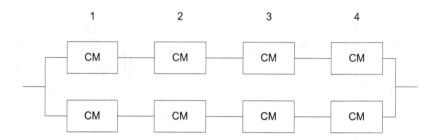

Fig. 2. Equivalent circuit of ODCS reservation on subsystem level

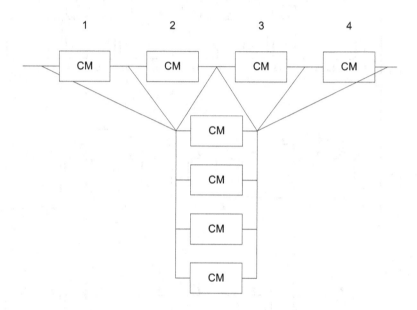

Fig. 3. Equivalent circuit of ODCS reservation on CFM level

distributes functional software, which consist executable task in ODCS, among serviceable CFM. After booting functional software from MSM a command to initialize the execution of functional software is on. At each functional software execution cycle there is test conducted functional CFM operation. Test performs as background task onboard. Node of the test depends of the status of a single command automatic control, which is an external signal for ODCS. If the signal is on, the test performs in advanced mode, if the signal is off, the test performs in standard mode. In advanced mode, the contextual data (data about the current flight parameters) of each CFM enter into a non-volatile ROM and test performs. It is believed that the test time is negligible compared with the significant change in the navigation parameters during the flight. The test includes: a full test of RAM, a checksum ROM test, CPU test (simple task of

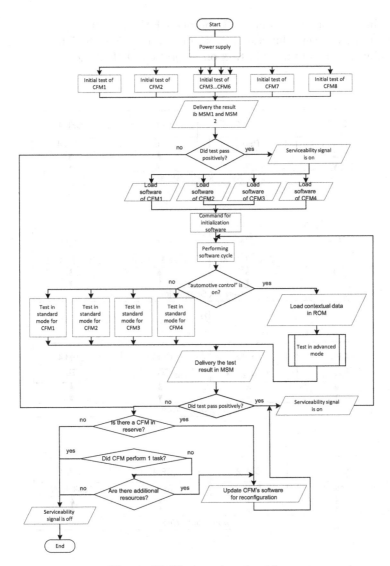

Fig. 4. ODCS operation algorithm

calculation with known result), the interrupt system test and input/output test (loop control). In standard mode, the test includes: a RAM test (only selected unused during task area), a checksum ROM test and processor command system test. After the test is completed the test result is passed in MSM, where it is stored in a dedicated memory sector. If test completed with success the integral serviceability signal forms. If test completed with CFM failure, the reconfiguration procedure begins. The operation algorithm during reconfiguration procedure depends on how functional tasks are distributed between CFM. Possible options are presented in fig. 5:

- appointment one functional tasks on one CFM (fig.5,a);
- appointment several functional tasks on several CFM (fig.5,b);
- appointment one functional task on several CFM (fig.5,c,d).

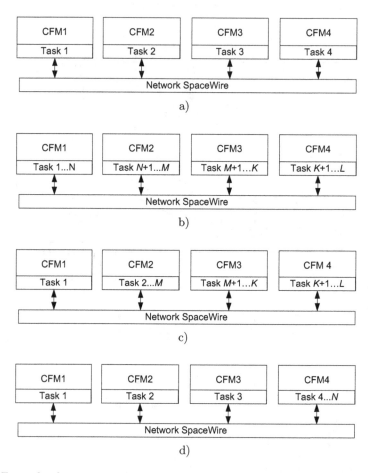

Fig. 5. Example of assignment functional tasks to CPM: a) each of the tasks performs on its own CFM; b) each CFM executes multiple tasks; c, d) several tasks are performed by individual CFM

Assignment supplication on CFM realizes by ODCS hardware and software by organization logical interaction protocols between CFM by internal links SpaceWire. Reconfiguration procedure with situation in fig.5,a (one task is on one CFM) is available when there is some CFM in reserve. If a CFM is available, MSM transmit functional software of failure CFM in its RAM. At same time functional software of other CFM is modified by changing ODCS configuration and device addresses. Then ODCS returns to the its original problem. If there is not any CFM in reserve, ODCS is considered faulty and integral serviceability signal is off. Reconfiguration procedure with situation in fig.5,b (several tasks are on several CFM) is available not only when there is a CFM in reserve, but also when there are additional (unused) resources to perform some additional tasks on used CFM. In this case, tasks of failed CFM can be distributed between serviceable and partially loaded modules. Any other case of assignment tasks (fig.5,c,d) ability of reconfiguration procedure depends on what CFM failed. If it is CFM, which performed one big task, then reconfiguration can be when there is a CFM in reserve, like with fig.5,a. If it is CFM, which performed many small tasks, then reconfiguration can be performed when there are available additional resources on CFM, which are already used. Then the failed CFM task is distributed between these CFM. According to the practice in aviation industry functional software algorithm includes calculation at fast and slow cycles. Mathematical calculation of parameters, which have rate of change comparable to the rate calculation, performs by quick calculation cycle. Calculation of parameters, which are not critical to the aircraft flight dynamics, performs by slow calculation cycle. The ODCS testing performs at slow cycle. Moreover, the developers provide 10-20% margin of time and resources that can be used for software modernization for the purpose to complicating the algorithm or to do reconfiguration procedure. After reconfiguration procedure operation ODCS algorithm is next: if the result of reconfiguration is positive, then serviceability signal generates, CFMs functional software updates and ODCS returns to performing functional software; if the result of reconfiguration is negative, then serviceability signal is off and ODCS is considered as faulty.

4 Conclusion

ODCS operation algorithm in case of reconfiguration depends of assignment tasks on CFMs computing resources during flight. Reconfiguration procedure is possible only if there are additional computing resources in ODCS by introduction instrumental redundancy. For example, in the initial moment only three of four CFM are used, then the fourth CFM is in reserve and in case in CFM failure this CFM begins to perform tasks of failure module.

References

1. Gamati, A., Brunette, C., Delamare, R., Gautier, T., Talpin, J.: A Modeling Paradigm for Integrated Modular Avionics Design. In: 32nd EUROMICRO Conference on Software Engineering and Advanced Applications (2006)
2. Lopez, J., Royo, P., Barrado, C., Pastor, E.: Modular avionics for seamless reconfigurable UAS missions. Department of Computer Architecture, Technical University of Catalonia, 08860 Castelldefels, Barcelona, Spain
3. Lpez, J., Royo, P., Pastor, E., Barrado, C., Santamaria, E.: A Middleware architecture for unmanned aircraft avionics. In: Proceedings of the 8th ACM/IFIP/USENIX International Conference on Middleware, Newport Beach, California (2007)

4. Littlefield, J., Viswanathan, R.: Advancing open standards in Integrated Modular Avionics: An industry analysis. In: IEEE/AIAA 26th Digital Avionics Systems Conference, DASC 2007, Dallas, TX (October 2007)
5. Lipari, G., Bini, E.: Resource partitioning among real-time applications. In: Proc. of the 15th Euromicro Conference on Real-Time Systems, pp. 151–158 (2003)
6. Garside, R., Pighetti, F.J.: Integrating Modular Avionics: A new role emerges. In: IEEE/AIAA 26th Digital Avionics System Dallas, TX (October 2007)
7. Lee, Y.-H., Kim, D., Younis, M., Zhou, J.: Scheduling tool and algorithm for integrated modular avionics systems. In: Proc. of the 19th Digital Avionics Systems Conference (DASC) (2000)
8. Vanek, B.: Future Trends in UAS Avionics. In: Proc. 10th Int. Symp. of Hungarian Researchers on Computational Intelligence and Informatics, Budapest, pp. 791–802 (2009)
9. Ananda, C.M., Venkatanarayana, K.G., Preme, M., Raghu, M.: Avionics Systems, Integration, and Technologies of Light Transport Aircraft. Defense Science Journal 61(4), 289–298 (2011)
10. Butz, H.: Open integrated modular avionic (IMA): State of the art and future development road map at Airbus deutschland. Technical report, Department of Avionic Systems at Airbus Deutschland GmbH (2010)
11. Prisaznuk, P.J.: ARINC 653 role in integrated modular avionics. In: 27th Digital Avionics Systems Conference, St. Paul, Minnesota (2008)
12. Hammarberg, J., Nadjm-Tehrani, S.: Formal verification of fault tolerance in safetycritical reconfigurable modules. International Journal of Software Tools for Technology Transfer (STTT) 7(3) (2005)

TCP Reno Congestion Window Size Distribution Analysis*

Vladimir Kokshenev and Sergey Suschenko

National Research Tomsk State University
vladimir_finf@mail.ru,
ssp@inf.tsu.ru

Abstract. Analysis of congestion window size distribution for TCP Reno sender is presented. The data for analysis are gathered from numerical results of an analytical model of Reno congestion control procedure based on Discrete-Time Markov Chain. The model was presented in [1] and as it is shown in this paper it provides a way to estimate congestion window distribution as a function of round trip time and loss rate for bulk transfer TCP flow. Presented results consider slow start, congestion avoidance and fast recovery phases, and fast retransmit, cumulative and selective acknowledgments, timeouts with exponential back-off and appropriate byte counting features of TCP. This paper also presents comparison of congestion window size distribution for selective and cumulative acknowledgments.

Keywords: TCP, Reno, Congestion Window Distribution, Analytical Model, Discrete-Time Markov Chain.

1 Introduction

TCP is the most widely used transport layer protocol in the Internet [2] and Reno is the most widely implemented congestion control procedure [3]. TCP Reno and NewReno are only congestion control algorithms that reached Standard Track category in RFC series [4, 5]. Reno is a part of TCP/IP stack of modern operating systems and networking equipment. Thus the speed and performance of many Internet services and applications are influenced by TCP characteristics.

RFC793 [6] defines window as a flow control mechanism of a transport connection. Receiver advertises a window size and guarantees that it has enough buffer space to accept the full window of data. This gives a receiver the ability to influence throughput of TCP connection. But there is no guarantee that the network has enough resources to serve offered data rate. As a result a series of congestion collapses had happened in 80s which led to development of congestion control procedures for TCP.

Congestion control is a feature of TCP which helps to avoid network overloading by excessive traffic. The main idea of congestion control procedures is

* This work is performed under the state order No. 1.511.2014/K of the Ministry of Education and Science of the Russian Federation.

A. Dudin et al. (Eds.): ITMM 2014, CCIS 487, pp. 205–213, 2014.

to change sending rate depending on network conditions. Congestion control bases decisions on different types of feedback received from a network such as loss rate, propagation delay or explicit congestion signaling. A lot of congestion control algorithms were developed to address different network types and conditions: Tahoe, Reno, NewReno, Vegas, Hybla, CTCP, Illinois, BIC, CUBIC, Westwood+, H-TCP, High Speed TCP, Scalable TCP, Veno, YeAH, FAST, Woodside, CHD, CDG, FIT, LP, MulTCP and others.

Each congestion control algorithm has unique footprint of window management strategy. A lot of researches have been made to model and estimate average TCP connection throughput of different congestion control procedures. But congestion window distribution was rarely analyzed. Though average throughput estimation can be referenced as the main goal of modeling, congestion window distribution is also important metric as it shows connection throughput variation.

In this paper, we present an extended analysis of congestion window distribution for TCP Reno sender. The analysis is based on a Discrete-Time Markov Chain analytical model of Reno congestion control procedure presented in [1]. The model implements slow start, congestion avoidance and fast recovery TCP phases, and covers fast retransmit, timeouts with exponential back-off, appropriate byte counting, cumulative and selective acknowledgments features of TCP. Analysis is based on numerical results of the analytical model. Comparison of congestion window distribution for selective and cumulative acknowledgments is also shown.

2   The Reno Protocol

Reno is most widely used congestion control procedure. It is based on three congestion window (CWND) management phases [3, 4]: Slow Start (SS), Congestion Avoidance (CA) and Fast Recovery (FR).

TCP increases congestion window by one for each positive acknowledgment received when SS phase is active. As a result CWND doubles each round trip time (RTT) cycle. Slow start threshold variable (SSTHRESH) defines when to switch from SS to CA mode. Protocol acts in SS mode if CWND is less than SSTHRESH.

Protocol switches into CA when CWND reaches or exceeds SSTHRESH value. Being in CA state, TCP sender increments congestion window by $1/CWND$ for each positive acknowledgment received. A window size of CWND segments will generate at most CWND acknowledgments in one RTT, so therefore CWND will be increased by at most one segment in one RTT [7].

TCP Reno uses two loss detection mechanisms. Timeout (TO) based loss detection can be counted as conservative option and duplicate acknowledgment (DupACK) based analysis in context of Reno congestion control can be referenced as liberal loss detection method.

Sender resets the already running retransmission timer and starts a new one each time a new segment is transmitted. The timer is set for the retransmission timeout (RTO) value supplied by the RTT estimation procedure. All unacknowledged segments are counted as lost if this timer expires. In response to RTO timer expiration TCP performs several actions:

- SSTHRESH parameter is set to $CWND/2$,
- CWND variable is set to 1,
- the protocol performs retransmission,
- slow start is re-initiated.

The other option to detect losses is to analyze incoming acknowledgments and track number of DupACKs. When the number of consecutive duplicate acknowledgments reaches value of 3 the closest unacknowledged segment is counted as lost. When a loss is detected by DupACKs TCP takes remediation actions:

- the protocol performs immediate retransmission, which is also called Fast Retransmission,
- SSTHRESH variable is set to $CWND/2$,
- CWND variable is set to $(CWND/2 + 3)$,
- the protocol switches to fast recovery mode.

Being in FR mode, TCP temporary increases CWND by one segment for each duplicate acknowledgment received and transmits new segment if general rules allow [3, 4]. When acknowledgment for the retransmitted segment arrives, CWND is deflated back to SSTHRESH, and protocol switches from FR to CA mode.

Most TCP implementations in Windows, Linux, FreeBSD operating systems and different networking equipment in addition to modes and features specified above use Karn algorithm [8, 9], selective acknowledgments [10–12], appropriate byte counting [13] and other advanced techniques.

1. Karn algorithm suggests RTO exponential back-off in case of retransmission timer expiration. The back-off is canceled when acknowledgment for retransmitted data arrives. This measure improves TCP adaptation to sudden RTT changes.
2. Traditional TCP uses cumulative acknowledgments. As a result TCP may experience poor performance in case of multiple losses in single window of data [10]. Selective acknowledgments (SACK) allow the receiver to acknowledge discontinuous blocks of segments that were received correctly.
3. TCP increases congestion window based on the number of bytes acknowledged by the arriving ACK when Appropriate Byte Counting (ABC) is used [13]. The algorithm mitigates the impact of delayed acknowledgments feature on connection throughput.

3   Analytical Models of TCP Reno

Reno algorithm is frequently referenced as Standard TCP, and it is the most widely implemented congestion control method [3]. A lot of analytical models of

Reno protocol were proposed in past twenty years ([1, 14–22] and many others). These researches differ in methods: some of them use renewal theory [15, 16] or Markov chains to model TCP sender behavior [1, 17, 19, 20, 22], others are based on fixed-point method [19–21] or other theories.

Though every mentioned model provides estimates of average TCP connection throughput, rare researches offer estimates for congestion window size distribution. Models [19, 20, 22] theoretically allow to obtain probabilities for different CWND values, and in [18, 21] congestion window distribution is explicitly analyzed.

The challenge of obtaining congestion window size distribution properly lies in complexity of CWND management strategy. Therefore in order to succeed one should use a method that theoretically allows to create precise models of complex TCP behavior. As it was mentioned in [23] Markov chains allow to create complex models of TCP sender and mimic CWND dynamics.

This research uses DTMC-based model proposed in [1] to get CWND size distribution. The model counts all Reno phases and several additional features. As it was shown in [1] the model produces accurate estimates for average connection throughput and closely matches with real TCP traces.

As it is not a trivial task to compare gathered results with models proposed in [18, 21] we present limited comparison only.

4   TCP Reno Analytical Model

We use two-dimensional Discrete-Time Markov Chain based analytical model proposed in [1] to model TCP sender behavior. It is assumed that statistical cycle duration is equal to one round trip time. We suggest that RTT is constant and each segment has the same size. It is also suggested that time required to send the whole receiver's advertised window (W) is less than RTT.

The first dimension of DTMC is used for CWND variable tracking, and the second dimension tracks SSTHRESH value changes.

The model doesn't count losses in reverse direction. Thus it is assumed that TCP ACKs are never lost. We denote the probability to transfer a segment successfully from sender to receiver as F. The model suggests uncorrelated losses.

In order to model timeout phase we use S as a value of retransmission timeout expressed in cycles. S is calculated as floored ratio of RTO and RTT values expressed in seconds:

$$S = \left\lfloor \frac{RTO}{RTT} \right\rfloor, \tag{1}$$

We use additional parameter MK to model Karn exponential back-off and additional parameter ML to model selective acknowledgments. MK denotes the upper limit for RTO exponential back-off, so the maximum timeout value is $2^{MK}S$. ML denotes the maximum number of losses that can be detected during SS and CA phases and recovered in FR.

Further details and transition probabilities (π_{in}^{jm}) for the DTMC are presented in [1].

Suggested Markov Chain is too complex to get analytical solution for state probabilities P_{ij}, but it is possible to solve a system of linear equations (2) numerically.

$$
\begin{cases}
P_{jm} = \sum_{i=1}^{W+2MK} \sum_{n=2}^{\lfloor W/2 \rfloor + 2^{MK}S} P_{in} \pi_{in}^{jm}, \quad j = \overline{1, W + 2MK}; \\
\hspace{9cm} m = \overline{2, \lfloor W/2 \rfloor + 2^{MK}S} \quad (2) \\
\sum_{j=1}^{W+2MK} \sum_{m=2}^{\lfloor W/2 \rfloor + 2^{MK}S} P_{jm} = 1
\end{cases}
$$

$Pcwnd_0$ denotes the probability of timeout phase when protocol doesn't send new data and awaits timeout expiration or acknowledgment arrival:

$$
Pcwnd_0 = \sum_{i=1}^{W+2MK} \sum_{n=\lfloor W/2 \rfloor + 1}^{\lfloor W/2 \rfloor + 2^{MK}S} P_{in}. \quad (3)
$$

Technically protocol has other than zero CWND value in case when full window of data has been sent and ACKs haven't been received yet. But it is convenient to use $Pcwnd_0$ to denote a protocol state when no data is sent and zero throughput in produced.

We denote the probability of congestion window size equal to i as $Pcwnd_i$ where $i = \overline{1, W}$:

$$
Pcwnd_i = \sum_{n=1}^{\lfloor W/2 \rfloor} P_{in}. \quad (4)
$$

5   Numerical Results and Analysis

Congestion window size distributions ($Pcwnd_i$) gathered numerically are shown in figure 1. The figure presents results for cumulative acknowledgments in different network conditions: $W = 12/23/45$, $S = 5$ ($RTO = 1000$ msec, $RTT = 200$ msec), $F = 0, 9/0, 95/0, 97/0, 99$.

Analysis of numerical results shows several noticeable facts which can also be observed in figure 1.

1. Higher packet loss leads to higher time spent in timeout phase when protocol does not send new data and simply waits timeout expiration or acknowledgment arrival. Thus $Pcwnd_0$ is greater for higher packet loss environments.
2. Interval of CWND values $\overline{1,3}$ has multiple inflection points. This is especially visible for high packet loss environments, i.e. $F = 0, 90$. In cases when congestion window is small enough to trigger packet loss detection by DupACKs, fast retransmit and fast recovery options are not available for TCP sender. This behavior differs from what is defined for CWND values $\overline{4, W}$, so therefore CWND distribution looks different too.

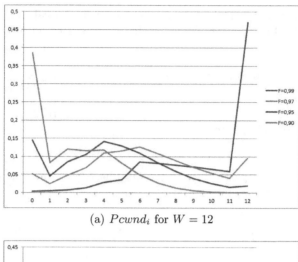

(a) $Pcwnd_i$ for $W = 12$

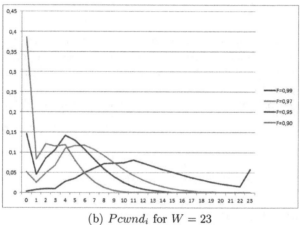

(b) $Pcwnd_i$ for $W = 23$

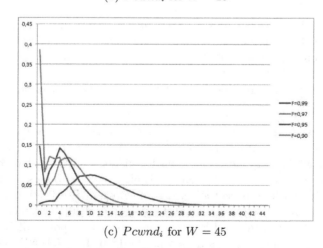

(c) $Pcwnd_i$ for $W = 45$

Fig. 1. Congestion window size distribution $Pcwnd_i$: cumulative acknowledgments, $S = 5$ (RTO=1000 msec, RTT=200 msec)

3. Congestion window distribution may have a jump in a point where sender's congestion window size value equal to receiver's advertised window ($CWND = W$). This is visible on figure 1(b) and especially visible on figure 1(a). In situation with infinite advertised window size there are no any jumps. But when the W is limited $Pcwnd_W$ stores the weight of the whole tail of the distribution. Also in this case congestion window distribution has a jump in the point where $CWND = \lfloor W/2 \rfloor$. This additional jump is caused by the previously mentioned jump and the nature of fast recovery phase: the protocol halves CWND after successful retransmission.

Figure 2 presents congestion window distribution for $S = 20$. Obviously $Pcwnd_0|_{S=20} > Pcwnd_0|_{S=5}$. Analysis of numerical results for different values of S shows that if sender follows RFC6298 [9] recommendations and sets timeout value not less than 1000 msec it will spent a lot of time awaiting timeout in cases of relatively low RTT.

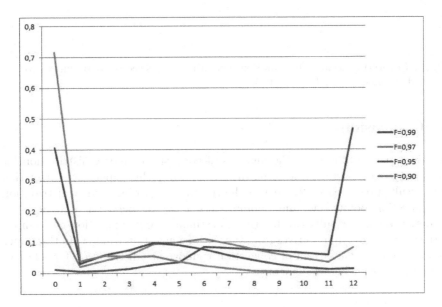

Fig. 2. Congestion window size distribution $Pcwnd_i$: cumulative acknowledgments, $S = 20$ (RTO=1000 msec, RTT=50 msec)

Comparative analysis of selective and cumulative acknowledgments is shown on the figure 3.

Model predicts that SACK provides potential for TCP throughput increase of up to 6-7.5%. It can be concluded that networks with higher loss rate and lower round-trip time benefit from SACK more[1]. SACK allows to detect losses more effectively, less timeout events mean less slow start re-initiations and more loss recoveries in FR mode. Thus congestion window distribution for selective acknowledgments has lower probabilities for $CWND = \overline{1,3}$ and higher probabilities for $CWND = \overline{4,W}$ in comparison to cumulative ACKs.

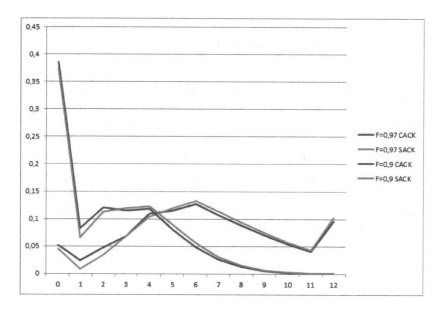

Fig. 3. Congestion window distribution comparison for selective and cumulative acknowledgements. $S = 5$, $F = 0,9/0,97$.

6   Conclusions

In this paper we presented the analysis of congestion window distribution for Reno congestion control procedure. Results shown in this paper have correlation with results presented in [21]. It was shown that congestion window distribution has inflection points and jump points caused by the behavior of Reno TCP sender. Presented results can be used to estimate TCP connection throughput variation as throughput derives from the size of congestion window. In this paper we also showed comparison of congestion window size distribution for selective and cumulative acknowledgments.

References

1. Kokshenev, V., Suschenko, S.: Analytical Model of the TCP Reno Congestion Control Procedure through a Discrete-Time Markov Chain. In: Vishnevsky, V., Kozyrev, D., Larionov, A. (eds.) DCCN 2013. CCIS, vol. 279, pp. 124–135. Springer, Heidelberg (2014)
2. Lee, D.J., Carpenter, B.E., Brownlee, N.: Media Streaming Observations: Trends in UDP to TCP Ratio. International Journal on Advances in Systems and Measurements 3(3-4) (2010)
3. Fall, K., Stevens, R.: TCP/IP Illustrated, 2nd edn. The Protocols, vol. 1. Addison-Wesley Professional Computing Series (2012)
4. Allman, M., Paxson, V., Blanton, E.: TCP Congestion Control. Internet RFC 5681 (September 2009)

5. Floyd, S., Henderson, T., Gurtov, A., Nishida, Y.: The NewReno Modification to TCPs Fast Recovery Algorithm. RFC6582 (April 2012)
6. Postel, J.: Transmission Control Protocol. Internet RFC 0793/STD 0007 (September 1981)
7. Van Jacobson, M.: Karels: Congestion avoidance and control. In: SIGCOMM 1988 (November 1988)
8. Karn, P., Partridge, C.: Improving Round-Trip Time Estimates in Reliable Transport Protocols. In: SIGCOMM 1987 (1987)
9. Paxson, V., Allman, M., Chu, J., Sargent, M.: Computing TCPs Retransmission Timer. RFC 6298 (June 2011)
10. Mathis, M., Mahdavi, J., Floyd, S., Romanow, A.: TCP Selective Acknowledgement Options. Internet RFC 2018 (October 1996)
11. Floyd, S., Mahdavi, J., Mathis, M., Podolsky, M.: An Extension to the Selective Acknowledgement (SACK) Option for TCP. Internet RFC 2883 (July 2000)
12. Blanton, E., Allman, M., Fall, K., Wang, K.: A Conservative Selective Acknowledgment (SACK)-based Loss Recovery Algorithm for TCP. RFC 3517 (April 2003)
13. Allman, M.: TCP Congestion Control with Appropriate Byte Counting (ABC). RFC 3465 (February 2003)
14. Lakshman, T.V., Madhow, U.: The performance of TCP/IP for networks with high bandwidth-delay products and random loss. ACM/IEEE Trans. on Networking 5, 336–350 (1997)
15. Padhey, J., Firoiu, V., Towsley, D., Kurose, J.: Modeling TCP Throughput: A simple Model and Its Empirical Validation. UMASS CMPSI Tech Report TR98-008 (February 1998)
16. Kumar, A.: Comparative Performance Analysis of versions of TCP in a Local Network with a Lossy Link. ACM/IEEE Trans. of Networking 6, 485–498 (1998)
17. Padhey, J., Firoiu, V., Towsley, D.: A stochastic model of TCP Reno congestion avoidance and control. Tech. Rep. UMASS-CS-TR-1999-02 (1999)
18. Misra, A., Baras, J., Ott, T.: Window Distribution of Multiple TCPs with Random Loss Queues. In: Proceedings of Global Telecommunications Conference, GLOBECOM 1999, pp. 1714–1729 (1999)
19. Casetti, C., Meo, M.: An analytical framework for the performance evaluation of TCP Reno connections. Computer Networks 37, 669–682 (2001)
20. Wierman, A., Osogami, T., Olsen, J.: A Unified Framework for Modeling TCP-Vegas, TCP-SACK, and TCP-Reno. In: Proceedings of the 11th IEEE/ACM International Symposium on Modeling, Analysis and Simulation of Computer Telecommunications Systems (MASCOTS 2003), pp. 1526–7539 (2003)
21. Kassa, D.F.: Analytic Models of TCP Performance. PhD Thesis, University of Stellenbosch, p. 199 (2005)
22. Ewald, N., Kemp, A.: Analytical Model of TCP NewReno through a CTMC. In: Bradley, J.T. (ed.) EPEW 2009. LNCS, vol. 5652, pp. 183–196. Springer, Heidelberg (2009)
23. Olsen, Y.: Stochastic modeling and simulation of the TCP protocol. Uppsla Dissertations in Mathematics 28, 94 (2003)

Optimization of the Road Capacity and the Public Transportation Frequency Which Are Based on Logit-Model of Travel Mode Choice

Mark Koryagin and Alexandra Dekina

Kemerovo State Agricultural Institite,
The Laboratory of Modeling of Social, Economics and Industrial Systems
Markovzeva Street, 5, 650056 Kemerovo, Russia
{markkoryagin,dekina-alexandra}@yandex.ru
http://www.ksai.ru

Abstract. An urban passenger transportation problem is researched. Municipal authorities and passengers are regarded as participants in the transportation system. The municipal authorities have to optimize road capacity and public transport frequency. Passengers travel mode choice is based on logit model. Traffic congestion is described by Greenshields equation. The existence of Nash equilibrium between municipal authorities and passengers is proved. The numerical example characterizing the influence of the parameters on the problem solution is given.

Keywords: Game theory, bus corridor, travel mode choice, traffic congestion.

1 Introduction

Most of cities in the developing nations are facing with the transport problems. The traffic is getting heavier, but the infrastructure is not developing so fast. Municipal authorities have to develop public transport and road infrastructure for private vehicles. But usually authorities decision based on current demand. Thus public transportation is under pressure in this situation: round trip time increases, expenditure grows and passenger flow declines. The optimal solution must be based on Transportation Demand Management (municipal authorities create conditions for optimal travel mode choice).

Decisions of passengers and authorities mutually influence each other (Fig. 1), therefore the game theory is used as a model. In the review [3] a lot of papers were considered in which game theoretic models of transport system were classified according to the set of participants (private company and travelers, authorities and travelers, private company and authorities, authorities between themselves, and private companies among themselves). The paper [6] takes into consideration the dependence between the traffic and the travel time.

A. Dudin et al. (Eds.): ITMM 2014, CCIS 487, pp. 214–222, 2014.
© Springer International Publishing Switzerland 2014

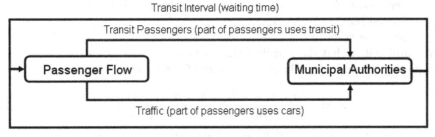

Fig. 1. Public transport management system

But this classification is not enough for the solution of traffic problems. On the other hand, many researchers of traffic [9] used either a set of participants: traveler, node and arc, which presented a road user, an intersection and a roadway segment.

The peculiarity of this paper is construction of the conflict model between the authorities and passengers with the traffic congestion taken into account. In this paper the traffic congestion is modeled by the well-known Greenshields equation [7].

Municipal authorities have to optimize road capacity and public transport frequency (or intervals) which are both depending on passengers travel mode choice (Fig.1). The authorities solution has to determine the travel time by car and by public transport. This very information is required for passengers decision-making. So vice versa the information about passengers travel mode choice passes to authorities.

The given article is about mathematical modeling with concern as to how municipal authorities make their decisions in conditions of passengers travel mode choice. This complicated problem is being solved on the game theory basis.

2   Passenger Flow

Nowadays the travel mode choice (TMC) is researched well enough [4]. There are two main modes of traveling for long distance: a private car and public transport. The passengers look for a better proportion between using a private car and the public transportation.

Travel mode choice theory describes passengers decision-making which depends on a lot of parameters. Passengers decision model may be based on objective function (total spending of time and money) [5]. But logit and probit distributions are usually used for TMC models [4], which are taken to represent solution without objective function.

The model passenger decision-making consists of the following parameters:

t_t – average travel time on public transport;

t – average travel time by car (not including driving time);
c_t – fare on public transport;
c – car travel costs $(c > c_t)$;
p – probability that the travelers choose a car.
Simple logit model calculates probability of car using as follows

$$p_c = \frac{exp\{\,a_0 + a_t(t_t - t) + a_c(c_t - c)\,\}}{1 + exp\{\,a_0 + a_t(t_t - t) + a_c(c_t - c)\,\}}, \tag{1}$$

where parameters a_0, a_t, a_c defined of influence of travel time and travel cost on passenger decision. The parameters of logit function ($a_0 = -1.88$, $a_t = 0.86$, $a_c = -0.6$) were calculated in [8].

The logit models describe decisions made by the passengers in accordance with survey data, but these models dont directly describe objective functions. Therefore we offer to use criterion, that based on quadratic deviation of passenger decision from optimal.

$$G(p) = (p - p_c)^2 = \left(p - \frac{exp\{\,a_0 + a_t(t_t - t) + a_c(c_t - c)\,\}}{1 + exp\{\,a_0 + a_t(t_t - t) + a_c(c_t - c)\,\}} \right)^2 \to \min_p \tag{2}$$

Evidently, that solution of (1) is $p = p_c$. Function $G(p)$ is convex upward in the parameter t, because the second derivative is

$$G''(p) = 2 > 0. \tag{3}$$

3 Municipal Authorities

For optimal city development its necessary to save two components: time of passengers [1] (value of time) and costs of transportation. Costs of transportation include not only direct expenditures but also road maintenance expenditures and ecological damage from transport operation.

The number of lanes (or road capacity) must be optimized, as the road capacity influences velocity. The basic Greenshields model [7] determines velocity as

$$\nu = \nu_0 \left(1 - \frac{\lambda_0 \rho_j}{\nu n} \right), \tag{4}$$

where n the number of lanes, l – length of the road, ν_0 – free speed, ρ_j jam density per lane, λ_0 – average rate of flow of vehicles. Note that the total flow of personal and public transport can be expressed as follows

$$\lambda_0 = \lambda\rho + \frac{1}{2t_\omega}. \tag{5}$$

The first summand of (5) is car flow intensity and the second summand is public transport frequency (or intensity of flow) with average passenger waiting time t_ω.

The travel time $t = \frac{l}{v}$ is used as variable, so we express the number of lanes n, depending on the variable t

$$n = \frac{\left(\lambda\rho + \frac{1}{2t_w}\right)\rho_j t^2 v_0}{l(v_0 t - l)}. \tag{6}$$

Function (6) is convex downward in the parameter t, because the second derivative of (6) is as follows

$$\frac{2\lambda_0 \rho_j v_0 l}{(v_0 t - l)^3} > 0. \tag{7}$$

The function of municipal losses of the consists of several components:

1) time losses of passengers connected with travelling by public transport and cars;

2) road costs;

3) transportation expenditures.

Passengers losses consist of travel time on cars and the public transport

$$p\lambda t\gamma + (1 - p)\lambda t_t \gamma = p\lambda t\gamma + (1 - p)\lambda(t + \Delta t + t_w)\gamma, \tag{8}$$

where γ – average value of time; t_w – average waiting time for public transport, and Δt time loss in bus stops and travel time from origin to transit station and travel time from station exit to final destination.

The road cost (construction and maintenance) depends on the number of lanes, the road length and the coefficient c_r

$$nlc_r. \tag{9}$$

Interval between buses is $2t_w$. Public transportation frequency is inversely related to bus interval. Therefore all spending on public transport is

$$c_p \frac{1}{(2t_w)}, \tag{10}$$

where c_p – spending on public transportation (per trip).

Then the losses of the city will be as follows

$$F = \lambda\gamma pt + c\lambda p + \lambda\gamma(1-p)[t_w + t + \Delta t] + \frac{c_p}{2t_w} + \frac{\left(\lambda p + \frac{1}{2t_w}\right)c_r \rho_j t^2 v_0}{(v_0 t - l)} \rightarrow \min_{t_w, v}. \tag{11}$$

Its simple to proof that (13) is convex downward in the parameters t, t_w.

4   Statement of the Problem

First of all we must describe the set of strategies. For the passengers its probability of car using $P = [0, 1]$. For the municipal authorities waiting time t_w and travel time t must be less then the upper bound $\bar{t}$ ($T = \lfloor 0, \bar{t} \rfloor$).

4.1 First Statement of the Problem (Equilibrium between Interests of Passengers and Municipal Authorities)

Passenger minimizes quadratic deviation of decision from logit model. The municipal authorities minimize total losses (11) by variation of the travel time and the public transport frequency.

The decision of municipal authorities depends on passengers decisions. Also passengers use the authorities decision (public transport interval and car travel time) for their own decision-making. The interference between decision-making of passengers and authorities leads to game theoretical formulation of the statement of the problem.

$$G \to \min_{p}; \tag{12}$$

$$F \to \min_{t_w, t}. \tag{13}$$

Its simple to proof that the game $\langle P, T^2, -G, -F \rangle$ has a Nash equilibrium, therefore all conditions of theorem [2] about set of strategies and objective functions are performed.

4.2 Second Statement of the Problem (Optimal Solution for Municipal Authorities)

We can use passengers solution $p = p_c$ in (11). Therefore single-criterion (non-game) statement of the problem follows:

$$F_0 = \frac{(\lambda \gamma t + c\lambda) exp\{a_0 + a_t(t_t - t) + a_c(c_t - c)\}}{1 + exp\{a_0 + a_t(t_t - t) + a_c(c_t - c)\}} +$$
$$+ \frac{\lambda \gamma(t_w + t + \Delta t)}{1 + exp\{a_0 + a_t(t_t - t) + a_c(c_t - c)\}} + \frac{c_p}{2t_w} + \tag{14}$$
$$+ \left(\lambda \frac{exp\{a_0 + a_t(t_t - t) + a_c(c_t - c)\}}{1 + exp\{a_0 + a_t(t_t - t) + a_c(c_t - c)\}} + \frac{1}{2t_w} \right) \frac{c_r \rho_j t^2 v_0}{v_0 t - l} \to \min_{t_w, t}.$$

Objective function F_0 is not convex on t, t_w. But numerical methods allow to solve for two variables.

5 Numerical Example

In this part of paper the solutions of two tasks are being compared. Values of input parameters were $v_0 = 60$ kilometers per hour; $\Delta t = 0.12$h; $c = 40$ roubles; $c_t = 12$ roubles; $\lambda = 5000$ travelers; $\rho_j = 20$ meters per car; $c_r = 50000$ roubles per lane; $\gamma = 150$ roubles per hour; $c_p = 500$ roubles per trip.

Thus we have two statements of the problem: game statement (solution is equilibrium) and single-criterion one (optimal for municipal authorities).

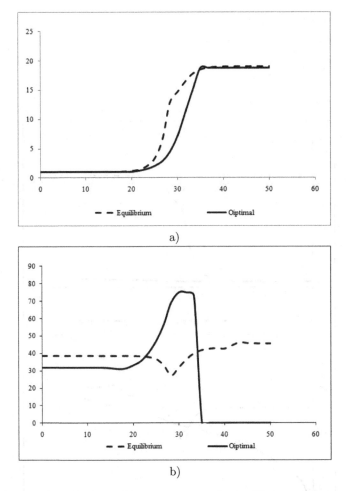

a)

b)

Fig. 2. Influence of the fare at public transport on the equilibrium situation a) number of lanes, b) frequency of public transport

Fig. 2a shows that number of lanes must be increased simultaneously with fare at the public transport. On the other hand public transport frequency (Fig. 2b) for game statement of the problem more stable than optimal solution (for some fare municipal authorities try to keep passengers on public transport, but for big fare its useless).

Fig. 3b shows that the growth of passenger flow intensity leads to the growth of frequency of public transport. But optimal solution needs of minimal number of lanes (Fig. 3a).

Under a very low level of life (value of time depends on income level) passengers have practically no choice (one lane is enough). On the other hand, under

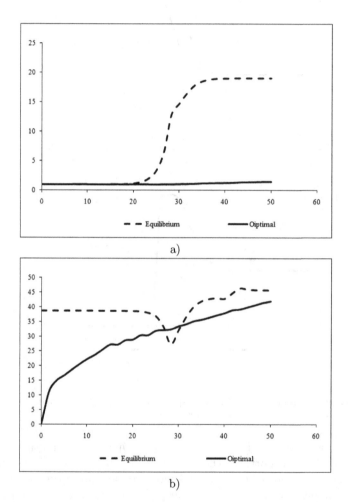

Fig. 3. Influence of the intensity of passenger flow on the equilibrium situation a) number of lanes, b) frequency of public transport

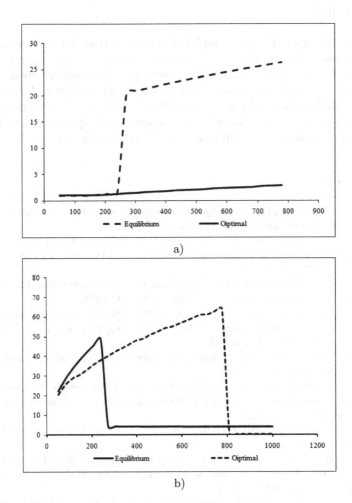

Fig. 4. Influence of the value of time on the equilibrium situation a) number of lanes, b) frequency of public transport

a high income level of population its more difficult to attract passengers to public transport and it degrades (Fig. 4b). Deference between game and non-game models increases when value of time varied from 300 to 800, but results for extreme values are similar.

6   Conclusion

In the paper the game theoretic model of transportation system is represented. The model has Nash equilibrium between interests of passengers and municipal authorities. The present research will be developed for real-sized road network. All road segments and passengers flows (between each origin and destination) will be included in the model as players. Thus model will allow to find optimal capacity of road segments and public transportation frequency for each route, because existence of Nash equilibrium will be accordingly proved.

References

1. Dodgson, J.S., Katsoulacos, Y.: Quality competition in bus services. Journal of Transport Economics and Policy 22, 263–281 (1988)
2. Glicksberg, I.L.: A Further Generalization of the Kakutani Fixed Point Theorem, with Application to Nash Equilibrium. Proceedings of the American Mathematical Society 3(1), 170–174 (1952)
3. Hollander, Y., Prashker, J.N.: The Applicability of Non-Cooperative Game Theory in Transport Analysis. Transportation 33(5), 481–496 (2006)
4. Horowitz, J.L., Koppelman, F.S., Lerman, S.R.: A Self Instructing Course in Dis-aggregate Mode Choice Modeling. Technology Sharing Program. U. S. Department of Transportation, Washington (1986)
5. Koryagin, M.E.: Competition of public transport flows. Automation and Remote Control 69(8), 1380–1389 (2008)
6. Bell, M.G.H., Wichiensin, M.: Road Use Charging and Inter-Modal User Equilib-rium: The Downs-Thompson Paradox Revisited. In: Energy, Transport, the Envi-ronment, pp. 373–383 (2012)
7. Greenshields, B.D.: A Study of Traffic Capacity. Highway Research Board Proceed-ings 14, 448–477 (1935)
8. Koryagin, M., Nesterova, A.: The influence of the availability of automobiles in a family on the choice of the method of transportation. Vestnik of Irkutsk State Technical University 48(1), 104–108 (2011)
9. Zhang, L.L., David, M., Zhu, S.: Agent-Based Model of Price Competition, Capacity Choice, and Product Differentiation on Congested Networks. Journal of Transport Economics and Policy 42(3), 435–461 (2008)

Nonparametric Estimation of Net Premium Functionals for Different Statuses in Collective Life Insurance[*]

Gennady Koshkin and Yaroslav Lopukhin

National Research Tomsk State University,
Lenin Avenue, 36, 634050
Tomsk, Russia
kgm@mail.tsu.ru,
yarl@mfu.tsu.ru

Abstract. The nonparametric estimators of net premiums in collective models of insurance for the different statuses of collective life insurance are proposed. The asymptotic normality and mean square convergence of the proposed estimates are proved. The main parts of asymptotic mean square errors (MSEs) for estimators of net premiums are found. The results of simulations show that the nonparametric estimates are just as good in practice.

Keywords: net premium, survival function, collective life insurance, k-survivor status, mixed status, nonparametric estimate.

1 Introduction

One of the main problem in actuarial mathematics is to find the "right" ratio between premiums and benefits. In this case, the calculation of net premiums allows to cover damages and to give the zero average income of the insurance company. Note that in the section, devoted to this area in the known book "Actuarial Mathematics" [1], is used the calculation of net premiums on the base of mortality tables. Interesting results by this approach have been presented in [2]-[8]. At present, the theory and practice of insurance is strongly required the using complex mathematical models and processes. Such results are obtained in [9]-[15]. In this paper, we develop the ideas from [16]-[20]. The nonparametric estimators of the net premium functionals for the different statuses of collective life insurance are constructed, and the asymptotic properties of the proposed estimators are studied.

---

[*] Supported by Russian Foundation for Basic Research, projects 13-08-00744, Tomsk State University Competitiveness Improvement Program, and the project "Goszadanie Minobrnauki Rossii".

A. Dudin et al. (Eds.): ITMM 2014, CCIS 487, pp. 223–233, 2014.

2   The Whole Life Insurance

In the long-term insurance the calculations of tariff rates take into account change of money value because the sum of S dollars after t years turns to the sum $S\,e^{\delta t}$ dollars, where δ is instantaneous interest rate.

The whole life insurance is example of the long-term insurance; in this situation the person pays p dollars to the insurance company, and the company pays b dollars to successors of the insured after his death. Though the premium p is less, than b, the company will receive the necessary sum b, since the premium is paid at the moment of the conclusion of the contract, and the payment is done great later.

We will use designations of actuarial mathematics later on. Let a random variable X denote the future lifetime, x be the age of the insured at the moment of policy issue, $T(x) = X - x$ denote the residual time of life. In time $T(x)$ premium, p, will turn in the sum, $p\,e^{\delta T(x)}$, and in this case the income of the company will be equal to

$$p\,e^{\delta T(x)} - b.$$

To have the required sum b dollars at the moment of client death, the insurance company must receive $b\,e^{-\delta T(x)}$ dollars at the time of policy issue. In economic terms, the sum $b\,e^{-\delta T(x)}$ expresses discounted value of the future insurance payment. As the above mentioned, this sum is a random variable, so it is natural to take as a net premium its average value

$$b\,\mathbf{E}\{e^{-\delta T(x)}\},$$

where $\mathbf{E}$ is the symbol of the expectation. In actuarial science, the benefit b is accepted as a unit payment, that is, $b = 1$, and the net premium of the whole life insurance $\overline{A}_x$ is equal to $\mathbf{E}\{e^{-\delta T(x)}\}$:

$$\overline{A}_x = E\{e^{-\delta T(x)}\} = \int_0^\infty e^{-\delta t} f_x(t)\,dt =$$

$$= \frac{1}{S(x)} \int_0^\infty e^{-\delta t} f(x+t)\,dt = \frac{1}{S(x)} \int_0^\infty e^{-\delta t}\,dF(x+t) \ , \tag{1}$$

where $F(x) = P(X \le x)$ is the distribution function of a random variable X, $S(x) = 1 - F(x) = P(X > x)$ is the survival function, $f(x) = -S'(x) = F'(x)$ is the curve of death or probability density function of random variable X.

Changing variables in the last integral of the formula (1), we have

$$\overline{A}_x = \frac{1}{S(x)} \int_x^\infty e^{-\delta(u-x)}\,dF(u) = \frac{\Phi(x,\delta)}{S(x)}, \tag{2}$$

where $\Phi(x,\delta) = \int_x^\infty e^{-\delta(u-x)}\,dF(u) = \int_0^\infty e^{-\delta(u-x)}\,I(x < u)\,dF(u)$, $I(A)$ is the indicator of a set A [1].

3   Collective Life Insurance

According to [1], the concept of status is the useful abstraction in the collective life insurance. Consider m members of ages $(x_1, x_2, \ldots, x_m)$ who desire to buy insurance policy. Let us denote the future lifetime of the k-th individual by $T(x_k) = X - x_k$. Let us put in a correspondence a status U with its future lifetime $T(U)$ and a set of numbers $T(x_1), T(x_2), \ldots, T(x_m)$ [16].

In [17]-[20], there were considered the cases of the joint-life status and the last-survivor status. In this paper, we consider the general case of the k-survivor status, which is denoted

$$U := \frac{k}{x_1 : x_2 : \ldots : x_m}$$

and exists as long as at least alive a k among the m individuals $(x_1), (x_2), \ldots, (x_m)$, i.e., it is considered destroyed upon the occurrence of the $(m - k + 1)$ death. It is clear that the joint-life status and last-survivor status are the special cases of the k-survivor status.

Also, separately consider the case of the $[k]$-deferred survivor status

$$U := \frac{[k]}{x_1 : x_2 : \ldots : x_m}.$$

Here, if a k of the m individuals $(x_1), (x_2), \ldots, (x_m)$ are exactly alive, i.e., the status starts at the $(m - k)$-th death and lasts until the $(m - k + 1)$-th death. This status is widely used in the calculation of sequences payments of limited duration [16].

Note that a new status can be also combined from the above base statuses. For example, the mixed status is a condition in which basis is a combination of statuses, and at least one of them asked for more than one individual.

Consider the following mixed statuses:

Case a: $(\overline{x_1 : x_2} : \overline{x_3 : x_4})$; the condition of this status persists if there are alive at least one of (x_1) and (x_2) and at least one of (x_3) and (x_4). The destruction moment of status $(\overline{x_1 : x_2} : \overline{x_3 : x_4})$ is defined as

$$T(U) = min\{max\{T(x_1), T(x_2)\}, max\{T(x_3), T(x_4)\}\}.$$

Case b: $\left(\overline{x_1 : x_2} : (x_3 : x_4)\right)$; here the condition persists if there are alive at least two of four, namely, (x_3) and (x_4), or when only one alive, and that either (x_1), or (x_2). The destruction moment of status $\left(\overline{x_1 : x_2} : (x_3 : x_4)\right)$ is defined as

$$T(U) = max\{max\{T(x_1), T(x_2)\}, min\{T(x_3), T(x_4)\}\}.$$

Case c: $(x_1 : x_2 : \overline{x_3 : x_4})$. In this case, the condition persists if there are alive (x_1), (x_2) and when one is alive, and it is either (x_3), or (x_4). The destruction moment of status $(x_1 : x_2 : \overline{x_3 : x_4})$ is defined as

$$T(U) = min\{T(x_1), T(x_2), max\{T(x_3), T(x_4)\}\}.$$

Similarly, the fracture point may be found for a combination of statuses.

4   Functionals of Net Premiums in Collective Life Insurance

Consider the random variables $Z_i = X_i - x_i$, $i = \overline{1, m}$. Order them in ascending and obtain the order statistics $Z_{(i)}$, $i = \overline{1, m}$. Then

$$\overline{A}_{x_1:x_2:\ldots:x_m} = \frac{\int_0^\infty e^{-\delta t} d\mathbf{P}\{Z_{(1)} < t\}}{\mathbf{P}\{Z_{(1)} > 0\}}$$

and

$$\overline{A}_{\overline{x_1:x_2:\ldots:x_m}} = \frac{\int_0^\infty e^{-\delta t} d\mathbf{P}\{Z_{(m)} < t\}}{\mathbf{P}\{Z_{(1)} > 0\}}.$$

By analogy with the individual cases (formula (1) and (2), we have

$$\overline{A}_{\underset{x_1:x_2:\ldots:x_m}{k}} = \frac{\int_0^\infty e^{-\delta t} d\mathbf{P}\{T(U) < t\}}{S(x_1, x_2, \ldots, x_m)} = \frac{\int_0^\infty e^{-\delta t} d\mathbf{P}\{Z_{(m-k+1)} < t\}}{\mathbf{P}\{Z_{(1)} > 0\}} =$$

$$= \frac{\Phi(x_1, x_2, \ldots, x_m, \delta, k)}{S(x_1, x_2, \ldots, x_m)}. \tag{3}$$

In the case of the $[k]$-deferred survivor status

$$\mathbf{P}\left\{Z_{(m-k)} \le t < Z_{(m-k+1)}\right\} = \mathbf{P}\left\{t < Z_{(m-k+1)}\right\} - \mathbf{P}\left\{t < Z_{(m-k)}\right\} =$$

$$= 1 - \mathbf{P}\left\{Z_{(m-k+1)} < t\right\} - 1 + \mathbf{P}\left\{Z_{(m-k)} < t\right\} =$$

$$= \mathbf{P}\left\{Z_{(m-k)} < t\right\} - \mathbf{P}\left\{Z_{(m-k+1)} < t\right\},$$

and the net premium

$$\overline{A}_{\underset{x_1:x_2:\ldots:x_m}{[k]}} = \frac{\int_0^\infty e^{-\delta t} d\mathbf{P}\left\{Z_{(m-k)} < t\right\} - \int_0^\infty e^{-\delta t} d\mathbf{P}\left\{Z_{(m-k+1)} < t\right\}}{\mathbf{P}\{Z_{(1)} > 0\}} =$$

$$= \frac{\Phi(x_1, x_2, \ldots, x_m, \delta, k - 1) - \Phi(x_1, x_2, \ldots, x_m, \delta, k)}{S(x_1, x_2, \ldots, x_m)} =$$

$$= \overline{A}_{\underset{x_1:x_2:\ldots:x_m}{k-1}} - \overline{A}_{\underset{x_1:x_2:\ldots:x_m}{k}}.$$

Similarly, the net premiums functionals for mixed cases can be written as

$$\overline{A}_{\overline{x_1:x_2}:\overline{x_3:x_4}} = \frac{1}{S(x_1, x_2, x_3, x_4)} \int_0^\infty e^{-\delta t} \times$$

$$\times d\mathbf{P}\left\{min\left\{max\left(X_1 - x_1, X_2 - x_2\right), max\left(X_3 - x_3, X_4 - x_4\right)\right\} \le t\right\},$$

$$\overline{A}_{\overline{x_1:x_2}:(x_3:x_4)} = \frac{1}{S(x_1, x_2, x_3, x_4)} \int_0^\infty e^{-\delta t} \times$$

$$\times d\mathbf{P}\left\{max\left\{max\left(X_1 - x_1, X_2 - x_2\right), min\left(X_3 - x_3, X_4 - x_4\right)\right\} \le t\right\},$$

$$\overline{A}_{x_1:x_2:\overline{x_3:x_4}} = \frac{1}{S(x_1, x_2, x_3, x_4)} \int_0^\infty e^{-\delta t} \times$$

$$\times d\mathbf{P}\left\{min\left\{X_1 - x_1, X_2 - x_2, max\left(X_3 - x_3, X_4 - x_4\right)\right\} \le t\right\}.$$

5 Nonparametric Estimates of Net Premiums in Collective Life Insurance

Let $(Z_{11}, \ldots, Z_{m1})$, $(Z_{12}, \ldots, Z_{m2}), \ldots,$ $(Z_{1n}, \ldots, Z_{mn})$ be an m-dimensional random sample of a size n. Denote the corresponding ordered set as

$(Z_{(1)1}, \ldots, Z_{(m)1}); (Z_{(1)2}, \ldots, Z_{(m)2}), \quad \ldots, \quad (Z_{(1)n}, \ldots, Z_{(m)n})$.

Estimate the distribution function $\mathbf{P}\left\{Z_{(m-k+1)} < t\right\}$ and the survival function $\mathbf{P}\left\{Z_{(1)} > 0\right\}$ by the simple nonparametric estimates

$\frac{1}{n}\sum_{i=1}^{n} I\left(Z_{(m-k+1)i} < t\right)$ and $\frac{1}{n}\sum_{i=1}^{n} I\left(Z_{(1)i} > 0\right)$.

Then nonparametric estimate of (3) is equal to

$$\hat{\bar{A}}_{\underset{x_1:x_2:\ldots:x_m}{k}} = \frac{\Phi_n(x_1, x_2, \ldots, x_m, \delta, k)}{S_n(x_1, x_2, \ldots, x_m)} =$$

$$= \frac{1}{S_n(x_1, x_2, \ldots, x_m)} \int_0^\infty e^{-\delta t} \left(\frac{1}{n}\sum_{i=1}^{n} I(Z_{(m-k+1)i} \leq t)\right)' dt =$$

$$= \frac{1}{S_n(x_1, x_2, \ldots, x_m)} \frac{1}{n} \sum_{i=1}^{n} \int_0^\infty e^{-\delta t} \bar{\delta}\left(t - Z_{(m-k+1)i}\right) dt =$$

$$= \frac{\frac{1}{n}\sum_{i=1}^{n} e^{-\delta\left(Z_{(m-k+1)}\right)} I(Z_{(m-k+1)} > 0)}{\frac{1}{n}\sum_{i=1}^{n} I(Z_{(1)} > 0)}, \tag{4}$$

where $\bar{\delta}(u)$ is the Dirac delta-function. Obtaining the formula (4), we used the filtering property of the delta-function: $\int_{-\infty}^{+\infty} \varphi(x)\bar{\delta}(x - a)dx = \varphi(a)$.

In the case of the $[k]$-deferred survivor status, the nonparametric plug-in estimate of net premium is defined as

$$\hat{\bar{A}}_{\underset{x_1:x_2:\ldots:x_m}{[k]}} = \hat{\bar{A}}_{\underset{x_1:x_2:\ldots:x_m}{k-1}} - \hat{\bar{A}}_{\underset{x_1:x_2:\ldots:x_m}{k}} =$$

$$= \frac{\Phi_n(x_1, x_2, \ldots, x_m, \delta, k - 1) - \Phi_n(x_1, x_2, \ldots, x_m, \delta, k)}{S_n(x_1, x_2, \ldots, x_m)}.$$

Similarly, the nonparametric estimates of the net premiums functionals for mixed cases can be found. For the status $\overline{x_1 : x_2} : \overline{x_3 : x_4}$, we have

$$\hat{\bar{A}}_{\overline{x_1:x_2}:\overline{x_3:x_4}} = \frac{1}{n\,S_n(x_1, x_2, x_3, x_4)}.$$

$$\cdot \sum_{i=1}^{n} e^{-\delta \min\{\max(X_{1i}-x_1, X_{2i}-x_2), \max(X_{3i}-x_3, X_{4i}-x_4)\}} \times$$

$$\times min\left\{max\left(X_{1i} - x_1, X_{2i} - x_2\right), max\left(X_{3i} - x_3, X_{4i} - x_4\right)\right\},$$

$$\hat{A}\overline{_{x_1:x_2:(x_3:x_4)}} = \frac{1}{n\,S_n\,(x_1,x_2,x_3,x_4)} \cdot$$

$$\cdot \sum_{i=1}^{n} e^{-\delta \max\{\max(X_{1i}-x_1,X_{2i}-x_2),\min(X_{3i}-x_3,X_{4i}-x_4)\}} \times$$

$$\times \max\left\{\max\left(X_{1i}-x_1,X_{2i}-x_2\right), \min\left(X_{3i}-x_3,X_{4i}-x_4\right)\right\},$$

$$\hat{A}_{x_1:x_2:\overline{x_3:x_4}} = \frac{1}{n\,S_n\,(x_1,x_2,x_3,x_4)} \cdot$$

$$\cdot \sum_{i=1}^{n} e^{-\delta \min\{X_{1i}-x_1,X_{2i}-x_2,\max(X_{3i}-x_3,X_{4i}-x_4)\}} \times$$

$$\times \min\left\{X_{1i}-x_1, X_{2i}-x_2, \max\left(X_{3i}-x_3,X_{4i}-x_4\right)\right\} \quad .$$

6   Asymptotic Properties of Functions of Statistics

Present the auxiliary results, which will be used below.

Let the function $H(\phi) : R^s \to R^1$, and $\phi = \phi(x) = (\phi_1(x),\ldots,\phi_s(x))$ be an s-dimensional bounded function; $H_j(\phi) = \frac{\partial H(\phi)}{\partial \phi_j}$, $j = \overline{1,s}$, $\nabla H(\phi) = (H_1(\phi),\ldots,H_s(\phi))$; the symbol T denote the transpose; $t_n = (t_{1n},\ldots,t_{sn})$ be an s-dimensional statistic, $t_{jn} = t_{jn}(x) = t_{jn}(x,X_1,\ldots,X_n)$, $j = \overline{1,s}$; $\|t_n\| = \sqrt{t_{1n}^2 + t_{2n}^2 + \ldots + t_{sn}^2}$ be the Euclidean norm of t_n; $t = t(x) = (t_1(x),\ldots,t_s(x))$ be an s-dimensional bounded function. Denote by $\Rightarrow N\{\mu,\sigma\}$ the symbol of weak convergence of sequence of random variables to the s-dimensional normal random variable with mean $\mu = (\mu_1,\mu_2,\ldots,\mu_s)$ and symmetric covariance matrix $\sigma = \|\sigma_{ij}\|$, $0 < \sigma_{jj} = \sigma_{jj}(x) < \infty, j = \overline{1,s}$. Also, let N and N^+ be sets of integers and even integers, accordingly.

Definition. Function $H(\cdot) \in N_{\nu,s}(t)$ if $H(z) : R^s \to R^1$ and function $t = t(x)$ at a point x has an ε-neighborhood $\{z : |z_i - t_i| < \varepsilon; i = \overline{1,s}\}$, in which $H(z)$ and its derivatives $\frac{\partial H(z)}{\partial z_j}$ up to the order ν are continuous and bounded.

Theorem 1. [21]. Let: 1)$H(z), \{H(t_n)\} \in N_{2,s}(t), 2)\mathbf{E}\|t_n - t\|^i = O\left(d_n^{-i/2}\right)$. Then for any $k \in N$

$$\left|\mathbf{E}\left[H(t_n) - H(t)\right]^k - \mathbf{E}\left[\nabla H(t)(t_n - t)^T\right]^k\right| = O\left(d_n^{-(k+1)/2}\right) \quad . \qquad (5)$$

Theorem 2. [21]. If a random vector $q_n(t_n - \phi) \Rightarrow N_s\{\mu,\sigma\}$ for some number sequence $q_n \uparrow \infty$, the function $H(t) \in N_{1,s}(\phi), \nabla H(\phi) \neq 0$, then the random variable

$$q_n\left(H(t_n) - H(t)\right) \Rightarrow N_1\{\nabla H(\phi)\mu^T, \nabla H(\phi)\sigma\nabla H^T(\phi)\} \quad . \qquad (6)$$

7   Asymptotic Properties of Estimates of Net Premiums

Formulate the result on asymptotic properties of nonparametric estimate (4).

Theorem 3. *If the survival function* $S(x_1, x_2, \ldots, x_m) \neq 0$, *then*

$$\sqrt{n}[\hat{\overline{A}}_{\overline{x_1:x_2:\ldots:x_m}}^{k} - \overline{A}_{\overline{x_1:x_2:\ldots:x_m}}^{k}] \Rightarrow$$

$$\Rightarrow N_1 \left\{ 0. \frac{\Phi(x_1, x_2, \ldots, x_m, 2\delta, k)\, S(x_1, x_2, \ldots, x_m) - \Phi^2(x_1, x_2, \ldots, x_m, \delta, k)}{n\, S^3(x_1, x_2, \ldots, x_m)} \right\},$$

and the MSE of nonparametric estimate (4) is equal to

$$u^2(\hat{\overline{A}}_{\overline{x_1:x_2:\ldots:x_m}}^{k}) =$$

$$= \frac{\Phi(x_1, x_2, \ldots, x_m, 2\delta, k)\, S(x_1, x_2, \ldots, x_m) - \Phi^2(x_1, x_2, \ldots, x_m, \delta, k)}{n\, S^3(x_1, x_2, \ldots, x_m)} +$$

$$+ O\left(\frac{1}{n^{3/2}}\right).$$

Proof. In the notation of Theorem 1, we have: $s = 2$, $t_n = (t_{1n}, t_{2n})$, $t_{1n} = \Phi_n(x_1, x_2, \ldots, x_m, \delta, k)$, $t_{2n} = S_n(x_1, x_2, \ldots, x_m)$,

$$H(t_n) = \frac{t_{1n}}{t_{2n}} = \frac{\Phi_n(x_1, x_2, \ldots, x_m, \delta, k)}{S_n(x_1, x_2, \ldots, x_m)} = \hat{\overline{A}}_{\overline{x_1:x_2:\ldots:x_m}}^{k},$$

$$t = (t_1, t_2) = (\Phi_n(x_1, x_2, \ldots, x_m, \delta, k), S(x_1, x_2), \ldots, x_m),$$

$$H(t) = \frac{t_1}{t_2} = \overline{A}_{\overline{x_1:x_2:\ldots:x_m}}^{k}, \quad d_n = n.$$

Then, by Lemma 3.1 from [22], the following expressions are hold:

$$\mathbf{E}|\Phi_n(x_1, x_2, \ldots, x_m, \delta, k) - \Phi(x_1, x_2, \ldots, x_m, \delta, k)|^i = O\left(n^{-\frac{i}{2}}\right),$$

$$\mathbf{E}|S_n(x_1, x_2, \ldots, x_m) - S_n(x_1, x_2, \ldots, x_m)|^i = O\left(n^{-\frac{i}{2}}\right).$$

If in formula (5) $k = 1$, then

$$\mathbf{E}\{\hat{\overline{A}}_{\overline{x_1:x_2:\ldots:x_m}}^{k}\} = \overline{A}_{\overline{x_1:x_2:\ldots:x_m}}^{k} + \frac{1}{t_2}\mathbf{E}\{t_{1n} - t_1\} - \frac{t_1}{t_2^2}\mathbf{E}\{t_{2n} - t_2\} + O\left(\frac{1}{n}\right).$$

As functions $t_1 = t_1(x_1, x_2, \ldots, x_m)$, $t_2 = t_2(x_1, x_2, \ldots, x_m)$ are continuous, then $\mathbf{E}\{t_{1n}\} = t_1$, $\mathbf{E}\{t_{2n}\} = t_2$, and $\mathbf{E}\{\hat{\overline{A}}_{\overline{x_1:x_2:\ldots:x_m}}^{k}\} = \overline{A}_{\overline{x_1:x_2:\ldots:x_m}}^{k} + O\left(\frac{1}{n}\right)$, i.e., $\hat{\overline{A}}_{\overline{x_1:x_2:\ldots:x_m}}^{k}$ is asymptotically unbised estimate.

Now, putting $k = 2$ and taking into account unbiasedness t_{1n}, t_{2n}, we have

$$u^2(\hat{A}_{\overline{x_1:x_2:\ldots:x_m}}) = \frac{1}{t_2^2} \mathbf{D}\{t_{1n}\} + \frac{t_1^2}{t_2^4} \mathbf{D}\{t_{2n}\} - 2\frac{t_1}{t_2^3} cov\{t_{1n}, t_{2n}\} + O\left(\frac{1}{n^{3/2}}\right). \quad (7)$$

Taking into account the randomness of sample
$(Z_{11}, \ldots, Z_{m1}), (Z_{12}, \ldots, Z_{m2}), \ldots, (Z_{1n}, \ldots, Z_{mn})$,
we find the corresponding variances and covariance:

$$\mathbf{D}\{t_{1n}\} = \mathbf{D}\left\{\frac{1}{n}\sum_{i=1}^{n} \varphi_i(x_1, x_2, \ldots, x_m, \delta, k)\right\} = \frac{1}{n}\mathbf{D}\{\varphi_1(x_1, x_2, \ldots, x_m, \delta, k)\} =$$

$$= \frac{1}{n}\left(\mathbf{E}\{\varphi_1(x_1, x_2, \ldots, x_m, 2\delta, k)\} - \mathbf{E}^2\{\varphi_1(x_1, x_2, \ldots, x_m, \delta, k)\}\right) =$$

$$= \frac{1}{n}\left(\Phi(x_1, x_2, \ldots, x_m, 2\delta, k) - \Phi^2(x_1, x_2, \ldots, x_m, \delta, k)\right),$$

$$\mathbf{D}\{t_{2n}\} = \mathbf{D}\left\{\frac{1}{n}\sum_{i=1}^{n} s_i(x_1, x_2, \ldots, x_m)\right\} =$$

$$= \frac{1}{n} S(x_1, x_2, \ldots, x_m)(1 - S(x_1, x_2, \ldots, x_m)),$$

$$cov\{t_{1n}, t_{2n}\} = \frac{1}{n} cov\{\varphi_1(x_1, x_2, \ldots, x_m, \delta, k), s_1(x_1, x_2, \ldots, x_m))\} =$$

$$= \frac{1}{n}\Phi(x_1, x_2, \ldots, x_m, \delta, k)(1 - S(x_1, x_2, \ldots, x_m)).$$

Substituting the found expressions in (7), we obtain the second assertion of Theorem 3.

In the notation of Theorem 2 we have:

$$\nabla H(\varphi) = (H_1, H_2), \quad H_1 = \frac{1}{S(x_1, x_2, , x_m)},$$

$H_2 = -\frac{\Phi(x_1, x_2, , x_m, \delta, k)}{S^2(x_1, x_2, , x_m)}$, $\quad q_n = \sqrt{n}$. As $S(x_1, x_2, , x_m) \neq 0$,
function $H(t) \in N_{1,2}(t)$.
And so,

$$\mathbf{E}\{S_n(x_1, x_2, , x_m)\} = S(x_1, x_2, , x_m),$$
$$\mathbf{E}\{\Phi_n(x_1, x_2, , x_m, \delta, k)\} = \Phi(x_1, x_2, , x_m, \delta, k),$$

i.e., $\mu^T = 0$; $\quad \sigma_{11} = \Phi(x_1, x_2, , x_m, 2\delta, k) - \Phi^2(x_1, x_2, , x_m, \delta, k)$,

$$\sigma_{12} = \sigma_{21} = \Phi(x_1, x_2, , x_m, \delta, k)(1 - S(x_1, x_2, , x_m)),$$
$$\sigma_{22} = S(x_1, x_2, , x_m)(1 - S(x_1, x_2, , x_m)).$$

That is why substituting the found expressions in (6), we obtain the first assertion of Theorem 3:

$$\nabla H(t)\mu^T = 0,$$
$$\nabla H(t)\sigma \nabla H(t)^T = S^{-3}(x_1, x_2, \ldots, x_m)\cdot$$
$$\cdot\left(\Phi(x_1, x_2, \ldots, x_m, 2\delta, k)S(x_1, x_2, \ldots, x_m) - \Phi^2(x_1, x_2, \ldots, x_m, \delta, k)\right).$$

Theorem 3 is proved.

8   Synthesis of Nonparametric Estimators of Net Premiums in Collective Life Insurance for Other Forms of Insurance

The above considered estimates of net premiums were constructed for the whole insurance; here we consider other forms of insurance.

8.1   The p-Years Term Life Insurance

In this case, the benefit to pay if the insured will die during of the contract validity. The company does not pay the benefit if the insured has lived the p years. Then

$$\hat{\bar{A}}_{\underline{x_1:x_2:\ldots:x_m}^{[k]}:p]} = \frac{\Phi_n(x_1, x_2, \ldots, x_m, \delta, k, p)}{S_n(x_1, x_2, \ldots, x_m)} =$$

$$= \frac{\frac{1}{n} \sum_{i=1}^{n} e^{-\delta\left(Z_{(m-k+1)}\right)} I(0 < Z_{(m-k+1)} \leq p)}{\frac{1}{n} \sum_{i=1}^{n} I(Z_{(1)} > 0)}.$$

8.2   The p-Years Endowment Life Insurance

Such form of insurance provides for a payment either following the death of the insured or upon his survival to the end of the p-years term. The given form of insurance accumulates the client's capital. The nonparametric estimate of net premium is expressed by the formula

$$\hat{\bar{A}}^s_{\underline{x_1:x_2:\ldots:x_m}^{[k]}:p]} = \frac{S_n(x_1, x_2, \ldots, x_m) - S_n(x_1 + p, x_2 + p, \ldots, x_m + p)}{S_n(x_1, x_2, \ldots, x_m)} \times$$

$$\times \hat{\bar{A}}_{\underline{x_1:x_2:\ldots:x_m}^{[k]}:p]} + \frac{S_n(x_1 + p, x_2 + p, \ldots, x_m + p)}{\overline{S}_n(x_1, x_2, \ldots, x_m)} e^{-\delta p}.$$

8.3   The r Years Deferred Life Insurance

This form of insurance provides for a benefit following the death of the insured when he dies at least r years following policy issue. The net premium is expressed in the form

$$_{r|}\hat{\bar{A}}_{\underline{x_1:x_2:\ldots:x_m}^{[k]}} = \frac{\Phi_n(x_1, x_2, \ldots, x_m, \delta, k, r)}{S_n(x_1, x_2, \ldots, x_m)} = \frac{\frac{1}{n} \sum_{i=1}^{n} e^{-\delta\left(Z_{(m-k+1)}\right)} I(r < Z_{(m-k+1)})}{\frac{1}{n} \sum_{i=1}^{n} I(Z_{(1)} > 0)}.$$

The properties of these nonparametric estimators are similar to the properties of (4).

9   Conclusion

The nonparametric estimates show their adaptability if the distribution is changed and exceed parametric estimates, oriented on the best result only for its own distributions. Often, the MSEs of nonparametric estimates are less than the MSEs of parametric estimates in 2-3 times. The main simulations results are obtained by making use of data from the Makeham and the de Moivre distributions.

References

1. Bowers, N.L., Gerber, H.U., Hickman, J.C., Jones, D.A., Nesbitt, C.J.: Actuarial Mathematics, 624p. The Society of Actuaries, Itasca (1986)
2. Bloink, R.: Premium Financed Surprises: Cancellation of Indebtedness Income and Financed Life Insurance. Tax Lawyer 63(3), 223–227 (2010)
3. Buhlmann, H.: The general economic premium principle. ASTIN Bulletin 14, 13–21 (1984)
4. Dong, Y.: Fair Valuation of Life Insurance Contracts under a Correlated Jump Diffusion Model. ASTIN Bulletin 41(2), 429–447 (2011)
5. Faust, R., Schmeiser, H., Zemp, A.: A Performance Analysis of Participating Life Insurance Contracts. Insurance: Mathematics and Economics 51(1), 158–171 (2013)
6. Maurer, R., Rogalla, R., Siegelin, I.: Participating Payout Life Annuities: Lessons from Germany. ASTIN Bulletin 43(2), 159–187 (2013)
7. Schmeiser, H., Wagner, J.: A Joint Valuation of Premium Payment and Surrender Options in Participating Life Insurance Contracts. Insurance: Mathematics and Economics 49(3), 580–596 (2011)
8. Sliwinski, A., Michalski, T., Roszkiewicz, M.: Demand for Life Insurance–An Empirical Analysis in the Case of Poland. The Geneva Papers on Risk and Insurance 38, 62–87 (2013)
9. Aase, K.K.: Premiums in a dynamic model of a reinsurance market. Scandinavian Actuarial Journal, 134–160 (1993)
10. Arora, N., Arora, P.: Insurance Premium Optimization: Perspective of Insurance Seeker and Insurance Provider. Journal of Management and Science 4(1), 43–53 (2014)
11. Bohnert, A., Gatzert, N.: Analyzing Surplus Appropriation Schemes in Participating Life Insurance from the Insurer's and the Policyholder's Perspective. Insurance: Mathematics and Economics 50(1), 64–78 (2012)
12. Gatzert, N.: Asset Management and Surplus Distribution Strategies in Life Insurance: An Examination with Respect to Risk Pricing and Risk Measurement. Insurance: Mathematics and Economics 42(2), 839–849 (2008)
13. Gatzert, N., Wesker, H.: The Impact of Natural Hedging on a Life Insurer's Risk Situation. Journal of Risk Finance 13(5), 396–423 (2012)
14. Huang, H., Lee, Y.-T.: Optimal Asset Allocation for a General Portfolio of Life Insurance Policies. Insurance: Mathematics and Economics 46(2), 271–280 (2010)
15. Rocio, P., Aguilar, C., Xu, C.: Design Life Insurance Participating Policies using Optimization Techniques, International Journal of Innovative Computing. Information and Control 6(4), 1655–1666 (2010)
16. Gerber, H.U.: Principles of Premium Calculation and Reinsurance. In: Transactions of the 21st International Congress of Actuaries, pp. 137–142 (1980)

17. Koshkin, G.M.: Nonparametric Estimation of Moments of Net Premium in Life Insurance. In: Aivazyan, S.A., Kharin, Y.S. (eds.) Computer Data Analysis and Modeling: Proceedings of the Fifth International Conference, June 8-12. A-M, vol. 1, pp. 148–153. BSU, Minsk (1998)
18. Koshkin, G.M., Lopukhin, Y.N.: On estimation of net premium in collective life insurance. In: Proceedings of the 5th Korea-Russian International Symposium on Science and Technology, KORUS 2001, June 26-July 3, vol. 2, pp. 296–299. Tomsk Polytechnic University, Tomsk (2001)
19. Koshkin, G.M., Lopukhin, Y.N.: Estimation of Net Premiums in Collective Models of Life Insurance. In: XIth Annual International AFIR Colloquium, September 6-7, vol. 2, pp. 447–457. Canadian Institute of Actuaries, Toronto (2001)
20. Koshkin, G.M., Lopukhin, Y.N.: Nonparametric Estimation of Net Premiums in Collective Insurance. In: Aivazyan, S., Kharin, Y., Rieder, H. (eds.) Computer Data Analysis and Modeling: Robustness and Computer Intensive Methods: Proceedings of the Sixth International Conference, Minsk, September 10-14. A-K, vol. 1, pp. 236–241. BSU, Minsk (2001)
21. Dobrovidov, A.V., Koshkin, G.M., Vasiliev, V.A.: Non-parametric State Space Models, Heber, UT 84032, USA, p. 501. Kendrick Press, Inc. (2012)
22. Ibragimov, I.A., Khas'minski, R.Z.: Asymptotic Theory of Estimation, p. 528. Nauka, Moskva (1979) (in Russian)

Tasks with Various Types of Answers in Systems Based on Mixed Diagnostic Tests*

Yuri Kostyuk and Vladimir Razin

National Research Tomsk State University, Tomsk, Russia
kostyuk_y_l@sibmail.com

Abstract. Using of tasks with different types of answers in intelligent teaching-testing systems based on mixed diagnostic tests is discussed. Tasks with formula-type answers are discussed in detail. A simple, intuitive language for writing formulae is described. A technology to verify answer formula compliance with a formula prepared in advance by the test developer is proposed. Description of a checking algorithm using LL(1)-parser, Reverse Polish Notation string generator and an interpreter to calculate the answer formula value for a prearranged set of input data is provided.

Keywords: teaching-testing system, mixed diagnostic tests, context-free language, LL(1)-parser, Reverse Polish Notation.

1 Introduction

Implementing mixed diagnostic tests [1] at different steps of testing usually requires using test tasks with different forms of answers in order to increase tests' efficiency. Traditional tests assume the presence of tasks with answers belonging to one of the following types [2]:

- answer in closed form, requiring a choice of one option, either correct or incorrect one;
- answer in form of multiple choice, requiring a choice of several options from some list;
- answer that requires establishing correspondence between two sequences;
- answer that requires establishing the correct sequence;
- answer in open form that can be represented as a number, a word or a text in the arbitrary form.

When the computer is used to check the answers it is easy to implement testing programs intended to check the answers automatically for almost all types of answers. The only exception is an open-form answer tasks that have to accept a wide variety of possible variations of a correct answer. As a rule, an answer in open form has to be written in some strict form (e.g. a numerical value in a

---

* Supported by Russian Foundation for Basic Research, projects no. 13-07-98037-r_sibir_a.

A. Dudin et al. (Eds.): ITMM 2014, CCIS 487, pp. 234–241, 2014.

predefined format), but such a requirement can significantly lower test applicability. To overcome this difficulty, some computer systems for testing students' knowledge [3] allow answers in open form described by regular expression. Such an answer is considered to be correct if it corresponds to some regular expression defined by the test developer in advance. However, this way is rather specific and doesn't cover all the variants of tasks with open answer.

Another kind of answers in open form is an answer in the form of essay. Such an answer can be processed, for example, by using Automated Essay Scoring and Calibrated Peer Review, both described in [4].

A usage of a new kind of tasks requiring an answer in the form of algebraic formula is proposed in this paper. Such tasks can be effectively used in tests for different subjects. Let's assume that such a formula answer includes values specified in the task itself, as long as common constant values and operations. In order to automate the checkup of such an answer one has to solve a problem of establishing equality of two formulae: the formula given by test developers and the formula provided as an answer. However, there exists no uniform method to solve this problem. For that reason, a particular method for solving this problem that can be convenient in use and effective in practice is suggested in this paper.

This paper describes a simple language for writing formulae, an analyzer for that language and Reverse Polish Notation (RPN) generator. Unlike most well-known translators, an RPN generator described in this paper has simple tabular structure and can be easily implemented in a testing system.

2   General Principles of Formula Type Answer Check Up

The algebraic formulae involving the calculation of one or several variables are to be considered. In general, a programming language can be used for formulae notation, however, the learner is supposed not just to know the language as it is, but also to be able to program in it. If the knowledge of the subject other than Computer Science is tested, such a requirement is definitely excessive. For that reason we introduce an intuitively understandable simplified language for the formulae notation, in which only the variables that are defined in the task itself are allowed. The formulae given in this language should be written in one or several lines with all the operations clearly signified. The notation of variables and constants is restricted by one or several Latin letters as well as Arabic numbers.

The algorithm of answer checkup is executed as follows:

1. Formula-answer (or a set of formulae) written by the learner is sent to the testing system and then analyzed. After that it is translated into a RPN. If some formal (syntactic) errors in the notation are found, the testing system must immediately inform the student that he has to correct the errors and give the right answer. This is of great importance since the testing is aimed at the knowledge checkup in the particular field rather than the language of formulae notation.

2. If there are no errors in the formula notation, RPN received by the testing system is transferred to the interpreter that makes the appropriate computation, giving the input data prepared in advance to its input. Matching of the results obtained is made using set of values prepared by the developers. If they are in full agreement, the formula is considered to be right. If approximate numbers are involved, matching should be made in terms of the computational error.

Therefore, in developing the test of this kind it is necessary not only to state the task distinctly but also to provide several sets of input data and the relevant results which the learner must get when applying the appropriate formula.

3   Formula Notation Language

The simple language of formulae is proposed. In a formula, variables predefined in the task, variables defined by the formula and numerical constants, are used. The variables are written according to the rules worked out in programming languages, i.e. they can consist in several Latin (lower- and upper-case) letters and numbers and they are to begin with the letter. The constants are written in numerical values, they can be integers and real numbers, fixed-point numbers and/or decimal exponent, either signed or unsigned. In the latter case, the letter e goes before the exponent. For some mathematical constants various alphabetical characters, e.g. pi, can be reserved. If one must use specific constants (for example, the physical ones), their notation should be given in the task.

Each formula notation is restricted by one line, if there is more than one formula, a semicolon and/or a new line should be used. The notation starts with the variable that must be computed, after that goes the equal mark, then the expression with variables, constants and round brackets. Operations are grouped by their priorities. Of highest priority is raising to the power operation (sign $\wedge$). Next are unary operations requiring only one operand. Among them are unary plus and minus, as well as operations which define the standard mathematical functions: square root (sqrt), sinus (sin), cosinus (cos), and so on. After that go multiplication (sign *) and division (sign /) operations. Of lowest priority are addition and subtraction operations (with two operands). Any part of the expression can be restricted with round brackets either on the right or on the left so as to change the order of the execution of operations.

The syntax of the language like this one can be written as the set of productions of context-free grammar [5]. For brevity sake, nonterminals (syntactic notions) are to be written with upper case Latin letters. Right and left parts of the production will be divided by the arrow, and terminals (by groups) written in the following way: raising to the power operation - sign $\wedge$, unary operations (apart from plus and minus) letter s, multiplication and division operations - sign *, plus and minus - sign +, variables letter a, constants letter k, brackets round brackets, formula separator (semicolon or new line) sign;.

Strictly speaking, the notation of variables and constants and operation designations, which require more than one character are not grammar terminals.

However, at the preliminary processing of the formula being analyzed the lexer [5] can be of use. It recognizes such grammar structures (lexical units) and at the output translates them into terminals. Lexer makes it possible to facilitate and speed up the further analysis.

The productions of the formulae language, with vertical line dividing different right parts for the group of rules, whose left parts are the same, are given below.

$$P \rightarrow P; A|A|A;$$
$$A \rightarrow a = S$$
$$S \rightarrow S + T|T$$
$$T \rightarrow T * F|F \qquad (1)$$
$$F \rightarrow W| + F|sF$$
$$W \rightarrow W \wedge G|G$$
$$G \rightarrow (S)|a|k$$

In these productions the initial nonterminal symbol is P.

4 LL(1)-Parsing of Formulae

One of the most effective methods of the analysis of character strings produced by context- free grammar is LL(1)-parsing [6]. In order to apply this method of parsing it is necessary to transform it to Greibach normal form when the right parts can be either empty or start with terminal character in all the productions. The grammar of this kind admits determinative LL(1)- parsing if for each group of productions with the same nonterminal in the left part, the right parts will be distinguishable by the first terminal.

Let entry string of characters be always completed with the boundary character $\perp$. In order the LL(1)-analyzer to work, it is necessary to draw a table whose columns are designated with terminals, the boundary character $\perp$ included, and strings are nonterminal of the transformed grammar. For all productions of grammar given as

$$A \rightarrow \alpha\gamma,$$

where A is a nonterminal, α – a terminal, γ – a string of both terminals and nonterminals, the right part of the rule $\alpha\gamma$ is placed on the intersection of the line marked with nonterminal A with the column designated as terminal a. If the production is in the form of

$$A \rightarrow \lambda,$$

where λ is the empty string, all the cells designated with nonterminal A and free of the other rules are written as λ. Before the work starts, first the boundary character $\perp$ is put into the stack of the parser, then goes the initial nonterminal symbol. At each stage the parser gets yet another terminal from the lexer and executes one of the two actions:

1. if there is a nonterminal on top of the stack, depending on what another entry terminal is, this nonterminal is changed in the stack by the right part characters of the corresponding production, with the characters written in the reverse order. If for another entry terminal character, written in the table is , a non-terminal is re-moved from the stack. If there is an empty cell in the table, the parser fixes the error in the input string;
2. if there is a terminal on top of the stack, it is compared with the another entry character. If there is a match, a terminal is removed from the stack and the transition to the following character in the input string is made. Lack of match gets the parser to fix the error.

The work of the parser ends when the entry string of characters turns out to be looked over. If, in this case, the stack is empty, the entry string of characters is considered correct, if it is not empty, the string is erroneous.

RPN generation is carried out while LL(1)-parser is at work. It is done in the following way. The second stack is necessary for RPN generator, its work being done simultaneously with that of the recognizer with its stack. Written in the second stack is the sequence of semantic characters that denote the actions generated by RPN elements. The actions on RPN generation will be done when the characters are removed from the second stack.

For the implementation of RPN generator together with the main table of LL(1)-parser it is necessary to give a semantic table. Its size fits with that of the main table. Moreover, for each non-empty cell of the main table, where there is a right part of any production, put into the semantic table is the sequence of operations, the number of which is equal to the length of the right part of this production.

Since terminal characters in the context free grammar are, indeed, lexical units, recognized by the lexer, such lexical units as variables names and constants contain additional semantic information links to tables of variables or tables of constants , i.e. the same terminals can be different semantically in the context free grammar.

RPN generation also involves forming the two tables: the table of constants and the table of variables. Besides, the constants and the variables that are required for the computations to be made in accordance with the conditions must be written in the table prior to the action of the LL(1) parser. Those constants and variables should be given by the test developer.

Let's consider the semantic actions on RPN generation for the formula grammar. Included in the sequence are the actions indicated by the following characters:

- a — writing in a RPN variable;
- k — writing in a RPN constant;
- = — writing in a RPN assignment operation ;
- + — writing in a RPN binary addition operation;
- — writing in a RPN binary subtraction operation ;
- * — writing in a RPN multiplication operation ;
- / — writing in a RPN division operation ;

— writing in a RPN unary subtraction operation;
- s — writing in a RPN standard function operation , for example, sqrt;
- ˆ — writing in a RPN raising to power operation;
- □ — empty action.

While a variable is entered into RPN, search is made in the table of variables. If there is a variable with the same name in the table, the link to it is entered into RPN. Otherwise, a new variable is put into the table of constants and the link to it is written in RPN. The actions indicated by the characters given above will be executed simultaneously with the pop of corresponding characters from the second stack.

In Table 1 one can find the parser's table which overlaps with the semantic table of RPN generator where there is only one column marked with terminal s. In the actual table for each operation computing a standard function there must be a column like that. Productions put into the table are derived from grammar rules (1) by their transformation to Greibach normal form.

Table 1. Integrated table of Parser and RPN Generator

	+	-	*	/	^	s	(	)	a	k	=	;	⊥
P	λ	λ	λ	λ	λ	λ	λ	λ	a = SB a□□ =	λ	λ	;P □□	λ
B	λ	λ	λ	λ	λ	λ	λ	λ	λ	λ	λ	;P □□	λ
S	+FVU □□□□	-FVU □□□□				sFVU □s□□	(S)WVU □□□□□□		aWVU a□□□□	kWVU k□□□□			
U	+TU □□+	λ	λ	λ	λ	λ	λ	λ	λ	λ	λ	λ	λ
T	+FV □□□	-FV □□[−]				sFV □s□	(S)WV □□□□□		aWV a□□	kWV k□□			
V	λ	λ	*FV □□*	/FV □□/	λ	λ	λ	λ	λ	λ	λ	λ	λ
F	+F □□	-F □ □□[−]				sF□ □□s	(S)W □□□□		aW a □	kW k□			
W	λ	λ	λ	λ	^GW □□^	λ	λ	λ	λ	λ	λ	λ	λ
G							(S) □□□		a a	k k			

Consider the work of parser and RPN generator. Let the input formula with a, b, c variables and numerical constants be as follows:

$$a = -\left(b + c\right) * b\hat{\ }3 / \left(b - 2.5\right) \qquad (2)$$

Formed at the output are the table of (a, b, c), variables and the table of (3, 2.5) constants and RPN in which the links to the table of variables are in curly brackets while the links to the table of constants are in square brackets.

$$\{1\}\{2\}\{3\} + [-]\,\{b\}\,[1]^{\wedge} * \{2\}\,[2] - / \qquad (3)$$

5   Formula Computation Using Reverse Polish Notation

Computation with predefined RPN can be made by the interpreter that uses a additional stack [6]. For the formulae language under consideration the structure of the stack is as follows. The stack consists in cells, each cell having two parts: 1) content type, 2) content that can serve either as a link to the variable in the table or a numerical value. Besides, for each variable in the table there must be space to store the current value as well as the tolerated error of computation since in the general case numerical values can be approximated.

For each input variable the test developers should predefine numerical values and put them into additional table where input data are presented as sets. Moreover, for each input variable, which must be defined by formulae, the tolerated error must be given.

Algorithm for computation is usually executed recursively in the following way. The number of loop iterations is the same as the number of sets of input data. Loop execution starts with writing the input values of the current set of data into the table and the tolerated error for the output variables. From here on, sequential scanning of generated RPN and actions with the stack are made according to the following rules:

– if the current element of RPN is a variable, the link to it is copied from RPN to the stack, the type of cell in the stack being defined in "link":
– if the current element of RPN is a constant, by the link the value of that constant is copied from the table of constants to the stack, the type of cell being defined in "value"; if the current element of RPN is a unary operation, the top element is removed from the stack, if
– it is of "link" type, the value by that link is read in the table of variables and if it is of "value" type, it is used, after that the unary operation is executed and its result is put into the stack, the type of cell being defined in "value";
– if the current element of RPN is a binary operation (apart from operation =), then , like in the case of unary operation, only one value is removed either from the stack or the table of variables and (unlike the unary operation) the other value is removed either from the stack or the table of variables whereupon the unary operation is executed and its result is put into the stack again;
– if the current element of RPN is operation =, like in the case of the unary operation, the value is removed from the stack or from the table of variables, then the link is removed from the stack, after that the retrieved value is put into the table of variables by the link.

After the computation has ended for the current set of input data matching of the values of input data put into the table is made using prepared by the developers true values in terms of the computational error. If for each set of data matching is successful, the formula type answer is considered correct.

6   Conclusion

Use of tasks with algebraic formula type answer improves the quality of testing in different subjects especially when the mixed diagnostic tests are implemented. The proposed method makes it possible to computerize the checkup of those answers, the knowledge of complicated programming languages is not required.

References

1. Yankovskaya, A.E., Fuks, I.L., Dementyev, Y.N.: Mixed Diagnoctic Tests in Construction Technology of the Training and Testing Systems. IJEIT 3, 169–174 (2013)
2. Avanesov, B.S.: Composition of test tasks: Textbook, 3rd edn., p. 240. Testing centre (2002)
3. Naprasnik, S.V., Tsimbaluk, E.S., Shkill, A.S.: Computer System for Testing Learners Knowledge. OpenTEST 2.0. In: Materials of 10 International Conference UADO Education and Virtuality 2006, pp. 454–461. KhNURE, Kharkov (2006)
4. Balfour, S.P.: Assessing Writing in MOOCs: Automated Essay Scoring and Calibrated Peer Review. Research & Practice in Assessment. Special Issue: MOOCs & Technology 8, 40–48 (2013)
5. Aho, A.V., Ullman, J.D.: The Theory of Parsing, Translation and Compiling. Parsing, vol. 1. Printice-Hall, Inc., Englewood Cliffs (1972)
6. Aho, A.V., Ullman, J.D.: The Theory of Parsing, Translation and Compiling. Compiling, vol. 2. Printice-Hall, Inc., Englewood Cliffs (1973)

Probability Density Function of a Non-profit Fund Surplus Under Hysteresis Surplus Control

Klimentii Livshits, Alexey Shkurkin, and Konstantin Yakimovich

National Research Tomsk State University and Branch of KemSU in
Anzhero-Sudzhensk
Tomsk, Anzhero-Sudzhensk, Russia
{kim47,shkurkin}@mail.ru,
konstantin.yakimovich@gmail.com

Abstract. The equations for the non-profit fund's capital distribution density under assumptions that the control of fund's capital is arbitrary hysteresis are obtained. The solution of this equations for the case of the small premium load are found.

Keywords: Non-profit fund, hysteresis control, distribution density of fund's capital, small premium load.

1 Introduction

By a non-profit fund we understand an organization created for the only purpose of collecting and distributing funds without making any profit. Among examples of non-profit funds are, in particular, state non-budget funds of the Russian Federation. Construction of non-profit fund models and related analysis can be found, for example, in [1]-[4] . In the articles the mentioned funds' features are analyzed under assumption of either threshold or combined threshold-hysteresis surplus control. In our work, we generalize the results of [4] and consider a general case of arbitrary hysteresis surplus control.

2 Mathematical Model for Fund's Surplus Change over Time

The main characteristic of the fund's state is its surplus $S(t)$ at time t. In the following we assume that the surplus can change due to several reasons:

1. The fund receives financial resourses. We assume that inflow times comprise a Poisson process with intensity λ. The financial resourses inflows (premiums) ξ are independent and identically distributed random variables with probability density $\varphi(x)$, mean $M\{\xi\} = a$ and second moment $M\{\xi^2\} = a_2$.

2. The fund spends premiums. We assume that premiums are expensed continuously at the rate $b(S)$, so that in time Δt a total of $b(S)\Delta t$ is paid out. We also assume that the spending process is controlled in a following fashion. Two

A. Dudin et al. (Eds.): ITMM 2014, CCIS 487, pp. 242–250, 2014.

boundary values S_1 and S_2 are set, with $S_1 < S_2$. For $S(t) < S_1$ $b(S) = b_l$, while for $S(t) > S_2$ $b(S) = b_u$. Since the fund's goals do not include earning profit, it is natural to assume that

$$b_l < \lambda a, \quad b_u > \lambda a .$$
(1)

Thus, for $S < S_0$ the fund on average spends less than it collects, while for $S > S_0$ it spends on average more than it collects.

For surplus values $S_1 \leq S \leq S_2$ the expense rate is set at $b(S) = b_0(S)$ or $b(S) = b_1(S)$ depending on how the surplus process $S(t)$ entered this domain. If the process crossed the lower bound S_1 up, then $b(S) = b_0(S)$, if it crossed the upper bound S_2 down, then $b(S) = b_1(S)$. It is assumed that functions $b_0(s)$ and $b_1(s)$ are continuous, monotonous and satisfy the following conditions:

$$b_0(S_1) = b_1(S_1) = b_l; \ b_0(S_2) = b_1(S_2) = b_u; \ b_0(S) < b_1(S) .$$
(2)

The domain $S_1 \leq S \leq S_2$ is in fact what we call the domain of hysteresis surplus control.

Finally, we assume that when $S < 0$ the fund continues its activities, but enters the state of insolvency, and its obligations are met as premiums flow in. This assumption defines the main distinction of our model from various dual risk models [5], [6], [7], [8].

3   Fund's Surplus Probability Density Function

We now present the equations that define the surplus probability density function $P(S)$ in each of the abovementioned domains in a steady state. Since the sum of the received premiums is a compound Poisson process [9], the surplus process in each domain is the sum of its nonstochastic component and its compound-Poisson component. Thus probability density function $P(S)$ exists and can only be non-continuous in S_1 and S_2. For convenience, let us move the reference point to $S = S_1$ and denote $S_0 = S_2 - S_1$. The lower bound will now be $S_1 = 0$.

We start with the domain where $S < 0$. Here we denote the probability density function $P(S)$ as $P_0(S)$. Consider two close points in time t and $t + \Delta t$. During time interval Δt the fund's surplus can be subject to the following changes. With probability $1 - \lambda \Delta t + o(\Delta t)$ the fund did not receive any premiums and hence its surplus decreased by $b_l \Delta t$. With probability $\lambda \Delta t + o(\Delta t)$ the fund received a random premium x and its surplus increased by $x - b_l \Delta t$. Other events have probability of order $o(\Delta t)$. Using the total probability formula, we get:

$$P_0(S) = (1 - \lambda \Delta t) P_0(S + b_l \Delta t) + \lambda \Delta t \int_0^\infty P_0(S - x)\varphi(x) \, dx + o(\Delta t).$$

As $\Delta t \to 0$ we arrive at the following equation:

$$b_l \dot{P}_0(S) = \lambda P_0(S) - \lambda \int_0^\infty P_0(S - x)\varphi(x) \, dx .$$
(3)

The solution for (3) must satisfy the following boundary condition: $P_0(-\infty) = 0$.

We now consider the domain $0 \le S \le S_0$. There are two possibilities in this case: $b(s) = b_0(s)$ and $b(s) = b_1(s)$.

We first consider the case of $b(s) = b_0(s)$. Denote as

$$g_0(S) = P\{S < S(t) \le S + dS, b(S) = b_0(S)\}/dS .$$

There are two possible ways to arrive at surplus value S at time $t + \Delta t$. Either at time t the surplus was equal to $S + b_0(S)\Delta t$ and during time Δt no premiums arrived. Or at time t the surplus was equal to $S + b_0(S)\Delta t - x$ and during time Δt a random premium x arrived. Writing down the probabilities of respective events and using the total probability formula we have:

$$\begin{aligned} g_0(S) = (1 - \lambda\Delta t)\, g_0(S + b_0(S)\Delta t) + \lambda\Delta t \int_0^S g_0(S - x)\varphi(x)\,dx + \\ + \lambda\Delta t \int_S^\infty P_0(S - x)\varphi(x)\,dx + o(\Delta t). \end{aligned} \tag{4}$$

As $\Delta t \to 0$ we get

$$b_0(S)\dot{g}_0(S) = \lambda g_0(S) - \lambda \int_0^S g_0(S - x)\varphi(x)\,dx - \lambda \int_S^\infty P_0(S - x)\varphi(x)\,dx . \tag{5}$$

The solution for (5) must satisfy the boundary condition $g_0(S_0) = 0$, which follows from the fact that when $b(s) = b_0(s)$ this bound can only be crossed from below and hence, when $S = S_0$ the first term on the right hand side of equation (4) is absent.

Now denote

$$g_1(S) = P\{S < S(t) \le S + dS, b(S) = b_1(S)\}/dS .$$

Similar to the above we get that the function $g_1(S)$ satisfies equation

$$b_1(S)\dot{g}_1(S) = \lambda g_1(S) - \lambda \int_0^S g_1(S - x)\varphi(x)\,dx . \tag{6}$$

Finally, we consider the domain $S > S_0$. Denote the probability density function $P(S)$ as $P_2(S)$. Function $P_2(S)$ satisfies equation

$$\begin{aligned} b_u\dot{P}_2(S) = \lambda P_2(S) - \lambda \int_0^{S-S_0} P_2(S - x)\varphi(x)\,dx - \\ - \lambda \int_{S-S_0}^S [g_0(S - x) + g_1(S - x)]\varphi(x)dx - \lambda \int_S^\infty P_0(S - x)\varphi(x)\,dx \end{aligned} \tag{7}$$

with boundary conditions $P_2(S_0) = g_1(S_0)$ and $P_2(\infty) = 0$.

The solution of (3), (5)–(7) must satisfy normalizing condition

$$\int_{-\infty}^0 P_0(S)dS + \int_0^{S_0} [g_0(S) + g_1(S)]dS + \int_{S_0}^\infty P_2(S)dS = 1 \tag{8}$$

and an additional condition

$$\dot{P}_2(S_0) = \dot{g}_1(S_0) + \dot{g}_0(S_0) , \tag{9}$$

which can be derived by comparing expressions (5)–(7) at S_0.

4 Probability Density Function of Fund's Surplus for Small Values of Premium Loading

It is impossible to derive the exact solution for the system (3), (5)–(7) in a general case of arbitrary premium distribution $\varphi(x)$. However, one can arrive at an approximate solution under some additional assumptions. Introduce parameter θ, where $0 < \theta < 1$, and assume that

$$b_l = (1 - \theta)\lambda a, \quad b_u = (1 + \theta)\lambda a . \tag{10}$$

Parameter θ is similar to insurance premium safety loading in modeling insurance risk processes [10]. We further consider an asymptotic case when $\theta \ll 1$. In practice, it means that for any value of surplus S the fund spends almost the same amount it receives in premiums. It is natural to assume that bounds S_1 and S_2, which define the domain of hysteresis surplus control, depend on safety loading θ. More strictly, we assume that for $\theta \to 0$ the difference between bounds $S_0(\theta) = S_2(\theta) - S_1(\theta) \to \infty$, but there is a finite $z_0 = \lim\limits_{\theta \to 0} \theta S_0(\theta)$. Finally, we represent spending rates in the form $b_i(s) = b_i(\theta s, \theta)$ and assume that the following limit exists:

$$c_i(z) = \lim_{\theta \to 0} \frac{b_i(z, \theta) - \lambda a}{\theta} . \tag{11}$$

This assumption is natural and follows from the following. Let $b_0(s)$ for example be linearly dependent on s, i.e.

$$b_0(s) = b_l + (b_u - b_l)\frac{s - S_1}{S_2 - S_1} .$$

By moving the reference point into $s = S_1$ and accounting for (10), we get

$$b_0(s) = \lambda a - \theta\lambda a(1 - \frac{\theta s}{\theta S_0}) .$$

Wherefrom

$$c_0(z) = -\lambda a(1 - \frac{z}{z_0}) .$$

Consider first the case of $S < 0$. In this domain we will be looking for the solution of (3) in the following form:

$$P_0(S) = \theta f_0(\theta S, \theta), \tag{12}$$

where $f_0(z, \theta)$ is some function. Substituting (12) into (3) and changing variables $\theta S = z$, we get an equation with respect to $f_0(z, \theta)$

$$\theta b_l \dot{f}(z, \theta) = \lambda f(z, \theta) - \lambda \int_0^\infty f_0(z - \theta x, \theta)\varphi(x)\,dx . \tag{13}$$

Assuming that $f_0(z, \theta)$ is twice differentiable in z and is uniformly continuous in θ, and taking Taylor expansion of the integrand with respect to the first argument and considering the first three terms of the sum, we get, accounting for (10), that

$$\ddot{f}_0\left(z,\theta\right) - \omega_0 \dot{f}_0\left(z,\theta\right) + \frac{o\left(\theta^2\right)}{\theta^2} = 0 \ ,$$

where

$$\omega_0 = \frac{2a}{a_2}.$$

Denote

$$f_0\left(z\right) = \lim_{\theta \to 0} f_0\left(z,\theta\right).$$

As $\theta \to 0$ we get the following equation with respect to $f_0\left(z\right)$

$$\ddot{f}_0\left(z\right) - \omega_0 \dot{f}_0\left(z\right) = 0 \ .$$

Wherefrom

$$f_0\left(z\right) = A_1 + A_2 e^{\omega_0 z} \ .$$

Since the boundary condition is $P_0\left(-\infty\right) = 0$ we have that

$$f_0\left(z\right) = A e^{\omega_0 z} \ , \tag{14}$$

where the constant A is defined by matching conditions.

Consider now the case of $0 \leq S \leq S_0$. We will look for the solution of (5) in this domain in the following form:

$$g_0\left(s\right) = \theta \psi_0\left(\theta s, \theta\right), \tag{15}$$

where $\psi_0\left(z,\theta\right)$ is assume to be twice continuously differentiable with respect to z and uniformly continuous with respect to θ. Substituting (12) and (15) into (5), we get, after substituting variables $\theta S = z$

$$\theta b_0(z,\theta)\dot{\psi}_0\left(z,\theta\right) = \lambda \psi_0\left(z,\theta\right) - \lambda \int\limits_0^\infty \psi_0\left(z - \theta x, \theta\right)\varphi\left(x\right) dx +$$

$$+\lambda \int\limits_{\frac{z}{\theta}}^\infty \psi_0\left(z - \theta x, \theta\right)\varphi\left(x\right) dx - \lambda \int\limits_{\frac{z}{\theta}}^\infty f_0\left(z - \theta x, \theta\right)\varphi\left(x\right) dx \ .$$

Taking Taylor expansion of $\psi_0\left(z - \theta x, \theta\right)$ with respect to the first argument and considering the first three terms we get

$$\frac{\lambda a_2}{2}\ddot{\psi}_0\left(z,\theta\right) + \frac{b_0(z,\theta)-\lambda a}{\theta}\dot{\psi}_0\left(z,\theta\right) - \frac{\lambda}{\theta^2}\int\limits_{\frac{z}{\theta}}^\infty \psi_0\left(z - \theta x, \theta\right)\varphi\left(x\right) dx+$$

$$+\frac{\lambda}{\theta^2}\int\limits_{\frac{z}{\theta}}^\infty f_0\left(z - \theta x, \theta\right)\varphi\left(x\right) dx + \frac{o\left(\theta^2\right)}{\theta^2} = 0 \ . \tag{16}$$

Function $\psi_0(z, \theta)$ is differentiable and hence bounded. Thus

$$\frac{1}{\theta^2} \int\limits_{\frac{z}{\theta}}^{\infty} \psi_0(z - \theta x)\varphi(x)\, dx \leq \max_y \psi_0(y, \theta) \frac{1}{z^2}\frac{z^2}{\theta^2} \int\limits_{\frac{z}{\theta}}^{\infty} \varphi(x)\, dx \leq$$

$$\leq \max_y \psi_0(y, \theta) \frac{1}{z^2} \int\limits_{\frac{z}{\theta}}^{\infty} x^2 \varphi(x)\, dx \to 0_{\theta \to 0},$$

since the second moment $M\{\xi^2\} = a_2$ exists by the model setup. The second integral in (16) can be evaluated similarly.

Denote

$$\psi_0(z) = \lim_{\theta \to 0} \psi_0(z, \theta). \tag{17}$$

By taking the limit as $\theta \to 0$ in (18) we arrive at equations for $\psi_0(z)$

$$\frac{\lambda a_2}{2}\ddot{\psi}_0(z) + c_0(z)\dot{\psi}_0(z) = 0 . \tag{18}$$

Wherefrom $\psi_0(z) = B_1 + B_2 \int_0^z \pi_0(x)dx$,
where

$$\pi_0(x) = \exp(-\frac{2}{\lambda a_2} \int\limits_0^x c_0(y)dy) . \tag{19}$$

Boundary condition $g_0(S_0) = 0$ now yields $\psi_0(z_0) = 0$. Hence

$$\psi_0(z) = B \int\limits_z^{z_0} \pi_0(x)dx . \tag{20}$$

In derivation of (20) we implicitly assumed that $S \neq 0$. Let now $S = 0$. Then from (5) we have

$$\theta b_0(0, \theta)\dot{\psi}_0(0, \theta) = \lambda \psi_0(0, \theta) - \lambda \int\limits_0^{\infty} f_0(-\theta x, \theta)\, \varphi(x)\, dx .$$

As $\theta \to 0$, we see that $\psi_0(0) = f_0(0)$. From this we obtain the relationship between constants A and B:

$$A = B \int\limits_0^{z_0} \pi_0(x)dx . \tag{21}$$

We will look for the solution of (6) relative to $g_1(s)$ in the form

$$g_1(s) = \theta \psi_1(\theta s, \theta). \tag{22}$$

Function $\psi_1(z,\theta)$ satisfies the following equation

$$\theta b_1(z,\theta)\dot{\psi}_1(z,\theta) = \lambda\psi_1(z,\theta) - \lambda\int\limits_0^{\frac{z}{\theta}} \psi_1(z-\theta x,\theta)\,\varphi(x)\,dx \ .$$

By once again assuming that $\psi_1(z,\theta)$ is twice continuously differentiable with respect to z and uniformly continuous in θ and taking the Taylor expansion of the integrand with respect to the first argument, and denoting

$$\psi_1(z) = \lim_{\theta\to 0}\psi_1(z,\theta),$$

we obtain after taking the limit as $\theta\to 0$ the equation relative to $\psi_1(z)$

$$\frac{\lambda a_2}{2}\ddot{\psi}_1(z) + c_1(z)\dot{\psi}_1(z) = 0 \ ,$$

the solution of which has the following form

$$\psi_1(z) = C_1 + C_2 \int\limits_0^z \pi_1(x)dx \ ,$$

where

$$\pi_1(z) = \exp\left(-\frac{2}{\lambda a_2}\int\limits_0^x c_1(y)dy\right). \tag{23}$$

On the lower bound $S = 0$ from (6) we have

$$\theta b_1(0,\theta)\dot{\psi}_1(0,\theta) = \lambda\psi_1(0,\theta) \ .$$

Hence, as $\theta\to 0$, we arrive at $\psi_1(0) = 0$. From this $C_1 = 0$ and

$$\psi_1(z) = C\int\limits_0^z \pi_1(x)dx \ . \tag{24}$$

Consider the domain $S > S_0$. The surplus probability density function must now satisfy (7). The solution for (7) takes the form

$$P_2(S) = \theta f_2(\theta S,\theta), \tag{25}$$

and it is easy to see that $f_2(z,\theta)$ satisfies

$$\theta b_u\dot{f}_2(z,\theta) = \lambda f_2(z,\theta) - \lambda \int\limits_0^{\frac{z-z_0}{\theta}} f_2(z-\theta x,\theta)\varphi(x)\,dx-$$

$$-\lambda \int\limits_{\frac{z-z_0}{\theta}}^{\frac{z}{\theta}} [\psi_0(z-\theta x,\theta) + \psi_1(z-\theta x,\theta)]\varphi(x)\,dx - \lambda \int\limits_{\frac{z}{\theta}}^{\infty} f_0(z-\theta x,\theta)\varphi(x)\,dx \ .$$

Assuming that $f_2(z, \theta)$ is twice continuously differentiable with respect to z and uniformly continuous in θ, taking Taylor expansion of the integrand with respect to the first argument and imposing $\theta \to 0$ we ultimately obtain the equation relative to

$$f_2(z) = \lim_{\theta \to 0} f_2(z, \theta)$$

for $z > z_0$:

$$\ddot{f}_2(z) + \omega_0 \dot{f}_2(z) = 0 .$$

Thus

$$f_2(z) = D_1 + De^{-\omega_0 z} .$$

Boundary conditions $P_2(+\infty) = 0$ and $P_2(S_0) = g_1(S_0)$ now yield $f_2(+\infty) = 0$ and $f_2(z_0) = \psi_1(z_0)$. Hence $D_1 = 0$,

$$D = Ce^{\omega_0 z_0} \int_0^{z_0} \pi_1(x) dx$$

and

$$f_2(z) = C \int_0^{z_0} \pi_1(x) dx e^{-\omega_0(z - z_0)} . \tag{26}$$

To obtain the relationship between constants B and C we look at equation (9), which for $\theta \to 0$ yields

$$\dot{f}_2(z_0) = \dot{\psi}_1(z_0) + \dot{\psi}_2(z_0) .$$

It follows from this that

$$B\pi_0(z_0) = C(\pi_1(z_0) + \omega_0 \int_0^{z_0} \pi_1(x) dx) . \tag{27}$$

Finally from normalizing condition (8) as $\theta \to 0$ we have

$$\int_{-\infty}^0 f_0(z) \, dz + \int_0^{z_0} [\psi_0(z) + \psi_1(z)] dz + \int_{z_0}^{+\infty} f_2(z) dz = 1 .$$

Wherefrom

$$B \int_0^{z_0} (1 + \omega_0 x) \pi_0(x) dx + C \int_0^{z_0} (1 + \omega_0(z_0 - x)) \pi_1(x) dx = \omega_0 . \tag{28}$$

Taking into consideration relations (12), (14), (22), (25), we have that for $\theta \ll 1$ the fund's surplus probability density function $P(S)$ is given by

$$P(S) = \begin{cases} \theta B \int\limits_{0}^{\theta S_0} \pi_0(x)dx \, e^{\theta \omega_0 S} + o\left(\theta\right), & S < 0 \ , \\ \theta(B \int\limits_{\theta S}^{\theta S_0} \pi_0(x)dx + C \int\limits_{0}^{\theta S} \pi_1(x)dx) + o\left(\theta\right), & 0 \le S \le S_0 \ , \\ \theta C \int\limits_{0}^{\theta S_0} \pi_1(x)dx \, e^{-\theta \omega_0 (S - S_0)} + o\left(\theta\right), & S > S_0 \ , \end{cases} \quad (29)$$

where B and C are defined by equations (27) and (28).

The approximation (29) for the solution of the system (3), (5)–(7) can be improved by considering higher-order terms in the expansions of functions $f_i(z, \theta)$ and $\psi_i(z, \theta)$ with respect to θ.

5   Conclusion

In this article we have found the probability density function of a non-profit fund surplus under arbitrary hysteresis surplus control and under additional assumption of small premium loading. The proposed method can be used to analyze other non-profit fund models where the premium loading is assumed to be small.

References

1. Zmeev, O.A.: Mathematical model of the social insurance fund with deterministic expensis for social programmes. Russian Physics Journal 3, 83–87 (2003)
2. Kitaeva, A.V., Terpugov, A.F.: Control of the social insurance fund's surplus. Tomsk State University Journal 280, 185–187 (2003)
3. Livshits, K.I., Bublik, Y.S.: Distribution density of non-profit fund's capital for Poisson's model under hysteresis control. Tomsk State University Journal of Control and Computer Science 3(12), 12–21 (2010)
4. Livshits, K.I., Bublik, Y.S.: Distribution density of non-profit fund's capital under hysteresis control. Bulletin of Tomsk Polytechnic University 315(5), 174–177 (2009)
5. Lin, X.S., Sendova, K.P.: The compound Poisson risk model with multiple thresholds. Insurance: Mathematics and Economics 42(2), 617–627 (2008)
6. Wen, Y.: On a Class of Dual Model with Diffusion. Int. J. Contemp. Math. Sciences 6(16), 793–799 (2011)
7. Ng, A.C.: On a dual model with a dividend threshold. Insurance: Mathematics and Economics 44(2), 315–324 (2009)
8. Li, B., Wu, R.: The dividend function in the jump-diffusion dual model with barrier dividend strategy. Applied Mathematics and Mechanics 29, 1239–1249 (2008)
9. Feller, W.: An Introduction to Probability Theory and its Applications. M.: Mir, vol. 1, p. 498 (1967)
10. Gluhova, E.V., Zmeev, O.A., Livshits, K.I.: Mathematical Models of Insurance, p. 180. Izd-vo TGU, Tomsk (2004)

Cramér-Lundberg Model with Stochastic Premiums and Continuous Non-insurance Costs

Klimentii Livshits and Konstantin Yakimovich

National Research Tomsk State University, Tomsk, Russia
kim47@mail.ru,
konstantin.yakimovich@gmail.com

Abstract. The goal of this paper is to estimate the probability of ruin and the moment-generating function of time to ruin an insurance company in a setting with insurance claims and premiums governed by compound Poisson processes and in the presence of continuous non-insurance costs.

Keywords: probability of ruin, moment-generating function of time to ruin, insurance premium safety loading.

1 Introduction

The article considers a mathematical model of an insurance company in a setting with insurance premiums that the company receives and claims that the company pays out both governed by independent compound Poisson processes. The main difference of the model in question from the well-known Cramér-Lundberg model with stochastic premiums [1],[2] is that in our model we additionally account for operating costs that the company bears in order to function.

2 Equation for Probability of Ruin

We assume that the flow of insurance premiums is a Poisson process with intensity λ, premiums ξ are independent and identically distributed random variables with probability density function $\varphi(x)$ and moments $M\{\xi\} = a$ and $M\{\xi^i\} = a_i$, $i = 2, 3$. Insurance claims η are assumed to be governed by a Poisson process with intensity μ, claims are also assumed to be i.i.d. random variables with probability density function $\psi(x)$ and moments $M\{\eta\} = b$ and $M\{\eta^i\} = b_i$, $i = 2, 3$. The company bears some additional costs that are not linked to claims. We further assume that these costs are incurred continuously so that for a time interval t the company incurs costs of ct.

Let $\bar{S}(t)$ be the insurer's mean surplus at time t. In the model considered mean surplus at time t will be given by the following expression:

$$\bar{S}(t) = S(0) + (\lambda a - \mu b - c)t \ . \tag{1}$$

A. Dudin et al. (Eds.): ITMM 2014, CCIS 487, pp. 251–260, 2014.

From (1) it follows that the insurer's surplus is monotonically increasing if

$$\lambda a = (1 + \theta)(\mu b + c) ,\tag{2}$$

where $\theta > 0$. If $\theta < 0$ the company becomes bankrupt. The parameter θ is the insurance premium safety loading.

Let $T = \inf\{t : t \geq 0, S(t) < 0\}$ and $T = \infty$ if $S(t) \geq 0 \, \forall t$. Random variable T is the time of ruin [3]. Denote $P(S) = \Pr\{T < \infty\}$ as the infinite time ruin probability with initial surplus S. It can be shown that the ruin probability $P(S)$ satisfies the following equation

$$(\lambda + \mu)P(S) = -cP'(S) + \lambda \int_0^\infty P(S + x)\varphi(x)dx +$$
$$+\mu \int_0^S P(S - x)\psi(x)dx + \mu \int_S^\infty \psi(x)dx .\tag{3}$$

The solution for (3) must satisfy the boundary conditions:

$$P(\infty) = 0, \quad P(0) = 1 .\tag{4}$$

Accordingly, in the model considered the ruin probability does not depend on the safety loading in case of zero initial surplus unlike in the classical model [3],[4], Cramér-Lundberg model with stochastic premiums [1],[2] and Markovian arrival risk model [5], [6], [7].

It is impossible to derive the exact solution of equation (3) in general case. Hence, it is useful to provide estimates of the ruin probability $P(S)$ under some additional conditions.

3   Upper Bound for Ruin Probability

We further show that under certain conditions we can arrive at the following inequality for the probability of ruin that is similar to Cramér inequality [1],[2].

Theorem 1. *Let $\theta > 0$ and let equation*

$$\lambda \int_0^\infty e^{-kx}\varphi(x)dx + \mu \int_0^\infty e^{kx}\psi(x)dx = \lambda + \mu - kc\tag{5}$$

have a root $k > 0$. Then equation (3) has a solution $P(S)$ that satisfies the following conditions:
    1)
$$P(S) \leq e^{-kS} ;\tag{6}$$

2) $P(S)$ is monotonically decreasing continuous function.

This solution can be obtained as a limit of a sequence of functions $P_n(S)$, $n = 1, 2 \ldots$, defined recursively by

$$P_{n+1}(S) = \frac{1}{c} \int\limits_0^S e^{-\frac{(\lambda+\mu)}{c}(S-u)} [\lambda \int\limits_0^\infty P_n(u+x)\varphi(x)dx + \mu \int\limits_0^u P_n(u-x)\psi(x)dx +$$
$$+ \mu \int\limits_u^\infty \psi(x)dx]du + e^{-\frac{(\lambda+\mu)S}{c}} \ ,$$

(7)

and the initial approximation $P_0(S)$ satisfies conditions 1 – 2.

Proof. Equation (3) can be represented as follows:

$$P(S) = \frac{1}{c} \int\limits_0^S e^{-\frac{(\lambda+\mu)}{c}(S-u)} [\lambda \int\limits_0^\infty P(u+x)\varphi(x)dx + \mu \int\limits_0^u P(u-x)\psi(x)dx +$$
$$+ \mu \int\limits_u^\infty \psi(x)dx]du + e^{-\frac{(\lambda+\mu)S}{c}} \ .$$

(8)

We solve equation (8) by using successive approximations according to (7) and by letting $P_n(S) \leq e^{-kS}$, where k is defined by equation (5). Then

$$P_{n+1}(S) \leq \frac{1}{c} \int\limits_0^S e^{-\frac{(\lambda+\mu)}{c}(S-u)} [\lambda e^{-ku} \int\limits_0^\infty e^{-kx}\varphi(x)dx + \mu e^{-ku} \int\limits_0^u e^{kx}\psi(x)dx +$$
$$+ \mu e^{-ku} \int\limits_u^\infty e^{-ku}\psi(x)dx]du + e^{-\frac{(\lambda+\mu)}{c}S} \leq$$
$$\leq \frac{1}{c} \int\limits_0^S e^{-\frac{(\lambda+\mu)}{c}(S-u)} e^{-ku}(\lambda + \mu - kc)du + e^{-\frac{(\lambda+\mu)}{c}S} = e^{-kS} \ .$$

We will now show that when $P_n(S)$ is monotonically decreasing and continuous, $P_{n+1}(S)$ is also monotonically decreasing and continuous. We first consider function

$$f_n(u) = \lambda \int\limits_0^\infty P_n(u+x)\varphi(x)dx + \mu \int\limits_0^u P_n(u-x)dx + \mu \int\limits_u^\infty \psi(x)dx \ .$$

(9)

For $u_2 > u_1$

$$f_n(u_2) - f_n(u_1) = \int\limits_0^\infty (P_n(u_2 + x) - P_n(u_1 + x))\varphi(x)dx +$$
$$+ \mu \int\limits_0^{u_1} (P_n(u_2 - x) - P_n(u_1 - x))\psi(x)dx +$$
$$+ \mu \int\limits_{u_1}^{u_2} P_n(u_2 - x)\psi(x)dx \ - \ \mu \int\limits_{u_1}^{u_2} \psi(x)dx \leq \mu \int\limits_{u_1}^{u_2} (e^{-k(u_2-x)} - 1)\psi(x)dx \leq 0 \ ,$$

since the first two terms are non-positive and due to monotonicity of $P_n(S)$. Further, for $S_2 > S_1$

$$P_{n+1}(S_2) - P_{n+1}(S_1) = \frac{1}{c} \int\limits_0^{S_1} e^{-\frac{\lambda+\mu}{c}u}(f_n(S_2 - u) - f_n(S_1 - u))du +$$
$$+ \frac{1}{c} \int\limits_{S_1}^{S_2} e^{-\frac{\lambda+\mu}{c}u} f_n(S_2 - u)du - (e^{-\frac{\lambda+\mu}{c}S_1} - e^{-\frac{\lambda+\mu}{c}S_2}) \ .$$

First term is non-positive. According to integral mean-value theorem

$$\frac{1}{c}\int_{S_1}^{S_2} f_n(S_2 - u)e^{-\frac{\lambda+\mu}{c}u}du = f_n(S_2 - u_0)\frac{1}{\lambda+\mu}(e^{-\frac{\lambda+\mu}{c}S_1} - e^{-\frac{\lambda+\mu}{c}S_2}) \ ,$$

where $u_0 \in (S_1, S_2)$. From equalities (9) and (5) it follows that

$$f_n(S_2 - u_0) \le e^{-k(S_2-u_0)}(\lambda\int_0^\infty e^{-kx}\varphi(x)dx + \mu\int_0^\infty e^{kx}\psi(x)dx)$$
$$= (\lambda + \mu - kc)e^{-k(S_2-u_0)} \le \lambda + \mu - kc \ .$$

Thus

$$P_{n+1}(S_2) - P_{n+1}(S_1) \le -\frac{kc}{\lambda+\mu}(e^{-\frac{\lambda+\mu}{c}S_1} - e^{-\frac{\lambda+\mu}{c}S_2}) \le 0 \ .$$

It follows from the above that operator (7) maps the domain of metric space defined by conditions 1-2 into itself. Let us prove that transformation (7) is a contraction mapping. Consider a function

$$F(v) = \lambda\int_0^\infty e^{-vx}\varphi(x)dx + \mu\int_0^\infty e^{vx}\psi(x)dx \ .$$

Function $F(v)$ has the following properties: $F(0) = \lambda + \mu$;
$F'(0) = -(\lambda a - \mu b) < 0$, as it is assumed that (2) holds, and, hence, is some neighborhood of $v = 0$ function $F(v)$ is decreasing. Further, $F''(v) > 0$ and $F(v) \to \infty$ as $v \to \infty$. Thus, $F(v)$ has a minimum at some point $v = k_0$. Moreover,

$$F(k_0) < \lambda + \mu - k_0c = \gamma(\lambda + \mu - k_0c), \quad 0 < \gamma < 1 \ ,$$

and, since $k_0 < k$, then $P_n(S) \le e^{-kS} \le e^{-k_0S}$.

Consider the following sequence of differences

$$|P_{n+1}(S) - P_n(S)|e^{k_0S} =$$
$$= \frac{1}{c}|\int_0^S e^{-\frac{\lambda+\mu}{c}(S-u)+k_0(S-u)}[\lambda\int_0^\infty (P_n(u+x) - P_{n-1}(u+x))e^{k_0(u+x)}e^{-k_0x}\varphi(x)dx +$$
$$+\mu\int_0^u (P_n(u-x) - P_{n-1}(u-x))e^{k_0(u-x)}e^{k_0x}\psi(x)dx]du| \le$$
$$\le \frac{1}{c}\int_0^S e^{-\frac{\lambda+\mu}{c}(S-u)+k_0(S-u)}\lambda\int_0^\infty e^{-k_0x}\varphi(x)dx +$$
$$\mu\int_0^\infty e^{k_0x}\psi(x)dx]du\max_s|P_n(S) - P_{n-1}(S)|e^{k_0S} \ .$$

From this,

$$|P_{n+1}(S) - P_n(S)|e^{k_0S} \le$$
$$\le \gamma\frac{\lambda+\mu-k_0c}{c}\int_0^S e^{\frac{\lambda+\mu-k_0c}{c}u}du \cdot \max_s|P_n(S) - P_{n-1}(S)|e^{k_0S} \le$$
$$\le \gamma\max_s|P_n(S) - P_{n-1}(S)|e^{k_0S} \ .$$

Because of the above as $n \to \infty$ $|P_{n+1}(S) - P_n(S)| \to 0$. According to contraction mapping principle the sequence $P_n(S)$ converges to the solution of equation (8), which satisfies conditions 1–2.

Example 1. Consider for illustrative purposes the simplest case of exponentially-distributed claims and premiums

$$\varphi(s) = \frac{1}{a} \exp(-\frac{s}{a}), \ \psi(s) = \frac{1}{b} \exp(-\frac{s}{b}) \ . \tag{10}$$

Solving (3) and accounting for (4) and (10), we arrive in case of $\theta > 0$ at

$$P(S) = \frac{z_2(1 + bz_1)}{z_2 - z_1} e^{z_1 S} + \frac{z_1(1 + bz_2)}{z_1 - z_2} e^{z_2 S} \ , \tag{11}$$

where

$$z_{1,2} = -\frac{(\lambda + \mu)ab + c(a - b)}{2abc} \pm \sqrt{\frac{((\lambda + \mu)ab + c(a - b))^2}{4a^2b^2c^2} - \frac{\theta(b\mu + c)}{abc}} \ . \tag{12}$$

For $k < 1/b$ equation (10) can be expressed as follows

$$\frac{\lambda}{1 + ka} + \frac{\mu}{1 - kb} = \lambda + \mu - kc \ .$$

If condition (2) holds, the positive root of this equation satisfying $k < 1/b$ is defined by the following expression

$$k = \frac{(\lambda + \mu)ab + c(a - b)}{2abc} - \sqrt{\frac{((\lambda + \mu)ab + c(a - b))^2}{4a^2b^2c^2} - (\lambda a - \mu b - c)} \ .$$

Figure 1 shows the dependence of ruin probability $P(S)$(solid lines) on initial surplus calculated according to expressions (11) and (12), and the dependence of its upper boundary $\hat{P}(S)$ (dashed line), calculated according to expression (11). Parameter values are $a = 1$, $\lambda = 10$, $b = 10$. Parameters $\mu = (1 - \varepsilon)\lambda a/(1 + \theta)b$, $c = \varepsilon\lambda a/(1 + \theta)$. Parameter $\varepsilon = 0.1$. As suggested by Fig. 1 the accuracy of approximation increases for larger values of initial surplus S and lower values of safety loading θ.

For small values of safety loading θ an approximate explicit expression can be derived for root k of equation (5). From expression (2) the equation (5) can be rewritten as follows

$$\Phi(k, \theta) = \lambda \int_0^\infty e^{-kx} \varphi(x)dx + \mu \int_0^\infty e^{kx} \psi(x)dx + (\frac{\lambda a}{1 + \theta} - \mu b)k - (\lambda + \mu) = 0 \ .$$

If $\theta = 0$ equation $\Phi(k, 0) = 0$ has a single root $k = 0$. Representing function $\Phi(k, \theta)$ as a Taylor series centered at $k = \theta = 0$ and considering only first three terms we will get the following approximation

$$\Phi(k, \theta) = (\lambda a_2 + \mu b_2)k^2 - 2\lambda a\theta k = 0 \ .$$

Hence when θ is small

$$k = \frac{2\lambda a\theta}{\lambda a_2 + \mu b_2} + o(\theta) = \frac{2(\mu b + c)\theta}{\lambda a_2 + \mu b_2} + o(\theta) \ . \tag{13}$$

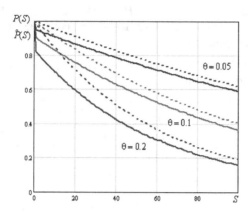

Fig. 1. Dependence of the ruin probability and its approximation on initial surplus

4   Ruin Probability for Small Values of Insurance Premium Safety Loading

The approximation of the ruin probability (6) can be improved when the safety loading θ is small. Define function

$$f(S,\theta) = P(\frac{(1-\theta)^{\kappa}}{\theta}S) \ , \tag{14}$$

where $\kappa \geq 0$. Function $f(S,\theta)$ satisfies the equation

$$(\lambda + \mu)f(S,\theta) = -c\frac{\theta}{(1-\theta)^{\kappa}}f'(S,\theta) + \lambda \int_0^{\infty} f(S + \frac{\theta}{(1-\theta)^{\kappa}}x, \theta)\varphi(x)dx +$$
$$+ \mu \int_0^{\infty} f(S - \frac{\theta}{(1-\theta)^{\kappa}}x, \theta)\psi(x)dx + R(S,\theta) \ , \tag{15}$$

where

$$R(S,\theta) = \mu \int_{\frac{S(1-\theta)^{\kappa}}{\theta}}^{\infty} \psi(x)dx - \mu \int_{\frac{S(1-\theta)^{\kappa}}{\theta}}^{\infty} f(S - \frac{\theta}{(1-\theta)^{\kappa}}x, \theta)\psi(x)dx$$

with boundary conditions

$$f(0,\theta) = 1, \quad \lim_{S \to \infty} f(S,\theta) = 0 \ . \tag{16}$$

Theorem 2. *If function $f(S,\theta)$ is at least three times differentiable in S and uniformly continuous in θ, then as $\theta \ll 1$*

$$f(S,\theta) = (1 - \frac{2(\mu b + c)}{(\lambda a_2 + \mu b_2)}(\frac{2(\mu b + c)(\lambda a_3 - \mu b_3)}{3(\lambda a_2 + \mu b_2)^2} - \kappa)\theta S)e^{-\frac{2(\mu b + c)}{\lambda a_2 + \mu b_2}S} + o(\theta) \ . \tag{17}$$

Proof. In can be shown that $R(S,\theta) = o(\theta^3)$ when $\theta \ll 1$.

Hence we will omit the last term in (15) in our further calculations.

Computing Taylor expansion of $f(S \pm \frac{\theta}{(1-\theta)^\kappa}x, \theta)$ with respect to the first argument and considering first three terms we will arrive from (15) at the following equation:

$$\frac{\lambda a_2 + \mu b_2}{2} f_S''(S,\theta) + (\mu b + c)f_S'(S,\theta) + \frac{o(\theta^2)}{\theta^2} = 0 . \tag{18}$$

Denote

$$f(S) = \lim_{\theta \to 0} f(S,\theta) . \tag{19}$$

As $\theta \to 0$ in (18) we have the following equation with respect to $f(S)$

$$\frac{\lambda a_2 + \mu b_2}{2} f_s''(s) + (\mu b + c)f_s'(s) = 0 . \tag{20}$$

Its solution, subject to boundary conditions (16), takes the form

$$f(S) = e^{-\frac{2(\mu b+c)}{\lambda a_2 + \mu b_2}S} . \tag{21}$$

Let us rewrite $f(S,\theta)$ as

$$f(S,\theta) = f(S) + \theta f_1(S,\theta) . \tag{22}$$

Substituting (22) into (15), taking Taylor expansions of $f(S \pm \frac{\theta}{(1-\theta)^\kappa}x)$ and $f_1(S \pm \frac{\theta}{(1-\theta)^\kappa}x, \theta)$ and considering terms of order θ^3 we have, taking into account (20), that

$$\frac{\lambda a_2 + \mu b_2}{2} f_{1S}''(S,\theta) + (\mu b + c)f_{1S}'(S,\theta) + \frac{\lambda a_3 - \mu b_3}{6} f^{(3)}(S) + \\ + \kappa \frac{\lambda a_2 + \mu b_2}{2} f^{(2)}(S) + \frac{o(\theta^3)}{\theta^3} = 0 . \tag{23}$$

Denote

$$f_1(S) = \lim_{\theta \to 0} f_1(S,\theta) . \tag{24}$$

As $\theta \to 0$ in (23) we arrive at the following equation with respect to $f_1(S)$

$$f_1''(S) + \frac{2(\mu b + c)}{\lambda a_2 + \mu b_2} f_1'(S) = \frac{4(\mu b + c)^2}{(\lambda a_2 + \mu b_2)^2} \left(\frac{2(\mu b + c)(\lambda a_3 - \mu b_3)}{3(\lambda a_2 + \mu b_2)^2} - \kappa \right)e^{-\frac{2(\mu b+c)}{\lambda a_2 + \mu b_2}S} . \tag{25}$$

Its solution satisfies conditions $f_1(0) = 0$, $f_1(\infty) = 0$, which follow from (16), and takes the form

$$f_1(S) = -\frac{2(\mu b + c)}{(\lambda a_2 + \mu b_2)} \left(\frac{2(\mu b + c)(\lambda a_3 - \mu b_3)}{3(\lambda a_2 + \mu b_2)^2} - \kappa \right)Se^{-\frac{2(\mu b+c)}{\lambda a_2 + \mu b_2}S} .$$

Hence the relation (17) is fulfiled.

Thus, we arrive at the following approximation for ruin probability $P(S)$

$$P(S) =$$
$$= \left(1 - \frac{2(\mu b+c)}{(\lambda a_2 + \mu b_2)} \left(\frac{2(\mu b+c)(\lambda a_3 - \mu b_3)}{3(\lambda a_2 + \mu b_2)^2} - \kappa \right)\frac{\theta^2}{(1-\theta)^\kappa}S \right)e^{-\frac{2(\mu b+c)\theta}{(\lambda a_2 + \mu b_2)(1-\theta)^\kappa}S} + o(\theta) .$$

5   Generating Function of Time of Ruin for Small Values of Premium Safety Loading

Let (Ω, F, P) be the probability space where trajectories of the insurer's capital process $S(t)$ are defined. Let the insurer's initial surplus at time zero be S. We distinguish two classes of the surplus process trajectories starting in this point: define trajectories leading to ruin as $\{S_\omega(t), \omega \in \Omega(S)\}$ and define trajectories leading to survival as $\{S_\omega(t), \omega \in \overline{\Omega}(S)\}$. Denote $t(s, \omega)$ as time to ruin on a trajectory leading to ruin. Denote

$$\Phi(s, u) = \int_{\Omega(S)} e^{-ut(S,\omega)} P(d\omega). \tag{26}$$

Since

$$P(S) = \int_{\Omega(S)} P(d\omega), \tag{27}$$

it follows that

$$\varphi(S, u) = \frac{\Phi(S, u)}{P(S)} \tag{28}$$

is the generating function of the conditional time to ruin for initial surplus value S.

Function $\Phi(S, u)$ must satisfy the following boundary conditions:

$$\Phi(S, 0) = P(S), \qquad \lim_{S \to \infty} \Phi(S, u) = 0, \quad \Phi(0, u) = 1. \tag{29}$$

The second condition follows from the fact that as $S \to \infty$ the integration domain in (26) becomes $\Omega(S) \to \emptyset$.

It can be shown that the function $\Phi(S, u)$ satisfies the following equation

$$(\lambda + \mu + u)\Phi(S, u) = -c\frac{\partial \Phi(S,u)}{\partial S} + \lambda \int_0^\infty \Phi(S + x, u)\varphi(x)\, dx +$$
$$+\mu \int_0^S \Phi(S - x, u)\psi(x)dx + \mu \int_S^\infty \psi(x)\, dx. \tag{30}$$

An explicit solution for (30) cannot be obtained. Let us further consider an asymptotic case when insurance premium safety loading is $\theta \ll 1$.

Define function

$$f(S, u, \theta) = \Phi(\frac{S}{\theta}, \theta^2 u). \tag{31}$$

Theorem 3. *If function $f(S, u, \theta)$ is at least twice differentiable with respect to S and uniformly continuous with respect to u and θ then as $\theta \ll 1$*

$$f(S, u, \theta) = e^{\chi_1(u)S} + O(\theta), \tag{32}$$

where

$$\chi_1(u) = -\frac{\mu b + c}{\lambda a_2 + \mu b_2} - \sqrt{\frac{(\mu b + c)^2}{(\lambda a_2 + \mu b_2)^2} + \frac{2u}{\lambda a_2 + \mu b_2}} \ . \tag{33}$$

The proof of theorem 3 is analogous to the proof of theorem 2. Thus,

$$\Phi(S, u) = e^{x_1(\frac{u}{\theta^2})\theta S} + O(\theta) . \tag{34}$$

Moments of conditional time to ruin are given by expressions

$$T_k(S) = (-1)^k \frac{\partial^k \Phi(S, u)}{\partial u^k}|_{u=0} / P(S) \ .$$

Therefore, given (2), for $\theta \ll 1$ the mean of time to ruin is given by

$$t_1(s) = \frac{S}{\lambda a \theta} + O(1) \ . \tag{35}$$

and the variance of time to ruin is

$$D(S) = \frac{S(\lambda a_2 + \mu b_2)}{\theta^3 \lambda^3 a^3} + O\left(\frac{1}{\theta^2}\right) . \tag{36}$$

Probability density function of time to ruin for initial surplus S would be given by

$$g(t, S) = \frac{S}{\sqrt{2\pi(\lambda a_2 + \mu b_2)}t\sqrt{t}} \exp\left\{-\frac{(S - \lambda a \theta t)^2}{2(\lambda a_2 + \mu b_2)}\right\} + O(\theta) . \tag{37}$$

Assume that as $\theta \to 0$ the insurer's initial surplus $S \to \infty$. Denote

$$m = \frac{1}{\lambda a \theta}, \qquad \sigma = \sqrt{\frac{\lambda a_2 + \mu b_2}{\theta^3 \lambda^3 a^3}} \ .$$

Theorem 4. *If* $\theta S(\theta) \to \infty$ *as* $\theta \to 0$ *then the random variable* $z = \frac{t - mS}{\sigma\sqrt{S}}$ *has a standard normal distribution.*

Proof. Generating function of z is

$$\varphi_z(S, u) = e^{\frac{m}{\sigma}\sqrt{S}u}\varphi\left(S, \frac{u}{\sigma\sqrt{S}}\right) =$$

$$= \exp\left\{\gamma u\sqrt{\theta S} + \gamma^2 \theta S(1 - \sqrt{1 + \frac{2u}{\gamma\sqrt{\theta S}}}\right\} + O(\theta) ,$$

where $\gamma = \sqrt{\frac{\lambda a}{\lambda a_2 + \mu b_2}}$. Since for $x \ll 1$ $\sqrt{1 + \alpha x} = 1 + \frac{\alpha x}{2} - \frac{\alpha^2 x^2}{8} + \frac{\alpha^3 x^3}{16} + o(x^3)$, it follows that

$$\varphi_z(S, u) = \exp\left\{\frac{u^2}{2} - \frac{u^3}{2\gamma\sqrt{\theta S}} + o(\frac{1}{\sqrt{\theta S}})\right\} + O(\theta).$$

As $\theta \to 0$ and as $\theta s \to \infty$, we have that $\lim_{\theta \to 0} \varphi_z(S, j\omega) = e^{-\frac{\omega^2}{2}}$.

Thus for $\theta \ll 1$ and $\theta s \gg 1$ random variable z is asymptotically normally distributed.

6 Conclusion

In this article, we calculate the main characteristics of the insurance company in a Cramér-Lundberg model with stochastic premiums and continuous costs required for insurer to function. We find the ruin probability and the distribution of time of ruin under additional assumption of small insurance premium safety loading. The proposed method can be used to analyze other insurance models where the safety loading is assumed to be small.

References

1. Livshits, K.I.: Probability of Ruin of an Insurance Company for the Poisson Model. Russian Physics Journal 42(4), 394–399 (1999)
2. Boykov, A.V.: Cramer-Lundberg Model with Stochastic Premiums. Theory of Probability and Its Applications 47(3), 549–553 (2002)
3. Panjer, H.Y., Willmont, G.E.: Insurance Risk Models, p. 442. Society of Actuaries (1992)
4. Grandell, J.: Simple approximation of ruin probabilities. Insurance: Mathematics and Economics 26, 157–173 (2000)
5. Wu, Y.: Bounds for the Ruin Probability Under a Markovian Modulated Risk Model. Commun. Statist. Stohastic Models. 15(1), 125–136 (1999)
6. Cheung, E.C.K., Feng, R.: A unified analysis of claim costs up to ruin in a Markovian arrival risk model. Insurance: Mathematics and Economics 53, 98–109 (2013)
7. Lu, Y., Li, S.: On the Probability of Ruin in a Markov-modulated Risk Model. Insurance: Mathematics and Economics 37(3), 522–532 (2005)

The Non-Markov Adaptive Retrial Queueing System with the Incoming MMPP-Flow of Requests

Tatyana Lyubina and Irina Garayshina

Branch of Kemerovo State University in Anzhero-Sudzhensk,
Anzhero-Sudzhensk, Russia
{lyubina_tv,irina_garayshina}@mail.ru

Abstract. In the paper, a non-Markov adaptive retrial queue system with the incoming MMPP-flow of requests on the condition of a highly-loaded RQ-system is considered. A system of equations for finding throughput capacity S, the γ-value, state probability distribution of a service device and the values of Markov chain managing the incoming MMPP-flow, is obtained.

Keywords: retrial queue system, highly-loaded RQ-system, throughput capacity.

1 Introduction

Retrial Queueing Systems (RQ-systems) have been widely used to model telephone networks, local area networks with random multiple access protocols, broadcasting and cellular radio networks, technological and transport systems and many other things.

Different RQ-systems have been investigated by J. R. Artalejo [1], [2], B. D. Choi [3], G. I. Falin [4], [5], I. I. Khomichkov, A. N. Dudin [6], [7], A. A. Nazarov [8], [9] and others. In RQ-systems the requests entering the system and finding the service device busy, do not leave the system but join the orbit to retry to occupy the service device later.

In the article, we study a non-Markov RQ-system with the incoming Markov Modulate Poisson Process (MMPP-flow) [10], managed by the adaptive access protocol.

2 Mathematical Model

The input process is Markov Modulate Poisson Process defined by a scalar matrix $\rho\mathbf{\Lambda}$ of arbitrary intensity $\rho\lambda_n$ and matrix $\mathbf{Q}$ of infinitesimal characteristics $q_{\nu n}$ of a Markov chain $n(t)$, managing the MMPP-flow. If the service device is free at the time of a request arrival the request occupies it to be served for some

A. Dudin et al. (Eds.): ITMM 2014, CCIS 487, pp. 261–268, 2014.

random amount of time having an arbitrary distribution function $B(x)$. Having been successfully served the request leaves the service device. If at the time of a request being served one more request arrives this new request joins the orbit. From the orbit after some random delay the request with the intensity $1/T(t)$, where $T(t)$ is the adapter condition at the current time, retries for service. If the service device is free the request is served, if the service device is occupied the request comes back to the orbit.

The task is to find the throughput capacity, stationary state probability distribution and the values of a Markov chain managing the incoming MMPP-flow of requests [11].

The system state at the time t is defined by a Markov process

$$\{k(t),\ z(t), n(t),\ i(t),\ T(t)\},$$

where $k(t)$ defines the service device state as follows: $k(t) = 0$ if the service device is free, and $k(t) = 1$ if the service device is busy servicing the request; $z(t)$ – the remaining service time for the request at the service device at the time t; $n(t)$ – the value of Markov chain, managing MMPP-flow; $i(t)$ – the number of requests in the orbit; the adapter during the time t changes its states $T(t)$ as follows:

$$T(t + \Delta t) = \begin{cases} T(t) - \alpha \Delta t, & \text{if } k(t) = 0, \\ T(t) + \beta \Delta t, & \text{if } k(t) = 1, \end{cases}$$

where $\alpha > 0$, $\beta > 0$ are the adapter parameters, which values are given.

Let us denote probability distributions

$$P_0(n, i, T, t) = \frac{\partial P\{k(t) = 0, n(t) = n, i(t) = i, T(t) < T\}}{\partial T},$$

$$P_1(z, n, i, T, t) = \frac{\partial P\{k(t) = 1, z(t) < z, n(t) = n, i(t) = i, T(t) < T\}}{\partial T},$$

which satisfy the following system of Kolmogorov equations for stationary probability distribution $P_0(n, i, T, t) = P_0(n, i, T)$ and $P_1(z, n, i, T, t) = P_1(z, n, i, T)$:

$$
\begin{cases}
-\alpha \dfrac{\partial P_0(n, i, T)}{\partial T} = -\left(\rho\lambda_n + \dfrac{i}{T}\right) P_0(n, i, T) + \dfrac{\partial P_1(0, n, i, T)}{\partial z} + \\
+ \sum\limits_{\nu} q_{\nu n} P_0(\nu, i, T), \\
\beta \dfrac{\partial P_1(z, n, i, T)}{\partial T} = \dfrac{\partial P_1(z, n, i, T)}{\partial z} - \dfrac{\partial P_1(0, n, i, T)}{\partial z} - \rho\lambda_n P_1(z, n, i, T) + \\
+ \dfrac{i+1}{T} B(z) P_0(n, i+1, T) + \rho\lambda_n B(z) P_0(n, i, T) + \\
+ \rho\lambda_n P_1(z, n, i-1, T) + \sum\limits_{\nu} q_{\nu n} P_1(z, \nu, i, T).
\end{cases}
$$

$$(1)$$

Let us denote the vectors

$$\begin{aligned}
\mathbf{P}_0(i, T) &= \{P_0(1, i, T), P_0(2, i, T), ..., P_0(N, i, T)\}, \\
\mathbf{P}_1(z, i, T) &= \{P_1(z, 1, i, T), P_1(z, 2, i, T), ..., P_1(z, N, i, T)\},
\end{aligned} \tag{2}$$

and define partial characteristic functions

$$\mathbf{H}_0(u_1, u_2) = \sum_i e^{-u_1 i} \int_0^\infty e^{-u_2 T} \mathbf{P}_0(i, T) dT,$$

$$\mathbf{H}_1(z, u_1, u_2) = \sum_i e^{-u_1 i} \int_0^\infty e^{-u_2 T} \mathbf{P}_1(z, i, T) dT. \tag{3}$$

Taking into account (2) and (3) we can rewrite the system (1) as follows

$$\begin{cases}
\mathbf{H}_0(u_1, u_2)(\mathbf{Q} - \rho\mathbf{\Lambda} + \alpha u_2\mathbf{I}) + \int_{u_2}^\infty \dfrac{\partial\mathbf{H}_0(u_1, x)}{\partial u_1} dx + \dfrac{\partial\mathbf{H}_1(0, u_1, u_2)}{\partial z} = 0, \\
\mathbf{H}_0(u_1, u_2) B(z)\rho\mathbf{\Lambda} - e^{u_1} \int_{u_2}^\infty \dfrac{\partial\mathbf{H}_0(u_1, x)}{\partial u_1} dx B(z) + \\
+ \mathbf{H}_1(z, u_1, u_2)\left(\mathbf{Q} - \left(1 - e^{-u_1}\right)\rho\mathbf{\Lambda} - \beta u_2\mathbf{I}\right) + \\
+ \dfrac{\partial\mathbf{H}_1(z, u_1, u_2)}{\partial z} - \dfrac{\partial\mathbf{H}_1(0, u_1, u_2)}{\partial z} = 0.
\end{cases} \tag{4}$$

We assume that for the parameters $\rho\mathbf{\Lambda}$, $\mathbf{Q}$ and b of the adaptive RQ-system there works the condition $(\mathbf{R}\,\mathbf{\Lambda}\,\mathbf{E})\,b = 1$, where $b = \int_0^\infty x dB(x)$ is the mean service time value.

3 Studying a RQ-System under a Heavy Load Condition

We will study the system (4) under a heavy load condition [12], defining the throughput capacity S of an adaptive RQ-system as a supremum of those ρ-values for which there exist stationary function regimes of an adaptive RQ-system, and considering the boundary condition $\rho \uparrow S$ fulfilled. The system (4) will be studied on the condition that $\varepsilon = S - \rho$ and $\varepsilon \to 0$.

Substituting

$$\rho = S - \varepsilon, \quad u_1 = \varepsilon w_1, \quad u_2 = \varepsilon w_2,$$

$$\mathbf{H}_0(u_1, u_2) = \mathbf{F}_0(w_1, w_2, \varepsilon), \quad \mathbf{H}_1(z, u_1, u_2) = \mathbf{F}_1(z, w_1, w_2, \varepsilon),$$

we get the system (4) in the form:

$$
\begin{cases}
\mathbf{F}_0\left(w_1, w_2, \varepsilon\right)\left(\mathbf{Q} - \left(S - \varepsilon\right)\mathbf{\Lambda} + \alpha\varepsilon w_2\mathbf{I}\right) + \\
+ \displaystyle\int_{w_2}^{\infty} \frac{\partial \mathbf{F}_0\left(w_1, x, \varepsilon\right)}{\partial w_1}dx + \frac{\partial \mathbf{F}_1\left(0, w_1, w_2, \varepsilon\right)}{\partial z} = 0, \\[3mm]
\mathbf{F}_0\left(w_1, w_2, \varepsilon\right)\left(S - \varepsilon\right)B(z)\mathbf{\Lambda} - e^{\varepsilon w_1}\displaystyle\int_{u_2}^{\infty}\frac{\partial \mathbf{F}_0\left(w_1, x, \varepsilon\right)}{\partial w_1}dx B(z) + \\[3mm]
+ \dfrac{\partial \mathbf{F}_1\left(z, w_1, w_2, \varepsilon\right)}{\partial z} + \\
+ \mathbf{F}_1\left(z, w_1, w_2, \varepsilon\right)\left(\mathbf{Q} - \left(1 - e^{-\varepsilon w_1}\right)\left(S - \varepsilon\right)\mathbf{\Lambda} - \beta\varepsilon w_2\mathbf{I}\right) - \\
- \dfrac{\partial \mathbf{F}_1\left(0, w_1, w_2, \varepsilon\right)}{\partial z} = 0 \ .
\end{cases}
\tag{5}
$$

Theorem 1. *The γ-value and the throughput capacity S of a non-Markov adaptive RQ-system with incoming MMPP-flow of requests is defined by a system of equations*

$$
\begin{cases}
\gamma\mathbf{R}_0\left(S, \gamma\right)\mathbf{E} - S\mathbf{R}_1\left(S, \gamma\right)\mathbf{\Lambda}\mathbf{E} = 0 \ , \\
\alpha\mathbf{R}_0\left(S, \gamma\right)\mathbf{E} - \beta\mathbf{R}_1\left(S, \gamma\right)\mathbf{E} = 0 \ ,
\end{cases}
\tag{6}
$$

where an adapter parameters α, β are given, $\mathbf{R}_k\left(S, \gamma\right)$ – state probability distribution of the service device and values of a Markov chain managing the incoming MMPP-flow, is defined by the equations

$$
\begin{aligned}
\mathbf{R}_0\left(S, \gamma\right)\left\{\left(S\mathbf{\Lambda} + \gamma\mathbf{I}\right)b + \mathbf{I}\right\}\mathbf{E} = 1 \ , \\
\mathbf{R}_0\left(S, \gamma\right) + \mathbf{R}_1\left(S, \gamma\right) = \mathbf{R} \ ,
\end{aligned}
\tag{7}
$$

where $b = \displaystyle\int_0^{\infty} x dB(x)$ – is the mean service time value, $\mathbf{R}$ – stationary probability distribution of Markov chain values $n(t)$, defined by the system $\mathbf{R}\mathbf{Q} = 0$ and $\mathbf{R}\mathbf{E} = 1$.

Proof. There are two stages of proving.

Stage 1:

Let $\lim\limits_{\varepsilon \to 0}\mathbf{F}_0(w_1, w_2, \varepsilon) = \mathbf{F}_0(w_1, w_2)$, $\lim\limits_{\varepsilon \to 0}\mathbf{F}_1(z, w_1, w_2, \varepsilon) = \mathbf{F}_1(z, w_1, w_2)$,

fulfilling this limiting transition in (5) we obtain a system

$$
\begin{cases}
\mathbf{F}_0\left(w_1, w_2\right)\left(\mathbf{Q} - S\mathbf{\Lambda}\right) + \displaystyle\int_{w_2}^{\infty}\frac{\partial \mathbf{F}_0\left(w_1, x\right)}{\partial w_1}dx + \frac{\partial \mathbf{F}_1\left(0, w_1, w_2\right)}{\partial z} = 0, \\[3mm]
\mathbf{F}_0\left(w_1, w_2\right)B(z)S\mathbf{\Lambda} - \displaystyle\int_{u_2}^{\infty}\frac{\partial \mathbf{F}_0\left(w_1, x\right)}{\partial w_1}dx B(z) + \mathbf{F}_1\left(z, w_1, w_2\right)\mathbf{Q} + \\[3mm]
+ \dfrac{\partial \mathbf{F}_1\left(z, w_1, w_2\right)}{\partial z} - \dfrac{\partial \mathbf{F}_1\left(0, w_1, w_2\right)}{\partial z} = 0 \ .
\end{cases}
\tag{8}
$$

We will seek for the solution of the system $\mathbf{F}_0(w_1, w_2)$ and $\mathbf{F}_1(z, w_1, w_2)$ as

$$\begin{aligned}
\mathbf{F}_0(w_1, w_2) &= \mathbf{R}_0\,(S, \gamma)\,\Phi(w_1, w_2) = \mathbf{R}_0\,(S, \gamma)\,\phi(w_2 + w_1\gamma)\ , \\
\mathbf{F}_1(z, w_1, w_2) &= \mathbf{R}_1(z, S, \gamma)\Phi(w_1, w_2) = \mathbf{R}_1(z, S, \gamma)\phi(w_2 + w_1\gamma)\ ,
\end{aligned} \tag{9}$$

where γ – some positive constant, its value will be defined later, and a function $\phi(w)$ at infinity is equal 0, then

$$\int\limits_{w_2}^{\infty} \frac{\partial \mathbf{F}_0(w_1, x)}{\partial w_1}\,dx = \int\limits_{w_2}^{\infty} \mathbf{R}_0\,(S, \gamma)\,\frac{\partial \phi(x + w_1\gamma)}{\partial w_1}\,dx =$$

$$= \gamma \mathbf{R}_0\,(S, \gamma)\int\limits_{w_2}^{\infty} \phi'(x + \gamma w_1)dx = -\gamma \mathbf{R}_0\,(S, \gamma)\,\phi(w_2 + \gamma w_1)\ ,$$

so the system (8) will be written in the form

$$\begin{cases}
\mathbf{R}_0\,(S, \gamma)\,(\mathbf{Q} - S\mathbf{\Lambda} - \gamma\mathbf{I}) + \dfrac{\partial \mathbf{R}_1(0, S, \gamma)}{\partial z} = 0, \\
\mathbf{R}_0\,(S, \gamma)\,(S\mathbf{\Lambda} + \gamma\mathbf{I})\,B(z) + \mathbf{R}_1(z, S, \gamma)\mathbf{Q} + \dfrac{\partial \mathbf{R}_1(z, S, \gamma)}{\partial z} - \dfrac{\partial \mathbf{R}_1(0, S, \gamma)}{\partial z} = 0\ .
\end{cases} \tag{10}$$

By applying the Laplace-Stieltjes transform to the system (10) we have

$$B^*(\eta) = \int\limits_0^{\infty} e^{-\eta z} dB(z), \quad \int\limits_0^{\infty} e^{-\eta z} d\mathbf{R}_1(z, S, \gamma) = \mathbf{R}_1^*(\eta, S, \gamma)\ ,$$

$$\int\limits_0^{\infty} e^{-\eta z} d\frac{\partial \mathbf{R}_1(z, S, \gamma)}{\partial z} = -\frac{\partial \mathbf{R}_1(0, S, \gamma)}{\partial z} + \eta \mathbf{R}_1^*(\eta, S, \gamma)\ ,$$

then the system(10) has the form

$$\begin{cases}
\mathbf{R}_0\,(S, \gamma)\,(\mathbf{Q} - S\mathbf{\Lambda} - \gamma\mathbf{I}) + \dfrac{\partial \mathbf{R}_1(0, S, \gamma)}{\partial z} = 0, \\
\mathbf{R}_0\,(S, \gamma)\,(S\mathbf{\Lambda} + \gamma\mathbf{I})\,B^*(\eta) + \mathbf{R}_1^*(\eta, S, \gamma)\mathbf{Q} + \eta \mathbf{R}_1^*(\eta, S, \gamma) - \dfrac{\partial \mathbf{R}_1(0, S, \gamma)}{\partial z} = 0\ .
\end{cases}$$

Thus it is not difficult to get the following equation:

$$\mathbf{R}_1^*(\eta, S, \gamma)\,(\eta\mathbf{I} + \mathbf{Q}) = \mathbf{R}_0\,(S, \gamma)\,\{(S\mathbf{\Lambda} + \gamma\mathbf{I})\,(1 - B^*(\eta)) - \mathbf{Q}\}\ . \tag{11}$$

With $\eta = 0$, denoting $\mathbf{R}_1^*\,(0, S, \gamma) = \mathbf{R}_1\,(S, \gamma)$, we can rewrite the equation as:

$$(\mathbf{R}_0\,(S, \gamma) + \mathbf{R}_1\,(S, \gamma))\,\mathbf{Q} = 0\ ,$$

which means $\mathbf{RQ} = 0$, where

$$\mathbf{R} = \mathbf{R}_0\left(S, \gamma\right) + \mathbf{R}_1\left(S, \gamma\right) \tag{12}$$

satisfies the condition of normalizing $\mathbf{RE} = 1$.

According to the equation (12) to find the vectors $\mathbf{R}_0\left(S, \gamma\right)$ and $\mathbf{R}_1\left(S, \gamma\right)$ it is enough to find one of them. We can rewrite the equation (11) as

$$\mathbf{R}_1^*(\eta, S, \gamma) = \mathbf{R}_0\left(S, \gamma\right)\left\{(S\Lambda + \gamma\mathbf{I})\left(1 - B^*(\eta)\right) - \mathbf{Q}\right\}(\eta\mathbf{I} + \mathbf{Q})^{-1}. \tag{13}$$

As

$$\mathbf{R}_1\left(S, \gamma\right) = \lim_{\eta \to 0} \mathbf{R}_1^*(\eta, S, \gamma) = $$
$$= \mathbf{R}_0\left(S, \gamma\right) \lim_{\eta \to 0}\left\{(S\Lambda + \gamma\mathbf{I})\left(1 - B^*(\eta)\right) - \mathbf{Q}\right\}(\eta\mathbf{I} + \mathbf{Q})^{-1}\ , \tag{14}$$

we can define the value of the limit (14), which is not equal to the product of limits, as there is no limit of the matrix $(\eta\mathbf{I} + \mathbf{Q})^{-1}$. We shall expand the matrix $(\eta\mathbf{I} + \mathbf{Q})^{-1}$ in the form of

$$(\eta\mathbf{I} + \mathbf{Q})^{-1} = \frac{1}{\eta}\mathbf{A} + \mathbf{B} + O(\eta)\ , \tag{15}$$

where the matrices $\mathbf{A}$ and $\mathbf{B}$ will be defined below.

The equation (15) can be rewritten as

$$\mathbf{I} = (\eta\mathbf{I} + \mathbf{Q})\left(\frac{1}{\eta}\mathbf{A} + \mathbf{B} + O(\eta)\right) = \frac{1}{\eta}\mathbf{QA} + \mathbf{QB} + \mathbf{A} + O(\eta)\ ,$$

or

$$\mathbf{I} = \left(\frac{1}{\eta}\mathbf{A} + \mathbf{B} + O(\eta)\right)(\eta\mathbf{I} + \mathbf{Q}) = \frac{1}{\eta}\mathbf{AQ} + \mathbf{BQ} + \mathbf{A} + O(\eta)\ ,$$

according to it for the matrix $\mathbf{A}$ we get the equation $\mathbf{AQ} = \mathbf{QA} = \mathbf{0}$, with its solution

$$\mathbf{A} = \mathbf{ER}\ . \tag{16}$$

Matrix $\mathbf{B}$ is the solution of the equation $\mathbf{QB} = \mathbf{BQ} = \mathbf{I} - \mathbf{A} = \mathbf{I} - \mathbf{ER}$.

Let us consider the limit in (14). Because of (16) we have

$$\lim_{\eta \to 0}\left\{(S\Lambda + \gamma\mathbf{I})\left(1 - B^*(\eta)\right) - \mathbf{Q}\right\}(\eta\mathbf{I} + \mathbf{Q})^{-1} = $$
$$= \lim_{\eta \to 0}\left\{(S\Lambda + \gamma\mathbf{I})\left(1 - B^*(\eta)\right) - \mathbf{Q}\right\}\left(\frac{1}{\eta}\mathbf{A} + \mathbf{B} + O(\eta)\right) = $$
$$= \lim_{\eta \to 0}\left\{(S\Lambda + \gamma\mathbf{I})\frac{1 - B^*(\eta)}{\eta}\mathbf{A} - \frac{1}{\eta}\mathbf{QA}+ \right.$$
$$\left. + (S\Lambda + \gamma\mathbf{I})\left(1 - B^*(\eta)\right)\mathbf{B} - \mathbf{QB} + O(\eta)\right\} = $$
$$= (S\Lambda + \gamma\mathbf{I})\, b\mathbf{A} - \mathbf{QB} = (S\Lambda + \gamma\mathbf{I})\, ba + \mathbf{A} - \mathbf{I}\ ,$$

where $b = \lim\limits_{\eta \to 0}\dfrac{1 - B^*(\eta)}{\eta} = \lim\limits_{\eta \to 0}\left(-B^{*\prime}(\eta)\right) = -B^{*\prime}(0) = \int\limits_0^\infty x\, dB(x) = b$.

It follows that we get the equation

$$\mathbf{R}_1\left(S, \gamma\right) = \mathbf{R}_0\left(S, \gamma\right)\left\{(S\Lambda + \gamma\mathbf{I})\, b\mathbf{A} + \mathbf{A} - \mathbf{I}\right\}\ ,$$

because of (16) it can be rewritten as

$$\mathbf{R}_1\,(S,\gamma) + \mathbf{R}_0\,(S,\gamma) = \mathbf{R}_0\,(S,\gamma)\,[(S\mathbf{\Lambda} + \gamma\mathbf{I})\,b + \mathbf{I}]\,\mathbf{A} =$$
$$= \mathbf{R}_0\,(S,\gamma)\,[(S\mathbf{\Lambda} + \gamma\mathbf{I})\,b + \mathbf{I}]\,\mathbf{ER}\ .$$

As $\mathbf{R}_0\,(S,\gamma) + \mathbf{R}_1\,(S,\gamma) = \mathbf{R}$, the equality of the vectors

$$\mathbf{R} = \{\mathbf{R}_0\,(S,\gamma)\,[(S\mathbf{\Lambda} + \gamma\mathbf{I})\,b + \mathbf{I}]\,\mathbf{E}\}\,\mathbf{R}$$

results in that for the first multiplier in the right-hand member of this equation the following equality is satisfied

$$\mathbf{R}_0\,(S,\gamma)\,[(S\mathbf{\Lambda} + \gamma\mathbf{I})\,b + \mathbf{I}]\,\mathbf{E} = 1\ ,$$

which defines the components of the vector $\mathbf{R}_0\,(S,\gamma)$ and is the same as (7).

Let us find one more condition which defines the components of the vector $\mathbf{R}_0\,(S,\gamma)$. The Laplace-Stieltjes transform $\mathbf{R}_1^*(\eta, S, \gamma)$ exists for all η with positive real parts, but because of the equality of (13) and

$$(\eta\mathbf{I} + \mathbf{Q})^{-1} = \frac{1}{|\eta\mathbf{I} + \mathbf{Q}|}\mathbf{D}(\eta),$$

where $\mathbf{D}(\eta)$ is the transpose to the matrix, composed of algebraic cofactors to the matrix elements, $\eta\mathbf{I} + \mathbf{Q}$, we may have $\mathbf{R}_1^*(\eta, S, \gamma)$ in the form

$$\mathbf{R}_1^*(\eta, S, \gamma) = \mathbf{R}_0\,(S,\gamma)\,\{(S\mathbf{\Lambda} + \gamma\mathbf{I})\,(1 - B^*(\eta)) - \mathbf{Q}\}\,\frac{\mathbf{D}(\eta)}{|\eta\mathbf{I} + \mathbf{Q}|}\ . \qquad (17)$$

With $\eta = \eta_l$, where η_l is roots of an equation, the denominator in (17) turns into zero, but the numerator is also zero, so for all η_l we have the equations

$$\mathbf{R}_0\,(S,\gamma)\,\{(S\mathbf{\Lambda} + \gamma\mathbf{I})\,(1 - B^*(\eta_l)) - \mathbf{Q}\}\,\mathbf{D}(\eta_l) = 0\ ,$$

defining the vector components $\mathbf{R}_0\,(S,\gamma)$.

Stage 2. To find S and γ we sum up all the equations of the system (5) over k and n. With $B(z = \infty) = B(\infty) = 1$ and sending $\varepsilon \to 0$, we have:

$$\mathbf{F}_0\,(w_1, w_2)\,\alpha w_2\mathbf{E} - w_1\int\limits_{w_2}^{\infty}\frac{\partial\mathbf{F}_0\,(w_1, x)}{\partial w_1}\mathbf{E}dx - \mathbf{F}_1\,(z, w_1, w_2)\,(w_1 S\mathbf{\Lambda} + \beta w_2\mathbf{I})\,\mathbf{E} = 0.$$

$$(18)$$

By applying solutions (9) to (18) we get the following equation

$$\alpha w_2\mathbf{R}_0\,(S,\gamma)\,\mathbf{E} + \gamma w_1\mathbf{R}_0\,(S,\gamma)\,\mathbf{E} - \mathbf{R}_1\,(S,\gamma)\,(w_1 S\mathbf{\Lambda} + \beta w_2\mathbf{I})\,\mathbf{E} = 0\ . \qquad (19)$$

We rewrite the equation (19) as follows:

$$w_1\,\{\gamma\mathbf{R}_0\,(S,\gamma)\,\mathbf{E} - S\mathbf{R}_1\,(S,\gamma)\,\mathbf{\Lambda}\mathbf{E}\} + w_2\,\{\alpha\mathbf{R}_0\,(S,\gamma)\,\mathbf{E} - \beta\mathbf{R}_1\,(S,\gamma)\,\mathbf{E}\} = 0\ ,$$

To turn it into an identity over w_1 and w_2 it is enough to have the following

$$\begin{cases} \gamma\mathbf{R}_0\,(S,\gamma)\,\mathbf{E} - S\mathbf{R}_1\,(S,\gamma)\,\mathbf{\Lambda}\mathbf{E} = 0\ , \\ \alpha\mathbf{R}_0\,(S,\gamma)\,\mathbf{E} - \beta\mathbf{R}_1\,(S,\gamma)\,\mathbf{E} = 0\ , \end{cases}$$

which coincides with (6), and because of (7) they set an equation system for two indeterminates S and γ, defined by the system. **The theorem is proved.**

4 Conclusions

Thus, an adaptive RQ-system MMPP|GI|1 under heavy load condition is investigated. As a result a system of equations (6)-(7) to find throughput capacity S and the γ-value, and also the probability distribution $\mathbf{R}_k\,(S,\gamma)$ of a service device state and values of a Markov chain managing the incoming MMPP-flow, is obtained.

References

1. Artalejo, J.R., Gomez-Corral, A.: Retrial Queueing Systems. A Computational Approach. Springer (2008)
2. Artalejo, J.R.: Accessible Bibliography on Retrial Queues. Math. Comput. Modelling 30, 1–6 (1999)
3. Choi, B.D., Choi, K.B., Lee, Y.W.: M/G/1 retrial queueing systems with two types of calls and finite capacity. Queueing Systems 19, 215–229 (1995)
4. Falin, G.I., Templeton, J.G.C.: Retrial queues. Chapman & Hall, London (1997)
5. Falin, G.I.: A Survey of Retrial Queues. Queuing Systems 7, 127–167 (1990)
6. Dudin, A., Klimenok, V.: Queuing System BMAP|G|1 with Repeated Calls. Mathematical and Computer Modeling 30, 115–128 (1999)
7. Dudin, A.N., Klimenok, V.I.: The state dependent M|M|1 retrial system. In: Choi, B.D. (ed.) Proceedings of the Fifth International Workshop on Retrial Queues, Seoul, pp. 81–88 (2004)
8. Kuznetsov, D.Y., Nazarov, A.A.: Adaptive network random access. Deltaplan, Tomsk (2002) (in Russian)
9. Sudyko, E.A., Nazarov, A.A.: Research of Markov RQ-system with conflicts applications and simplest incoming stream. Vestnik of Tomsk state University. Control, Computer Engineering and Computer Science 3(12), 97–106 (2010)
10. Garayshina, I.R., Moiseeva, S.P., Nazarov, A.A.: Research methods correlated streams and special systems of mass service. NTL, Tomsk (2010) (in Russian)
11. Nazarov, A.A., Lyubina, T.V.: The non-Markov dynamic RQ-system with the incoming MMP-flow of requests. Automation and Remote Control 74(7), 1132–1143 (2013)
12. Nazarov, A.A., Moiseeva, S.P.: The Asymptotical Analysis Method in Queueing Theory. NTL, Tomsk (2006) (in Russian)

Performance Evaluation of Integrated Wireless Networks with Virtual Partition of Channels

Agassi Melikov[1], Mehriban Fattakhova[2], Gulnara Velidzanova[2], and János Sztrik[3]

[1] National Aviation Academy, Azerbaijan
agassi.melikov@rambler.ru
[2] Institute of Cybernetics, National Academy of Sciences, Azerbaijan
meri-fattax@mail.ru, gulnaravelicanova@rambler.ru
[3] University of Debrecen, Hungary
sztrik.janos@inf.unideb.hu

Abstract. A new access scheme in integrated wireless networks which is based on virtual partition of channels among voice and data calls is proposed. In this scheme a voice call occupies a free channel in its own zone and if there is no available channel in the given zone a handover voice call searches for an idle channel in another zone. A threshold for number of handover voice calls in zone of channels for data calls is defined. To determine the access scheme of new data calls a state-dependent threshold based rule is introduced. An effective method to calculate the QoS metrics of the defined access scheme is developed. Some sample results illustrating the numerical experiments are collected and analyzed.

Keywords: integrated networks, voice and data calls, partition of channels, quality of service metrics, calculation method.

1 Introduction

In the last few decades the teletraffic theory has become a very important and effective scientific discipline representing a set of probabilistic methods to solve problems of designing and optimization of telecommunication systems. Information technology solutions require analytical, numerical, approximate, simulation and hybrid techniques. The current state of mathematical theory of the teletraffic problems has been collected in a detailed review [1]. That paper and other ones justify that performance evaluation of wireless networks plays a central role, see for example [2], [4], [5], [8], [13]. Developing effective methods to calculate QoS metrics of integrated cellular networks under various Call Admission Control (CAC) schemes are important, see [10], [12].

Main goals of any CAC when determining the rules is to use the scarce resources (frequencies, time slots, codes and their combinations). These rules are necessary to prevent (or minimization) the conflict situations due to employment of specified resources as well as to satisfy the desired QoS level for heterogeneous calls.

A. Dudin et al. (Eds.): ITMM 2014, CCIS 487, pp. 269–276, 2014.

In an integrated cellular network calls of real-times (e.g. voice calls) and non-real times (e.g. data calls) are distinguished. In such networks either CAC based on guard channels scheme or CAC based on cut-off scheme are used. In both schemes all channels are available for calls of any type.

To reduce the possibility of conflict situations the schemes which are based on the partition of pool of channels between heterogeneous calls are more useful. Literature review shows that models of integrated cellular networks with such kind of access schemes are insufficiently investigated.

Note that fixed (rigid or isolated) partition of channels is not effective one as noted by [6] and thus other schemes are required. It should be underlined that non-isolated schemes of partition of channels in networks with single traffic (a network of the second generation) have been offered in [9] and in Chapter 1 of the book [11] (pp. 18 and 19). The main contribution is that in these schemes the partition of channels is not rigid, i.e. the scheme of virtual partition of channels (Virtual Partitioning, VP) is suggested.

In the present paper a multi-parametric VP-scheme for partition of channels in integrated cellular networks is proposed. Exact formulas to calculate QoS metrics of such CAC scheme are developed. Figures generated by results of several sample numerical experiments are shown and analyzed.

2   VP-scheme of Partition

A base station of integrated cellular network contains of $N > 1$ radio channels. These channels are divided into two parts: a number of N_v channels is assigned for voice calls only and the remaining $N_{vd} = N - N_v$ channels are used by both voice and data calls. In other words, the pool of channels is divided into individual zone with N_v channels (for voice calls only) and common one with N_{vd} channels (both for voice and data calls).

Virtual partition means the following: on termination of processing of the v-call in a v-zone, the released channel is transferred into the vd-zone if there is a v-call in that zone and simultaneously the channel in a vd-zone which process a v-call, is transferred into the v-zone.

In this network four types of Poisson-type arrival traffics, i.e. new (ov-calls) and handover voice calls (hv-calls), furthermore new (od-calls) and handover data calls (hd-calls) are assumed. Intensity of x-calls is $\lambda_x, x \in \{ov, hv, od, hd\}$, respectively.

Distribution function of channel holding time for both kind of calls is supposed to be exponential the mean for voice calls (new or handover) is $1/\mu_v$, and the corresponding parameter for data calls (new or handover) is $1/\mu_d$. Identity of channel holding times for new and handover calls of both types is explained by the memoryless property of the exponential distribution.

Access of v-calls is specified by the following rules:

– If upon arrival an ov-call, there is free channel in v-zone, then it is accepted; otherwise, it is rejected.

- If upon arrival an hv-call, there is free channel in v-zone, then it is accepted; otherwise, free channel is searched in vd-zone. At that limit to maximum number of hv-calls in vd-zone is defined, i.e. maximum number of hv-calls in vd-zone is R_{hv}, $1 \leq R_{hv} \leq N_{vd}$. If at the moment of arriving an hv-call, number of hv-calls in vd-zone is equal R_{hv}, then it is rejected. Note that the average channel holding time for hv-calls in vd-zone is $1/\mu_v$, too.

Access of d-calls is specified by the following rules:

- If upon arrival an hd-call, there is free channel in vd-zone, then it is accepted; otherwise, it is rejected.
- If upon arrival an od-call, the number of d-calls in vd-zone is less than R_{od}, $1 \leq R_{od} \leq N_{vd} - 1$, then it is accepted; otherwise, it is rejected.

The primary goal of our investigation is to find the main QoS metrics of this system, namely, the loss probabilities of calls for each type.

3   Method to Solve the Problem

The state of a cell is described by a two-dimensional vector $n = (n_d, n_v)$ where n_d and n_v denote the total number of data calls and voice calls, respectively. Then the state space of the corresponding two-dimensional Markov chain (2-D MC) is defined as follows:

$$S = \{n : n_d = 0, 1, \ldots, N_{vd}; \quad n_v = 0, 1, \ldots, N_v + R_{hv}; \quad n_d + n_v \leq N\}. \quad (1)$$

According to the introduced access scheme, non-negative elements of generating matrix (Q-matrix) are determined from the following relationships:

$$q(n, n') = \begin{cases} \lambda_d & \text{if } n_d < R_{od}, \quad n' = n + e_1, \\ \lambda_{hd} & \text{if } n_d \geq R_{od}, \quad n' = n + e_1, \\ \lambda_v & \text{if } n_v < N_v, \quad n' = n + e_2, \\ \lambda_{hv} & \text{if } N_v \leq n_v < N_v + R_{hv}, \quad n' = n + e_2, \\ n_d \mu_d & \text{if } n' = n - e_1, \\ n_v \mu_v & \text{if } n' = n - e_2, \\ 0 & \text{in other cases,} \end{cases} \quad (2)$$

where $\lambda_v = \lambda_{ov} + \lambda_{hv}$, $\lambda_d = \lambda_{od} + \lambda_{hd}$, $e_1 = (1, 0)$, $e_2 = (0, 1)$.

It is easy to show that given finite 2-D is irreducible, so in this chain equilibrium regime exists. Let $p(n)$ denote the stationary probability of state $n \in S$.

The above-mentioned QoS metrics are determined as appropriate marginal distributions of the defined 2-D MC.

Let P_x be the loss probability of x-calls, $x \in \{hv, ov, hd, od\}$. Taking into account described above rules for accepting of heterogeneous calls and by using PASTA-theorem [14] we obtain the following formulas to calculate the QoS metrics of the network:

$$P_{ov} = \sum_{n \in S} p(n)I(n_v \geq N_v); \tag{3}$$

$$P_{od} = \sum_{n \in S} p(n)I(n_d \geq R_{od}); \tag{4}$$

$$P_{hd} = \sum_{n \in S} p(n)\delta(n_d + n_v, N_{vd}); \tag{5}$$

$$P_{hv} = \sum_{n \in S} p(n)\Big(\delta(n_v, R_{hv})\big(1 - \delta(n_d + n_v, N_{vd})\big) +$$

$$+ \big(1 - \delta(n_v, R_{hv})\big)\delta(n_d + n_v, N_{vd})\Big); \tag{6}$$

where $\delta(i,j)$ are Kronecker's symbols, $I(A)$ is the indicator function of event A. Thus, to calculate these QoS metrics, as usual the solution of system of global balance equations (SGBE) is required. In general, due to the large state space we face to the state space explosion problem. However, in this case the considered SGBE has analytical solution in multiplicative form. Namely, we have

Proposition. The stationary distribution has the following multiplicative form:
Case $R_{od} \leq N_{vd} - R_{hv}$:

$$p(i,j)$$

$$= \begin{cases} \frac{v_d^i \, v_v^j}{i! \, j!}p(0,0) & \text{if } 0 \leq i \leq R_{od}, 0 \leq j \leq N_v, \\[2mm] \left(\frac{v_d}{v_{hd}}\right)^{R_{od}} \frac{v_{hd}^i \, v_v^j}{i! \, j!}p(0,0) & \text{if } R_{od} + 1 \leq i \leq N_{vd}, 0 \leq j \leq N_v, \\[2mm] \left(\frac{v_v}{v_{hv}}\right)^{N_v} \frac{v_d^i \, v_{hv}^j}{i! \, j!}p(0,0) & \text{if } 0 \leq i \leq R_{od}, N_v + 1 \leq j \leq N_v + R_{hv}, \\[2mm] \left(\frac{v_d}{v_{hd}}\right)^{R_{od}} \left(\frac{v_v}{v_{hv}}\right)^{N_v} \frac{v_{hd}^i \, v_{hv}^j}{i! \, j!}p(0,0) & \text{if } R_{od} + 1 \leq i \leq N_{vd} - 1, \\ & \quad N_v + 1 \leq j \leq min(N_v + R_{hv}, N - i); \end{cases} \tag{7}$$

Case $R_{od} > N_{vd} - R_{hv}$:

$$p(i,j)$$

$$= \begin{cases} \frac{v_d^i \, v_v^j}{i! \, j!}p(0,0) & \text{if } 0 \leq i \leq R_{od}, 0 \leq j \leq N_v, \\[2mm] \left(\frac{v_d}{v_{hd}}\right)^{R_{od}} \frac{v_{hd}^i \, v_v^j}{i! \, j!}p(0,0) & \text{if } R_{od} + 1 \leq i \leq N_{vd}, 0 \leq j \leq N_v, \\[2mm] \left(\frac{v_v}{v_{hv}}\right)^{N_v} \frac{v_d^i \, v_{hv}^j}{i! \, j!}p(0,0) & \text{if } 0 \leq i \leq R_{od}, N_v + 1 \leq j \leq \\ & \quad min(N_v + R_{hv}, N - i), \\[2mm] \left(\frac{v_d}{v_{hd}}\right)^{R_{od}} \left(\frac{v_v}{v_{hv}}\right)^{N_v} \frac{v_{hd}^i \, v_{hv}^j}{i! \, j!}p(0,0) & \text{if } R_{od} + 1 \leq i \leq N_{vd} - 1, \\ & \quad N_v + 1 \leq j \leq N - i; \end{cases} \tag{8}$$

In both formulas as usual $p(0,0)$ is determined by the normalizing condition.

Proof of this fact is based on Kolmogorov's theorem about reversibility of 2-D MC, see for example [7]. Indeed, it is easily shown that there is no circulation between states $n, n+e_1, n+e_2, n+e_1+e_2$ of the state diagram of the underlying 2-D MC. Indeed, in both cases $R_{od} \le N_{vd} - R_{hv}$ and $R_{od} > N_{vd} - R_{hv}$ circulation flow among the indicated four states in both directions (clockwise and counter clockwise) is equals $\lambda_d \lambda_v (n_d + 1) \mu_d (n_v + 1) \mu_v$. In other words, system of local balance equations (SLBE) is fulfilled, i.e. there is a general solution of the SLBE for state probabilities. Thus by choosing the path $(0,0), (1,0), , (i,0), (i,1), , (i,j)$ from state $(0,0)$ to state (i,j) we find that multiplicative solution (7) (or (8)) is hold. Note that in this proof scheme it is required take into account four cases in formulas (7) and (8) which are indicated in the right sides of the indicated formulas.

Now we are ready to obtain the respective loss probabilities in the following explicit formulas

$$P_{ov} = \sum_{i=0}^{N_{vd}-R_{hv}} \sum_{j=N_v}^{N_v-R_{hv}} p(i,j) + \sum_{i=N_{vd}-R_{hv}+1}^{N_{vd}} \sum_{j=N_v}^{N-i} p(i,j); \tag{9}$$

$$P_{hv} = \sum_{i=0}^{N_{vd}-R_{hv}} p(i, N_v + R_{hv}) + \sum_{i=N_{vd}-R_{hv}+1}^{N_{vd}} p(i, N - i); \tag{10}$$

$$P_{od} = \sum_{i=R_{od}}^{N_{vd}} \sum_{j=0}^{min(N_v+R_{hv}, N-i)} p(i,j); \tag{11}$$

$$P_{hd} = \sum_{i=0}^{N_v-1} p(N_{vd}, i) + \sum_{i=N_{vd}-R_{hv}}^{N_{vd}} p(i, N - i). \tag{12}$$

4 Numerical Results

The developed above explicit formulas allow us to investigate behavior of QoS metrics of the proposed partition scheme over any range of change of values of loading parameters of heterogeneous calls and number of channels. First of all, here it is assumed that allocation of entire pool of channels between zones is fixed and only regulated parameters are R_{hv} and R_{od}. It is clear that the increase in value of one of the parameters R_{hv} and R_{od} (in an admissible area) favorably influences the QoS metric of calls of the corresponding type only.

The initial data for total number of channels and loading parameters of heterogeneous calls are as in [3], i.e.

$$N = 30, \quad \lambda_o + \lambda_h = 0.15, \quad \lambda_{od} + \lambda_{hd} = 0.3, \quad \mu_v^{-1} = 2, \quad \mu_d^{-1} = 120.$$

Below assume that $N_v = 12$, $N_{vd} = 18$ and 30 % of the total intensity of voice calls are handover voice calls and 80 % of the total intensity of data calls are new data calls.

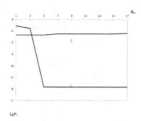

Fig. 1. P_v vs R_{hv}; $1 - P_{hv}$, $2 - P_{ov}$

Consider the results of numerical experiments for the model with VP-scheme for partition of channels. In Fig. 1 the dependency of QoS metrics on the parameter R_{hv} is shown. It is seen from Fig. 1 that function P_{hv} decreases in small values of parameter R_{hv} with high speed, thereafter it becomes almost constant; function P_{ov} increases with insignificant speed in small values of indicated parameter, thereafter it becomes almost constant also. Almost constants are both functions P_{od} and P_{hd} versus R_{hv} (see Fig. 2). Such behavior of functions P_{od} and P_{hd} is explained via small intensity of handover voice calls.

Dependency of QoS metrics on the parameter R_{od} are shown in Figs. 3 and 4. Here both functions P_{ov} and P_{hv} increases with insignificant speed in small values of indicated parameter, thereafter it becomes almost constant (see Fig. 3). However, function P_{od} decreases with significant speed versus R_{od} while function P_{hd} is almost constant one (see Fig. 4).

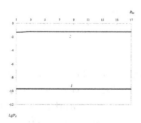

Fig. 2. P_d vs R_{hv}; $1 - P_{hd}$, $2 - P_{od}$

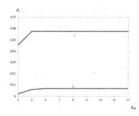

Fig. 3. P_v vs R_{od}; $1 - P_{hv}$, $2 - P_{ov}$

It should be noted that as these numerical results show that all QoS metrics have monotony property. These facts allow us to develop the algorithms to find the set of effective values in order to satisfy the given QoS level.

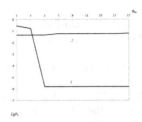

Fig. 4. P_d vs R_{od}; $1 - P_{od}$, $2 - P_{hd}$

5   Conclusions

In this paper, a virtual scheme to partition of entire pool of channels of isolated cell in integrated wireless networks was proposed. In accordance to this scheme all channels are virtually distributed between voice and data calls and there are limits to the number of handover voice calls and new data calls in zone of channels for data calls. The indicated limits are state-dependent parameters.

Explicit formulas to calculate loss probabilities of the network under given partition scheme were developed. The obtained formulas allow us to solve the problems related to satisfy desired QoS level of heterogeneous calls. These problems are subject of future research.

Acknowledgments. The work of A. Melikov, M.Fattakhova and G. Velidzanova was supported by Science Development Foundation under the President of the Republic of Azerbaijan - Grant No. EIF-RITN-MQM-2/IKT-2-2013-7(13)-29/01/1.

The work of János Sztrik was realized in the frames of TÁMOP 4.2.4. A/2-11-1-2012-0001 National Excellence Program - Elaborating and operating an inland student and researcher personal support system. The project was subsidized by the European Union and co-financed by the European Social Fund.

References

1. Basharin, G.P., Samouylov, K.E., Yarkina, N.V., Gudkova, I.A.: A new stage in mathematical teletraffic theory. Automation Remote Control 70(12), 1954–1964 (2009)
2. Bérczes, T., Almási, B., Kuki, A., Sztrik, J.: The effect of RF unit breakdowns in sensor communication networks. Infocommunications Journal 5, 11–16 (2013)
3. Carvalho, G.H.S., Martins, V.S., Frances, C.R.L., Costa, J.C.W.A., Carvalho, S.V.: Performance analysis of multi-service wireless network: An approach integrating cac, scheduling, and buffer management. Comput. Electric. Eng. 34, 346–356 (2008)

4. Daigle, J.N.: Queuing theory with applications to packet telecommunication. Springer, New York (2005)
5. Van Do, T., Chakka, R., Sztrik, J.: Spectral expansion solution methodology for QBD-M processes and applications in Future Internet engineering. In: Nguyen, N.T., van Do, T., Thi, H.A. (eds.) ICCSAMA 2013. SCI, vol. 479, pp. 131–142. Springer, Heidelberg (2013)
6. Feng, W., Kowada, M.: Performance analysis of wireless mobile networks with queueing priority and guard channels. International Transactions on Operational Research 15, 481–508 (2008)
7. Kelly, F.P.: Reversibility and stochastic networks. John Wiley & Sons, New York (1979)
8. Lakatos, L., Szeidl, L., Telek, M.: Introduction to queuing systems with telecommunication applications. Springer, Heidelberg (2013)
9. Melikov, A.Z., Fattakhova, M.I., Babayev, A.T.: Investigation of cellular communication networks with private channels for service of handover calls. Automatic Control and Computer Sciences 39(3), 61–69 (2005)
10. Oh, Y., Kim, C.S., Melikov, A.Z., Fattakhova, M.I.: Numerical analysis of multi-parameter strategy of access in multiservice cellular communication networks. Automation and Remote Control 71(12), 2558–2572 (2010)
11. Ponomarenko, L., Kim, C.S., Melikov, A.: Performance analysis and optimization of multi-traffic on communication networks. Springer, Heidelberg (2010)
12. Schneps-Schneppe, M., Iversen, V.B.: Call admission control in cellular networks. In: Ortiz, J.H. (ed.) Mobile Networks, pp. 111–136. Intech (2012)
13. Stasiak, M., Glabowski, M., Wishniewski, A., Zwierzykowski, P.: Modeling and dimensioning of mobile networks. John Wiley, Chichester (2011)
14. Wolff, R.W.: Poisson arrivals see time averages. Operations Research 30(2), 223–231 (1992)

Router Speed Analysis*

Pavel Mikheev

Tomsk State University,
Lenina str., 36, 634050 Tomsk, Russia
doka.patrick@gmail.com

Abstract. This paper describes a model of transit node of network for data transmission, which distributes an input flow in several outgoing directions. In addition the paper presents investigations of quality influence of communication channels, distribution of incoming traffic in outgoing directions and strategy of limited buffer memory sharing of transit node between communication channel queue on the throughput of network segments with various speed of incoming and outgoing interfaces.

Keywords: star network, memory lock, throughput, mathematical model, Markov chain, buffer memory sharing strategy, traffic split.

1 Introduction

One of the main factors characterizing operational parameters of network are limited buffer memory locks of the switching (2nd level of network architecture) [1] and routing nodes (3rd level of network architecture) [2]. The throughput of the network fragment is mainly defined by the capacity of buffer memory of transit node. When distributing incoming traffic for transit node in the outgoing directions, the volume of the missed flow considerably depends on strategy of limited buffer memory sharing between output interfaces. The main problem for the architects to solve is how to divide the shared buffer space for storing transit data packages between output communication channels [3–5]. One of the first buffer memory sharing research was performed in [6]. Originally the classification of various schemes of buffer memory division was offered in [7]. As the operation of computer networks has a significantly discrete character [8, 9], ref. [10] describes studying the effect of buffer memory locks on speed of network fragments using queuing systems with finite storage and discrete time. The research results were developed and reviewed by author: the comparative analysis for the three strategies of limited buffer memory division in terms of queuing systems with discrete time of the served load index was performed. This paper is further development of this research. Effect of buffer memory locks of transit node on the share of the served loading by a star-like network fragment with various speed of incoming and outgoing interfaces is analyzed.

---

* This work is performed under the state order No. 1.511.2014/K of the Ministry of Education and Science of the Russian Federation.

A. Dudin et al. (Eds.): ITMM 2014, CCIS 487, pp. 277–286, 2014.

2   Discrete Model of Star-Like Network Segment with Traffic Distribution

Consider a star-like network segment, including $M + 1$ links of data transmission where the information flow arrives into the central transit node through the one incoming communication channel and is distributed along the M outgoing directions. Assuming that all incoming communication channels have identical physical speeds for data transmission, and data transmission speed in the incoming channel is S times higher. Besides, consider that time of package processing during receiving and sending in sending and receiving nodes is the same. Then time of the full transfer cycle of package t will be identical for all incoming links of the considered fragment, and the time for an incoming link is t/S. Let us consider that the package which arrived in transit node in the current cycle t, will be transferred through the output channel only in the next cycle. Note that during the time $t - S$ packages can arrive into the transit node through incoming channel whereas only one package can leave through each of the outgoing directions. Assume that error-free transfer of data package in the incoming and outgoing channel is defined by F and F_m, $m = \overline{1, M}$ probabilities, respectively. All packet flow entering the transit node is distributed in m-th output channel with B_m, $\sum_{m=1}^{M} B_m = 1$ probability. B_m values can be determined as the shares of incoming flow sent to m-th outgoing channel. It is not difficult to see that the time of error-free packet transfer through the each internodal connection is a random value. This value has the geometrical distribution law with parameter F and F_m, $m = \overline{1, M}$ in incoming and outgoing channels, respectively.

The pool of shared buffer memory of volume K for packet storing in queues to output interfaces of transit node. The queue size q_m to each m-th outgoing channel is limited by the limiting value of $N_m \leq K$ determined by strategy of division of buffer memory between outgoing channels. For each incoming packet sent to the certain incoming channel the buffer is allocated if the output queue q_m for the given direction do not exceed the maximum size of $q_m < N_m$. Besides, queues to outgoing communication channels restricted by $\sum_{m=1}^{M} q_m < K$. Obviously, that in each case for division of buffer pool between outgoing directions the queue size to m-th channel q_m does not exceed the value of Q_m which meets the following conditions: $Q_m \leq N_m$ and $\sum_{m=1}^{M} Q_m = K$.

Generally there are five strategies for division of buffer memory between outgoing communication channels [7]. Two of them are extreme: full-accessible and full division strategies. The other three strategies are intermediate and allow various implementations [7]. Let us analyze three strategies: full-accessible ($N_m = K$), full division ($N_m = K/M$) and one intermediate ($K/M < N_m < K$, $\sum_{m=1}^{M} Q_m = K$).

Behavior of the considered network fragment is represented as the Markov queuing system with discrete time, finite storage and the M servers [11]. The incoming flow is defined by the quality of the incoming channel F and parameter of incoming connection speed S, and service time is defined by the quality of m-th incoming channel F_m. Distribution of incoming requests of queuing system along M servers is set by probabilities B_m, $m = \overline{1,M}$. Queue dynamics to output communication channels of this queuing system under stationary conditions is described by Markov chain in the M-dimensional space. The set of possible conditions of Markov chain is determined by division of buffer memory strategy between incoming channels and does not exceed the value of $N_m + 1$.

In Markov discrete chain with finite number of states describing considered queuing system in the established mode, we will define transitional probabilities π_I^J from state I to state J. Here I and J — M-bit numbers according to initial and changed states with value ranges of each bit from 0 to N_m: $I = i_1, \ldots, i_M$; $i_m = \overline{0, N_m}$; $J = j_1, \ldots, j_M$; $j_m = \overline{0, N_m}$; $m = \overline{1, M}$.

Denote state probabilities of M-dimensional Markov chain by $P_{i_1,\ldots,i_M}$, $i_m = \overline{0, Q_m}$, $m = \overline{1, M}$. It is obvious that the record $P_{i_1,\ldots,i_M}$ is equivalent to the record P_I. One of the main characteristics of storage-limited queuing system is throughput [11]:

$$
Z(M, K, F, \boldsymbol{F}, \boldsymbol{B}) = \sum_{m=1}^{M} F_m \sum_{i_1=0}^{Q_1} \cdots \sum_{i_{m-1}=0}^{Q_{m-1}} \sum_{i_m=1}^{Q_m} \sum_{i_{m+1}=0}^{Q_{m+1}} \cdots \sum_{i_M=0}^{Q_M} P_{i_1,\ldots,i_M}, \quad (1)
$$

where $\boldsymbol{F} = \{F_1, \ldots, F_M\}$ is vector of values F_m, $m = \overline{1, M}$. In case of uniformity of all outgoing directions this vector is denoted as F_*. $\boldsymbol{B} = \{B_1, \ldots, B_M\}$ — is vector of values B_m, $m = \overline{1, M}$.

3 Full-Accessible Buffer Memory Sharing Strategy

Full-accessible buffer memory sharing strategy provides full availability of the whole buffer pool for any outgoing direction of transit node. In this case the package queue for any outgoing channels can occupy all buffer memory of transit node.

Consider the operation of star-like network fragment at full-accessible buffer memory sharing as the fragment with the following parameters: two outgoing channels ($M = 2$) divide the buffer pool of transit node with volume $K = 2$, speed parameter of incoming channel is $S = 2$. In this case $N_m = 2$. Table 1 shows transition probabilities grouped by identity of changed status J.

The equilibrium equations describing employment of buffer space of transit node by packets to various outgoing communication channels under stationary conditions can be written as follows:

$$
P_{00}F(2 - F) = P_{10}F_1(1 - F)^2 + P_{01}F_2(1 - F)^2 + P_{11}F_1F_2(1 - F)^2;
$$

$$
P_{10}\left[F(2 - F) + F_1(1 - F)^2 - 2F_1FB_1(1 - F)\right] = 2P_{00}FB_1(1 - F) + 2P_{01}F_2 \times
$$

$$
\times FB_1(1 - F) + P_{11}\left[F_2(1 - F_1)(1 - F)^2 + 2F_2F_1FB_1(1 - F)\right] + P_{20}F_1(1 - F)^2;
$$

$$
P_{01}\left[F(2 - F) + F_2(1 - F)^2 - 2F_2FB_2(1 - F)\right] = 2P_{00}FB_2(1 - F) + 2P_{10}F_1 \times
$$

$$
\times FB_2(1 - F) + P_{11}\left[F_1(1 - F_2)(1 - F)^2 + 2F_1F_2FB_2(1 - F)\right] + P_{02}F_2(1 - F)^2;
$$

$$P_{11}\big[F_1 + F_2 - F(2 - F)(F_1B_1 + F_2B_2) - F_1F_2(1 - F)^2 - 2F^2F_1F_2B_1B_2\big] =$$
$$= 2P_{00}F^2B_1B_2 + P_{10}\big[(1 - F_1)FB_2(2 - F) + 2F_1F^2B_1B_2\big] + P_{01}\big[(1 - F_2)F\times$$
$$\times B_1(2 - F) + 2F_2F^2B_1B_2\big] + P_{20}F_1FB_2(2 - F) + P_{02}F_2FB_1(2 - F);$$
$$P_{20}\big[F_1 - F_1FB_1(2 - F)\big] = P_{00}F^2B_1^2 + P_{10}\big[(1 - F_1)FB_1(2 - F) + F_1F^2B_1^2\big] +$$
$$+ P_{01}F_2F^2B_1^2 + P_{11}\big[F_2(1 - F_1)FB_1(2 - F) + F_2F_1F^2B_1^2\big];$$
$$P_{02}\big[F_2 - F_2FB_2(2 - F)\big] = P_{00}F^2B_2^2 + P_{01}\big[(1 - F_2)FB_2(2 - F) + F_2F^2B_2^2\big] +$$
$$+ P_{10}F_1F^2B_2^2 + P_{11}\big[F_1(1 - F_2)FB_2(2 - F) + F_1F_2F^2B_2^2\big].$$

Table 1. Transition probabilities

$\pi_{i_1 i_2}^{j_1 j_2}$	i_1	i_2	j_1	j_2
$(1 - F)^2$	0	0	0	0
$F_1(1 - F)^2$	1	0	0	0
$F_2(1 - F)^2$	0	1	0	0
$F_1F_2(1 - F)^2$	1	1	0	0
$2FB_1(1 - F)$	0	0	1	0
$(1 - F_1)(1 - F)^2 + 2F_1FB_1(1 - F)$	1	0	1	0
$2F_2FB_1(1 - F)$	0	1	1	0
$F_2(1 - F_1)(1 - F)^2 + 2F_2F_1FB_1(1 - F)$	1	1	1	0
$F_1(1 - F)^2$	2	0	1	0
$2FB_2(1 - F)$	0	0	0	1
$2F_1FB_2(1 - F)$	1	0	0	1
$(1 - F_2)(1 - F)^2 + 2F_2FB_2(1 - F)$	0	1	0	1
$F_1(1 - F_2)(1 - F)^2 + 2F_1F_2FB_2(1 - F)$	1	1	0	1
$F_2(1 - F)^2$	0	2	0	1
$2F^2B_1B_2$	0	0	1	1
$(1 - F_1)FB_2(2 - F) + 2F_1F^2B_1B_2$	1	0	1	1
$(1 - F_2)FB_1(2 - F) + 2F_2F^2B_1B_2$	0	1	1	1
$(1 - F_1)(1 - F_2) + F_1(1 - F_2)FB_1(2 - F) +$ $+ F_2(1 - F_1)FB_2(2 - F) + 2F_1F_2F^2B_1B_2$	1	1	1	1
$F_1FB_2(2 - F)$	2	0	1	1
$F_2FB_1(2 - F)$	0	2	1	1
$F^2B_1^2$	0	0	2	0
$(1 - F_1)FB_1(2 - F) + F_1F^2B_1^2$	1	0	2	0
$F_2F^2B_1^2$	0	1	2	0
$F_2(1 - F_1)FB_1(2 - F) + F_2F_1F^2B_1^2$	1	1	2	0
$1 - F_1 + F_1FB_1(2 - F)$	2	0	2	0
$F^2B_2^2$	0	0	0	2
$F_1F^2B_2^2$	1	0	0	2
$(1 - F_2)FB_2(2 - F) + F_2F^2B_2^2$	0	1	0	2
$F_1(1 - F_2)FB_2(2 - F) + F_1F_2F^2B_2^2$	1	1	0	2
$1 - F_2 + F_2FB_2(2 - F)$	0	2	0	2

The solution of the simultaneous equations for arbitrary F, F_1, F_2, B_1, B_2 is rather complex, that is why the values of probability state and throughput (1) for reliable input channel ($F = 1$) are given here: $P_{00} = P_{10} = P_{01} = 0$; $P_{11} = EF_1F_2B_1B_2$; $P_{20} = EF_2^2B_1^2(1 - F_1B_2)$; $P_{02} = EF_1^2B_2^2(1 - F_2B_1)$; $Z(2,2,1,\boldsymbol{F},\boldsymbol{B}) = E(F_1F_2^2B_1 + F_2F_1^2B_2 - F_1^2F_2^2B_1B_2)$; $E = \left[F_1F_2B_1B_2(1 - F_2B_1 - F_1B_2) + F_1^2B_2^2 + F_2^2B_1^2\right]^{-1}$.

Fig. 1,a and 1,b illustrate effects of the missed flow on the traffic distribution structure for network fragment with parameters $S = 2$, $M = 2$, $K = 2$, and $S = 2$, $M = 2$, $K = 4$ respectively. Fig. 1 shows that the throughput considerably is determined by structure of traffic distribution on outgoing channels and has optimal parameter set B_m, $m = \overline{1, M}$. As capacity of the buffer memory becomes higher, the throughput increases and its value becomes more sensitive to optimal parameter set B_m, $m = \overline{1, M}$. It should be noted that maximum throughput is reached at uniform traffic distribution ($B_m = 1/M$, $m = \overline{1, M}$) with statistically uniform communication channels $F_m = F_*$, $m = \overline{1, M}$.

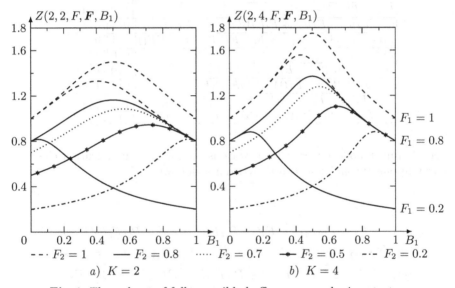

Fig. 1. Throughput of full-accessible buffer memory sharing strategy

4 Full Memory Sharing Strategy

Full memory sharing strategy allocates a certain amount of buffer memory to each of the outgoing directions. In this case directions are not locked in case of buffer overfilling for one of the directions. Inefficient use of memory of transit node at low loading of one of the directions is possible.

Consider the previous network segment ($S = 2$, $M = 2$, $K = 2$) using full memory sharing strategy ($N_m = 1$). Transition probabilities for this case are presented in Table 2. The equilibrium equations describing employment of buffer

Table 2. Transition probabilities

$\pi_{i_1 i_2}^{j_1 j_2}$	i_1	i_2	j_1	j_2
$(1-F)^2$	0	0	0	0
$F_1(1-F)^2$	1	0	0	0
$F_2(1-F)^2$	0	1	0	0
$F_1 F_2(1-F)^2$	1	1	0	0
$FB_1(2-F-FB_2)$	0	0	1	0
$(1-F_1)(1-FB_2)^2 + F_1 FB_1(2-F-FB_2)$	1	0	1	0
$F_2 FB_1(2-F-FB_2)$	0	1	1	0
$F_2(1-F_1)(1-FB_2)^2 + F_2 F_1 FB_1(2-F-FB_2)$	1	1	1	0
$FB_2(2-F-FB_1)$	0	0	0	1
$F_1 FB_2(2-F-FB_1)$	1	0	0	1
$(1-F_2)(1-FB_1)^2 + F_2 FB_2(2-F-FB_1)$	0	1	0	1
$F_1(1-F_2)(1-FB_1)^2 + F_1 F_2 FB_2(2-F-FB_1)$	1	1	0	1
$2F^2 B_1 B_2$	0	0	1	1
$(1-F_1)FB_2(2-FB_2) + 2F_1 F^2 B_1 B_2$	1	0	1	1
$(1-F_2)FB_1(2-FB1) + 2F_2 F^2 B_1 B_2$	0	1	1	1
$(1-F_1)(1-F_2) + F_1(1-F_2)FB_1(2-FB_1)+$ $+F_2(1-F_1)FB_2(2-FB_2) + 2F_1 F_2 F^2 B_1 B_2$	1	1	1	1

space of transit node by packets to various outgoing communication channels under stationary conditions can be written as follows:

$$P_{00}F(2-F) = P_{10}F_1(1-F)^2 + P_{01}F_2(1-F)^2 + P_{11}F_1 F_2(1-F)^2;$$
$$P_{10}\left[2FB_2 - F^2 B_2^2 + F_1(1-F)^2\right] = P_{00}FB_1(2-F-FB_2)+$$
$$+P_{01}F_2 FB_1(2-F-FB_2) + P_{11}F_2\left[(1-FB_2)^2 - F_1(1-F)^2\right];$$
$$P_{01}\left[2FB_1 - F^2 B_1^2 + F_2(1-F)^2\right] = P_{00}FB_2(2-F-FB_1)+$$
$$+P_{10}F_1 FB_2(2-F-FB_1) + P_{11}F_1\left[(1-FB_1)^2 - F_2(1-F)^2\right];$$
$$P_{11}\left[F_1(1-FB_1)^2 + F_2(1-FB_2)^2 - F_1 F_2(1-F)^2\right] = 2P_{00}F^2 B_1 B_2+$$
$$+P_{10}\left[(1-F_1)FB_2(2-FB_2) + 2F_1 F^2 B_1 B_2\right]+$$
$$+P_{01}\left[(1-F_2)FB_1(2-FB_1) + 2F_2 F^2 B_1 B_2\right].$$

Taking into account the normalization condition we find relations for finite probabilities of Markov chain:

$$P_{00} = \frac{F_1 F_2 (1-F)^2 (1 - X_1 X_2 X_3)}{A};$$

$$P_{10} = P_{00} \frac{FB_1(2-FB_1)X_2[1 - X_3(1-F)^2] + F_1[X_1 X_2 - (1-F)^2]}{F_1(1-F)^2(1-X_1X_2X_3)};$$

$$P_{01} = P_{00} \frac{FB_2(2-FB_2)X_1[1 - X_3(1-F)^2] + F_2[X_1 X_2 - (1-F)^2]}{F_2(1-F)^2(1-X_1X_2X_3)};$$

$$P_{11} = P_{00}\left\{2F^2 B_1 B_2 (3 - 2F + F^2 B_1 B_2)[1 - X_3(1-F)^2] - (F_1 + F_2 + $$

$$ + F(2-F)X_3)[X_1 X_2 - (1-F)^2]\right\}\Big/ F_1 F_2 (1-F)^2 (1-X_1X_2X_3);$$

$$X_1 = (1 - FB_1)^2; \quad X_2 = (1 - FB_2)^2; \quad X_3 = (1-F_1)(1-F_2);$$

$$A = F_1 F_2 (1-F)^2(1 - X_1X_2X_3) + [F_2 FB_1(2-FB_1)X_2 + F_1 FB_2 \times$$

$$ \times (2 - FB_2)X_1 + 2F^2 B_1 B_2 (3 - 2F + F^2 B_1 B_2)][1 - X_3(1-F)^2] + $$

$$ + [2F_1 F_2 - F_1 - F_2 - F(2-F)X_3][X_1 X_2 - (1-F)^2].$$

The throughput (1) for random levels of data transmission validity of the considered fragment is written as follows:

$$Z(2,2,F,\boldsymbol{F},\boldsymbol{B}) = \left\{[F_1 F_2 FB_1(2-FB_1)X_2 + F_1 F_2 FB_2(2-FB_2)X_1 + \right.$$

$$ + 2F^2 B_1 B_2(F_1 + F_2)(3 - 2F + F^2 B_1 B_2)][1 - X_3(1-F)^2] + (F_1 + F_2)\times$$

$$ \left. \times [F_1 F_2 - F_1 - F_2 - F(2-F)X_3][X_1 X_2 - (1-F)^2]\right\}\Big/ A. \qquad (2)$$

Fig. 2,a and Fig. 2,b show the dependences of the throughput from traffic distribution of network fragment with parameters $S = 2$, $M = 2$, $K = 2$, and $S = 2$, $M = 2$, $K = 4$, respectively. Curves in Fig. 2 demonstrate that when using full division or full-accessible strategies the throughput has a maximum by parameters of traffic distribution B_m, $m = \overline{1, M}$, but in this case the insignificant deviation from an optimum parameter set of distribution slightly reduces the value of the throughput.

5   Intermediate Strategy for Buffer Memory Division

Both full-accessible and full buffer memory division strategies are two contrasts in division of buffer memory of the transit node between queues to the outgoing directions. The point of intermediate strategy is to allocate the part of buffer memory for individual needs of each outgoing interface and to make other memory equally accessible for all outgoing directions. Fig. 3 presents the results of numerical research of network segment with parameters $S = 2$, $M = 2$, $K = 4$ and intermediate strategy of buffer memory sharing of the transit node between the outgoing directions (one individual buffer for each direction and two common buffers). In this case $N_m = 3$.

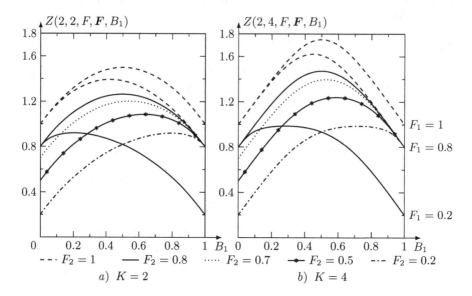

Fig. 2. Throughput of full memory sharing strategy

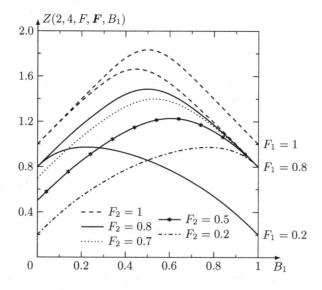

Fig. 3. Throughput of intermediate strategy

6 Comparative Analysis of the Strategies for Buffer Memory Division

The dependence of the throughput for considered strategies of buffer memory division from reliability of data transmission in uniform incoming channels is given in fig. 4. The presented curves show that there are parametric areas of network

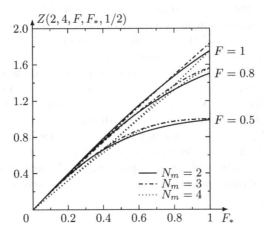

Fig. 4. Throughput of full memory sharing ($N_m = 2$), full-accessible ($N_m = 4$) and intermediate ($N_m = 3$) strategies for $M = 2$, $K = 4$

fragment for which each considered strategy is more preferable than competing strategy. As the speed parameter S is increased the equivalence point of full memory sharing and full-accessible memory sharing strategies shifts towards (to the right) reduction of error level of communication channels. For $S = 2$ strategy equivalence is reached in $F_1 = F_2 = F$ (see fig. 4), while for $S = 1$ equality of throughput for full-accessible and full memory division strategies is reached in $F_1 = F_2 = F/2$. Thus the advantage of full memory division strategy is increases, and this advantage is rather essential. For example, in case of $S = 2$, $M = 2$, $K = 2$, $F = 1$, $F_1 = F_2 = 0.5$ at $B_1 = B_2$ full division strategy increases level of the throughput by 22% in comparison with full-accessible strategy of memory division. At the same time the intermediate strategy slightly concedes to strategy of full memory division at high error level in outgoing channels, and provides the best indicators of the throughput (fig. 4) when increasing the quality of outgoing channels. The similar case takes place for three outgoing channels.

Numerical research of fragments with two and three outgoing channels show that the throughput is considerably defined by traffic distribution along the outgoing channels and has the optimum parameter set of B_m, $m = \overline{1, M}$ for all memory division strategies (see fig. 1–3). It is easy to see that at significantly non-uniform quality of outgoing communication channels the missed flow has a pronounced maximum from distribution B_1. It is also noted that as volume of buffer memory increases the maximum of throughput is displaced in area of traffic distribution with an increasing share of the flow sent to the outgoing channel of low quality. The maximum of the throughput is reached at uniform traffic distribution for outgoing communication channels uniform in quality.

7   Conclusion

This paper presents the model of transit node of the data transmission, distributing incoming information flow in several incoming directions. Various strategies of buffer memory division of transit node between queues to outgoing communication channels are analyzed. The router model with uniform speed of service of incoming and outgoing interfaces is postponed for a case when the speed of incoming channel surpasses the outgoing directions of data transmission. The analysis of numerical results demonstrates that for buffer memory of volume $K \geq M$ throughput has a maximum traffic distribution in outgoing communication channels B_m, $m = \overline{1, M}$ for any of the considered strategy of division. In general extreme nature of dependence of the missed flow from traffic distribution needs to be considered when implementing of routing algorithms with information flow distribution between the set couple of corresponding nodes on several various routes.

References

1. Froom, R., Sivasubramanian, B., Frahim, E.: Implementing Cisco IP Switched Networks (SWITCH) Foundation Learning Guide. Cisco Press, Indianapolis (2010)
2. Teare, D.: Implementing Cisco IP Routing (ROUTE) Foundation Learning Guide. Cisco Press, Indianapolis (2010)
3. Empson, S.: CCNA Routing and Switching Portable Command Guide. Cisco Press, Indianapolis (2013)
4. Oppenheimer, P.: Top-Down Network Design, 3rd edn. Cisco Press, Indianapolis (2012)
5. Wilkins, S.R.: Designing for Cisco Internetwork Solutions (DESGN) Foundation Learning Guide, 3rd edn. Cisco Press, Indianapolis (2012)
6. Irland, M.: Buffer management in a packet switch. IEEE Trans. Commun. 26(3), 328–337 (1978)
7. Kamoun, F., Kleinrock, L.: Analysis of shared finite storage in a computer network node environment under general traffic conditions. IEEE Trans. Commun. 28(7), 992–1003 (1980)
8. Ivanovskiy, V.B.: On properties of output flows in digital service systems. Autom. Remote Control. 45(11), 1413–1419 (1984)
9. Kleynrock, L.: Queueing Systems, vol. II. Wiley (1975)
10. Suschenko, S.P.: The influence of buffer overfilling on the speed of synchronous data-transmission control procedures. Autom. Remote Control. 60(10), 1460–1468 (1999)
11. Kleynrock, L.: Queueing Systems: Theory, vol. I. Wiley (1975)

The First Jump Separation Technique for the Tandem Queueing System $GI/(GI/\infty)^{K\star}$

Alexander Moiseev

Tomsk State University
Tomsk, Russia
moiseev.tsu@gmail.com

Abstract. Application of the first jump separation technique for analysis of the tandem queueing system with high-intensive renewal arrival process, infinite number of servers and general service time distribution is presented in the paper. An equation for characteristic function of the multi-dimensional joint distribution of the number of customers at the system stages is derived. The equation is solved under an asymptotic condition of the infinite growth of the arrival rate. It is shown that the distribution under study can be approximated by the multi-dimensional Gaussian distribution. Numerical example shows the range of the approximation applicability.

Keywords: tandem queueing system, high-intensive arrival process, renewal process.

1 Introduction

Tandem queueing systems [1] are important to study of processes in the data processing systems [2], manufacturing and other practical fields [3]. They also are applicable for queueing network decomposition [4]. Usually, the researches in this filed are devoted to solve some concrete practical problems or consider the tandem systems of special configuration (e.g.,[5,6]). In this paper, we provide the analysis of the tandem model of quite general form.

In section 3, we have obtained the integral equation for the characteristic function of the number of customers at the system stages and solve it under an asymptotic condition of infinite growth of the arrivals' rate. This condition can be considered as one of the well-known heavy-traffic conditions [7]. A solution of the equation under this condition (see section 5) allows to build an approximation for the multi-dimensional distribution of the system states. In the section 6, we consider a concrete numerical example and draw a conclusions about the range of the approximation applicability.

---

* This work is performed under the state order No. 1.511.2014/K of the Ministry of Education and Science of the Russian Federation.

A. Dudin et al. (Eds.): ITMM 2014, CCIS 487, pp. 287–300, 2014.

2   Mathematical Model

Let's consider a tandem (multi-stage) queueing system with K stages, renewal arrival process, an infinite number of servers at each stage and i.i.d. service times. Let $A(x)$ be a cumulative distribution function of the inter-arrival intervals, $B_k(x)$ be a cumulative distribution function of the service time at the k-th stage $(k = 1, \ldots, K)$. After a service is completed for some customer at one stage, this customer moves to the next stage for the further service until service will be completed at the final, K-th, stage.

Let $i_k(t)$ be a number of customers at the k-th service stage at the moment t. Denote a vector $\boldsymbol{i}^{\mathrm{T}}(t) = \{i_1(t), \ldots, i_K(t)\}$. The goal of the research is to obtain the characteristics of the K-dimensional stochastic process $\boldsymbol{i}(t)$ at arbitrary moment t.

3   The First Jump Equations

Denote by $P(\boldsymbol{i}, t)$ the probability $P(\boldsymbol{i}, t) = \mathrm{P}\{\boldsymbol{i}(t) = \boldsymbol{i}\}$. To obtain the equations for probabilities $P(\boldsymbol{i}, t)$, we use a first jump separation technique which was presented in the works [8], [9]. We adapt this method for the tandem queues as follows.

Let a customer come to the empty system at the moment $t_0 = 0$. This customer will be special for us and we call it as *the first customer*. Denote by $S_k(t)$ a probability that at the moment $t > 0$ the first customer is serviced at the k-th stage. It is obvious that the value $S_0(t) = 1 - \sum_{k=1}^{K} S_k(t)$ is a probability that the first customer has left the system before the moment t. Let's suggest here that the probabilities $S_k(t)$ are known. We will calculate them later, in the section 4.

Applying a technique by [8], [9], we obtain the following equations for probabilities $P(\boldsymbol{i}, t)$:

$$P(\boldsymbol{0}, t) = S_0(t) \int_0^t P(\boldsymbol{0}, t - x)\, dA(x) + S_0(t)\,[1 - A(t)] \ , \tag{1}$$

$$P(\boldsymbol{e}_k, t) = S_0(t) \int_0^t P(\boldsymbol{e}_k, t - x)\, dA(x) +$$

$$+\, S_k(t)\,[1 - A(t)] + S_k(t) \int_0^t P(\boldsymbol{0}, t - x)\, dA(x) \text{ for } k = 1, \ldots, K \ . \tag{2}$$

Here $\boldsymbol{0}$ is a zero vector, $\boldsymbol{e}_k$ is a vector with all zero entries except an entry number k which is equal to 1.

For each vector $\boldsymbol{i}$, which is not equal to $\boldsymbol{0}$ or to $\boldsymbol{e}_k$ $(k = 1, \ldots, K)$, we can write the set of equations:

$$P(\boldsymbol{i}, t) = S_0(t) \int_0^t P(\boldsymbol{i}, t - x) \, dA(x) + \sum_{k=1}^K S_k(t) \int_0^t P(\boldsymbol{i} - \boldsymbol{e}_k, t - x) \, dA(x) \quad (3)$$

where entries of the vector $\boldsymbol{i}$ are running from 0 to ∞. We call the system of equations (1)–(3) as *the first jump equations* for probability distribution $P(\boldsymbol{i}, t)$ of the process $\boldsymbol{i}(t)$.

Consider a characteristic function $H(\boldsymbol{u}, t)$ of the distribution of the K-dimensional stochastic process $\boldsymbol{i}(t)$ at the moment t

$$H(\boldsymbol{u}, t) = \sum_{i_1=0}^\infty \cdots \sum_{i_K=0}^\infty e^{ju_1 i_1 + \ldots + ju_K i_K} P(i_1, \ldots, i_K, t) = \sum_{i=0}^\infty e^{j\boldsymbol{u}^\mathrm{T} \cdot \boldsymbol{i}} P(\boldsymbol{i}, t) \ . \quad (4)$$

Here $j = \sqrt{-1}$ is an imaginary unit. The function $H(\boldsymbol{u}, t)$ has vector argument $\boldsymbol{u}^\mathrm{T} = \{u_1, \ldots, u_K\}$.

Theorem 1. *The characteristic function $H(\boldsymbol{u}, t)$ satisfies the first jump equation as follows:*

$$H(\boldsymbol{u}, t) = \left[S_0(t) + \sum_{k=1}^K S_k(t) e^{ju_k} \right] \left[1 - A(t) + \int_0^t H(\boldsymbol{u}, t - x) \, dA(x) \right] \ . \quad (5)$$

Proof. From (1)–(4), we obtain the expression

$$H(\boldsymbol{u}, t) = S_0(t) \int_0^t e^{j\boldsymbol{u}^\mathrm{T} \cdot \boldsymbol{0}} P(\boldsymbol{0}, t - x) \, dA(x) + S_0(t) e^{j\boldsymbol{u}^\mathrm{T} \cdot \boldsymbol{0}} \left[1 - A(t) \right] +$$

$$+ \sum_{k=1}^K S_0(t) \int_0^t e^{j\boldsymbol{u}^\mathrm{T} \cdot \boldsymbol{e}_k} P(\boldsymbol{e}_k, t - x) \, dA(x) + \sum_{k=1}^K S_k(t) e^{j\boldsymbol{u}^\mathrm{T} \cdot \boldsymbol{e}_k} \left[1 - A(t) \right] +$$

$$+ \sum_{k=1}^K S_k(t) \int_0^t e^{j\boldsymbol{u}^\mathrm{T} \cdot \boldsymbol{e}_k} P(\boldsymbol{0}, t - x) \, dA(x) + \sum_{i > e_k} S_0(t) \int_0^t e^{j\boldsymbol{u}^\mathrm{T} \cdot \boldsymbol{i}} P(\boldsymbol{i}, t - x) \, dA(x) +$$

$$+ \sum_{i > e_k} \sum_{k=1}^K S_k(t) \int_0^t e^{j\boldsymbol{u}^\mathrm{T} \cdot \boldsymbol{i}} P(\boldsymbol{i} - \boldsymbol{e}_k, t - x) \, dA(x) \ .$$

Suggesting that $P(\boldsymbol{i}, t) = 0$ for vectors $\boldsymbol{i}$ with negative entries, we can reduce this formula to the form (5).

4   Calculation of the Probabilities $S_k(t)$

Let's obtain the probabilities $S_k(t)$ that the first customer is serviced at the stage k at the moment t $(k = 1, \ldots, K)$.

It is obvious that for the first stage

$$S_1(t) = 1 - B_1(t) \ .$$

Next, denote by τ_k the service time of the first customer at the k-th stage under the condition that the first customer was moved to this stage. So, we can write the following expression for $k = 2$:

$$S_2(t) = \mathrm{P}\{\tau_1 < t < \tau_1 + \tau_2\} = \int_0^t \mathrm{P}\{\tau_1 < t < \tau_1 + \tau_2 | \tau_1 = x\} \, dB_1(x) =$$

$$= \int_0^t \mathrm{P}\{t < x + \tau_2\} \, dB_1(x) = \int_0^t \mathrm{P}\{\tau_2 > t - x\} \, dB_1(x) =$$

$$= \int_0^t [1 - B_2(t - x)] \, dB_1(x) = B_1(t) - \int_0^t B_2(t - x) \, dB_1(x) =$$

$$= B_1(t) - (B_1 * B_2)(t) \ . \tag{6}$$

Here $(B_1 * B_2)(x)$ is a convolution of the functions $B_1(x)$ and $B_2(x)$. Using denotations $B_k^*(t) = (B_1 * \ldots * B_k)(t)$ for $k > 1$ and $B_1^*(t) = B_1(t)$, we can reduce expression (6) to the following form:

$$S_2(t) = B_1^*(t) - B_2^*(t) \ . \tag{7}$$

For every value k from 3 to K we can write:

$$S_k(t) = \mathrm{P}\left\{ \sum_{\nu=1}^{k-1} \tau_\nu < t < \sum_{\nu=1}^{k} \tau_\nu \right\} \ .$$

Using denotation $\xi_k = \sum_{\nu=1}^{k-1} \tau_\nu$, we can reduce this expression to the form

$$S_k(t) = \mathrm{P}\{\xi_k < t < \xi_k + \tau_k\} \ .$$

Performing for this expression the same transforms as in the formulas (6)–(7) and taking into account that $\mathrm{P}\{\xi_k < t\} = B_{k-1}^*(t)$, we obtain the following expression for the probabilities $S_k(t)$:

$$S_k(t) = B_{k-1}^*(t) - B_k^*(t) \tag{8}$$

for $k \geq 2$. Using the denotation $B_0^*(t) = 1$, we can extend this formula to be right for every value of the number k from 1 up to K.

5 Investigation of the Stochastic Characteristics for the Tandem Queue under a Condition of the High Arrival Rate

Equation (5) and the formulas (8) provide a tool for analysis of the tandem queueing systems of the type $GI/(GI/\infty)^K$. Here we consider an asymptotic [10] analysis for this system under a condition of the high arrival rate.

Let's obtain asymptotic expressions for function $H(u, t)$ under a condition of an infinite growth of arrival rate [11]. Rate of the arrival process is represented in a form $N\lambda$, where $N > 0$ is a parameter which gets large values (asymptotically $N \to \infty$), and value of λ is defined as

$$\lambda = \frac{1}{\int\limits_0^\infty [1 - A(x)] \, dx} = \frac{1}{a} \ .$$

Here a is an expected value of the random variable defined by distribution function $A(x)$.

In the paper [11] we have considered the renewal process with the rate $N\lambda$. Because a value of the parameter N is large, such process were named as *High-Intensive* (or *HI-processes*). It was shown that inter-arrival intervals of the renewal process with the rate $N\lambda$ are independent random variables distributed accordingly the distribution function $A(Nx)$. So, for the tandem queueing system $GI/(GI/\infty)^K$ with high arrival rate equation (5) gets a form

$$H(u, t) = \left[S_0(t) + \sum_{k=1}^K S_k(t) e^{ju_k} \right] \left[1 - A(Nt) + \int_0^t H(u, t - x) \, dA(Nx) \right] .$$

$$(9)$$

In the following subsections we will obtain an approximation for the multi-dimensional law of the distribution of the number of customers at the system stages.

5.1 The First-Order Asymptotic Form

Performing the following changes of variables in the equation (9)

$$\frac{1}{N} = \varepsilon, \qquad u = \varepsilon w, \qquad H(u, t) = F(w, t, \varepsilon) \ , \qquad (10)$$

we obtain the equation

$$F(w, t, \varepsilon) = \left[S_0(t) + \sum_{k=1}^K S_k(t) e^{jw_k\varepsilon} \right] \left[1 - A\left(\frac{t}{\varepsilon}\right) + \int_0^t F(w, t - x, \varepsilon) \, dA\left(\frac{x}{\varepsilon}\right) \right] .$$

$$(11)$$

Let's prove the following statement for an asymptotic approximation $F(w, t) = \lim\limits_{\varepsilon \to 0} F(w, t, \varepsilon)$.

Theorem 2. *The expression for the function $F(\boldsymbol{w}, t)$ has the following form:*

$$F(\boldsymbol{w}, t) = \exp\left\{ \lambda j \boldsymbol{w}^{\mathrm{T}} \int_0^t \boldsymbol{S}(\tau)\, d\tau \right\} \tag{12}$$

where $\boldsymbol{S}(\tau) = \{S_1(t), \ldots, S_K(t)\}^{\mathrm{T}}$.

Proof. Making a substitution $z = x/\varepsilon$ in the subintegral expression at formula (11) we obtain the equation

$$F(\boldsymbol{w}, t, \varepsilon) = \left[S_0(t) + \sum_{k=1}^K S_k(t) e^{j w_k \varepsilon} \right] \left[1 - A\left(\frac{t}{\varepsilon}\right) + \int_0^{\frac{t}{\varepsilon}} F(\boldsymbol{w}, t - z\varepsilon, \varepsilon)\, dA(z) \right].$$

Let's use the following expansions:

$$e^{j w_k \varepsilon} = 1 + j w_k \varepsilon + \mathrm{O}\left(\varepsilon^2\right) \ ,$$

$$F(\boldsymbol{w}, t - z\varepsilon, \varepsilon) = F(\boldsymbol{w}, t, \varepsilon) - z\varepsilon \frac{\partial F(\boldsymbol{w}, t, \varepsilon)}{\partial t} + \mathrm{O}\left(\varepsilon^2\right)$$

where $\mathrm{O}\left(\varepsilon^2\right)$ is infinitesimal which has order of ε^2. We obtain the following equation

$$F(\boldsymbol{w}, t, \varepsilon) = \left[S_0(t) + \sum_{k=1}^K S_k(t)\left(1 + j w_k \varepsilon\right) \right] \times$$

$$\times \left[1 - A\left(\frac{t}{\varepsilon}\right) + \int_0^{\frac{t}{\varepsilon}} \left\{ F(\boldsymbol{w}, t, \varepsilon) - z\varepsilon \frac{\partial F(\boldsymbol{w}, t, \varepsilon)}{\partial t} \right\} dA(z) \right] + \mathrm{O}\left(\varepsilon^2\right) \ .$$

Performing here an asymptotic transition $\varepsilon \to 0$, we reduce this formula to the form

$$F(\boldsymbol{w}, t) = \lim_{\varepsilon \to 0} F(\boldsymbol{w}, t, \varepsilon) =$$

$$= \lim_{\varepsilon \to 0} \left\{ \left[1 + \sum_{k=1}^K S_k(t) j w_k \varepsilon \right] \left[\int_0^\infty F(\boldsymbol{w}, t, \varepsilon)\, dA(z) - \int_0^\infty z\varepsilon \frac{\partial F(\boldsymbol{w}, t, \varepsilon)}{\partial t}\, dA(z) \right] +$$

$$+ \mathrm{O}\left(\varepsilon^2\right) \right\} = \lim_{\varepsilon \to 0} \left\{ \left[1 + \sum_{k=1}^K S_k(t) j w_k \varepsilon \right] \left[F(\boldsymbol{w}, t, \varepsilon) - \varepsilon \frac{\partial F(\boldsymbol{w}, t, \varepsilon)}{\partial t} a \right] + \mathrm{O}\left(\varepsilon^2\right) \right\} \ .$$

As a result, we obtain the following differential equation:

$$\frac{\partial F(\boldsymbol{w}, t, \varepsilon)}{\partial t} = \lambda F(\boldsymbol{w}, t) \sum_{k=1}^K S_k(t) j w_k \ .$$

Solving it with an initial condition $F(\boldsymbol{w}, 0) = 1$, we obtain the following expression for the function $F(\boldsymbol{w}, t)$:

$$F(\boldsymbol{w}, t) = \exp\left\{\lambda \sum_{k=1}^{K} jw_k \int_0^t S_k(\tau)\, d\tau\right\} = \exp\left\{\lambda j\boldsymbol{w}^{\mathrm{T}} \int_0^t \boldsymbol{S}(\tau)\, d\tau\right\}.$$

The theorem is proved.   □

Let's implement the inverse changes of variables for expressions (10) in the formula (12). We obtain the following approximation of the characteristic function $H(\boldsymbol{u}, t)$:

$$H(\boldsymbol{u}, t) \approx \exp\left\{\lambda j N \boldsymbol{u}^{\mathrm{T}} \int_0^t \boldsymbol{S}(\tau)\, d\tau\right\}$$

when N has large value. So, the average numbers of customers at the k-th stage of the considering tandem system at the moment t can be approximated by values $\lambda N \int_0^t S_k(\tau)\, d\tau$ for $k = 1, \ldots, K$ where $S_k(t)$ is defined by expressions (8).

5.2   Second-Order Asymptotic Form

Denote by $H_2(\boldsymbol{u}, t)$ the function which is determined by the following formula

$$H(\boldsymbol{u}, t) = H_2(\boldsymbol{u}, t)\exp\left\{\lambda N \sum_{k=1}^{K} ju_k \int_0^t S_k(\tau)\, d\tau\right\}. \tag{13}$$

Substituting this expression into equation (9), we obtain the following equation

$$H_2(\boldsymbol{u}, t)\exp\left\{\lambda N \sum_{k=1}^{K} ju_k \int_0^t S_k(\tau)\, d\tau\right\} = \left[S_0(t) + \sum_{k=1}^{K} S_k(t)e^{ju_k}\right] \times$$

$$\times \left[1 - A(Nt) + \int_0^t H_2(\boldsymbol{u}, t - x)\exp\left\{\lambda N \sum_{k=1}^{K} ju_k \int_0^{t-x} S_k(\tau)\, d\tau\right\} dA(Nx)\right]. \tag{14}$$

Performing here the following changes of variables

$$\frac{1}{N} = \varepsilon^2, \qquad \boldsymbol{u} = \varepsilon\boldsymbol{w}, \qquad H_2(\boldsymbol{u}, t) = F_2(\boldsymbol{w}, t, \varepsilon), \tag{15}$$

we obtain the equation

$$F_2(\boldsymbol{w}, t, \varepsilon)\exp\left\{\frac{\lambda}{\varepsilon^2} \sum_{k=1}^{K} j\varepsilon w_k \int_0^t S_k(\tau)\, d\tau\right\} = \left[S_0(t) + \sum_{k=1}^{K} S_k(t)e^{j\varepsilon w_k}\right] \times$$

$$\times \left[1 - A\left(\frac{t}{\varepsilon^2}\right) + \int_0^t F_2(\boldsymbol{w}, t-x, \varepsilon) \exp\left\{\frac{\lambda}{\varepsilon^2}\sum_{k=1}^K j\varepsilon w_k \int_0^{t-x} S_k(\tau)\, d\tau\right\} dA\left(\frac{x}{\varepsilon^2}\right)\right].$$
(16)

Let's prove the following statement for the asymptotic approximation $F_2(\boldsymbol{w}, t)$
$= \lim_{\varepsilon \to 0} F_2(\boldsymbol{w}, t, \varepsilon)$.

Theorem 3. *The expression for the function $F_2(\boldsymbol{w}, t)$ has the following form:*

$$F_2(\boldsymbol{w}, t) = \exp\left\{\lambda \sum_{k=1}^K \frac{(jw_k)^2}{2}\int_0^t S_k(\tau)\, d\tau + \frac{\kappa}{2}\sum_{k=1}^K\sum_{\nu=1}^K jw_k jw_\nu \int_0^t S_k(\tau) S_\nu(\tau)\, d\tau\right\}$$
(17)

where $\kappa = \lambda^3\left(\sigma^2 - a^2\right)$ and σ^2 is a variance of the random variable with distribution function $A(x)$.

Proof. Performing the substitution $z = x/\varepsilon^2$ in the integral on $dA(\cdot)$ in the equation (16), we reduce this equation to the form:

$$F_2(\boldsymbol{w}, t, \varepsilon) = \left[S_0(t) + \sum_{k=1}^K S_k(t) e^{j\varepsilon w_k}\right] \times$$

$$\times \left[\left\{1 - A\left(\frac{t}{\varepsilon^2}\right)\right\} \exp\left\{-\frac{\lambda}{\varepsilon}\sum_{k=1}^K jw_k \int_0^t S_k(\tau)\, d\tau\right\} + \right.$$

$$\left. + \int_0^{\frac{t}{\varepsilon^2}} F_2(\boldsymbol{w}, t-z\varepsilon^2, \varepsilon) \exp\left\{-\frac{\lambda}{\varepsilon}\sum_{k=1}^K jw_k \int_{t-z\varepsilon^2}^t S_k(\tau)\, d\tau\right\} dA(z)\right]. \quad (18)$$

Using an expansion $\int_{t-z\varepsilon^2}^t S_k(\tau)\, d\tau = z\varepsilon^2 S_k(t) + \mathrm{O}\left(\varepsilon^4\right)$, we get the relation

$$\exp\left\{-\frac{\lambda}{\varepsilon}\sum_{k=1}^K jw_k \int_{t-z\varepsilon^2}^t S_k(\tau)\, d\tau\right\} = \exp\left\{-\frac{\lambda}{\varepsilon}\sum_{k=1}^K jw_k\left[z\varepsilon^2 S_k(t) + \mathrm{O}\left(\varepsilon^4\right)\right]\right\} =$$

$$= \exp\left\{-z\lambda\sum_{k=1}^K j\varepsilon w_k S_k(t) + \mathrm{O}\left(\varepsilon^3\right)\right\} =$$

$$= 1 - z\lambda\sum_{k=1}^K j\varepsilon w_k S_k(t) + \frac{z^2\lambda^2}{2}\left[\sum_{k=1}^K j\varepsilon w_k S_k(t)\right]^2 + \mathrm{O}\left(\varepsilon^3\right). \quad (19)$$

Further we consider a case when the functions $F_2(\boldsymbol{w}, t, \varepsilon)$ and $A(x)$ have the following features:

$$\int_{\frac{t}{\varepsilon^2}}^\infty F_2(\boldsymbol{w}, t-z\varepsilon^2, \varepsilon)\, dA(z) = \mathrm{o}\left(\varepsilon^2\right),$$

$$\left\{1 - A\left(\frac{t}{\varepsilon^2}\right)\right\} \exp\left\{-\frac{\lambda}{\varepsilon}\sum_{k=1}^{K} jw_k \int_0^t S_k(\tau)\, d\tau\right\} = o\left(\varepsilon^2\right) \qquad (20)$$

where $o\left(\varepsilon^2\right)$ is an infinitesimal of the order greater than ε^2. Substituting expressions (20) and (19) into equation (18) and using the expansions

$$e^{j\varepsilon w_k} = 1 + j\varepsilon w_k + \frac{(j\varepsilon w_k)^2}{2} + O\left(\varepsilon^3\right) \ ,$$

$$F_2(\boldsymbol{w}, t - z\varepsilon^2, \varepsilon) = F_2(\boldsymbol{w}, t, \varepsilon) - z\varepsilon^2 \frac{\partial F_2(\boldsymbol{w}, t, \varepsilon)}{\partial t} + o\left(\varepsilon^2\right) \ ,$$

we obtain the following formula

$$F_2(\boldsymbol{w}, t, \varepsilon) = \left[S_0(t) + \sum_{k=1}^{K} S_k(t) + \sum_{k=1}^{K} S_k(t) j\varepsilon w_k + \sum_{k=1}^{K} S_k(t)\frac{(j\varepsilon w_k)^2}{2}\right] \times$$

$$\times \int_0^\infty \left\{\left[F_2(\boldsymbol{w}, t, \varepsilon) - z\varepsilon^2 \frac{\partial F_2(\boldsymbol{w}, t, \varepsilon)}{\partial t}\right] \times \right.$$

$$\times \left. \left[1 - z\lambda\sum_{k=1}^{K} j\varepsilon w_k S_k(t) + \frac{z^2\lambda^2}{2}\left\{\sum_{k=1}^{K} j\varepsilon w_k S_k(t)\right\}^2\right] dA(z)\right\} + o\left(\varepsilon^2\right) =$$

$$= \left[1 + \sum_{k=1}^{K} S_k(t) j\varepsilon w_k + \sum_{k=1}^{K} S_k(t)\frac{(j\varepsilon w_k)^2}{2}\right]\left[F_2(\boldsymbol{w}, t, \varepsilon) - F_2(\boldsymbol{w}, t, \varepsilon)\lambda a\sum_{k=1}^{K} j\varepsilon w_k S_k(t) +\right.$$

$$\left. + F_2(\boldsymbol{w}, t, \varepsilon)\frac{\lambda^2 a_2}{2}\left(\left\{\sum_{k=1}^{K} j\varepsilon w_k S_k(t)\right\}^2 - \varepsilon^2 \frac{\partial F_2(\boldsymbol{w}, t, \varepsilon)}{\partial t} a\right)\right] + o\left(\varepsilon^2\right) \ .$$

Here a_2 is the second initial moment of the random variable with distribution function $A(x)$. As a result, we obtain the equation

$$\varepsilon^2 \frac{\partial F_2(\boldsymbol{w}, t, \varepsilon)}{\partial t} a = F_2(\boldsymbol{w}, t, \varepsilon)\left[\sum_{k=1}^{K} S_k(t)\frac{(j\varepsilon w_k)^2}{2} + \left\{\frac{\lambda^2 a_2}{2} - 1\right\}\left\{\sum_{k=1}^{K} j\varepsilon w_k S_k(t)\right\}^2\right] + o\left(\varepsilon^2\right) \ .$$

Dividing each part of this equation by ε^2 and performing an asymptotic transition $\varepsilon \to 0$, we obtain the following differential equation for the function $F_2(\boldsymbol{w}, t) = \lim_{\varepsilon \to 0} F_2(\boldsymbol{w}, t, \varepsilon)$:

$$\frac{\partial F_2(\boldsymbol{w}, t)}{\partial t} = F_2(\boldsymbol{w}, t)\left[\lambda\sum_{k=1}^{K} \frac{(jw_k)^2}{2}S_k(t) + \frac{\kappa}{2}\sum_{k=1}^{K}\sum_{\nu=1}^{K} jw_k jw_\nu S_k(t)S_\nu(t)\right]$$

where $\kappa = \lambda^3\left(a_2 - 2a^2\right) = \lambda^3\left(\sigma^2 - a^2\right)$. Solving this equation with the initial condition $F_2(\boldsymbol{w}, 0) = 1$ we obtain the following expression for the function $F_2(\boldsymbol{w}, t)$:

$$F_2(\boldsymbol{w}, t) = \exp\left\{\lambda \sum_{k=1}^{K} \frac{(jw_k)^2}{2} \int_0^t \dot{S}_k(\tau)\, d\tau + \frac{\kappa}{2} \sum_{k=1}^{K} \sum_{\nu=1}^{K} jw_k jw_\nu \int_0^t S_k(\tau) S_\nu(\tau)\, d\tau\right\}.$$

The theorem is proved.   □

5.3   Multidimensional Gaussian Approximation

Let's perform the inverse substitutions for expressions (15) and (13) in the formula (17). Supposing that N is large enough, we obtain the following approximation

$$H(\boldsymbol{u}, t) \approx \exp\left\{\lambda N \sum_{k=1}^{K} ju_k s_k(t) + \lambda N \sum_{k=1}^{K} \frac{(ju_k)^2}{2} s_k(t) + \kappa N \sum_{k=1}^{K} \sum_{\nu=1}^{K} \frac{ju_k ju_\nu}{2} V_{k\nu}(t)\right\}.$$

Here we use denotations $s_k(t) = \int_0^t S_k(\tau)\, d\tau$ and $V_{k\nu}(t) = \int_0^t S_k(\tau) S_\nu(\tau)\, d\tau$ for k and ν running from 1 up to K.

Using matrix denotations $\boldsymbol{S}(t) = \text{diag}\{s_1(t), \ldots, s_K(t)\}$, $\boldsymbol{V}(t) = \{V_{k\nu}\}_{k,\nu=\overline{1,K}}$ and $\boldsymbol{e} = \{1, 1, \ldots, 1\}^{\mathrm{T}}$, we can write the following expression for the approximation $h(\boldsymbol{u}, t)$ of the characteristic function $H(\boldsymbol{u}, t)$:

$$h(\boldsymbol{u}, t) = \exp\left\{\lambda N j \boldsymbol{u}^{\mathrm{T}} \boldsymbol{S}(t) \boldsymbol{e} + \frac{1}{2} N j \boldsymbol{u}^{\mathrm{T}} [\lambda \boldsymbol{S}(t) + \kappa \boldsymbol{V}(t)]\, j\boldsymbol{u}\right\}. \qquad (21)$$

So, under a condition of the high arrival rate, the multi-dimensional distribution of the number of customers at the stages of the tandem queueing system $GI/(GI/\infty)^K$ at the moment t can be approximated by the multi-dimensional normal distribution with a vector of the means $\lambda N \boldsymbol{S}(t) \boldsymbol{e}$ and a covariance matrix $N[\lambda \boldsymbol{S}(t) + \kappa \boldsymbol{V}(t)]$.

To obtain the approximation $h(\boldsymbol{u}) = \lim_{t \to \infty} h(\boldsymbol{u}, t)$ for a characteristic function of the stationary distribution, we perform in the expression (21) the asymptotic transition $t \to \infty$. The result is as follows

$$h(\boldsymbol{u}) = \exp\left\{\lambda N j \boldsymbol{u}^{\mathrm{T}} \boldsymbol{S} \boldsymbol{e} + \frac{1}{2} N j \boldsymbol{u}^{\mathrm{T}} [\lambda \boldsymbol{S} + \kappa \boldsymbol{V}]\, j\boldsymbol{u}\right\}. \qquad (22)$$

Here $\boldsymbol{S} = \boldsymbol{S}(\infty)$ and $\boldsymbol{V} = \boldsymbol{V}(\infty)$. It is not difficult to show that the diagonal entries of the vector $\boldsymbol{S}$ are equal to the means of the service times at the stages of the system.

So, under the condition when N is large enough, the stationary joint distribution of the number of customers at the stages of the tandem system can be approximated by the multi-dimensional Gaussian distribution with a vector of the means $\lambda N \boldsymbol{S} \boldsymbol{e}$ and a covariance matrix $N[\lambda \boldsymbol{S} + \kappa \boldsymbol{V}]$.

6   Numerical Results and Applicability of the Approximation

Simulations of the system evolution for various numerical examples demonstrate a good accuracy of the obtained Gaussian approximation for the state distribution of the tandem systems $GI/(GI/\infty)^K$ in both the cases of stationary and non-stationary regimes (see expressions (22) and (21)) when a value of the arrival rate is large enough. Here we present one of the examples and we consider it only for the stationary distribution approximation. Parameters of the model are the following:

- *for arrival process* – distribution of the inter-arrival intervals is gamma distribution with the shape parameter $\alpha = 0.5$ and the rate parameter $\beta = 0.5 \cdot N$;
- *the number of stages* is 4;
- *service time distributions* are gamma distributions with the following parameters:

$$
\begin{array}{llll}
\text{for the first stage:} & \alpha_1 = 2, & \beta_1 = 1; \\
\text{for the second stage:} & \alpha_2 = 2, & \beta_2 = 2; \\
\text{for the third stage:} & \alpha_3 = 1.5, & \beta_3 = 1; \\
\text{for the fourth stage:} & \alpha_4 = 0.5, & \beta_4 = 1.
\end{array}
$$

Means of the service times at the stages are equal to $2, 1, 1.5, 0.5$. We can consider that the average service time at the stages is about 1. The mean of the inter-arrival intervals is equal to $1/N$ and, therefore, the rate of the arrival process is equal to N. So, the value of parameter N characterizes how much the arrival rate is large.

Simulations of the system were performed for the values of parameter N equal to 1, 10, 30, 60, 100 and 1000. To estimate the accuracy of the Gaussian approximation, we will use the Kolmogorov distances [12]

$$d_k = \sup_x |F_k(x) - G_k(x)|$$

for the marginal state distributions of each stage of the system. Here $F_k(x)$ is the cumulative distribution function of the Gaussian distribution and $G_k(x)$ is the empiric cumulative distribution function constructed based on results of the system evolution simulation, k is the number of the stage: $k = 1, 2, 3, 4$. For the purpose of more relevant estimation of the accuracy, we should consider the Kolmogorov distance between the multi-dimensional distributions. But the calculation of this distance is very hard computing procedure, e.g. for the value $N = 1000$ it is necessary to process about 10^{12} points. So, here we restrict ourselves to the considering the one-dimensional distributions.

Values of the distances are presented in the Table 1 for various values of the parameter N. The Figure 1 demonstrates the dynamics of the Kolmogorov distances when a value of N grows. It is easy to see that the growth of the arrival rate (parameter N) implies that the Gaussian approximation (22) becomes more accurate. Basing on our numerical experiments, we can draw a conclusion that this approximation is applicable when the value of the parameter N (ratio of the

Table 1. The Kolmogorov distances between the one-dimensional Gaussian approximations and the empiric cumulative distribution functions for each stage of the system and various values of the parameter N

N	d_1	d_2	d_3	d_4
1	0.1729	0.2320	0.2143	0.3691
10	0.0578	0.0815	0.0690	0.1150
30	0.0331	0.0475	0.0384	0.0677
60	0.0224	0.0339	0.0280	0.0488
100	0.0160	0.0288	0.0213	0.0395
1000	0.0121	0.0148	0.0128	0.0214

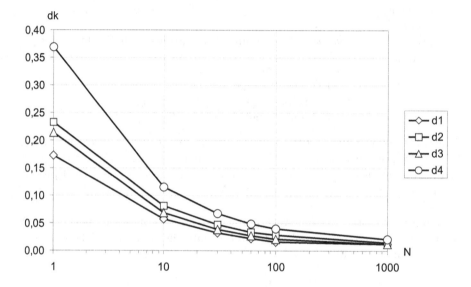

Fig. 1. Dynamics of the Kolmogorov distances d_k for each system stage when value of the parameter N (logarithmic scale) grows

arrivals rate to servicing rate) is about 30 or more. In this case the Kolmogorov distance is less than 0.05.

Marginal distributions of the number of customers at the stage 2 of the system are presented at the Figure 2 for the values of parameter N equal to 1, 10, 30, 100. These figures demonstrate the general trend of the approximation character.

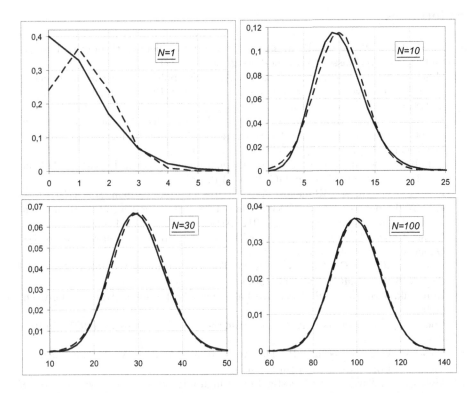

Fig. 2. Comparison of the analytical approximation (dashed line) and simulation results (solid line) of the probability distributions of the customers number at the second stage of the system for values of the parameter N equal to 1, 10, 30, 100

7   Conclusions

In the paper we have obtained the multi-dimensional Gaussian approximation for the stationary probability distribution of the number of customers at the stages of the tandem queueing system with high-intensive renewal arrival process.

The approximation is obtained under an asymptotic condition of the arrivals' rate growth. Numerical results show that this approximation for the tandem system under study is quite accurate when the rate of input is greater than the rate of service by 30 times or more.

The investigation was performed by means of the first jump separation technique. The results coincide with ones obtained by means of the original method of screened process [13].

References

1. Gnedenko, B.W., König, D. (eds.): Handbuch der Bedienungstheorie: II. Formeln und andere Ergebnisse. Akademie-Verlag, Berlin (1984)
2. Grachev, V.V., Moiseev, A.N., Nazarov, A.A., Yampolsky, V.Z.: Tandem queue as a model of the system of distributed data processing. Proc. of TUSUR 2(26), part 2, 248–251 (2012) (in Russian)
3. Balsamo, S., De Nitto Personè, V., Inverardi, P.: On using Queueing Network Models with finite capacity queues for Software Architectures performance prediction. Performance Evaluation 51(2-3), 269–288 (2002)
4. Heindl, A.: Decomposition of general tandem queueing networks with MMPP input. Performance Evaluation 44(1-4), 5–23 (2001)
5. Kim, C., Dudin, A., Klimenok, V., Taramin, O.: A Tandem $BMAP/G/1 \rightarrow \bullet/M/N/0$ Queue with Group Occupation of Servers at the Second Station. Mathematical Problems in Engineering 2012, Article ID 324604 (2012)
6. Gómez-Corral, A.: A Tandem Queue with Blocking and Markovian Arrival Process. Queueing Systems 41, 343–370 (2002)
7. Kingman, J.F.C.: On Queues in Heavy Traffic. Journal of the Royal Statistical Society. Series B 24(2), 383–392 (1962)
8. Korolyuk, V.S., Korolyuk, V.V.: Stochastic models of systems. Kluwer, Dordrecht (1999)
9. Bocharov, P.P., Pechinkin, A.V.: Queueing theory. RUDN, Moscow (1995) (in Russian)
10. Nazarov, A.A., Moiseeva, S.P.: The Asymptotical Analysis Method in Queueing Theory. NTL, Tomsk (2006) (in Russian)
11. Moiseev, A., Nazarov, A.: Investigation of high intensive general flow. In: Proc. of the IV International Conference, Problems of Cybernetics and Informatics (PCI 2012), pp. 161–163. IEEE, Baku (2012)
12. Kolmogorov, A.: Sulla determinazione empirica di una legge di distribuzione. Giornale Dell' Intituto Italiano Degli Attuari 4, 83–91 (1933)
13. Moiseev, A.N., Nazarov, A.A.: Asymptotic analysis of a multistage queuing system with a high-rate renewal arrival process. Optoelectronics, Instrumentation and Data Processing 50(2), 163–171 (2014)

Locally Optimal Control
for Discrete Time Delay Systems
with Interval Parameters*

Oksana Mukhina and Valery Smagin

Department of Applied Mathematics and Cybernetics,
Tomsk State University, 36 Lenin str., 634050 Tomsk, Russia
oksm7@sibmail.com, vsm@mail.tsu.ru

Abstract. The problem of locally optimal control for discrete time delay systems with interval parameters based on the probabilistic method is considered. The goal of control: systems output must track a reference input. Control is defined as a function of the measured variables and the reference input. The asymptotic behavior of the system is investigated. Example is given to illustrate the usefulness of the proposed approach.

Keywords: locally optimal control, interval parameters, time delay, tracking a reference input, asymptotic behavior.

1 Introduction

Locally optimal discrete control systems are a special case of the discrete model predictive control (MPC) with a forecast for 1 cycle. The main advantage of the method of locally optimal control is a significant simplification of the synthesis procedure. For last years field of MPC application and, accordingly, the method of locally optimal control has been considerably extended covering the problems of control for technical systems, production systems, inventory control and portfolio optimisation [1–7].

In this paper the problem of synthesis of locally optimal control on the tracking output for discrete systems with interval parameters and time delay is considered. This paper is a generalization of [8, 9] to the case of presence interval parameters in the model [10]. These parameters reflect the situation when the exact values of parameters are not defined and only intervals for parameters are known. In this situation, one of the most important tasks is to construct such a control that ensures desirable requirements needed for the system on the whole range of variable parameters of the object. To solve these problems with interval parameters methods interval the Cochran arithmetic, the theory of fuzzy sets [11, 12], the methods of linear matrix inequalities [13] and probabilistic methods are applied [14].

---

* Supported by Tomsk State University Competitiveness Improvement Program, and the project "Goszadanie Minobrnauki Rossii".

A. Dudin et al. (Eds.): ITMM 2014, CCIS 487, pp. 301–311, 2014.

We propose to realize synthesis of tracking system control on output with indirect measurements for discrete systems with interval parameters and time delay based on the probabilistic method [14]. Control is defined as a function of the measured variables and the reference input. Proposed tracking control has the property of robustness.

2 Problem Statement

Consider the following discrete-time system with time delay and interval parameters:

$$x(k+1) = (A + \sum_{i=1}^{r} A_i\theta_i)x(k) + (A' + \sum_{i=1}^{r} A'_i\theta_i)x(k-h) +$$
$$+ (B + \sum_{i=1}^{r} B_i\theta_i)u(k) + q(k),$$
$$x(\tau) = \gamma(\tau), \tau = -h, 1-h, 2-h, \dots, 0; k = 0, 1, 2, \dots, \tag{1}$$
$$y(k) = Sx(k) + \nu(k). \tag{2}$$

In (1), (2) $x(k) \in R^n$ is state vector, $h > 0$ is positive integer time delay, $u(k) \in R^m$ is control input, $y(k) \in R^l$ is observations vector, $A, A_i, A', A'_i, B,$ $B_i, i = \overline{1,r}$ are constant matrices of appropriate dimensions, S is matrix of observations channel, matrix B and S are of full rank, pairs of matrices (A, B) and (A', B) are controllable, pairs of matrices (S, A) and (S, A') are observable, x_0 are initial conditions $(M\{x_0 x_0^T\} = P_{x_0})$, $q(k)$ is Gaussian random sequence of input disturbances, $\nu(k)$ is Gaussian random sequence of observations errors with characteristics: $M\{q(k)\} = 0, M\{\nu(k)\} = 0, M\{q(k)\nu^T(j)\} = 0, M\{q(k)q^T(j)\} = Q(k)\delta_{kj}, M\{\nu(k)\nu^T(j)\} = V(k)\delta_{k,j}$ ($\delta_{k,j}$ is Kronecker delta, $Q(k) = Q^T(k) \geq 0, V(k) = V^T(k) \geq 0$ are nonnegative definite matrices), θ_i are uncertain parameters interval type $(-1 \leq \theta_i \leq 1)$.

Local criterion has the form:

$$I(k) = M\{(w(k+1) - z(k))^T C(w(k+1) - z(k)) + u^T(k)Du(k)\}, \tag{3}$$

where $w(k) = Hx(k)$ is controlled output of the system (H is matrix of the output of the system), $C = C^T, D = D^T \geq 0$ are weighting matrices, $z(k) \in R^n$ is the reference input which described by the equation:

$$z(k) = Fz(k) + q_z(k), z(0) = z_0, k = 0, 1, 2, \dots. \tag{4}$$

In (4) $q_z(k)$ is Gaussian random sequence with characteristics $M\{q_z(k)\} = 0,$ $M\{q_z(k)q^T(j)\} = 0, M\{q_z(k)\nu^T(j)\} = 0, M\{q_z(k)q_z^T(j)\} = Q_z(k)\delta_{k,j}, z_0$ are initial conditions $(M\{z_0 z_0^T\} = P_{z_0}, M\{z_0 x_0^T\} = P_{z_0 x_0}, M\{x_0 z_0^T\} = P_{x_0 z_0}), F$ is matrix of the dynamics model of the reference input.

It is required to construct a control of system (1), using observations (2) and minimizing the criterion (3).

The essence of the probabilistic approach is that the uncertain interval parameters are replaced by independent random sequences $\bar{\theta}(k)$, with uniform distribution law on the interval [-1, 1].

3   Locally Criterion Optimization

Let the control law of system (1) under observations (2) is defined as:

$$u(k) = K_1(k)y(k) + K_2(k)y(k-h) + K_3(k)z(k), \qquad (5)$$

where transfer coefficients are $K_1(k), K_2(k), K_3(k)$ to be determined.

Solution of the problem locally criterion optimization is given in the following theorem.

Theorem 1. If for system (1), observations (2) and local criterion (3) matrices

$$\overline{C} = (B^{\mathrm{T}}H^{\mathrm{T}}CHB + D + \frac{1}{3}\sum_{i=1}^{r} B_i^{\mathrm{T}}H_i^{\mathrm{T}}CHB_i) > 0,$$

$$\overline{P}(k) = \begin{pmatrix} SP_x(k)S^{\mathrm{T}} + V(k) & SP_x(k-h,k)S^{\mathrm{T}} & P_{zx}(k)S^{\mathrm{T}} \\ SP_x(k,k-h)S^{\mathrm{T}} & SP_x(k-h)S^{\mathrm{T}} + V(k-h) & P_{zx}(k,k-h)S^{\mathrm{T}} \\ SP_{xz}(k) & SP_{xz}(k-h,k) & P_z(k) \end{pmatrix} \quad (6)$$

are positive definite for all $k = 1, 2, \ldots$, then optimal in the sense of minimum criteria (3) transfer coefficients for control (5) are determined by the formulas:

$$K_1^*(k) = aK_2(k) + bK_3(k) + c, \qquad (7)$$

$$K_2^*(k) = [K_3(k)(bd + e) + cd + f][E - ad], \qquad (8)$$

$$K_3^*(k) = [(cd + f)(E - ad)^{-1}(ag + n) + cg + m][(1 - bg) \\ -(bd - e)(E - ad)^{-1}(ag + n)]^{-1}, \qquad (9)$$

where E is the identity matrix,

$$a = -SP_x(k-h,k)S^{\mathrm{T}}[SP_x(k)S^{\mathrm{T}} + V(k)]^{-1},$$

$$b = -P_{zx}(k)S^{\mathrm{T}}[SP_x(k)S^{\mathrm{T}} + V(k)]^{-1},$$

$$c = -\overline{C}^{-1}[\tfrac{1}{3}\sum_{i=1}^{r} B_i^{\mathrm{T}}H^{\mathrm{T}}C(HA_i'P_x(k-h,k) + HA_iP_x(k))S^{\mathrm{T}} \\ +B^{\mathrm{T}}H^{\mathrm{T}}C(HAP_x(k) + HA'P_x(k-h,k) \\ -P_{zx}(k))S^{\mathrm{T}}][SP_x(k)S^{\mathrm{T}} + V(k)]^{-1},$$

$$d = -SP_x(k,k-h)S^{\mathrm{T}}[SP_x(k-h)S^{\mathrm{T}} + V(k-h)]^{-1},$$

$$e = -P_{zx}(k,k-h)S^{\mathrm{T}}[SP_x(k-h)S^{\mathrm{T}} + V(k-h)]^{-1},$$

$$f = -\overline{C}^{-1}[\tfrac{1}{3}\sum_{i=1}^{r} B_i^{\mathrm{T}}H^{\mathrm{T}}C(HA_i'P_x(k-h) + HA_iP_x(k,k-h))S^{\mathrm{T}} \\ +B^{\mathrm{T}}H^{\mathrm{T}}C(HAP_x(k,k-h) + HA'P_x(k-h) \\ -P_{zx}(k,k-h))S^{\mathrm{T}}][SP_x(k-h)S^{\mathrm{T}} + V(k-h)]^{-1},$$

$$g = -SP_{xz}(k)P_z^{-1}(k), n = -SP_{xz}(k-h,k)P_z^{-1}(k),$$

$$m = -\overline{C}^{-1}[\tfrac{1}{3}\sum_{i=1}^{r} B_i^{\mathrm{T}}H^{\mathrm{T}}C(HA_i'P_{xz}(k-h,k) + HA_iP_{xz}(k)) \\ +B^{\mathrm{T}}H^{\mathrm{T}}C(HAP_{xz}(k) + HA'P_{xz}(k-h,k) + P_z(k))S^{\mathrm{T}}]P_z^{-1}(k). \qquad (10)$$

In (10) introduced the notation: $P_z(k) = M\{z(k)z^{\mathrm{T}}(k)\}$, $P_x(k) = M\{x(k) \times x^{\mathrm{T}}(k)\}$, $P_{zx}(k,r) = P_{xz}^{\mathrm{T}}(r,k) = M\{z(k)x^{\mathrm{T}}(r)\}$, $P_{xz}(k,r) = P_{zx}^{\mathrm{T}}(r,k)$

$= M\{x(k)z^{\mathrm{T}}(r)\}$, $P_x(k, r) = M\{x(k)x^{\mathrm{T}}(r)\}$, $P_{xz}(k) = P_{zx}^{\mathrm{T}}(k) = M\{x(k) \times z^{\mathrm{T}}(k)\}$, $P_{zx}(k) = P_{xz}^{\mathrm{T}}(k) = M\{z(k)x^{\mathrm{T}}(k)\}$, which are determined by a system of difference matrix equations with delays.

Proof. To calculate the local criterion we obtain the equation of state by substituting (5) in (1):

$$x(k+1) = (A + \textstyle\sum_{i=1}^{r} A_i\theta_i)x(k) + (A' + \textstyle\sum_{i=1}^{r} A'_i\theta_i)x(k-h)$$
$$+(B + \textstyle\sum_{i=1}^{r} B_i\theta_i)u(k) + q(k)$$
$$= (A + \textstyle\sum_{i=1}^{r} A_i\theta_i)x(k) + (A' + \textstyle\sum_{i=1}^{r} A'_i\theta_i)x(k-h)$$
$$+(B + \textstyle\sum_{i=1}^{r} B_i\theta_i)K_1(k)Sx(k) + (B + \textstyle\sum_{i=1}^{r} B_i\theta_i)K_1(k)\nu(k)$$
$$+(B + \textstyle\sum_{i=1}^{r} B_i\theta_i)K_2(k)Sx(k-h) + (B + \textstyle\sum_{i=1}^{r} B_i\theta_i)K_1(k)\nu(k-h)$$
$$+(B + \textstyle\sum_{i=1}^{r} B_i\theta_i)K_3(k)z(k) + q(k). \tag{11}$$

Taking into account (1), (2), (4), (5), (6), characteristics of random sequences $q(k)$ and $\nu(k)$ we can calculate the value of the local criterion (3):

$$I(k) = M\{(w(k+1) - z(k))^{\mathrm{T}}C(w(k+1) - z(k)) + u^{\mathrm{T}}(k)Du(k)\}$$
$$= tr(HA + HBK_1(k)S)^{\mathrm{T}}C(HA + HBK_1(k)S)P_x(k)$$
$$+\tfrac{1}{3}tr\textstyle\sum_{i=1}^{r}(HA_i + HB_iK_1(k)S)^{\mathrm{T}}C(HA_i + HB_iK_1(k)S)P_x(k)$$
$$+trS^{\mathrm{T}}K_1^{\mathrm{T}}(k)DK_1(k)SP_x(k) + tr(HA + HBK_1(k)S)^{\mathrm{T}}C(HA'$$
$$+HBK_2(k)S)P_x(k-h,k) + \tfrac{1}{3}tr\textstyle\sum_{i=1}^{r}(HA_i + HB_iK_1(k)S)^{\mathrm{T}}C(HA_i$$
$$+HB_iK_2(k)S)P_x(k-h,k) + trS^{\mathrm{T}}K_1^{\mathrm{T}}(k)DK_2(k)SP_x(k-h,k)$$
$$+tr(HA + BK_1(k)S)^{\mathrm{T}}C(HBK_3(k) - E)P_{zx}(k) + \tfrac{1}{3}tr\textstyle\sum_{i=1}^{r}(HA_i$$
$$+HB_iK_1(k)S)^{\mathrm{T}}CHB_iK_3(k)P_{zx}(k) + trS^{\mathrm{T}}K_1^{\mathrm{T}}(k)DK_3(k)P_{zx}(k)$$
$$+tr(HA' + HBK_2(k)S)^{\mathrm{T}}C(HA + HBK_1(k)S)P_x(k,k-h)$$
$$+\tfrac{1}{3}tr\textstyle\sum_{i=1}^{r}(HA'_i + HB_iK_2(k)S)^{\mathrm{T}}C(HA_i + HB_iK_1(k)S)P_x(k,k-h)$$
$$+trS^{\mathrm{T}}K_2^{\mathrm{T}}(k)DK_1(k)SP_x(k,k-h) + tr(HA' + BK_2(k)S)^{\mathrm{T}}C(HA'$$
$$+HBK_2(k)S)P_x(k-h) + \tfrac{1}{3}tr\textstyle\sum_{i=1}^{r}(HA'_i + HB_iK_2(k)S)^{\mathrm{T}}C(HA'_i$$
$$+HB_iK_2(k)S)P_x(k-h) + trS^{\mathrm{T}}K_2^{\mathrm{T}}(k)DK_2(k)SP_x(k-h) + tr(HA'$$
$$+HBK_2(k)S)^{\mathrm{T}}C(HBK_3(k) - E)P_{z}x(k,k-h) + tr(HBK_3(k) - E)^{\mathrm{T}}$$
$$\times C(HA + HBK_1(k)S)P_{xz}(k) + \tfrac{1}{3}tr\textstyle\sum_{i=1}^{r}K_3^{\mathrm{T}}(k)B_i^{\mathrm{T}}H^{\mathrm{T}}C(HA_i$$
$$+HB_iK_1(k)S)P_{xz}(k) + trSK_3^{\mathrm{T}}(k)DK_1(k)SP_{xz}(k) + tr(HBK_3(k) - E)^{\mathrm{T}}$$
$$\times C(HA' + HBK_2(k)S)P_{xz}(k-h,k)$$
$$+\tfrac{1}{3}tr\textstyle\sum_{i=1}^{r}K_3^{\mathrm{T}}(k)B_i^{\mathrm{T}}H^{\mathrm{T}}C(HA'_i + HB_iK_2(k)S)P_{xz}(k-h,k)$$
$$+trSK_3^{\mathrm{T}}(k)DK_2(k)SP_{xz}(k-h,k)$$
$$+tr(HBK_3(k) - E)^{\mathrm{T}}C(HBK_3(k) - E)P_z(k)$$
$$+\tfrac{1}{3}tr\textstyle\sum_{i=1}^{r}K_3^{\mathrm{T}}(k)B_i^{\mathrm{T}}H^{\mathrm{T}}CHB_iK_3(k)P_z(k) + trK_3^{\mathrm{T}}(k)DK_3(k)P_z(k)$$
$$+trK_1^{\mathrm{T}}(k)(B^{\mathrm{T}}H^{\mathrm{T}}CHB + D + \tfrac{1}{3}tr\textstyle\sum_{i=1}^{r} B_i^{\mathrm{T}}H^{\mathrm{T}}CHB_i)K_1(k)V(k)$$
$$+trK_2^{\mathrm{T}}(k)(B^{\mathrm{T}}H^{\mathrm{T}}CHB + D + \tfrac{1}{3}tr\textstyle\sum_{i=1}^{r} B_i^{\mathrm{T}}H^{\mathrm{T}}CHB_i)K_2(k)V(k-h)$$
$$+trCHQ(k). \tag{12}$$

The moments $P_x(k,j)$, $P_z(k,j)$, $P_{xz}(k,j)$, $P_{zx}(k,j)$ in (12) are defined by the following formulas:

$$P_x(k+1,j+1) = M\{\,x(k+1)x(j+1)^\mathrm{T}\,\}$$
$$= \xi(i)P_x(k,j)\xi^\mathrm{T}(j) + \xi(i)P_x(k-h,j)\xi_2^\mathrm{T}(j) + \tfrac{1}{3}\sum_{i=1}^r \xi_i(k)P_x(k,j)\xi_i^\mathrm{T}(j)$$
$$+\tfrac{1}{3}\sum_{i=1}^r \xi_i(k)P_x(k,j-h)\xi_{2_i}^\mathrm{T}(j) + \xi_2(k)P_x(k-h,j)\xi^\mathrm{T}(j)$$
$$+\xi_2(k)P_x(k-h,j-h)\xi_2^\mathrm{T}(j) + \tfrac{1}{3}\sum_{i=1}^r \xi_{2_i}(k)P_x(k-h,j)\xi_i^\mathrm{T}(j)$$
$$+\tfrac{1}{3}\sum_{i=1}^r \xi_{2_i}(k)P_x(k-h,j-h)\xi_{2_i}^\mathrm{T}(j) + Q_1(k,j),\, P_x(0) = P_{x_0}. \tag{13}$$
$$P_z(k+1,j+1) = M\{\,z(k+1)z(j+1)^\mathrm{T}\,\}$$
$$= FP_z(k,j)F^\mathrm{T} + Q_z(k,j)\delta_{k,j},\, P_z(0) = P_{z_0}. \tag{14}$$
$$P_{zx}(k+1,j+1) = M\{\,z(k+1)x(j+1)^\mathrm{T}\,\} = FP_{zx}(k,j)\xi^\mathrm{T}(j)$$
$$+FP_{zx}(k,j-h)\xi_2^\mathrm{T}(j) + FP_z(k,j)K_3^\mathrm{T}(j)B^\mathrm{T},\, P_{zx}(0) = P_{z_0 x_0}. \tag{15}$$
$$P_{xz}(k+1,j+1) = M\{\,x(k+1)z(j+1)^\mathrm{T}\,\} = \xi(k)P_{xz}(k,j)F^\mathrm{T}$$
$$+\xi_2(k)P_{xz}(k-h,j)F^\mathrm{T} + BK_3(k)P_z(k,j)F^\mathrm{T},\, P_{xz}(0) = P_{x_0 z_0}, \tag{16}$$

where:

$$\xi(k) = A + BK_1^*(k)S,\, \xi_2(k) = A' + BK_2^*(k)S,\, \xi_3(k) = BK_3^*(k) - E,$$
$$\xi_i(k) = A_i + B_iK_1^*(k)S,\, \xi_{2_i}(k) = A_i' + B_iK_2^*(k)S,$$
$$Q_1(k,j) = Q(k,j)\delta_{k,j} + \xi(k)P_{xz}(k,j)K_3^{*\mathrm{T}}(j)B^\mathrm{T}$$
$$+\tfrac{1}{3}\sum_{i=1}^r \xi_i(k)P_{xz}(k,j)K_3^{*\mathrm{T}}(j)B_i^\mathrm{T} + \xi_2(k)P_{xz}(k-h,j)K_3^{*\mathrm{T}}B^\mathrm{T}$$
$$+\tfrac{1}{3}\sum_{i=1}^r \xi_i(k)P_{xz}(k-h,j)K_3^{*\mathrm{T}}(j)B_i^\mathrm{T} + BK_3^*(k)P_{zx}(k,j)\xi^\mathrm{T}(j)$$
$$+BK_3^*(k)P_{zx}(k,j-h)\xi_2^\mathrm{T}(j) + BK_3^*(k)P_z(k,j)K_3^{*\mathrm{T}}(j)B^\mathrm{T}$$
$$+\tfrac{1}{3}\sum_{i=1}^r B_iK_3^*(k)P_z(k,j)K_3^{*\mathrm{T}}(j)B_i^\mathrm{T} + \tfrac{1}{3}\sum_{i=1}^r B_iK_3^*(k)P_{zx}(k,j)\xi_i^\mathrm{T}(j)$$
$$+\tfrac{1}{3}\sum_{i=1}^r B_iK_3^*(k)P_{zx}(k,j-h)\xi_{2_i}^\mathrm{T}(j) + \tfrac{1}{3}\sum_{i=1}^r B_iK_3^*(k)P_{zx}(k,j)\xi_i^\mathrm{T}(j)$$
$$+BK_1^*(k)V(k,j)\delta_{k,j}K_1^{*\mathrm{T}}(j)B^\mathrm{T} + BK_2^*(k)V(k-h,j-h)\delta_{k-h,j-h}K_2^{*\mathrm{T}}(j)B^\mathrm{T}$$
$$+\tfrac{1}{3}\sum_{i=1}^r B_iK_1^{*\mathrm{T}}(k)V(k,j)\delta_{k,j}K_1^{*\mathrm{T}}(j)B_i^\mathrm{T}$$
$$+\tfrac{1}{3}\sum_{i=1}^r B_iK_2^*(k)V(k-h,j-h)\delta_{k-h,j-h}K_2^{*\mathrm{T}}(j)B_i^\mathrm{T}. \tag{17}$$

We calculate the values of the gradients of the criterion (12) by $K_1(k)$, $K_2(k)$ and $K_3(k)$. Using rules for differentiating functions tr from product of the matrices by matrix argument [15], and equating them to zero, we obtain the formulas:

$$K_1(k) = -\overline{C}^{-1}[\overline{C}K_2(k)SP_x(k-h,k)S^\mathrm{T} + \overline{C}K_3(k)P_{zx}(k)S^\mathrm{T}$$
$$+\tfrac{1}{3}\sum_{i=1}^r B_i^\mathrm{T}H^\mathrm{T}CH(A_iP_x(k) + A_i'P_x(k-h,k))S^\mathrm{T}$$
$$+B^\mathrm{T}H^\mathrm{T}C(HAP_x(k) + HA'P_x(k-h,k)$$
$$-P_{zx}(k))S^\mathrm{T}][SP_x(k)S^\mathrm{T} + V(k)]^{-1}, \tag{18}$$
$$K_2(k) = -\overline{C}^{-1}[\overline{C}K_1(k)SP_x(k,k-h)S^\mathrm{T} + \overline{C}K_3(k)P_{zx}(k,k-h)S^\mathrm{T}$$
$$+\tfrac{1}{3}\sum_{i=1}^r B_i^\mathrm{T}H^\mathrm{T}CH(A_iP_x(k,k-h) + A_i'P_x(k-h))S^\mathrm{T}$$
$$+B^\mathrm{T}H^\mathrm{T}C(HAP_x(k,k-h) + HA'P_x(k-h)$$

$$-P_{zx}(k, k-h))S^{\mathrm{T}}][SP_x(k-h)S^{\mathrm{T}} + V(k-h)]^{-1}, \qquad (19)$$

$$K_3(k) = -\overline{C}^{-1}[\overline{C}K_1(k)SP_{xz}(k) + \overline{C}K_2(k)SP_{xz}(k-h, k)$$
$$+\tfrac{1}{3}\sum_{i=1}^{r} B_i^{\mathrm{T}}H^{\mathrm{T}}CH(A_iP_{xz}(k) + A_i'P_{xz}(k-h, k))$$
$$+B^{\mathrm{T}}H^{\mathrm{T}}C(HAP_{xz}(k) + HA'P_{xz}(k-h, k) + P_z(k))]P_z(k)^{-1}. \qquad (20)$$

After calculating $K_1(k)$, $K_2(k)$, $K_3(k)$, we rewrite (18)–(20) in the form of:

$$\overline{C}[K_1(k)(SP_x(k)S^{\mathrm{T}} + V(k)) + K_2(k)SP_x(k-h, k)S^{\mathrm{T}}$$
$$+K_3(k)P_{zx}(k)S^{\mathrm{T}}] = -B^{\mathrm{T}}H^{\mathrm{T}}C[HAP_x(k) + HA'P_x(k-h, k)$$
$$-P_{zx}(k)]S^{\mathrm{T}} - \tfrac{1}{3}\sum_{i=1}^{r} B_i^{\mathrm{T}}H^{\mathrm{T}}CH[A_iP_x(k) + A_i'P_x(k-h, k)], \qquad (21)$$
$$\overline{C}[K_1(k)SP_x(k, k-h)S^{\mathrm{T}} + K_2(k)(SP_x(k-h)S^{\mathrm{T}} + V(k-h))$$
$$+K_3(k)P_{zx}(k, k-h)S^{\mathrm{T}}] = -B^{\mathrm{T}}H^{\mathrm{T}}C[HAP_x(k, k-h) + HA'P_x(k-h)$$
$$-P_{zx}(k, k-h)]S^{\mathrm{T}} - \tfrac{1}{3}\sum_{i=1}^{r} B_i^{\mathrm{T}}H^{\mathrm{T}}CH[A_iP_x(k, k-h) + A_i'P_x(k-h)], (22)$$
$$\overline{C}[K_1(k)SP_{xz}(k) + K_2(k)SP_{xz}(k-h, k) + P_z(k)] =$$
$$-B^{\mathrm{T}}H^{\mathrm{T}}C[HAP_{xz}(k) + HA'P_{xz}(k-h, k) - P_z(k)]$$
$$-\tfrac{1}{3}\sum_{i=1}^{r} B_i^{\mathrm{T}}H^{\mathrm{T}}CH[A_iP_{xz}(k) + A_i'P_{xz}(k-h, k)]. \qquad (23)$$

We obtain system of matrix equations (21)–(23) in the following form:

$$\overline{C}[K_1(k)|K_2(k)|K_3(k)]\overline{P}(k) =$$
$$-[B^{\mathrm{T}}H^{\mathrm{T}}C(HAP_x(k) + HA'P_x(k-h, k) - P_{zx}(k))S^{\mathrm{T}}$$
$$-\tfrac{1}{3}\sum_{i=1}^{r} B_i^{\mathrm{T}}H^{\mathrm{T}}CH(A_iP_x(k) + A_i'P_x(k-h, k))|$$
$$B^{\mathrm{T}}H^{\mathrm{T}}C(HAP_x(k, k-h) + HA'P_x(k-h) - P_{zx}(k, k-h))S^{\mathrm{T}}$$
$$-\tfrac{1}{3}\sum_{i=1}^{r} B_i^{\mathrm{T}}H^{\mathrm{T}}CH(A_iP_x(k, k-h) + A_i'P_x(k-h))|$$
$$B^{\mathrm{T}}H^{\mathrm{T}}C(HAP_{xz}(k) + HA'P_{xz}(k-h, k) - P_z(k))$$
$$-\tfrac{1}{3}\sum_{i=1}^{r} B_i^{\mathrm{T}}H^{\mathrm{T}}CH(A_iP_{xz}(k) + A_i'P_{xz}(k-h, k))]. \qquad (24)$$

Taking into account (6), matrices $\overline{C}$ and $\overline{P}(k)$ are nonsingular for all $k = 0, 1, 2, \ldots$, consequently, the equation (24) is solvable relative to block matrix $[K_1(k)|K_2(k)|K_3(k)]$ and has the unique solution (7)–(9).

4   Asymptotic Behavior

Theorem 2. Let in description of system (1), observations (2), criterion (3) and model of reference input (4) matrices A, A_i, A', A_i', B, B_i, Q, S, V, C, D, $i = \overline{1, r}$ are constant, $F = E$, $q_z(k) = 0$. Then, if the condition (6) of theorem 1 is satisfied, there exist steady-state solution of equations (13), (15), (16), matrices $P_x = \lim_{k\to\infty} P_x(k) \geq 0$, $Q_1 = \lim_{k\to\infty} Q_1(k) \geq 0$, pair of matrix $(A, \sqrt{Q_1})$ is stabilizable, matrix of the dynamics of a closed-loop system $\xi = A + BK_1^*S$ asymptotically stable for $K_1^* = \lim_{k\to\infty} K_1^*(k)$.

Proof. If matrix $P_x \geq 0$, then from lemma 12.2 [16] on conditions that pair of matrices $(A, \sqrt{Q_1})$ is stabilized it follows that matrix ξ is asymptotically stable. Applying theorem 3.6 [16], we obtain that if pair of matrices $(A, \sqrt{Q_1})$ is stabilized, then pair of matrices $(\xi, \sqrt{Q_1})$ is also stabilized. This proves the justice of theorem 2.

Asymptotic tracking accuracy is defined by criterion:

$$J = \lim_{k \to \infty} M\{\, \|x(k) - z\|^2 \,\}, \tag{25}$$

where $\|.\|$ is Euclidean norm of vector, z is constant reference input. First construct evaluation for criterion $J = M\{\, \|x(k) - z\|^2 \,\}$. Then define condition $k \to \infty$ and find evaluation for criterion (25). At the same time assume that the conditions of theorem 2 are satisfied and $\|\xi\|_s = \alpha_1, \|\xi_2\|_s = \alpha_2, \|\varphi\|_s = \phi_1, \|\varphi_2\|_s = \phi_2$ (here $\|.\|$ is spectral norm of matrix, $K_1^* = \lim_{k \to \infty} K_1^*(k), K_2^* = \lim_{k \to \infty} K_2^*(k), K_3^* = \lim_{k \to \infty} K_3^*(k)$). Suppose $\alpha_1^2 + \phi^2 < 1$. We note that satisfaction of this condition ensures the asymptotic stability of the closed-loop system with state delay [17]. Taking into account (1), (2), (5) by transfer coefficients K_1^*, K_2^*, K_3^*, we calculate value of criterion (25) for k+1 step:

$$\begin{aligned}
J = M\{\, & x^{\mathrm{T}}(k)\xi^{\mathrm{T}}\xi x(k) + x^{\mathrm{T}}(k)\xi^{\mathrm{T}}\xi_2 x(k-h) + x^{\mathrm{T}}(k)\xi^{\mathrm{T}}\xi_3 z \\
& + x^{\mathrm{T}}(k-h)\xi_2^{\mathrm{T}}\xi x(k) + x^{\mathrm{T}}(k-h)\xi_2^{\mathrm{T}}\xi_2 x(k-h) + x^{\mathrm{T}}(k-h)\xi_2^{\mathrm{T}}\xi_3 z \\
& + z^{\mathrm{T}}(k)\xi_3^{\mathrm{T}}\xi x(k) + z^{\mathrm{T}}(k)\xi_3^{\mathrm{T}}\xi_2 x(k-h) + x^{\mathrm{T}}(k)\varphi^{\mathrm{T}}\varphi x(k) \\
& + x^{\mathrm{T}}(k)\varphi^{\mathrm{T}}\varphi_2 x(k-h) + x^{\mathrm{T}}(k-h)\varphi_2^{\mathrm{T}}\varphi x(k) + x^{\mathrm{T}}(k-h)\varphi_2^{\mathrm{T}}\varphi_2 x(k-h) \\
& + x^{\mathrm{T}}(k)\varphi^{\mathrm{T}}\tfrac{1}{3}\sum_{i=1}^{r} B_i \theta_i K_3^* z + x^{\mathrm{T}}(k-h)\varphi_2^{\mathrm{T}}\tfrac{1}{3}\sum_{i=1}^{r} B_i \theta_i K_3^* z \\
& + \sum_{i=1}^{r} z^{\mathrm{T}} K_3^{*\mathrm{T}} \theta^{\mathrm{T}} B^{\mathrm{T}} \varphi x(k) + \tfrac{1}{3}\sum_{i=1}^{r} z^{\mathrm{T}} K_3^{*\mathrm{T}} \theta^{\mathrm{T}} B^{\mathrm{T}} B_i \theta_i K_3^* z \\
& + \tfrac{1}{3}\sum_{i=1}^{r} z^{\mathrm{T}} K_3^{*\mathrm{T}} \theta^{\mathrm{T}} B^{\mathrm{T}} \varphi_2 x(k-h) \,\} + z^{\mathrm{T}}\xi_3^{\mathrm{T}}\xi_3 z + \mathrm{tr}\tilde{Q}, \tag{26}
\end{aligned}$$

where $\varphi = \sum_{i=1}^{r}(A_i \theta_i + B_i \theta_i K_1^* S)$, $\varphi_2 = \sum_{i=1}^{r}(A_i' \theta_i + B_i \theta_i K_2^* S)$, $\tilde{Q} = Q + BK_1^* V K_1^{*\mathrm{T}} B^{\mathrm{T}} + \tfrac{1}{3}\sum_{i=1}^{r} B_i K_1^* V K_1^{*\mathrm{T}} B_i^{\mathrm{T}} + BK_2^* V K_1^{*\mathrm{T}} B^{\mathrm{T}} + \tfrac{1}{3}\sum_{i=1}^{r} B_i K_2^* V K_1^{*\mathrm{T}} \times B_i^{\mathrm{T}}$.

From (26) by virtue Cauchy–Schwarz inequality we obtain the estimate:

$$\begin{aligned}
J(k+1) \leq\ & (\alpha_1^2 + \phi^2) J_1(k) + (\alpha_1 \alpha_2 + \phi\phi_2) J_2(k, k-h) \\
& + 2(\alpha_1 r_1 + \phi R) J_3(k) + (\alpha_1 \alpha_2 + \phi\phi_2) J_2(k-h, k) + (\alpha_2^2 + \phi_2^2) J_1(k-h) \\
& + 2(\alpha_2 r_1 + \phi_2 R) J_3(k-h) + r_1^2 + R^2 + \mathrm{tr}\tilde{Q}, \tag{27}
\end{aligned}$$

where $J_1(k) = M\{\, \|x(k)\|^2 \,\}$, $J_2(k, k-h) = M\{\, \|x(k)\|\|x(k-h)\| \,\}$, $J_3(k) = M\{\, \|x(k)\| \,\}$, $r_1 = \|\xi_3 z\|$, $R = \|\sum_{i=1}^{r} B_i \theta_i K_3^* z\|$, $\tilde{Q} = \lim_{k \to \infty} \tilde{Q}(k) = \lim_{k \to \infty} \tilde{Q}(k-1)$.

Then, considering that the trajectory of the closed-loop system is described by the equation:

$$\begin{aligned}
x(k) =\ & (\xi + \phi)x(k-1) + (\xi_2 + \phi_2)x(k-h-1) \\
& + (B + \sum_{i=1}^{r} B_i \theta_i)K_1^* \nu(k-1) + (B + \sum_{i=1}^{r} B_i \theta_i)K_2^* \nu(k-h-1) \\
& + (B + \sum_{i=1}^{r} B_i \theta_i)K_3^* z + q(k-1), \tag{28}
\end{aligned}$$

we calculate recurrence relations for criteria $J_1(k)$, $J_2(k, k-h)$, $J_3(k)$ which are part of (27):

$$
\begin{aligned}
J_1(k) \leq & (\alpha_1^2 + \phi^2)^k J_1(0) + (\alpha_1\alpha_2 + \phi\phi_2)\sum_{i=1}^{r}(\alpha_1^2 + \phi^2)^{k-j} \\
& \times J_2(j-1, j-h-1) + 2(\alpha_1 r_2 + \phi R)\sum_{i=1}^{r}(\alpha_1^2 + \phi^2)^{k-j}J_3(j-1) \\
& + (\alpha_1\alpha_2 + \phi\phi_2)\sum_{i=1}^{r}(\alpha_1^2 + \phi^2)^{k-j}J_2(j-h-1, j-1) \\
& + (\alpha_2^2 + \phi_2^2)\sum_{i=1}^{r}(\alpha_1^2 + \phi^2)^{k-j}J_1(j-h-1) \\
& + 2(\alpha_2 r_2 + \phi_2 R)\sum_{i=1}^{r}(\alpha_1^2 + \phi^2)^{k-j}J_3(j-h-1) \\
& + \frac{(\alpha_1^2+\phi^2)^k-1}{(\alpha_1^2+\phi^2)-1}(r_2^2 + R^2 + tr\tilde{Q}),
\end{aligned}
\tag{29}
$$

where $r_2 = \|BK_3^* z\|$. Recurrence relation for $J_2(k, k-h)$ takes the form:

$$
\begin{aligned}
J_2(k, k-h) \leq & (\alpha_1^2 + \phi^2)^k J_2(0, -h) \\
& + (\alpha_1\alpha_2 + \phi\phi_2)\sum_{i=1}^{r}(\alpha_1^2 + \phi^2)^{k-j}J_2(j-1, j-h-1) \\
& + (\alpha_1 r_2 + \phi R)\sum_{i=1}^{r}(\alpha_1^2 + \phi^2)^{k-j}J_3(j-1) \\
& + (\alpha_1\alpha_2 + \phi\phi_2)\sum_{i=1}^{r}(\alpha_1^2 + \phi^2)^{k-j}J_1(j-h-1) \\
& + (\alpha_2^2 + \phi_2^2)\sum_{i=1}^{r}(\alpha_1^2 + \phi^2)^{k-j}J_2(j-h-1, j-2h-1) \\
& + r_2(\alpha_1 + \alpha_2)\sum_{i=1}^{r}(\alpha_1^2 + \phi^2)^{k-j}J_3(j-h-1) \\
& + R(\phi + \phi_2)\sum_{i=1}^{r}(\alpha_1^2 + \phi^2)^{k-j}J_3(j-h-1) \\
& + (r_2\alpha_2 + R\phi_2)\sum_{i=1}^{r}(\alpha_1^2 + \phi^2)^{k-j}J_3(j-2h-1) \\
& + \frac{(\alpha_1^2+\phi^2)^k-1}{(\alpha_1^2+\phi^2)-1}(r_2^2 + R^2 + tr\tilde{Q}_1),
\end{aligned}
\tag{30}
$$

where $\tilde{Q}_1 = BK_1^* V K_2^{*\mathrm{T}} B^{\mathrm{T}} + \frac{1}{3}\sum_{i=1}^{r} B_i K_1^* V K_2^{*\mathrm{T}} B_i^{\mathrm{T}}$, $\tilde{Q}_1 = \lim_{k\to\infty}\tilde{Q}_1(k) = \lim_{k\to\infty}\tilde{Q}_1(k-1)$.

Recurrence relation for $J_3(k)$ is:

$$
J_3(k) \leq \alpha_1^k J_3(0) + \alpha_2\sum_{i=1}^{r}\alpha_1^{k-j}J_3(j-h-1) + \frac{\alpha_1^k-1}{\alpha_1-1}(r_2 + R).
\tag{31}
$$

Taking into account inequality (28)–(30) we construct the value of the criterion (25). Then in $k \to \infty$ from (27) we obtain:

$$
\begin{aligned}
J \leq & \left[\frac{(\alpha_1+\alpha_2)^2+(\phi^2+\phi_2)^2}{1-(\alpha_1^2+\phi_2^2)}\right](r_2^2 + R^2) + \left[\frac{\alpha_1^2+\alpha_2^2+\phi^2+\phi_2^2}{1-(\alpha_1^2+\phi_2^2)}\right]tr\tilde{Q} \\
& + 2\left[\frac{r_1(\alpha_1+\alpha_2)+R(\phi+\phi_2)}{1-\alpha_1}\right](r_2 + R) + R^2 + r_1^2 + 2\left[\frac{\alpha_1\alpha_2+\phi_1\phi_2}{1-(\alpha_1^2+\phi^2)}\right]tr\tilde{Q}_1.
\end{aligned}
\tag{32}
$$

From the equation (31) it is evident, that in almost natural restrictions on the class of dynamic systems the method of locally optimal tracking under indirect measurements with errors provides asymptotic tracking with accuracy determined by the intensity of additive disturbances and errors in the observations, dynamic characteristics of a closed-loop system, values of the parameters of the object and the transmission coefficients tracking control system.

5   Illustrative Example

Let the system (1) and local criterion (3) be described by the following matrices and vectors:

$$A = \begin{pmatrix} 0.05 & 1 \\ -0.025 & 1 \end{pmatrix}, A' = \begin{pmatrix} 0 & 0 \\ 0.03 & 0 \end{pmatrix}, B = \begin{pmatrix} 0.1 \\ 1 \end{pmatrix}, A_1 = \begin{pmatrix} 0.05 & 0 \\ 0 & 0 \end{pmatrix},$$

$$A_2 = \begin{pmatrix} 0 & 0 \\ 0.005 & 0 \end{pmatrix}, A_3 = \begin{pmatrix} 0 & 0.1 \\ 0 & 0 \end{pmatrix}, A_4 = \begin{pmatrix} 0 & 0 \\ 0 & 0.1 \end{pmatrix}, A_5 = A_6 = \begin{pmatrix} 0 & 0 \\ 0 & 0 \end{pmatrix},$$

$$A_1' = \begin{pmatrix} 0.03 & 0 \\ 0 & 0 \end{pmatrix}, A_2' = \begin{pmatrix} 0 & 0 \\ 0.005 & 0 \end{pmatrix}, A_3' = \begin{pmatrix} 0 & 0.01 \\ 0 & 0 \end{pmatrix}, A_4' = \begin{pmatrix} 0 & 0 \\ 0 & 0.002 \end{pmatrix},$$

$$A_5' = A_6' = \begin{pmatrix} 0 & 0 \\ 0 & 0 \end{pmatrix}, B_1 = B_2 = B_3 = B_4 = \begin{pmatrix} 0 \\ 0 \end{pmatrix}, B_5 = \begin{pmatrix} 0.05 \\ 0 \end{pmatrix}, B_6 = \begin{pmatrix} 0 \\ 0.25 \end{pmatrix},$$

$$Q = \begin{pmatrix} 0.02 & 0 \\ 0 & 0.02 \end{pmatrix}, S = \begin{pmatrix} 0 & 1 \end{pmatrix}, H = \begin{pmatrix} 1 & 0 \end{pmatrix}, D = 0.2, F = 1, h = 1, z = \begin{pmatrix} 10 \\ 10 \end{pmatrix}.$$

The quality of two control systems was compared. The first control system was simulated by optimum transmission coefficients with interval parameters. The second control system was simulated by optimum transmission coefficients which were calculated using the nominal values of the parameters.

One can see from the graphs that the optimal control system has the property of robustness and accuracy of tracking in such a control system is higher than in the system synthesized by nominal values of the parameters. As a criterion for assessing the quality of the convergence of the state vector $x(k)$ to the reference input $z(k)$ calculated the average estimation error:

$$e_i = \frac{\sum_{k=1}^{N} |x(k) - z(k)|}{N},$$

where $z(k)$ is the reference input. In the table the value of quality criteria for the convergence of two algorithms ($N = 100$) for 5 different sets of interval parameters θ_i is cited:

algorithm 1 is the proposed algorithm,
algorithm 2 is the control calculated using nominal values of the parameters [8].

Average error

Algorithm	e_1	e_2	e_3	e_4	e_5
1	0.261	0.829	1.599	0.666	0.488
2	0.407	0.994	2.106	1.004	0.697

We note that in the simulation, interval on which the criterion is calculated is shifted on the value of the transient process (15 steps).

The table shows, that average error deviation of the state vector $x(k)$ from tracking vector $z(k)$ by proposed algorithm is smaller than control using algorithm 2.

6 Conclusion

The problem of controlling the output of a discrete systems with interval parameters and time delay based on the synthesis of locally optimal linear tracking control system of discrete systems with indirect observations using the probabilistic method has been solved. The asymptotic behavior of the system has been analyzed. It is shown that the optimal control system with constant transfer coefficients has the property of robustness and provides better tracking accuracy than the control system, which was synthesized using the nominal parameter values.

References

1. Camacho, E.F., Bordons, C.: Model Predictive Control. Springer, London (2004)
2. Aggelogiannaki, E., Doganis, P., Sarimveis, H.: An Adaptive Model Predictive Control Configuration for Production-Inventory Systems. International Journal of Production Economics 114, 165–178 (2008)
3. Wang, W., Rivera, D.: A Novel Model Predictive Control Algorithm for Supply Chain Management in Semiconductor Manufacturing. In: 2005 American Control Conference, Portland, OR, pp. 1–6 (2005)
4. Stoica, C., Arahal, M.: Application of Robustified Model Predictive Control to a Production-Inventory System. In: 48th IEEE Conference on Decision and Control and 28th Chinese Control Conference Shanghai, China, pp. 3993–3998 (2009)
5. Henneta, J.-C.: A Globally Optimal Local Inventory Control Policy for Multistage Supply Chains. International Journal of Production Research 47(2), 435–453 (2009)
6. Dombrovskii, V.V., Dombrovskii, D.V.: Predictive Control of Random-Parameter Systems with Multiplicative Noise. Application to Investment Portfolio Optimization. Automation and Remote Control 66(4), 583–595 (2005)
7. Dai, L., Xia, Y., Fu, M., Mahmoud, M.: Discrete-Time Model Predictive Control. In: Advances in Discrete Time Systems, ch. 4, pp. 77–116. InTech (2012)
8. Tang, G., Sun, H., Liu, Y.: Optimal Tracking Control for Discrete Time-Delay Systems with Persistent Disturbances. Asian Journal of Control 8(8), 135–140 (2006)
9. Mukhina, O., Smagin, V.: Local-Optimal Output Control for Discrete Systems with State Delays. Tomsk State University Journal of Control and Computer Science 1(26), 4–13 (2014) (in Russian)
10. Patre, B.M., Bandyopadhyay, B.: Robust Control for Two-Time-Scale Discrete Interval Systems. Reliable Computing 12, 45–58 (2006)
11. Zadeh, L.A.: Fuzzy Sets. Information and Control 8, 338–353 (1965)
12. Lin, T.-S., Chan, S.-W.: Robust Adaptive Fuzzy Sliding Mode Control for a Class of Uncertain Discrete-Time Nonlinear Systems. International Journal of Innovative Computing, Information and Control 8(1(A)) (2012)
13. Capron, B.D.O., Uchiyama, M.T.: Linear Matrix Inequality-Based Robust Model Predictive Control for Time-Delayed Systems. IET Control Theory and Applications 6, 37–50 (2012)
14. Alon, N., Krivelevich, M.: Extremal and Probabilistic Combinatorics. In: Gowers, W.T. (ed.) Princeton Companion to Mathematics, pp. 562–575. Princeton University Press (2008)

15. Athans, M.: The Matrix Minimum Principle. Information and Control 11(5/6), 592–606 (1968)
16. Wonham, W.M.: Linear Multivariable Control: A Geometric Approach. Springer (1979)
17. Stojanovic, S., Debeljkovic, D.: On the Asymptotic Stability of Linear Discrete Time Delay Systems. Mechanical Engineering 2(1), 35–48 (2004)

The M/G/∞ Queue in Random Environment*

Anatoly Nazarov and Galina Baymeeva

National Research Tomsk State University, Tomsk, Russia
{nazarov.tsu,baymeevag}@gmail.com

Abstract. Infinite-server queues are often used to approximate the behavior of systems with sufficiently large number of servers such as banks, call-centers, supermarkets or digital distribution platforms. Such systems often become subjects to external influences which affect their performance, particularly their arrival rate and service-time distribution. In this article we consider a mathematical model of such a situation as an $M/G/\infty$ queue operating in a "random environment", for which the underlying process is a continuous-time Markov chain with finite number of states. The arrival rate and service-time distribution change randomly according to the environment state. Note that distribution of service-time of customers, which are currently being served, does not change until the service-time is finished. The approximate probability distribution of the number of customers under certain conditions is obtained.

Keywords: queueing theory, infinite-server queue, random environment, general service-time, method of screened flow, method of asymptotic analysis.

1 Introduction

Infinite-server queues are often used to approximate the behavior of systems with sufficiently large number of servers, such as banks, call-centers, supermarkets or digital distribution platforms. Such systems often become subjects to external influences which affect their performance, particularly, their arrival rate and service-time distribution. For instance, the change of bank rate set by the Central bank affects the conditions under which commercial banks give loans to their clients. These, in turn, significantly influence the intensity of clients' arrival. In this article we consider a mathematical model of such situation as an $M/G/\infty$ queue operating in a "random environment", for which the underlying process is a continuous-time Markov chain with finite number of states. The arrival rate and service-time distribution change randomly according to the environment state. Note that distribution of service-time customers which are currently being served does not change until the service-time is finished. Say the bank provided a credit to the client on certain conditions and during the repayment period there was a

---

* The work is performed under the state order of the Ministry of Education of the Russian Federation N 1.511.2014/K.

A. Dudin et al. (Eds.): ITMM 2014, CCIS 487, pp. 312–324, 2014.

change of bank rate. The client will continue to repay his debt on those initial conditions — as mentioned in a loan agreement.

Queues in a random environment have been intensively studied in the literature. In particular, a lot of research is conducted on infinite-sever queues operating in a random environment, either Markovian [1,2,3] or semi-Markovian [4,5,6]. In [7] stochastic decomposition formula is obtained for the number of customers in an $M/G/\infty$ system with service speeds depending on general ergodic process. In one of the oldest paper on random environment [8] the problem, very similar to ours, is studied. Three cases of service behavior at environment transition epochs are covered. The first one is considered in the present paper — service-time distribution stays the same while customer is in the system. In the second case when a change in environment occurs all customers in service are discharged from the system immediately. This situation is also covered in [9]. The last case considers customers in service moving to a secondary queue which is an infinite-server system with bulk arrivals. As a result, the steady-state mean number of customers in the secondary queue is obtained.

2   Problem Statement

We consider an $M/G/\infty$ queue operating in a "random environment", i.e. depending on the state of a continuous-time Markov chain $s(t)$. That is, when the process is in state $s = \overline{1, K}$, the queue operates as an $M(\lambda_s)/G(B_s(x))/\infty$ with arrival rate λ_s and service-time is distributed according to the probability distribution function $B_s(x)$ on each server. We study the case when service-time distribution of a customer which is currently being served does not change until the service is finished. As a model of such a system we use a 2-dimensional stochastic process $\{i(t), s(t)\}$, where $i(t)$ is a number of customers in the system. Apparently the process $\{i(t), s(t)\}$ is not a Markovian one. Let us denote as $P(i, s, t)$ the probability of system state:

$$P(i, s, t) = P\{i(t) = i, s(t) = s\}, i \geq 0, s = \overline{1, K}. \tag{1}$$

The objective of our research is to obtain the distribution of probabilities (1) as well as numerical characteristics such as expected value and variance of the number of customers in the system [10]. In this paper, we apply the original methods for queueing systems, such as method of screened flow and method of asymptotic analysis [11].

3   Method of Screened Flow

3.1   Method Description

We pick a moment T and mark on the time axis $(-\infty, T)$ the moments of event occurrences in the arrival flow. The customer will be referred to as "screened", with probability

$$S_s(t) = 1 - B_s(T - t), s = \overline{1, K}, \tag{2}$$

if it arrived in the system at time $t < T$ and was not fully serviced until the moment T. Thus, the customers of screened flow at time T will be in the system taking up its servers.

Let us denote as $n(t)$ the number of events of the screened flow which occurred until the moment t. If at some initial time $t_0 < T$ the system is empty, i.e. there are no customers under service, then for the time T the following equality holds:

$$i(T) = n(T). \tag{3}$$

Next, choose the initial time t_0 so that at all times $t < t_0$ events of screened flow do not occur, i.e.

$$S_s(t) = 1 - B_s(T - t) = 0, s = \overline{1, K}, t < t_0. \tag{4}$$

Since $B_s(x)$ is the distribution function, it is obvious enough to put $t_0 = -\infty$.

Equation (3) allows us to reduce the research problem of non-Markovian queue with infinite number of servers to the problem of non-stationary screened flow analysis which is defined by the process $n(t)$. Characteristics of the process $n(t)$ at time T coincide with the characteristics of value $i(T)$.

This "screening" method can also be used for the analysis of queueing networks. For example, the dynamic screening method is applied for $GI - (GI|\infty)^K$ queueing network in [12].

3.2   System of Kolmogorov Differential Equations

As stated above, it is legit to consider the process $\{s(t), n(t)\}$ in order to study our system. This two-dimensional process is a Markovian one. We write the possible transitions to the states of $n(t)$ and their probabilities assuming $n(t) = n$ as follows:

$$n(t + \Delta t) = \begin{cases} n + 1, & \text{with probability } \lambda_s \Delta t S_s(t) + o(\Delta t), s = \overline{1, K}, \\ n, & \text{with probability } 1 - \lambda_s \Delta t S_s(t) + o(\Delta t), s = \overline{1, K}. \end{cases}$$

Similarly to (1), we define the probabilities

$$P(s, n, t) = P\{s(t) = s, n(t) = n\}, s = \overline{1, K}, n \geq 0, \tag{5}$$

and according to the law of total probability [13] we write

$$P(s, n, t + \Delta t) = \{1 + q_{ss} \Delta t\}\{[1 - \lambda_s \Delta t S_s(t)]P(s, n, t) +$$

$$+ \lambda_s \Delta t S_s(t) P(s, n - 1, t)\} + \sum_{\nu \neq s} q_{\nu s} \Delta t P(\nu, n, t) + o(\Delta t), \tag{6}$$

$$n \geq 0, s = \overline{1, K}.$$

Performing simple operations on (6), we obtain the direct system of Kolmogorov differential equations:

$$\frac{\partial P(s, n, t)}{\partial t} = \lambda_s S_s(t)\{P(s, n - 1, t) - P(s, n, t)\} + \sum_{\nu=1}^{K} q_{\nu s} P(\nu, n, t), \tag{7}$$

$$n \geq 0, s = \overline{1, K}.$$

The initial condition for the solution $P(s, n, t)$ of this system of equations at time t_0 we define as

$$P(s, n, t_0) = \begin{cases} r(s), & \text{if } n = 0, \\ 0, & \text{if } n > 0, \end{cases} \qquad (8)$$

where $r(s)$ are steady-state probabilities of the environment states.

3.3 Characteristic Functions

We introduce partial characteristic functions as follows

$$H(s, u, t) = \sum_{n=0}^{\infty} e^{jun} P(s, n, t), s = \overline{1, K}, \qquad (9)$$

where $j = \sqrt{-1}$. We multiply the equations of the system (7) by e^{jun} and sum all the equations over n from 0 to ∞ and obtain the system of Kolmogorov differential equations which defines partial characteristic functions:

$$\frac{\partial H(s, u, t)}{\partial t} = \lambda_s S_s(t)(e^{ju} - 1)H(s, u, t) + \sum_{\nu=1}^{K} q_{\nu s} H(\nu, u, t), \qquad (10)$$

$$s = \overline{1, K}.$$

We rewrite this system as a vector-matrix equation, denoting

$$\mathbf{\Lambda} = \begin{bmatrix} \lambda_1 & 0 & \cdots & 0 \\ 0 & \lambda_2 & \cdots & 0 \\ \vdots & \vdots & \ddots & \vdots \\ 0 & 0 & \cdots & \lambda_K \end{bmatrix}, \mathbf{S}(t) = \begin{bmatrix} S_1(t) & 0 & \cdots & 0 \\ 0 & S_2(t) & \cdots & 0 \\ \vdots & \vdots & \ddots & \vdots \\ 0 & 0 & \cdots & S_K(t) \end{bmatrix},$$

$$\mathbf{Q} = \begin{bmatrix} q_{11} & q_{12} & \cdots & q_{1K} \\ q_{21} & q_{22} & \cdots & q_{2K} \\ \vdots & \vdots & \ddots & \vdots \\ q_{K1} & q_{K2} & \cdots & q_{KK} \end{bmatrix}, \mathbf{H}(u, t) = \begin{bmatrix} H(1, u, t) & H(2, u, t) & \cdots & H(K, u, t) \end{bmatrix}.$$

Here $\mathbf{\Lambda}$, $\mathbf{S}(t)$ are diagonal matrices containing the conditional intensities λ_s of the arrival flow and the probabilities $S_s(t)$ of the screened flow's event occurrences at time t, respectively, $\mathbf{Q}$ is the transition rate matrix of the environment and $\mathbf{H}(u, t)$ is a vector characteristic function. Now the system (10) can be rewritten as

$$\frac{\partial \mathbf{H}(u, t)}{\partial t} = \mathbf{H}(u, t)\{(e^{ju} - 1)\mathbf{\Lambda S}(t) + \mathbf{Q}\}. \qquad (11)$$

It is obvious that obtaining an explicit solution to the equation above is quite uneasy. We apply the method of asymptotic analysis to the equation (11) in order to obtain the explicit distribution under limiting conditions of high arrival intensity and frequent change of the environment states.

4   Method of Asymptotic Analysis

4.1   Method Description

Method of asymptotic analysis for queueing systems consists of analysis of the equations defining any characteristics of the system and allows to obtain the explicit distribution and numerical characteristics under some asymptotic condition. Returning to our problem we set

$$\tilde{\mathbf{\Lambda}} = \mathbf{\Lambda} N, \tilde{\mathbf{Q}} = \mathbf{Q} N. \tag{12}$$

Then the asymptotic condition is given as follows:

$$N \to \infty. \tag{13}$$

Thus, our goal is to solve the following equation:

$$\frac{1}{N} \frac{\partial \mathbf{H}(u,t)}{\partial t} = \mathbf{H}(u,t)\{(e^{ju} - 1)\mathbf{\Lambda S}(t) + \mathbf{Q}\} \tag{14}$$

with the initial condition derived from (8), as (13) takes place.

4.2   First Degree Asymptotic

In the equation (14), we denote $\varepsilon = \frac{1}{N}$ and make substitutions

$$u = \varepsilon w, \mathbf{H}(u,t) = \mathbf{F}_1(w,t,\varepsilon),$$

and then it can be rewritten as

$$\varepsilon \frac{\partial \mathbf{F}_1(w,t,\varepsilon)}{\partial t} = \mathbf{F}_1(w,t,\varepsilon)\{(e^{j\varepsilon w} - 1)\mathbf{\Lambda S}(t) + \mathbf{Q}\}. \tag{15}$$

Here w has the meaning of a scaled argument of the vector characteristic function.

In (15), we set $\varepsilon \to 0$ and denote

$$\lim_{\varepsilon \to 0} \mathbf{F}_1(w,t,\varepsilon) = \mathbf{F}_1(w,t). \tag{16}$$

This yields:

$$\mathbf{F}_1(w,t)\mathbf{Q} = \mathbf{0}. \tag{17}$$

Row-vector $\mathbf{0}$ above contains zeros. It then follows that $\mathbf{F}_1(w,t)$ can be represented as a product

$$\mathbf{F}_1(w,t) = \mathbf{r}\Phi_1(w,t). \tag{18}$$

Here $\Phi_1(w,t)$ is a scalar function, which will be defined below, and $\mathbf{r}$ is a row-vector of steady-state probability distribution of environment states:

$$\begin{cases} \mathbf{r}\mathbf{Q} = \mathbf{0}, \\ \mathbf{r}\mathbf{e} = 1, \end{cases} \tag{19}$$

where $\mathbf{e}$ is a column-vector containing units. We post-multiply (15) by $\mathbf{e}$ and divide by ε:

$$\frac{\partial \mathbf{F}_1(w,t,\varepsilon)}{\partial t}\mathbf{e} = \frac{e^{j\varepsilon w}-1}{\varepsilon}\mathbf{F}_1(w,t,\varepsilon)\boldsymbol{\Lambda}\mathbf{S}(t)\mathbf{e}. \tag{20}$$

In (20) we set $\varepsilon \to 0$ and substitute $\mathbf{F}_1(w,t)$ with product (18). This gives the following equation:

$$\frac{\partial \Phi_1(w,t)}{\partial t} = jw\Phi_1(w,t)\mathbf{r}\boldsymbol{\Lambda}\mathbf{S}(t)\mathbf{e}, \tag{21}$$

which is an ordinary differential equation separable in $\Phi_1(w)$ and w. Its solution considering (8) is given as follows:

$$\Phi_1(w,t) = exp\{jw\int_{-\infty}^{t} \mathbf{r}\boldsymbol{\Lambda}\mathbf{S}(\tau)\mathbf{e}d\tau\}. \tag{22}$$

Denote

$$\kappa_1(t) = \int_{-\infty}^{t} \mathbf{r}\boldsymbol{\Lambda}\mathbf{S}(\tau)\mathbf{e}d\tau. \tag{23}$$

Finally,

$$\mathbf{H}(u,t) = \mathbf{F}_1(w,t,\varepsilon) \approx \mathbf{F}_1(w,t) = \mathbf{r}\Phi_1(w,t) = \mathbf{r}exp\{jw\kappa_1(t)\}, \tag{24}$$

where $w = Nu$. It follows that

$$M\{e^{jun(t)}\} = \mathbf{H}(u,t)\mathbf{e} \approx h_1(u,t) = exp\{ju\kappa_1(t)N\}. \tag{25}$$

Since (3) takes place, we can finally conclude:

$$M\{e^{jui(T)}\} = M\{e^{jun(T)}\} = \mathbf{H}(u,T)\mathbf{e} \approx h_1(u,T) = exp\{ju\kappa_1(T)N\}. \tag{26}$$

We calculate the value $\kappa_1(T)$:

$$\begin{aligned}\kappa_1(T) &= \int_{-\infty}^{T} \mathbf{r}\boldsymbol{\Lambda}\mathbf{S}(t)\mathbf{e}dt = \int_{-\infty}^{T}\sum_{s=1}^{K}\mathbf{r}(s)\lambda_s S_s(t)dt = \\ &= \sum_{s=1}^{K}\mathbf{r}(s)\lambda_s\int_{-\infty}^{T}\{1 - B_s(T-t)\}dt.\end{aligned} \tag{27}$$

In the integral we make a substitution $\tau = T - t$, then (27) can be rewritten as:

$$\begin{aligned}\sum_{s=1}^{K}\mathbf{r}(s)\lambda_s\int_{-\infty}^{T}\{1-B_s(T-t)\}dt &= \\ = \sum_{s=1}^{K}\mathbf{r}(s)\lambda_s\int_{0}^{\infty}\{1-B_s(\tau)\}d\tau &= \sum_{s=1}^{K}\mathbf{r}(s)\lambda_s b_s,\end{aligned} \tag{28}$$

where b_s is the mean of the service time. Thus

$$\kappa_1(T) = \sum_{s=1}^{K} \mathbf{r}(s)\lambda_s \int_0^\infty \{1 - B_s(\tau)\}d\tau = \sum_{s=1}^{K} \mathbf{r}(s)\lambda_s b_s = \mathbf{r}\mathbf{\Lambda}\mathbf{B}\mathbf{e}, \qquad (29)$$

where $\mathbf{B}$ is a diagonal matrix containing service time means:

$$\mathbf{B} = \begin{bmatrix} b_1 & 0 & \cdots & 0 \\ 0 & b_2 & \cdots & 0 \\ \vdots & \vdots & \ddots & \vdots \\ 0 & 0 & \cdots & b_K \end{bmatrix}.$$

4.3   Second Degree Asymptotic

In the equation (14) we make a substitution

$$\mathbf{H}(u,t) = \mathbf{H}_2(u,t)exp\{ju\kappa_1(t)N\}. \qquad (30)$$

The function $\mathbf{H}_2(u,t)$ here has a meaning of centered characteristic function as the following relation takes place:

$$\mathbf{H}_2(u,t)\mathbf{e} = \mathbf{H}(u,t)e^{-ju\kappa_1(t)N}\mathbf{e} = M\{exp[ju(n(t) - \kappa_1(t)N)]\}. \qquad (31)$$

The substitution (30) yields an equation which defines function $\mathbf{H}_2(u,t)$:

$$\frac{1}{N}\frac{\partial \mathbf{H}_2(u,t)}{\partial t} = \mathbf{H}_2(u,t)\{(e^{ju} - 1)\mathbf{\Lambda}\mathbf{S}(t) + \mathbf{Q} - ju(\mathbf{r}\mathbf{\Lambda}\mathbf{S}(t)\mathbf{e})\mathbf{I}\}. \qquad (32)$$

The matrix $\mathbf{I}$ here is the identity matrix.

We define $\varepsilon^2 = \frac{1}{N}$ and make the following denotations:

$$u = \varepsilon w, \mathbf{H}_2(u,t) = \mathbf{F}_2(w,t,\varepsilon).$$

We then rewrite (32) as:

$$\varepsilon^2 \frac{\partial \mathbf{F}_2(w,t,\varepsilon)}{\partial t} = \mathbf{F}_2(w,t,\varepsilon)\{(e^{j\varepsilon w} - 1)\mathbf{\Lambda}\mathbf{S}(t) + \mathbf{Q} - j\varepsilon w(\mathbf{r}\mathbf{\Lambda}\mathbf{S}(t)\mathbf{e})\mathbf{I}\}. \qquad (33)$$

As $\varepsilon \to 0$, denoting

$$\lim_{\varepsilon \to 0} \mathbf{F}_2(w,t,\varepsilon) = \mathbf{F}_2(w,t), \qquad (34)$$

we obtain

$$\mathbf{F}_2(w,t)\mathbf{Q} = \mathbf{0}. \qquad (35)$$

It follows that

$$\mathbf{F}_2(w,t) = \mathbf{r}\Phi_2(w,t), \qquad (36)$$

where $\Phi_2(w,t)$ is a scalar function which will be defined later. The solution $\mathbf{F}_2(w,t,\varepsilon)$ for the equation (33) we rewrite as follows

$$\mathbf{F}_2(w,t,\varepsilon) = \Phi_2(w,t)\{\mathbf{r} + j\varepsilon w \mathbf{f}_2(t)\} + O(\varepsilon^2), \tag{37}$$

where $\mathbf{f}_2(t)$ is a row-vector, which will be defined below.

In the equation (33) we represent the function $e^{j\varepsilon w}$ as a first degree Taylor polynomial. This yields the following expression:

$$\mathbf{F}_2(w,t,\varepsilon)\{j\varepsilon w[\mathbf{\Lambda S}(t) - (\mathbf{r}\mathbf{\Lambda S}(t)\mathbf{e})\mathbf{I}] + \mathbf{Q}\} = O(\varepsilon^2) \tag{38}$$

In (38) we substitute $\mathbf{F}_2(w,t,\varepsilon)$ with approximation (37) and obtain the following system:

$$j\varepsilon w \Phi_2(w,t)\{\mathbf{r}\mathbf{\Lambda S}(t)[\mathbf{I} - \mathbf{er}] + \mathbf{f}_2(t)\mathbf{Q}\} = O(\varepsilon^2). \tag{39}$$

We divide both sides of it by ε and set $\varepsilon \to 0$. We then have:

$$\Phi_2(w,t)\{\mathbf{r}\mathbf{\Lambda S}(t)[\mathbf{I} - \mathbf{er}] + \mathbf{f}_2(t)\mathbf{Q}\} = 0. \tag{40}$$

Given $\Phi_2(w,t) \neq 0$, the following identity takes place:

$$\mathbf{f}_2(t)\mathbf{Q} = \mathbf{r}\mathbf{\Lambda S}(t)\{\mathbf{er} - \mathbf{I}\}. \tag{41}$$

Hence, the vector $\mathbf{f}_2(t)$ is defined by the inhomogeneous underdetermined linear system.

The solution $\mathbf{f}_2(t)$ of the system (41) we write as

$$\mathbf{f}_2(t) = c(t)\mathbf{r} + \mathbf{g}(t), \tag{42}$$

where $c(t)$ is an arbitrary scalar function and the row vector $\mathbf{g}(t)$ is any specific solution to the system (41) satisfying a certain condition, for example:

$$\mathbf{g}(t)\mathbf{e} = 0. \tag{43}$$

A solution $\mathbf{g}(t)$ to the system (41), (43) we write as

$$\mathbf{g}(t) = \mathbf{r}\mathbf{\Lambda S}(t)\mathbf{G}, \tag{44}$$

where $\mathbf{G}$ is a matrix solving the following system:

$$\begin{cases} \mathbf{GQ} = \mathbf{er} - \mathbf{I}, \\ \mathbf{Ge} = \mathbf{0}^T. \end{cases} \tag{45}$$

Thus, the function $\mathbf{f}_2(t)$ is written as follows:

$$\mathbf{f}_2(t) = c(t)\mathbf{r} + \mathbf{r}\mathbf{\Lambda S}(t)\mathbf{G}. \tag{46}$$

Let us now derive the explicit expression for the function $\Phi_2(w,t)$. To do this, we approximate the exponential function in (33) with the second degree Taylor polynomial and make a substitution (37). This yields the equation:

$$\varepsilon^2 \frac{\partial \Phi_2(w,t)}{\partial t}\mathbf{r} = \Phi_2(w,t)\{j\varepsilon w[\mathbf{r}\mathbf{\Lambda S}(t)(\mathbf{I} - \mathbf{er}) + \mathbf{f}_2(t)\mathbf{Q}] +$$
$$+ \frac{(j\varepsilon w)^2}{2}[\mathbf{r}\mathbf{\Lambda S}(t) + 2\mathbf{f}_2(t)\mathbf{\Lambda S}(t) - 2(\mathbf{r}\mathbf{\Lambda S}(t)\mathbf{e})\mathbf{f}_2(t)]\} + O(\varepsilon^3). \tag{47}$$

We then post-multiply both parts of the system (47) by vector $\mathbf{e}$. Due to (41), the equation can be rewritten as:

$$\varepsilon^2 \frac{\partial \Phi_2(w,t)}{\partial t} = \frac{(j\varepsilon w)^2}{2} \Phi_2(w,t)\{\mathbf{r} + 2\mathbf{f}_2(t)[\mathbf{I} - \mathbf{er}]\}\mathbf{\Lambda S}(t)\mathbf{e} + O(\varepsilon^3). \tag{48}$$

We divide both sides of (48) by ε^2 and set $\varepsilon \to 0$. This gives:

$$\frac{\partial \Phi_2(w,t)}{\partial t} = \frac{(jw)^2}{2} \Phi_2(w,t)\{\mathbf{r} + 2\mathbf{f}_2(t)[\mathbf{I} - \mathbf{er}]\}\mathbf{\Lambda S}(t)\mathbf{e}. \tag{49}$$

A solution for this ordinary differential equation, considering the initial condition derived from (8), is written as follows:

$$\Phi_2(w,t) = exp\{\frac{(jw)^2}{2}\kappa_2(t)\}, \tag{50}$$

where $\kappa_2(t)$ denotes the following expression:

$$\kappa_2(t) = \int\limits_{-\infty}^{t} \{\mathbf{r} + 2\mathbf{f}_2(\tau)[\mathbf{I} - \mathbf{er}]\}\mathbf{\Lambda S}(\tau)\mathbf{e}d\tau. \tag{51}$$

Thus, the expression for the centered characteristic function $\mathbf{H}_2(u,t)$ is obtained and is written as follows:

$$\mathbf{H}_2(u,t) = \mathbf{F}_2(w,t,\varepsilon) \approx \mathbf{F}_2(w,t) = \mathbf{r}\Phi_2(w,t) =$$
$$= \mathbf{r}exp\{\frac{(jw)^2}{2}\kappa_2(t)\} = \mathbf{r}exp\{\frac{(ju)^2}{2}\kappa_2(t)N\}. \tag{52}$$

It follows that

$$\mathbf{H}(u,t) = \mathbf{H}_2(u,t)e^{ju\kappa_1(t)N} \approx \mathbf{r}exp\{ju\kappa_1(t)N + \frac{(ju)^2}{2}\kappa_2(t)N\}, \tag{53}$$

$$M\{e^{jun(t)}\} = \mathbf{H}(u,t)\mathbf{e} \approx h_2(u,t) = exp\{ju\kappa_1(t)N + \frac{(ju)^2}{2}\kappa_2(t)N\}. \tag{54}$$

Considering (3), the following identities are true:

$$\mathbf{H}(u,T) \approx \mathbf{r}exp\{ju\kappa_1(T)N + \frac{(ju)^2}{2}\kappa_2(T)N\},$$
$$M\{e^{jui(T)}\} = M\{e^{jun(T)}\} = \mathbf{H}(u,T)\mathbf{e} \approx h_2(u,T) = \tag{55}$$
$$= exp\{ju\kappa_1(T)N + \frac{(ju)^2}{2}\kappa_2(T)N\}.$$

Let us specify the expression for $\kappa_2(T)$, considering (46) and the system (45):

$$\kappa_2(T) = \int_{-\infty}^{T} \{\mathbf{r} + 2\mathbf{f}_2(t)[\mathbf{I} - \mathbf{er}]\}\mathbf{\Lambda S}(t)\mathbf{e}d\tau =$$

$$= \int_{-\infty}^{T} \mathbf{r\Lambda S}(t)\mathbf{e}dt + 2\int_{-\infty}^{T} \{c(t)\mathbf{r} + \mathbf{r\Lambda S}(t)\mathbf{G}\}\{\mathbf{I} - \mathbf{er}\}\mathbf{\Lambda S}(t)\mathbf{e}dt =$$

$$= \kappa_1(T) + 2\int_{-\infty}^{T} \mathbf{r\Lambda S}(t)\mathbf{G\Lambda S}(t)\mathbf{e}dt =$$

$$= \mathbf{r\Lambda Be} + 2\sum_{s=1}^{K}\sum_{s'=1}^{K}\mathbf{r}(s)\lambda_s\lambda_{s'}\mathbf{G}_{ss'}\int_{-\infty}^{T} S_s(t)S_{s'}(t)dt =$$

$$= \mathbf{r\Lambda Be} + 2\mathbf{r\Lambda BG\Lambda e} - 2\mathbf{r\Lambda}(\mathbf{M} \times \mathbf{G})\mathbf{\Lambda e}.$$

The symbol $\times$ here denotes the Hadamard product for matrices. Note that

$$\int_{-\infty}^{T} S_s(t)S_{s'}(t)dt = \int_{0}^{\infty} (1 - B_s(x))(1 - B_{s'}(x))xdx = \tag{56}$$

$$= M\{min(\tau_s, \tau_{s'})\}, s, s' = \overline{1, K}.$$

The matrix $\mathbf{M}$ is defined as follows:

$$\mathbf{M} = \left[M\{min(\tau_s, \tau_{s'})\}\right], s, s' = \overline{1, K}.$$

Finally we can write

$$\kappa_2(T) = \mathbf{r\Lambda Be} + 2\mathbf{r\Lambda BG\Lambda e} - 2\mathbf{r\Lambda}(\mathbf{M} \times \mathbf{G})\mathbf{\Lambda e}. \tag{57}$$

It follows from the last equality that $\kappa_2(T)$ does not depend on the arbitrary function $c(t)$, which is present in (46).

After obtaining (55) it is clear that the probability distribution of the number of customers in the system is asymptotic normal and $\kappa_1(t)N$ and $\kappa_2(t)N$ are, respectively, the first and the second cumulants. It is known that

$$M\{i(T)\} \approx \kappa_1(T)N, D\{i(T)\} \approx \kappa_2(T)N. \tag{58}$$

Inverse Fourier transform of (55) gives the probability density function of the normally distributed random variable:

$$p(x) = \frac{1}{\sqrt{2\pi\kappa_2(T)N}}exp\left\{-\frac{(x - \kappa_1(T)N)^2}{2\kappa_2(T)N}\right\}. \tag{59}$$

It is necessary to switch from this continuous distribution to discrete as follows:

$$P(i) = Cp(i), i \geq 0, \tag{60}$$

where the constant value C is defined considering the normalizing condition:

$$\sum_{i=0}^{\infty} P(i) = C \sum_{i=0}^{\infty} p(i) = 1. \tag{61}$$

Due to (61), C is given as follows:

$$C = 1/\sum_{i=0}^{\infty} p(i) \tag{62}$$

Thus the asymptotic probability distribution of the number of customers in the system $M(\lambda_s)/G(B_s(x))/\infty$ is obtained using the methods of screened flow and asymptotic analysis. Given the concrete distribution $B_s(x)$ we can calculate the exact values for $\kappa_1(t)N$ and $\kappa_2(t)N$.

5 Example

In conclusion we apply derived formulas to a concrete example. Let us set the transition-rate matrix $\mathbf{Q}$ and diagonal matrix $\mathbf{\Lambda}$ as follows:

$$\mathbf{Q} = \begin{bmatrix} -5 & 2 & 3 \\ 0 & -3 & 3 \\ 2 & 1 & -3 \end{bmatrix}, \mathbf{\Lambda} = \begin{bmatrix} 4 & 0 & 0 \\ 0 & 3 & 0 \\ 0 & 0 & 5 \end{bmatrix}.$$

Given these values, row-vector $\mathbf{r}$ and matrix $\mathbf{G}$ are defined as:

$$\mathbf{r} = \begin{bmatrix} 0.2 & 0.3 & 0.5 \end{bmatrix}, \mathbf{G} = \begin{bmatrix} 0.127 & -0.043 & -0.083 \\ -0.073 & 0.157 & -0.083 \\ -0.006 & -0.077 & 0.083 \end{bmatrix}.$$

Let service-time be gamma-distributed with shape parameter α_s and rate parameter β_s. Given the expression for the gamma-distribution mean, the matrix $\mathbf{B}$ can be specified as follows:

$$\mathbf{B} = \begin{bmatrix} \frac{\alpha_1}{\beta_1} & 0 & 0 \\ 0 & \frac{\alpha_2}{\beta_2} & 0 \\ 0 & 0 & \frac{\alpha_3}{\beta_3} \end{bmatrix} = \begin{bmatrix} 2 & 0 & 0 \\ 0 & 0.375 & 0 \\ 0 & 0 & 0.500 \end{bmatrix}.$$

Finally, we specify the matrix $\mathbf{M}$ and its general term $M\{min(\tau_s, \tau_{s'})\} = = \frac{\alpha_s \alpha_{s'}}{\beta_s + \beta_{s'}}$:

$$\mathbf{M} = \begin{bmatrix} 2 & 0.667 & 0.667 \\ 0.667 & 0.562 & 0.300 \\ 0.667 & 0.300 & 0.250 \end{bmatrix}.$$

We now may calculate the values $\kappa_1(T)$ and $\kappa_2(T)$:

$$\kappa_1(T) = 2.275,$$
$$\kappa_2(T) = 1.185.$$

Using these parameters we can plot the probability distribution of the number of customers in the system, as N is reasonably large. The probabilities of i customers being serviced are defined by expression (60). Let $N = 10$. The asymptotic probability distribution graph is given in Figure 1.

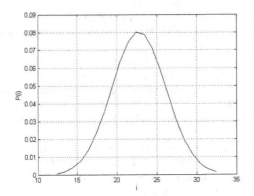

Fig. 1. Asymptotic probability distribution graph of the number of customers in the system

6   Conclusion

In this paper the research was conducted on the $M/G/\infty$ queue, operating in a Markovian random environment. Thus, it is necessary to analyze the 2-dimensional stochastic process, which is non-Markovian. We studied the case when the service-time distribution of customers which are currently being served does not change until the customer is fully serviced, even if the environment has jumped to another state. Using the method of screened flow we reduced the mentioned problem to the analysis of a 2-dimensional Markovian process. We then used the method of asymptotic analysis to find the limiting probability distribution of the number of clients in the system, which turned out to be Gaussian. We as well derived the expressions for its parameters — mean and variance.

Previously the work was done on analyzing the $M/M/\infty$ queue in random environment in case the arrival and service rate vary along with the environment state [14]. The solution to such a problem, namely approximate probability distribution of the queue length and exact formulas for the mean and variance, is obtained quite easily. On the other hand, if the service rate stays the same until the client is present in the system the complications arise as the service-time distribution is different at the infinite number of servers. Thus we obtain the solution for the more general problem as we have the proper tools to perform it.

In the future it is planned to build a simulation model for the queueing system in question and compare simulated results to the calculated using formulas (29) and (57). We will also consider the case when the random environment is semi-Markovian.

References

1. Baykal-Gursoy, M., Xiao, W.: Stochastic Decomposition in $M/M/\infty$ Queues with Markov Modulated Service Rates. Queueing Syst. 48, 75–88 (2004)
2. O'Cinneide, C.A., Purdue, P.: The $M/M/\infty$ Queue in a Random Environment. J. Appl. Prob. 23, 175–184 (1986)
3. Blom, J., Kella, O., Mandjes, M., Thorsdottir, H.: Markov-Modulated Infinite-Server Queues with General Service Times. Queueing Syst. 76, 403–424 (2014)
4. D'Auria, B.: $M/M/\infty$ queues in semi-Markovian random environment. Queueing Syst. 58, 221–237 (2008)
5. Falin, G.: The $M/M/\infty$ Queue in Random Environment. Queueing Syst. 58, 65–76 (2008)
6. Fralix, B.H., Adan, I.J.B.F.: An Infinite-Server Queue Influenced by a Semi-Markovian Environment. Queueing Syst. 61, 65–84 (2009)
7. D'Auria, B.: Stochastic Decomposition of the $M/G/\infty$ Queue in a Random Environment. Oper. Res. Lett. 35, 805–812 (2007)
8. Purdue, P., Linton, D.: An Infinite-Server Queue Subject to an Extraneous Phase Process and Related Models. J. Appl. Prob. 18, 236–244 (1981)
9. Linton, D., Purdue, P.: An $M/G/\infty$ Queue with m Customer Types Subject to Periodic Clearing. Opsearch 16, 80–88 (1979)
10. Nazarov, A.A., Terpugov, A.F.: Queueing Theory: The Study Guide. NTL, Tomsk (2004) (in Russian)
11. Nazarov, A.A., Moiseeva, S.P.: Method of Asymptotic Analysis in Queueing theory. NTL, Tomsk (2006) (in Russian)
12. Nazarov, A.A., Moiseev, A.N.: Analysis of an open non-Markovian $GI - (GI|\infty)^K$ queueing network with high-rate renewal arrival process. Probl. Inf. Transm. 49, 167–178 (2013)
13. Nazarov, A.A., Terpugov, A.F.: Theory of Probability and Stochastic Processes: The Study Guide. NTL, Tomsk (2010) (in Russian)
14. Baymeeva, G.V.: The Study of $M/M/\infty$ Queue in Random Environment. In: 18th All-Russian Conference "Young People's Scientific Creativity. Mathematics. Informatics", pp. 3–5. Tomsk University Publishing House, Tomsk (2014) (in Russian)

The Accuracy of Gaussian Approximations of Probabilities Distribution of States of the Retrial Queueing System with Priority of New Customers*

Anatoly Nazarov and Yana Chernikova

National Research Tomsk State University
Lenina, 36, Tomsk, Russia
nazarov.tsu@gmail.com,
evgenevna.92@mail.ru
http://www.tsu.ru

Abstract. In this paper, we study retrial queueing system with priority of new customers. We consider the weighted sum of exponential distribution and gamma distribution as an example of the numerical realization. Distribution of probabilities of the number of customers in the orbit is obtain with the aid of numerical algorithm. Two Gaussian approximations are constructed using both the moments of obtained distribution and the asymptotic semi-invariants. The third order approximation is presented. This approximation is more accurate than Gaussian approximation based on the moments of distribution.

Keywords: retrial queuing system, throughput, orbit, priority customer, asymptotic semi-invariant.

1 Introduction

Retrial queueing systems are characterized by the feature that arrivals who find the server unavailable are obliged to leave the service area and to try again for their customers in random order and at random intervals. Between trials a customer is called to be in "orbit". This feature plays a special role in several computer and communications networks. Queues with repeated attempts have been widely used to model many problems in telephone switching systems, communication systems and local area network problems. For recent bibliographies on retrial queues, see [1], [2]. Artalejo also provided extensive surveys of retrial queues.

Priority mechanism is an invaluable scheduling method that allows customers to receive different quality of service. Service priority is clearly today a main feature of the operation of any manufacturing system. Several authors including

---

* This work is performed under the state order No. 1.511.2014/K of the Ministry of Education and Science of the Russian Federation.

A. Dudin et al. (Eds.): ITMM 2014, CCIS 487, pp. 325–333, 2014.

Cobham [3], Phipps [4], Schrage [5], Jaiswal [6], Madan [7], Simon [8], Takagi [9], Choi and Chang [10], [11] have studied priority queues.

In this paper we study retrial queuing system M/G/1 with priority as primary customers and customers in the orbit.

2   Mathematical Model

Let us consider retrial queueing system M/G/1 with priority of the new customers. Structure of the system is depicted in Figure 1. We assume that arrival

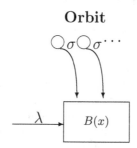

Fig. 1. Mathematical model

flow to the system is described by the stationary Poisson process with intensity λ. Customer, which finds the server be free, occupies it for service during a random time with distribution function $B(x)$. If the server is busy, then an arrived customer replaces the customer, which is in service, and occupies the server. The customer, which was in service, moves to a so customer's orbit where it performs a random delay with duration determined by exponential distribution with parameter σ. From the orbit, after the random delay, the customer occupies the device again. If the device is free then the customer occupies it for a random service time. If the server is busy then the customer from the orbit replaces the customer on service and occupies the server, while the customer which was on service goes to the orbit. After each service interruption, new service time also is characterized by distribution function $B(x)$. Let

$$B^*(\alpha) = \int\limits_0^\infty \exp(-\alpha x)\, dB(x)$$

be Laplace-Stieltjes transform of the distribution function $B(x)$. Let $i(t)$ be the number of customers in the orbit, $k(t)$ define S the server state in the following way:

$$k(t) = \begin{cases} 0, \text{ if server is free at moment } t, \\ 1, \text{ if server is busy at moment } t. \end{cases}$$

We would like to solve a problem of computation of stationary distribution of probabilities of the number of customers in the orbit and server state.

Since the process $\{k(t), i(t)\}$ is not Markovian then we analyze the process $\{k(t), i(t), z(t)\}$, where $z(t)$ is the residual service time of a customer in service, if any.

3   Kolmogorov's Equations

Let us denote by $P\{k(t) = 0, i(t) = i\} = P_0(i, t)$ a probability that, at the moment t, the server is in the state 0 and i customers stay in the orbit. Let $P\{k(t) = 1, i(t) = i, z(t) < z\} = P_1(i, z, t)$ is a probability that, at the moment t the server is in the state 1, residual service time is less than z and the number of customers in the orbit is equal to i.

Let us assume that the system is operating in a stationary mode, i.e. $P_0(i, t) \equiv P_0(i)$, $P_1(i, z, t) \equiv P_1(i, z)$.

Let us write the system of equations for stationary distribution:

$$\begin{cases} -\dfrac{\partial P_1(i, z)}{\partial z} + \dfrac{\partial P_1(i, 0)}{\partial z} = \lambda B(z) P_0(i) - (\lambda + i\sigma) P_1(i, z) + \\ \quad i\sigma B(z) P_1(i, \infty) + \lambda B(z) P_1(i - 1, \infty) + (i + 1)\sigma B(z) P_0(i + 1), \qquad (1) \\ \dfrac{\partial P_1(i, 0)}{\partial z} = (\lambda + i\sigma) P_0(i), \ i \geq 0, z > 0. \end{cases}$$

where we mean that

$$\frac{\partial P_1(i, 0, t)}{\partial z} = \left. \frac{\partial P_1(i, z, t)}{\partial z} \right|_{z=0}.$$

4   Numerical Realization

We got numerical algorithm, following which you can obtain a distribution of probability $P(i)$, where $P(i) = P_0(i) + P_1(i)$. This algorithm consists of the following steps:

1. Temporarily assume that $P_1(0) = 1$;
2. Select sufficiently large integer number N, and $P_1(i)$, $i = \overline{1, N}$ compute probabilities by

$$P_1(i) = \frac{[\lambda + i\sigma - \lambda B^*(\lambda + i\sigma)] \frac{\lambda}{i\sigma} - \lambda B^*(\lambda + i\sigma)}{(\lambda + i\sigma) B^*(\lambda + i\sigma)} \cdot P_1(i - 1), i \geq 1;$$

3. Using formulas

$$P_0(0) = \frac{\sigma}{\lambda} \cdot \frac{B^*(\lambda)}{1 - B^*(\lambda)} P_0(1)$$

and

$$P_0(i+1) = \frac{\lambda}{(i+1)\sigma} \cdot P_1(i), i \ge 0,$$

find the value of $P_0(0)$;

4. Using relation

$$P_0(i+1) = \frac{\lambda}{(i+1)\sigma} \cdot P_1(i), i \ge 0,$$

compute the values of $P_0(i)$ for all $i = \overline{1, N}$;

5. Compute the sum $d = \sum_{i=0}^{N} (P_0(i) + P_1(i))$;

6. Assume that probabilities $P_0(i)$, $P_1(i)$, $i \ge 0$ are equal to $\frac{1}{d} P_k(i)$, $i \ge 0$, $k = 0, 1$;

7. Stop algorithm if the computed value of $P_0(N) + P_1(N)$ is sufficiently small, for instance is equal to computer zero. Otherwise, increase value N and return to Step 2 of algorithm.

In particular, we analyze the weighted sum of gamma distribution and exponential distribution

$$B^*(u) = q \left(1 + \frac{u}{\beta}\right)^{-\alpha} + (1 - q) \left(1 + \frac{u}{\gamma}\right)^{-1}, \tag{2}$$

where α, β, γ and $0 \le q \le 1$ are positive parameters.

Note that density function $B'(u)$ of distribution having Laplace-Stieltjes transform (2) is given by

$$B'(u) = q \frac{\beta^\alpha}{\Gamma(\alpha)} u^{\alpha-1} e^{-\beta u} + (1 - q)\gamma e^{-\gamma u}. \tag{3}$$

For various parameter values in the range that $\alpha > 1, q < 1$, we compute distributions $P_k(i), k = 0, 1$ and $P(i), i \ge 0$ by means of the algorithm described above. Using distribution $P(i), i \ge 0$, we compute the first three moments of this distribution as:

$$a_1 = \sum_{i=0}^{N} i P(i), \ a_2 = \sum_{i=0}^{N} (i - a_1)^2 P(i), \ a_3 = \sum_{i=0}^{N} (i - a_1)^3 P(i), \tag{4}$$

E. g., if the parameters of distribution (3) are fixed as follow is:

$$q = 0.5; \ \alpha = 2; \ \beta = 5; \ \gamma = 5. \tag{5}$$

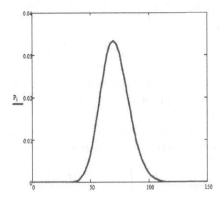

Fig. 2. Graph of distribution $P(i)$ for $\sigma = 0.02$

Graph of distribution $P(i)$, $i \geq 0$ computed by means of the numerical algorithm for parameters of servicing time distribution, which is given by (5), $\lambda = 1.5$
for $\sigma = 0.02$ is given in figure 2.

Looking at the graphs of distribution $P(i)$ it seems appropriate to compare
them to density function of Gaussian distribution, which is defined by the same
moments of a_1 and a_2 as distribution $P(i)$, $i \geq 0$.

Let us denote normal distribution function with moments a_1 and a_2 by $F(x)$,
$P_2(i)$ be discrete distribution of nonnegative quantity which is defined by

$$P_2(i) = [F(i+1) - F(i)](1 - F(0))^{-1}, i \geq 0. \tag{6}$$

We will call distribution $P_2(i)$, $i \geq 0$, as Gaussian's approximation or second-
order approximation of distribution $P(i)$, which is was computed by means of
the developed above numerical algorithm.

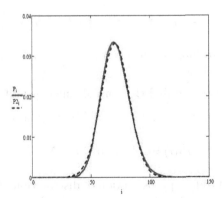

Fig. 3. Graphs of distribution $P(i)$ and Gaussian distribution for $\sigma = 0.02$

To estimate the accuracy of this approximation we will take the following value

$$\Delta_2 = \max_{0 \leq i \leq N} |\sum_{n=0}^{i} (P(n) - P_2(n))|, \tag{7}$$

which is called as Kolmogorov's distance between distributions.

As shown in Table 1 this approximation is quite accurate for $\sigma \leq 0.2$ since $\Delta_2 < 0.05$.

Now, let us try to construct even more accurate approximation.

Note that characteristic function $h_2(u)$ of normal distribution is given by

$$h_2(u) = \exp \left\{ jua_1 + \frac{(ju)^2}{2} a_2 \right\} \tag{8}$$

and it accounts the average value a_1 and variance a_2. But it does not account quite meaningful asymmetry of the graphs of distribution $P(i)$, $i \geq 0$. The coefficient of asymmetry of any distribution accounts the third moment, a_3, of distribution. So let us consider the function

$$h_3(u) = \exp \left\{ jua_1 + \frac{(ju)^2}{2} a_2 + \frac{(ju)^3}{6} a_3 \right\}, \tag{9}$$

which is not a characteristic function, but allows to construct the following discrete distribution.

The inverse Fourier transformation for function given by (9) by is defined

$$h_4(i) = \frac{1}{2\pi} \int_{-\infty}^{\infty} e^{-jui} h_3(u) du.$$

Next in $h_4(i)$ we reduce imaginary components and real parts in the following manner

$$h_3(i) = \frac{1}{4} \left\{ h_4(i) + \overline{h_4(i)} + | h_4(i) + \overline{h_4(i)} | \right\}.$$

where $\overline{h_4(i)}$ is conjugate number to $h_4(i)$ and $| h_4(i) + \overline{h_4(i)} |$ stands for the module of complex number.

We fix some integer N and find the normalizing constant $d = \sum_{i=0}^{n} h_3(i)$.

Let us write the following discrete probability distribution

$$P_3(i) = \frac{1}{d} h_3(i), 0 \leq i \leq N,$$

which we call as third-order approximation of distribution $P(i)$, $i \geq 0$.

Analogously to (7), to estimate the accuracy of this new approximation we find the value of Δ_3, substituting instead of $P_3(n)$ in (7).

Values of Δ_2 and Δ_3 with various values λ of intensity for some values of intensity σ are listed in the Table 1.

Table 1. Numerical result for values Δ_2 and Δ_3

σ		0.5	0.1	0.05	0.02	0.01	0.005
$\lambda = 1$	Δ_2	0.1773	0.0796	0.0436	0.0226	0.0160	0.0113
	Δ_3	0.0264	0.0173	0.0174	0.0042	0.0020	0.0010
$\lambda = 1.5$	Δ_2	0.1026	0.0441	0.0278	0.0175	0.0124	0.0088
	Δ_3	0.0831	0.0209	0.0068	0.0027	0.0013	0.0006
$\lambda = 2$	Δ_2	0.1291	0.0363	0.0256	0.0163	0.0115	0.0082
	Δ_3	0.0627	0.0123	0.0063	0.0024	0.0012	0.0006
$\lambda = 2.25$	Δ_2	0.1236	0.0385	0.0274	0.0337	0.0163	0.0060
	Δ_3	0.0501	0.0143	0.0074	0.0226	0.0142	0.0050

As is seen from table 1 it is divided into three parts:

In the first, right one, approximation it is not recommended to apply distributions $P(i)$ by the distributions of the second and third order, because Δ_2 and Δ_3 exceed permissible error of 0.03.

In the second part approximation by distributions by the distributions of the third order is accepted, but approximation by Gaussian distribution is not recommended.

In the right low part both approximation are accepted, and also approximation of the distribution by distribution of the third order is more correct than approximation by Gaussian distribution, because Δ_3 is 8-15 less than Δ_2.

To distribution (3) let's find three moments of asymptotic semi-invariants κ_1, κ_2 and κ_3:

$$a_1^{as} = \frac{\kappa_1}{\sigma}, \ a_2^{as} = \frac{\kappa_2}{\sigma}, \ a_3^{as} = \frac{\kappa_3}{\sigma}.$$

Now let's compare the graph of distribution $P(i)$ with Gaussian distribution that is defined by the moments a_1^{as} and a_2^{as}.

Let us denote normal distribution function with moments a_1^{as} and a_2^{as} by $F^{as}(x)$, $P^{as}(i)$ be discrete distribution of non-negative quantity which is defined by

$$P_2^{as}(i) = [F^{as}(i+1) - F^{as}(i)](1 - F^{as}(0))^{-1}, i \geq 1. \tag{10}$$

We will call distribution $P_2^{as}(i)$, $i \geq 1$, as Gaussian's approximation or second-order approximation of distribution $P(i)$, which is was computed by asymptotic semi-invariants.

To estimate the accuracy of this approximation we will take the following value

$$\Delta_2^{as} = \max_{0 \leq i \leq N} | \sum_{n=0}^{i} (P(n) - P_2^{as}(n)) |.$$

Note that characteristic function $h_2^{as}(u)$ of normal distribution is given by

$$h_2^{as}(u) = \exp \left\{ j u a_1^{as} + \frac{(ju)^2}{2} a_2^{as} \right\}.$$

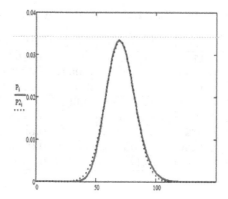

Fig. 4. Graphs of distribution $P(i)$ and Gaussian distribution with moments a_1^{as} and a_2^{as} for $\sigma = 0.02$

So let us consider the function

$$h_3^{as}(u) = \exp\left\{jua_1^{as} + \frac{(ju)^2}{2}a_2^{as} + \frac{(ju)^3}{6}a_3^{as}\right\},$$

which is not a characteristic function, but allows to construct the following discrete distribution

$$P_3^{as}(i) = \frac{1}{d}h_3^{as}(i), 0 \le i \le N.$$

We will call distribution $P_3^{as}(i)$, $i \ge 0$, as third-order approximation of distribution $P(i)$, which is was computed by asymptotic semi-invariants.

Analogously to (7), to estimate the accuracy of this new approximation we find the value of Δ_3^{as}. Values of Δ_2^{as} and Δ_3^{as} with various values λ of intensity for some values of intensity σ are listed in the Table 2.

Table 2. Numerical result for values Δ_2^{as} and Δ_3^{as}

σ		0.5	0.1	0.05	0.02	0.01	0.005
$\lambda = 1$	Δ_2^{as}	0.0947	0.0639	0.0277	0.0139	0.0096	0.0067
	Δ_3^{as}	0.0954	0.0175	0.0287	0.0120	0.0081	0.0056
$\lambda = 1.5$	Δ_2^{as}	0.0599	0.0252	0.0212	0.0133	0.0094	0.0066
	Δ_3^{as}	0.1049	0.0485	0.0288	0.0182	0.0128	0.0091
$\lambda = 2$	Δ_2^{as}	0.0693	0.0411	0.0291	0.0182	0.0128	0.0090
	Δ_3^{as}	0.0703	0.0638	0.0459	0.0293	0.0207	0.0147
$\lambda = 2.25$	Δ_2^{as}	0.0731	0.0571	0.0401	0.0251	0.0273	0.0112
	Δ_3^{as}	0.1353	0.0860	0.0623	0.0399	0.0321	0.0176

Analysing the second table it is possible to conclude that it is divided into 3 parts:

In the left part, the values Δ_2^{as} and Δ_3^{as} are greater than 0.03 that is an intolerable error. That means that it is highly impossible to use the approximations of the second and third orders.

In the second part it is not recommended to use the approximations of the third order. The approximations of the second order is quite possible.

In the right upper part, the values of the quantities Δ_2^{as} are not greater that 0.03. That means that it is possible to use the asymptotic of the second and third order, where asymptotic of the second order is better that the one of the third order.

5 Conclusion

In this paper we considered the weighted sum of gamma distribution and exponential distribution. Distribution of probabilities of the number of customers in the orbit was obtained with the aid of numerical algorithm. Approximation with the aim of Gaussian distribution with the moments that were found by this distribution and with the moments, that were found by asymptotic semi-invariants was carried on. The third order approximation was described. It was stated that approximation of the third order is more accurate than Gaussian approximation based on the moments of distribution $P(i)$. And also it was established that approximation of the third order are less appropriate that approximation of Gaussian distribution with the moments that were found with the aim of the asymptotic semi-invariants.

References

1. Falin, G.I.: A survey of retrial queues. Queueing Systems 7(2), 127–167 (1990)
2. Yang, T.J., Templeton, G.C.: A survey on retrial queues. Queueing Systems 2(3), 201–233 (1987)
3. Cobham, A.: Priority Assignments in Waiting Line Problems. Operations Research 2(1), 70–76 (1954)
4. Phipps, T.E.: Machine Repair as a Priority Waiting Line Problem. Operations Research 4(1), 76–85 (1956)
5. Schrage, L.E.: The Queue M/G/1 with Feedback to Lower Priority Queues. Management Science 13(7), 466–474 (1967)
6. Jaiswal, N.: Priority Queues. Academic Press, New York (1968)
7. Madan, C.: A Priority Queueing System with Service Interruptions. Statistica Neerlandica 27(3), 115–123 (1973)
8. Simon, B.: Priorty Queues with Feedback. Journal of the Association for Computing Machinery 31(1), 134–149 (1984)
9. Takagi, H.: Queueing Analysis: A Foundation of Performance Analysis. Vacation and Priority Systems, part 1, vol. 1. Elsevier Science Publishers B.V., Amsterdam (1991)
10. Choi, B.D., Chang, Y.: Single Server Retrial Queues with Priority Calls. Mathematical and Computer Modeling 30(3-4), 7–32 (1999)
11. Choi, B.D., Choi, K.B., Lee, Y.W.: M/G/1 retrial queueing systems with two types of calls and finite capacity. Queueing Systems 19(1-2), 215–229 (1995)

Asymptotic Analysis of Closed Markov Retrial Queuing System with Collision

Anatoly Nazarov[1], Anna Kvach[1], and Vladimir Yampolsky[2]

[1] National Research Tomsk State University
[2] National Research Tomsk Polytechnic University
Tomsk, Russia
nazarov.tsu@gmail.com,
kvach_as@mail.ru,
yampolsky@incom.tomsk.ru

Abstract. We consider a closed retrial queuing system M/M/1//N with collision of the customers. We assume that sources can be in two states: generating a primary customers and waiting for the end of successful service. Source which sends the customer for service, moves into the waiting state and stays in this state till the end of the service of this customer. This system is solved using the asymptotic method under conditions of infinitely increasing number of sources. We establish formulas for computing the prelimit distribution of the number of sources in "waiting" state. Also, we determine the range of applicability of the asymptotic results in preliminiting situation.

Keywords: closed queueing system, retrial queue, collision, asymptotic analysis.

1 Introduction

Retrial queue is a queuing system [1–3] characterized by the following feature: customers, who find server busy goes to the orbit and after random time repeat their demand. It is assumed that the orbit is infinitely large and every call retry its attempts until it is satisfied. The field of practical application of such system is very extensive. RQ-system can be applied for researching telecommunication and computer system, for engineering cellular mobile networks, computer networks, ets. For a detailed overviews of main results about retrial queues, we refer the reader to the excellent book of Falin and Templeton [4]. For an extensive bibliography, see [5]. As regards the closed retrial queuing with finite number of source, it is elaborately discussed by Almási B. et al [6, 7], Artalejo J.R. [8], and Dragieva V.I. [9, 10]. In this paper, we consider the M/M/1//N retrial queue with collision. In the papers Nazarov A.A., Lyubina T.V. are considered the various open retrial queuing systems with collision [11, 12].

2 Model Description

We consider a closed retrial queuing system of type M/M/1//N in Kendals notation with collision of the customers. This mean that the system has one

A. Dudin et al. (Eds.): ITMM 2014, CCIS 487, pp. 334–341, 2014.

server and N sources. Each one of them generated a primary customers according to a Poisson flow with rate λ/N. We assume that sources can be in two states: generating a primary customers and waiting for the end of successful service. Source which send the customer for service, moves into the "waiting" state and stays in this state till the end of the service of this customer. If a primary customer finds server idle, he enters into service immediately, during service time, which distributed exponentially with parameter μ. Otherwise, if server is busy, arriving customer involves into collision with servicing customer and they both moves into the orbit. Retrial customer repeat his demand for service with an exponential distribution with rate σ/N. We assume that primary customers, retrial customers and service time are mutually independent.

At time t let $i(t)$ be the number of sources locating in "waiting" state and $k(t)$ determines the server state

$$k(t) = \begin{cases} 0, & \text{if the server is free,} \\ 1, & \text{if the server is busy.} \end{cases}$$

Let us denote by $P\{k(t) = k, i(t) = i\} = P_k(i,t)$ the joint probability that at the time t there are i sources in "waiting" state and the server is in the "k" state. Under the above assumption the process $\{k(t), i(t)\}$ is a 2-dimentional Markov process with state space $\{0, 1, \ldots, N\} \times \{0, 1\}$

The differential Kolmogorov equations for probabilities $P_k(i,t)$ are

$$\frac{\partial P_0(0,t)}{\partial t} = -\lambda P_0(0,t) + \mu P_1(1,t) ,$$

$$\frac{\partial P_1(1,t)}{\partial t} = -\left(\lambda \frac{N-1}{N} + \mu\right) P_1(1,t) + \lambda P_0(0,t) + \frac{\sigma}{N} P_0(1,t) ,$$

$$\frac{\partial P_0(i,t)}{\partial t} = -\left(\lambda \frac{N-i}{N} + \sigma \frac{i}{N}\right) P_0(i,t) + \mu P_1(i+1,t) +$$
$$+ \lambda \frac{N-i+1}{N} P_1(i-1,t) + \sigma \frac{i-1}{N} P_1(i,t) ,$$

$$\frac{\partial P_1(i,t)}{\partial t} = -\left(\lambda \frac{N-i}{N} + \sigma \frac{i-1}{N} + \mu\right) P_1(i,t) +$$
$$+ \lambda \frac{N-i+1}{N} P_0(i-1,t) + \sigma \frac{i}{N} P_0(i,t) .$$

Note this system in steady state

$$-\lambda P_0(0) + \mu P_1(1) = 0 ,$$

$$-\left(\lambda \frac{N-1}{N} + \mu\right) P_1(1) + \lambda P_0(0) + \frac{\sigma}{N} P_0(1) = 0 ,$$

$$-\left(\lambda \frac{N-i}{N} + \sigma \frac{i}{N}\right) P_0(i) + \mu P_1(i+1) + \lambda \frac{N-i+1}{N} P_1(i-1) +$$
$$+ \sigma \frac{i-1}{N} P_1(i) = 0 , \qquad (1)$$

$$-\left(\lambda \frac{N-i}{N} + \sigma \frac{i-1}{N} + \mu\right) P_1(i) + \lambda \frac{N-i+1}{N} P_0(i-1) +$$
$$+ \sigma \frac{i}{N} P_0(i) = 0 .$$

The partial characteristic functions are denoted by

$$H_k(u) = \sum_{i=0}^{N} e^{jui} P_k(i) .$$

Then system (1) corresponds as

$$\frac{j}{N}(\sigma - \lambda)\frac{dH_0(u)}{du} + \frac{j}{N}\left(\lambda e^{ju} - \sigma\right)\frac{dH_1(u)}{du} - \lambda H_0(u)+$$
$$+ \left(\lambda e^{ju} + \mu e^{-ju} - \frac{\sigma}{N}\right) H_1(u) = 0 ,$$
$$\frac{j}{N}\left(\lambda e^{ju} - \sigma\right)\frac{dH_0(u)}{du} + \frac{j}{N}(\sigma - \lambda)\frac{dH_1(u)}{du} + \lambda e^{ju} H_0(u)+$$
$$+ \left(\frac{\sigma}{N} - \lambda - \mu\right) H_1(u) = 0 . \tag{2}$$

In order to solve this system, we use method of asymptotic analysis [13] under conditions of infinitely increasing number of sources $(N \to \infty)$.

3   Asymptotic of the First Order

Let us denote $\dfrac{1}{N} = \varepsilon$.

Introducing following substitute

$$u = \varepsilon w, \qquad H_k(u) = F_k(w, \varepsilon) ,$$

we can transform system (2) to the form:

$$j(\sigma - \lambda)\frac{\partial F_0(w,\varepsilon)}{\partial w} + j\left(\lambda e^{j\varepsilon w} - \sigma\right)\frac{\partial F_1(w,\varepsilon)}{\partial w} - \lambda F_0(w,\varepsilon)+$$
$$+ \left(\lambda e^{j\varepsilon w} + \mu e^{-j\varepsilon w} - \varepsilon\sigma\right) F_1(w,\varepsilon) = 0,$$
$$j\left(\lambda e^{j\varepsilon w} - \sigma\right)\frac{\partial F_0(w,\varepsilon)}{\partial w} + j(\sigma - \lambda)\frac{\partial F_1(w,\varepsilon)}{\partial w} + \lambda e^{j\varepsilon w} F_0(w,\varepsilon)+$$
$$+ (\varepsilon\sigma - \lambda - \mu) F_1(w,\varepsilon) = 0. \tag{3}$$

Theorem 1. *The limiting value* $F_0(w), F_1(w)$ *of function* $F_0(w,\varepsilon), F_1(w,\varepsilon)$ *(the solutions of the system (3)), are given by the formulas*

$$F_0(w) = R_0 e^{jw\kappa_1}, \quad F_1(w) = R_1 e^{jw\kappa_1},$$

where R_k *the stationary distributions of probabilities of the service state are defined as follows*

$$R_1 = \frac{\sigma(2\lambda + \mu) - \sqrt{\sigma^2(2\lambda - \mu)^2 + 8\sigma\mu\lambda^2}}{4\mu(\sigma - \lambda)},$$

$$R_0 = 1 - \frac{\sigma(2\lambda + \mu) - \sqrt{\sigma^2(2\lambda - \mu)^2 + 8\sigma\mu\lambda^2}}{4\mu(\sigma - \lambda)}, \tag{4}$$

and κ_1 is

$$\kappa_1 = \frac{2\mu R_1^2}{\sigma(1 - 2R_1)}.$$

4   Asymptotic of the Second Order

To find the asymptotic of the second order we must execute following substitute at system (2):

$$H_k(u) = H_k^{(2)}(u) \exp\{ju\kappa_1 N\}.$$

By putting $\dfrac{1}{N} = \varepsilon^2$, $u = \varepsilon w$, $H_k^{(2)}(u) = F_k^{(2)}(w, \varepsilon)$, we get

$$j\varepsilon (\sigma - \lambda) \frac{\partial F_0^{(2)}(w, \varepsilon)}{\partial w} + j\varepsilon \left(\lambda e^{j\varepsilon w} - \sigma\right) \frac{\partial F_1^{(2)}(w, \varepsilon)}{\partial w} -$$

$$- \left[\lambda + (\sigma - \lambda)\kappa_1\right] F_0^{(2)}(w, \varepsilon) +$$

$$+ \left[\lambda e^{j\varepsilon w}(1 - \kappa_1) + \mu e^{-j\varepsilon w} + \sigma \kappa_1 - \varepsilon^2 \sigma\right] F_1^{(2)}(w, \varepsilon) = 0,$$

$$j\varepsilon \left(\lambda e^{j\varepsilon w} - \sigma\right) \frac{\partial F_0^{(2)}(w, \varepsilon)}{\partial w} + j\varepsilon (\sigma - \lambda) \frac{\partial F_1^{(2)}(w, \varepsilon)}{\partial w}$$

$$+ \left[\lambda e^{j\varepsilon w}(1 - \kappa_1) + \sigma \kappa_1\right] F_0^{(2)}(w, \varepsilon) -$$

$$- \left[\lambda(1 - \kappa_1) + \mu + \sigma \kappa_1 - \varepsilon^2 \sigma\right] F_1^{(2)}(w, \varepsilon) = 0.$$

$$(5)$$

Theorem 2. *The limiting value $F_0^{(2)}(w)$, $F_1^{(2)}(w)$ of function $F_0^{(2)}(w, \varepsilon)$, $F_1^{(2)}(w, \varepsilon)$ (the solutions of the system (5)), are given by the formulas*

$$F_k^{(2)}(w) = R_k \Phi^{(2)}(w),$$

where

$$\Phi^{(2)}(w) = \exp\left\{\frac{(jw)^2}{2} \kappa_2\right\},$$

$$\kappa_2 = \mu R_1 \cdot \frac{1 + (R_1 - R_0) R_0}{\lambda - (\lambda - \sigma)(R_1 - R_0)^2}.$$

5   Asymptotic of the Third Order

To find the asymptotic of the third order we must execute following substitute at system (2):

$$H_k(u) = H_k^{(3)}(u) \exp\left\{ju\kappa_1 N + \frac{(ju)^2}{2} \kappa_2 N\right\}.$$

$$(6)$$

At system (2) make substitutions

$$\frac{1}{N} = \varepsilon^3, \quad u = \varepsilon w, \quad H_k^{(3)}(u) = F_k^{(3)}(w, \varepsilon), \tag{7}$$

and we obtain

$$
\begin{aligned}
& j\varepsilon^2 (\sigma - \lambda) \frac{\partial F_0^{(3)}(w, \varepsilon)}{\partial w} + j\varepsilon^2 \left(\lambda e^{j\varepsilon w} - \sigma\right) \frac{\partial F_1^{(3)}(w, \varepsilon)}{\partial w} + \\
& + \left[(\lambda - \sigma)(\kappa_1 + j\varepsilon w \kappa_2) - \lambda\right] F_0^{(3)}(w, \varepsilon) + \\
& + \left[(\sigma - \lambda e^{j\varepsilon w})(\kappa_1 + j\varepsilon w \kappa_2) + \lambda e^{j\varepsilon w} + \mu e^{-j\varepsilon w} - \varepsilon^3 \sigma\right] F_1^{(3)}(w, \varepsilon) = 0, \\
& j\varepsilon^2 \left(\lambda e^{j\varepsilon w} - \sigma\right) \frac{\partial F_0^{(3)}(w, \varepsilon)}{\partial w} + j\varepsilon^2 (\sigma - \lambda) \frac{\partial F_1^{(3)}(w, \varepsilon)}{\partial w} + \\
& + \left[(\sigma - \lambda e^{j\varepsilon w})(\kappa_1 + j\varepsilon w \kappa_2) + \lambda e^{j\varepsilon w}\right] F_0^{(3)}(w, \varepsilon) + \\
& + \left[(\lambda - \sigma)(\kappa_1 + j\varepsilon w \kappa_2) - \lambda - \mu + \varepsilon^3 \sigma\right] F_1^{(3)}(w, \varepsilon) = 0.
\end{aligned}
\tag{8}
$$

Theorem 3. *The limiting value $F_0^{(3)}(w)$, $F_1^{(3)}(w)$ of function $F_0^{(3)}(w, \varepsilon)$, $F_1^{(3)}(w, \varepsilon)$ (the solutions of the system (8)), are given by the formulas*

$$F_k^{(3)}(w) = R_k \Phi^{(3)}(w),$$

where

$$\Phi^{(3)}(w) = \exp\left\{\frac{(jw)^3}{3!} \kappa_3\right\},$$

$$
\kappa_3 = 2 \cdot \left\{ \frac{\left[(\lambda - \sigma)(R_0 - R_1)^2 \kappa_2 + \mu a\right] \cdot \left[\frac{1}{2} - ((R_0 - R_1)(\sigma - \lambda) + \lambda) \frac{\kappa_2}{\mu R_1}\right]}{\lambda + (\sigma - \lambda)(R_0 - R_1)^2} \right. +
$$

$$
+ \left. \frac{\lambda \kappa_2 \cdot \left[\frac{a}{R_0} + \frac{1}{2}\right] + \mu \left[\frac{1}{R_0} - \frac{a}{2}\right]}{\lambda + (\sigma - \lambda)(R_0 - R_1)^2} \right\},
$$

$$a = R_0 R_1 (R_1 - R_0).$$

We can find the characteristic function $h(u)$ of the number of sources is in "waiting" state. Using a substitution, reversing to the (7), and, considering (6), we get

$$h(u) = \left(H_0^{(3)}(u) + H_1^{(3)}(u)\right) \exp\left\{ j u \kappa_1 N + \frac{(ju)^2}{2} \kappa_2 N \right\},$$

where

$$H_0^{(3)}(u) = R_0 \exp\left\{\frac{(ju)^3}{3!}\kappa_3 N\right\}, \quad H_1^{(3)}(u) = R_1 \exp\left\{\frac{(ju)^3}{3!}\kappa_3 N\right\}.$$

Thus

$$h(u) = \exp\left\{ju\kappa_1 N + \frac{(ju)^2}{2}\kappa_2 N + \frac{(ju)^3}{3!}\kappa_3 N\right\}.$$

6   Computing Procedure

Consider now system M/M/1//N in prelimiting situation.

Theorem 4. *The joint distribution of the service and source state can be computed from the following steps:*

1. *Choose the model parameters* λ, μ, σ, N.
2. *Put* $P_1(0) = 0$.
3. *For* $i = 0$ *compute quantity* $\dfrac{P_1(1)}{P_0(0)}$ *from (1).*
4. *For* $1 \le i \le N-1$ *compute recursively* $\dfrac{P_0(i)}{P_0(0)}$ *and* $\dfrac{P_1(i)}{P_0(0)}$ *from the following*
   formulas

$$\frac{P_0(i)}{P_0(0)} = \frac{N}{i\sigma}\left\{\left(\lambda\frac{N-i}{N} + \sigma\frac{i-1}{N} + \mu\right)\frac{P_1(i)}{P_0(0)} - \lambda\left(\frac{N-i+1}{N}\right)\frac{P_0(i-1)}{P_0(0)}\right\},$$

$$\frac{P_1(i+1)}{P_0(0)} = \frac{1}{\mu}\left\{\lambda\frac{N-i}{N} + \sigma\frac{i}{N}\right\}\frac{P_0(i)}{P_0(0)} - \lambda\frac{(N-i+1)}{N}\frac{P_1(i-1)}{P_0(0)} - \sigma\frac{(i-1)}{N}\frac{P_1(i)}{P_0(0)},$$

   which are obtain from (1).
5. *For* $i = N$ *compute quantity* $\dfrac{P_0(N)}{P_0(0)}$

$$\frac{P_0(N)}{P_0(0)} = \frac{1}{\sigma}\left\{\left(\sigma\frac{N-1}{N} + \mu\right)\frac{P_1(N)}{P_0(0)} - \frac{\lambda}{N}\cdot\frac{P_0(N-1)}{P_0(0)}\right\}.$$

6. *The quantity* $P_0(0)$ *may be found with the help of the normalizing conditions*

$$P_0(0) = 1\bigg/\sum_{i=0}^{N}\left(\frac{P_0(i)}{P_0(0)} + \frac{P_1(i)}{P_0(0)}\right).$$

7. *Compute* $P_k(i)$ *from*

$$P_k(i) = \frac{P_k(i)}{P_0(0)}\cdot P_0(0).$$

   Now we can get one-dimensional distribution of the number of sources in
   "waiting" state

$$P(i) = P_0(i) + P_1(i).$$

7   Numerical Results

We assume that prelimit distribution $P(i)$ can be approximated by asymptotic distribution $P_\nu(i)$ of the ν-th order ($\nu = 2, 3$) in some domains of the system parameters. In order to compare distributions, we use the Kolmogorov distance:

$$\Delta_\nu = \max_{0 \le k \le N} \left| \sum_{i=0}^{k} P_\nu(i) - \sum_{i=0}^{k} P(i) \right|.$$

In numerical computation the model parameters λ, μ, σ are fixed, $\lambda = 5$, $\mu = 10$, $\sigma = 20$ and the number of sources N takes values $5, 10, 18, 25, 100$.
The results are reported in Table 1.

Table 1. Kolmogorov distance between prelimit distribution $P(i)$ and asymptotic distribution $P_\nu(i)$ of the ν-th order ($\nu = 2, 3$)

	$\lambda = 5, \ \mu = 10, \ \sigma = 20$				
N	5	10	18	25	100
Δ_2	0,184	0,091	0,039	0,028	0,012
Δ_3	0,165	0,080	0,029	0,019	0,008

Table 1 show that asymptotic approximation of the second order with a good degree of accuracy approximates prelimit distribution $P(i)$ for $N \ge 25$. At the same time, the Kolmogorov distance between distributions $P(i)$ and $P_3(i)$ becomes less then $0, 03$ for $N \ge 18$. Let us note, that accuracy of all approximations generally improves as N increases.

Our numerical experiment show, that accuracy increase with the growth of the order of approximation and determine the range of applicability of the asymptotic results in prelimiting situation.

8   Conclusion

In this paper, we research a closed retrial queuing system M/M/1//N with collision of the customers. Using the method of asymptotic analysis under conditions of infinitely increasing number of sources, we obtain a distribution of the number of sources in "waiting" state. Also, we obtain the probability distribution with help of numerical algorithm in prelimiting situation. Comparing there distributions we can conclude, that prelimit distribution can be approximated by asymptotic distribution for $N \ge 25$ (the second order approximation) and for $N \ge 18$ (the third order approximation). Therefore, those approximations have a high level of accuracy and can be used on practical engineering application where the relative error does not represent a severe constrain.

References

1. Nazarov, A.A., Terpugov, A.F.: The queuing theory. "NTL" Publishing House, Tomsk (2004) (in Russian)
2. Gnedenko, B.V., Kovalenko, I.N.: Introduction to queuing theory. "KomKniga" Publishing House, Moscow (2007) (in Russian)
3. Koening, D., Shtoyan, D.: Methods of the queuing theory. "Radio and Communications" Publishing House, Moscow (1981) (in Russian)
4. Falin, G.I., Templeton, J.G.C.: Retrial queues, p. 328. Chapman & Hall, London (1997)
5. Artalejo, J.R., Gomez-Corral, A.: Retrial Queueing Systems: A Computational Approach, p. 309. Springer (2008)
6. Almási, B., Roszik, J., Sztrik, J.: Homogeneous Finite-Source Retrial Queues with Server Subject to Breakdowns and Repairs. Mathematical and Computer Modeling 42, 673–682 (2005)
7. Sztrik, J., Almási, B., Roszik, J.: Heterogeneous finite-source retrial queues with server subject to breakdowns and repairs. Journal of Mathematical Sciences 132, 677–685 (2006)
8. Artalejo, J.R.: Retrial queues with a finite number of sources. J. Korean Math. Soc. 35, 503–525 (1998)
9. Dragieva, V.I.: Single-line queue with finite source and repeated calls. Problems of Information Transmission 30, 283–289 (1994)
10. Dragieva, V.I.: System State Distributions In One Finite Source Unreliable Retrial Queue, http://elib.bsu.by/handle/123456789/35903
11. Lyubina, T.V., Nazarov, A.A.: Research of the Markov dynamic retrial queue system with collision. Herald of Tomsk State University. Journal of Control and Computer Science 3(12), 73–84 (2010) (in Russian)
12. Lyubina, T.V., Nazarov, A.A.: Research of the non-Markov dynamic retrial queue system with collision. Herald of Kemerovo State University 1(49), 38–44 (2012) (in Russian)
13. Nazarov, A.A., Moiseeva, S.P.: Methods of asymptotic analysis in a queuing theory. "NTL" Publishing House, Tomsk (2006) (in Russian)

Optimal State Estimation in Modulated MAP Event Flows with Unextendable Dead Time

Luydmila Nezhelskaya

National Research Tomsk State University, Tomsk, Russia
ludne@mail.ru

Abstract. We consider the optimal estimation problem for the states of a modulated MAP event flow with two states; it is one of the mathematical models for an incoming stream of claims (events) in digital integral servicing networks. The observation conditions for this flow are such that each event generates a period of dead time during which other events from the flow are inaccessible for observation and do not extend the dead time period (unextendable dead time). We find an explicit form for posterior probabilities of the flow states. The decision about the flow state is made with the maximal a posteriori criterion.

Keywords: modulated, MAP event flows, unextendable dead time, optimal state estimation.

1   Introduction

Mathematical models of queueing theory are widely used to describe real physical, technical, and other systems and processes. Thanks to the fast development of computer hardware and information technologies, another important field of queueing theory applications has arisen, namely the design and creation of informational and computational networks, computer communication networks, satellite networks, telecommunication networks, etc.

In practice, parameters that determine the incoming flow of events change in time, and the changes are often random; the latter has led researchers to consider doubly stochastic flows of events. One of the first works in this direction was probably the paper [1] in which a doubly stochastic flow is defined as a flow whose intensity is a random process. Doubly stochastic flows can be divided into two classes: flows whose intensity is a continuous random process and flows whose intensity is a piecewise constant random process with a finite number of states. We emphasize that flows of the second class were introduced virtually at the same time in 1979, in [2]-[4]. In [2],[3], these flows were called MC (Markov Chain) flows; in [4], MVP (Markov Versatile Processes) flows. Starting from the end of the 1980s, the latter, especially after [5], have usually been called MAP (Markovian Arrival Process) event flows. We note that MAP-flows of events are especially characteristic for real telecommunication networks [6].

In the studies of event flows, we can distinguish two classes of problems: (1) estimating the states of an event flow; (2) estimating flow parameters.

A. Dudin et al. (Eds.): ITMM 2014, CCIS 487, pp. 342–350, 2014.

One of the distorting factors in our estimates of event flow states and parameters is the dead time of sensing devices [7] which results from a detected event. Other events that occur during a dead time period are inaccessible for observation (simply speaking, they are lost). We can assume that this period lasts for some fixed time (unextendable dead time). One example of such flows is given by the CSMA/CD protocol, a random multiple access protocol with conflict detection which is widely used in computer networks. At the moment a conflict is registered (detected) on the input of a certain network node, the "stub" ("plug") signal is broadcast in the network; while the "stub" signal is being sent out, claims arriving to this network node are refused service and are forwarded to callback source. Here the time during which the network node is closed for servicing claims that arrive there after a conflict is found can be treated as the dead time of the device that registers conflicts in the network node.

In the present work we solve the optimal state estimation problem for a modulated MAP flow under incomplete observability conditions (in our case, unextendable dead time). We propose an optimal state estimation algorithm in which the decision about a MAP flow state is made by maximizing the posterior distribution, which is the most complete characteristic of the flow state that we can get from a sample of observations. The criterion itself minimizes the total probability of error in making the decision [8].

2   Problem Setting

We consider a modulated MAP flow of events with intensity represented by a piecewise constant random process $\lambda(t)$ with two states: $\lambda(t) = \lambda_1$ or $\lambda(t) = \lambda_2$ ($\lambda_1 > \lambda_2$). The time during which process $\lambda(t)$ remains at the ith, $i = 1, 2$, state depends on two random values: 1) the first random value has exponential distribution function $F_i^{(1)}(t) = 1 - e^{-\alpha_i t}$, $i = 1, 2$; when the ith state ends process $\lambda(t)$ transits with probability equal one from the ith state to the jth, $i, j = 1, 2$ $(i \neq j)$; 2) the second random value has exponential distribution function $F_i^{(2)}(t) = 1 - e^{-\lambda_i t}$, $i = 1, 2$; when the ith state ends process $\lambda(t)$ transits with probability $P_1(\lambda_j|\lambda_i)$ from the ith state to the jth $(i \neq j)$ and a flow event occurs or process $\lambda(t)$ transits with probability $P_0(\lambda_j|\lambda_i)$ from the ith state to the jth $(i \neq j)$, but the flow event does not occur, or process $\lambda(t)$ transits from the ith state to the ith with probability $P_1(\lambda_i|\lambda_i)$ and a flow event occurs. Here $P_1(\lambda_j|\lambda_i) + P_0(\lambda_j|\lambda_i) + P_1(\lambda_i|\lambda_i) = 1$; $i, j = 1, 2$; $i \neq j$.

The first and the second random values are independent from each other. Under these assumptions, $\lambda(t)$ is a Markov process. The infinitesimal characteristics matrices for the process $\lambda(t)$ are as follows [6]:

$$
\begin{aligned}
\mathbf{D_0} &= \left\| \begin{array}{cc} -(\alpha_1 + \lambda_1) & \alpha_1 + \lambda_1 P_0(\lambda_2|\lambda_1) \\ \alpha_2 + \lambda_2 P_0(\lambda_1|\lambda_2) & -(\alpha_2 + \lambda_2) \end{array} \right\|, \\
\mathbf{D_1} &= \left\| \begin{array}{cc} \lambda_1 P_1(\lambda_1|\lambda_1) & \lambda_1 P_1(\lambda_2|\lambda_1) \\ \lambda_2 P_1(\lambda_1|\lambda_2) & \lambda_2 P_1(\lambda_2|\lambda_2) \end{array} \right\|;
\end{aligned} \tag{1}
$$

here matrix $\mathbf{D}_0$ describes the situation when on the semiinterval $[t, t + \Delta t)$, where Δt (here and in what follows) is a sufficiently small value, there is no flow event; matrix $\mathbf{D}_1$, when on the semiinterval $[t, t + \Delta t)$ a flow event occurs.

Note that the modulated MAP flow definition does not explicitly say in which process state $\lambda(t)$ flow event occurs when the process $\lambda(t)$ transits from the ith state to the jth $(i, j = 1, 2; i \neq j)$. This is inconsequential for state estimation since flow events and transitions of the process $\lambda(t)$ from the ith state to the jth $(i, j = 1, 2; i \neq j)$ occur instantaneously.

After each event registered at time t_k, there begins a time of fixed duration T (dead time) during which other events from the original modulated MAP flow are inaccessible for observation. When dead time is over, the first new event again gives rise to a period of dead time of duration T and so on. One possible scenario of the resulting situation is shown on Fig. 1, where $t_1, t_2, \ldots$ denote the moments when events occur in the observed flow; 1 and 2 are states of the random process $\lambda(t)$; black circles denote modulated MAP flow events inaccessible for observation; dashed lines denote dead time durations.

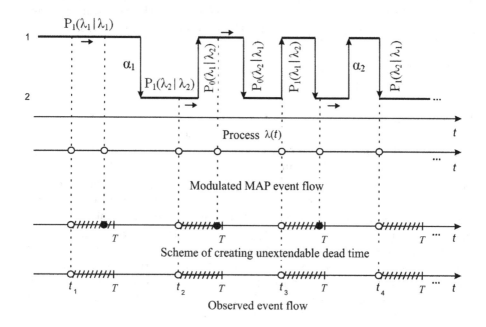

Fig. 1. Forming the observed flow

Since the process $\lambda(t)$ is unobservable in principle, and we can only observe time moments when events in the observed flow $t_1, t_2, \ldots$ occur, we have to estimate the state of the process $\lambda(t)$ (modulated MAP flow) when the observations stop.

We consider the stationary operation mode for the observed flow, so we disregard transition processes on the observation interval (t_0, t), where t_0 denotes the beginning of observations and t denotes the end of observations (decision making time). Then we can let $t_0 = 0$ without loss of generality. To make the decision regarding the state of the process $\lambda(t)$ at time moment t, we have to find posterior probabilities $w(\lambda_i|t) = w(\lambda_i|t_1, \ldots, t_m, t)$, $i = 1, 2$, of the fact that at time moment t, the value of the process $\lambda(t) = \lambda_i$ (where m is the number of observed events in time t), here $w(\lambda_1|t) + w(\lambda_2|t) = 1$. The decision regarding the state of the process $\lambda(t)$ is made by comparing the probabilities: if $w(\lambda_i|t) \geq w(\lambda_j|t)$, $i, j = 1, 2$; $i \neq j$, then we decide that $\hat{\lambda}(t) = \lambda_i$.

3   Optimal Estimation Algorithm for the States of the Modulated MAP Event Flow

We will consider a decision making moment t in the interval (t_k, t_{k+1}), $k = 1, 2, \ldots$, between neighboring events in the observed flow. For an initial interval (t_0, t_1), the time flow t lies between the beginning of observation t_0 and the first observed event in the flow. Consider the interval (t_k, t_{k+1}) with duration value $\tau_k = t_{k+1} - t_k$, $k = 0, 1, \ldots$. On the other hand, since the event observed at time moment t_k gives rise to a dead time period of duration T, we get that $\tau_k = T + \theta_k$, where θ_k is the duration value of the interval between the end of dead time $t_k + T$ and the moment t_{k+1}, i.e., the interval (t_k, t_{k+1}) breaks down into two adjacent intervals, the first semiinterval $(t_k, t_k + T]$ and the second interval $(t_k + T, t_{k+1})$. We emphasize that conditions of finding the posterior probability $w(\lambda_1|t)$ on the semiinterval $(t_k, t_k + T]$ and interval $(t_k + T, t_{k+1})$ are different in principle. Here we assume that the value of T is known exactly.

Lemma 1. *On time intervals (t_0, t_1) and $(t_k + T, t_{k+1})$, $k = 1, 2, \ldots$, when a modulated MAP event flow is observed, the posterior probability $w(\lambda_1|t)$ satisfies the differential equation*

$$
\begin{aligned}
w'(\lambda_1|t) = {}& [\lambda_1 - \lambda_2 - \lambda_1 P_0(\lambda_2|\lambda_1) + \lambda_2 P_0(\lambda_1|\lambda_2)] w^2(\lambda_1|t) - \\
& - [\alpha_1 + \alpha_2 + \lambda_1 - \lambda_2 + 2\lambda_2 P_0(\lambda_1|\lambda_2)] w(\lambda_1|t) + [\alpha_2 + \lambda_2 P_0(\lambda_1|\lambda_2)],
\end{aligned} \tag{2}
$$

$$
t_0 < t < t_1, \quad t_k + T < t < t_{k+1}, \quad k = 1, 2, \ldots.
$$

Proof. To derive the formulas for posterior probability $w(\lambda_1|t)$ we use a well-known method [8]: we first consider discrete observations divided by sufficiently small time intervals Δt and then make the limit transition as Δt tends to zero. First we suppose that the time is discrete and changes with step Δt: $t = n\Delta t$, $n = 0, 1, \ldots$. We introduce a two-dimensional process $(\lambda^{(n)}, r_n)$, where $\lambda^{(n)} = \lambda(n\Delta t)$ is the value of process $\lambda(t)$ at time moment $n\Delta t$ ($\lambda^{(n)} = \lambda_i, i = 1, 2$), and $r_n = r_n(\Delta t) = r(n\Delta t) - r((n-1)\Delta t)$ is the number of events in the flow observed on the interval $((n-1)\Delta t, n\Delta t)$ of length Δt, $r_n = 0, 1, \ldots$. We denote by $\mathbf{r}_m = (r_0, r_1, \ldots, r_m)$ the sequence of the numbers of events in time from zero to $m\Delta t$ on intervals $((n-1)\Delta t, n\Delta t)$ of length Δt ($n = \overline{0, m}$). Here r_0 is the number of events observed on the interval $(-\Delta t, 0)$. This number is undefined since there are no observations during this period, so we can set an arbitrary value to it, say $r_0 = 0$. We denote by $\lambda^{(m)} = \left(\lambda^{(0)}, \lambda^{(1)}, \ldots, \lambda^{(m)}\right)$ the sequence of unknown (unobservable) values of the process $\lambda(n\Delta t)$ at time moments $n\Delta t$ $(n = \overline{0, m})$; $\lambda^{(0)} = \lambda(0) = \lambda_i$, $i = 1, 2$. We denote by $w\left(\lambda^{(m)}|\mathbf{r}_m\right)$ the conditional probability of the value $\lambda^{(m)}$ given that we have observed a realization $\mathbf{r}_m$. Similarly $w\left(\lambda^{(m+1)}|\mathbf{r}_{m+1}\right)$. For the Markov random process $\left(\lambda^{(m)}, r_m\right)$, a recurrent relation is proved in [9] for the posterior probabilities $w\left(\lambda^{(m)}|\mathbf{r}_m\right)$ and $w\left(\lambda^{(m+1)}|\mathbf{r}_{m+1}\right)$:

$$
w\left(\lambda^{(m+1)}|\mathbf{r}_{m+1}\right) = \frac{\sum\limits_{\lambda^{(m)}=\lambda_1}^{\lambda_2} w\left(\lambda^{(m)}|\mathbf{r}_m\right)p\left(\lambda^{(m+1)}, r_{m+1}|\lambda^{(m)}, r_m\right)}{\sum\limits_{\lambda^{(m+1)}=\lambda_1}^{\lambda_2}\sum\limits_{\lambda^{(m)}=\lambda_1}^{\lambda_2} w\left(\lambda^{(m)}|\mathbf{r}_m\right)p\left(\lambda^{(m+1)}, r_{m+1}|\lambda^{(m)}, r_m\right)},
$$

$$(3)$$

where $p\left(\lambda^{(m+1)}, r_{m+1}|\lambda^{(m)}, r_m\right)$ is the transition probability for the process $\left(\lambda^{(n)}, r_n\right)$ in one step Δt from state $\left(\lambda^{(m)}, r_m\right)$ to state $\left(\lambda^{(m+1)}, r_{m+1}\right)$.

In the considered case of a modulated MAP flow, the random process $\left(\lambda^{(n)}, r_n\right)$, by our assumptions and by its constructions, will be a Markov process, so formula (3) holds.

The transition probability $p\left(\lambda^{(m+1)}, r_{m+1}|\lambda^{(m)}, r_m\right)$ for the modulated MAP flow in (3) can be written as

$$
\begin{aligned}
p\left(\lambda^{(m+1)}, r_{m+1}|\lambda^{(m)}, r_m\right) &= \\
= p\left(\lambda^{(m+1)}|\lambda^{(m)}\right) &p\left(r_{m+1}|\lambda^{(m)}, \lambda^{(m+1)}\right); \quad \lambda^{(m)}, \lambda^{(m+1)} = \lambda_1, \lambda_2.
\end{aligned}
\tag{4}
$$

Taking into account that $w\left(\lambda^{(m)}|\mathbf{r}_m\right) = w\left(\lambda^{(m)}|\mathbf{r}_m(t)\right) = w\left(\lambda^{(m)}|t\right)$, $w\left(\lambda^{(m+1)}|\mathbf{r}_{m+1}\right) = w\left(\lambda^{(m+1)}|t + \Delta t\right)$ and also (4) and letting in (3) $\lambda^{(m+1)} = \lambda_1$, we can rewrite (3) as

$$
w\left(\lambda_1|t + \Delta t\right) = \frac{\sum\limits_{s=1}^{2} w\left(\lambda_s|t\right)p\left(\lambda_1|\lambda_s\right)p\left(r_{m+1}|\lambda_s, \lambda_1\right)}{\sum\limits_{j=1}^{2}\sum\limits_{s=1}^{2} w\left(\lambda_s|t\right)p\left(\lambda_j|\lambda_s\right)p\left(r_{m+1}|\lambda_s, \lambda_j\right)}.
\tag{5}
$$

By the definition of the modulated MAP flow, the value r_{m+1} in (5) takes only two values: $r_{m+1} = 0$ or $r_{m+1} = 1$. Here we consider the behavior of probability $w\left(\lambda_1|t\right)$ on the interval $(t_k + T, t_{k+1})$ between the end of dead time $t_k + T$ and the moment t_{k+1} when the next event in the observed flow occurs, i.e., $t_k + T < t < t_{k+1}$, $t_k + T < t + \Delta t < t_{k+1}$. Then in (5) $r_{m+1} = 0$ and taking into account the matrix $\mathbf{D}_0$ in (1) the transition probabilities (4) on the subinterval $[t, t + \Delta t) = [m\Delta t, (m + 1)\Delta t)$ take the following form:

$$
\begin{aligned}
p\left(\lambda_s|\lambda_s\right)p\left(r_{m+1} = 0|\lambda_s, \lambda_s\right) &= 1 - (\alpha_s + \lambda_s)\Delta t + o(\Delta t), \quad s = 1, 2, \\
p\left(\lambda_j|\lambda_s\right)p\left(r_{m+1} = 0|\lambda_s, \lambda_j\right) &= \\
= \left[\alpha_s + \lambda_s P_0\left(\lambda_j|\lambda_s\right)\right]\Delta t + o(\Delta t), &\quad s, j = 1, 2, s \neq j.
\end{aligned}
\tag{6}
$$

Substituting (6) into (5), we find the numerator A_1 and denominator B_1 in (5):

$$A_1 = [1 - (\alpha_1 + \lambda_1)\,\Delta t]\,w\,(\lambda_1|t) + [\alpha_2 + \lambda_2 P_0\,(\lambda_1|\lambda_2)]\,\Delta t w\,(\lambda_2|t) + o\,(\Delta t)\,,$$

$$B_1 = 1 - \Delta t\,\{\lambda_1\,[1 - P_0\,(\lambda_2|\lambda_1)]\,w\,(\lambda_1|t) + \lambda_2\,[1 - P_0\,(\lambda_1|\lambda_2)]\,w\,(\lambda_2|t)\} + o\,(\Delta t)\,.$$

Substituting A_1 and B_1 into (5) and taking into account that

$$B_1^{-1} = 1 + \Delta t\,\{\lambda_1\,[1 - P_0\,(\lambda_2|\lambda_1)]\,w\,(\lambda_1|t) + \lambda_2\,[1 - P_0\,(\lambda_1|\lambda_2)]\,w\,(\lambda_2|t)\} + o\,(\Delta t)$$

(since $(1 - x)^{-1} = 1 + x + o\,(x)$ for $x > 0$ sufficiently small), we get:

$$w\,(\lambda_1|t + \Delta t) - w\,(\lambda_1|t) = \Delta t\,\{- (\alpha_1 + \lambda_1)\,w\,(\lambda_1|t) +$$
$$+ \lambda_1\,[1 - P_0\,(\lambda_2|\lambda_1)]\,w^2\,(\lambda_1|t) +$$
$$\lambda_2\,[1 - P_0\,(\lambda_1|\lambda_2)]\,w\,(\lambda_1|t)\,w\,(\lambda_2|t) + [\alpha_2 + \lambda_2 P_0\,(\lambda_1|\lambda_2)]\,w\,(\lambda_2|t)\} + o\,(\Delta t)\,.$$

Dividing the left- and the right-hand side by Δt, taking into account that $w\,(\lambda_2|t) = 1 - w\,(\lambda_1|t)$, and passing to the limit for $\Delta t \to 0$, we find (2). This completes the proof of Lemma 1.

Lemma 2. *The posterior probability $w\,(\lambda_1|t)$ at the time a modulated MAP flow event t_k, $k = 1, 2, \ldots$, occurs is given by the following formula:*

$$w\,(\lambda_1|t_k + 0) =$$
$$\frac{\lambda_2 P_1(\lambda_1|\lambda_2) + [\lambda_1 P_1(\lambda_1|\lambda_1) - \lambda_2 P_1(\lambda_1|\lambda_2)]w(\lambda_1|t_k - 0)}{\lambda_2[1 - P_0(\lambda_1|\lambda_2)] + [\lambda_1 - \lambda_2 - \lambda_1 P_0(\lambda_2|\lambda_1) + \lambda_2 P_0(\lambda_1|\lambda_2)]w(\lambda_1|t_k - 0)}, \tag{7}$$

$$k = 1, 2, \ldots \quad .$$

Proof. Suppose that on the interval $(t, t + \Delta t)$ at time moment t_k $(t < t_k < t + \Delta t)$ there occurs a flow event $(r_{m+1} = 1)$. We have two adjacent intervals (t, t_k) and $(t_k, t + \Delta t)$. The duration of the first interval is $\Delta t' = t_k - t$; the duration of the second is $\Delta t'' = t + \Delta t - t_k$. Then $w\,(\lambda_s|t) = w\,(\lambda_s|t_k - \Delta t')$, $s = 1, 2$; $w\,(\lambda_1|t + \Delta t) = w\,(\lambda_1|t_k + \Delta t'')$, and (5) becomes

$$w\,(\lambda_1|t_k + \Delta t'') = \frac{\sum\limits_{s=1}^{2} w\,(\lambda_s|t_k - \Delta t')p\,(\lambda_1|\lambda_s)\,p\,(r_{m+1} = 1|\lambda_s, \lambda_1)}{\sum\limits_{j=1}^{2}\sum\limits_{s=1}^{2} w\,(\lambda_s|t_k - \Delta t')p\,(\lambda_j|\lambda_s)\,p\,(r_{m+1} = 1|\lambda_s, \lambda_j)}. \tag{8}$$

Taking into account the matrix $\mathbf{D}_1$ in (1) on the interval $(t, t + \Delta t) = (m\Delta t, (m + 1)\,\Delta t)$, we can rewrite transition probabilities (4) as

$$p\,(\lambda_s|\lambda_s)\,p\,(r_{m+1} = 1|\lambda_s, \lambda_s) = \lambda_s \Delta t P_1\,(\lambda_s|\lambda_s) + o\,(\Delta t)\,, \quad s = 1, 2;$$
$$p\,(\lambda_j|\lambda_s)\,p\,(r_{m+1} = 1|\lambda_s, \lambda_j) = \lambda_s \Delta t P_1\,(\lambda_j|\lambda_s) + o\,(\Delta t)\,, \quad s, j = 1, 2, s \neq j. \tag{9}$$

Substituting (9) into (8), we get the numerator A_2 and the denominator B_2 in (8):

$$A_2 = \Delta t\,[\lambda_1 P_1\,(\lambda_1|\lambda_1)\,w\,(\lambda_1|t_k - \Delta t') + \lambda_2 P_1\,(\lambda_1|\lambda_2)\,w\,(\lambda_2|t_k - \Delta t')] + o\,(\Delta t)\,,$$

$$B_2 = \Delta t \left\{ \lambda_1 \left[1 - P_0 \left(\lambda_2 | \lambda_1 \right) \right] w \left(\lambda_1 | t_k - \Delta t' \right) + \right.$$
$$\left. + \lambda_2 \left[1 - P_0 \left(\lambda_1 | \lambda_2 \right) \right] w \left(\lambda_2 | t_k - \Delta t' \right) \right\} + o \left(\Delta t \right).$$

Substituting A_2 and B_2 into (8), dividing the numerator and denominator by Δt, taking into account that $w \left(\lambda_2 | t_k - \Delta t' \right) = 1 - w \left(\lambda_1 | t_k - \Delta t' \right)$, and passing to the limit for $\Delta t \to 0$ ($\Delta t'$ and $\Delta t''$ tend to zero simultaneously), we get (7). This completes the proof of Lemma 2.

Remark. At point t_k the probability $w \left(\lambda_1 | t \right)$ is discontinuous (there is a finite jump at this point). The probability $w \left(\lambda_1 | t_k + 0 \right)$ depends on the value $w \left(\lambda_1 | t_k - 0 \right)$, i.e., on the value of probability $w \left(\lambda_1 | t \right)$ at time moment t_k when $w \left(\lambda_1 | t \right)$ defined in (2) changes on the interval $(t_{k-1} + T, t_k)$ adjacent to the semiinterval $(t_k, t_k + T]$, $k = 2, 3, \ldots$. Thus, the value $w \left(\lambda_1 | t_k + 0 \right)$ "combines" the entire prehistory of our modulated MAP flow observations starting from the time moment $t_0 = 0$ until the moment t_k. As in initial condition $w \left(\lambda_1 | t_0 + 0 \right) = w \left(\lambda_1 | t_0 = 0 \right)$ on the semiinterval $[t_0, t_1)$ we take the prior final probability of the first state of the process $\lambda \left(t \right)$:

$$\pi_1 = \frac{\alpha_2 + \lambda_2 \left[1 - P_1 \left(\lambda_2 | \lambda_2 \right) \right]}{\alpha_1 + \alpha_2 + \lambda_1 \left[1 - P_1 \left(\lambda_1 | \lambda_1 \right) \right] + \lambda_2 \left[1 - P_1 \left(\lambda_2 | \lambda_2 \right) \right]}, \tag{10}$$

which is the decision of the differential equation

$$\pi_1' \left(t | t_0 \right) = \left[-\alpha_1 - \alpha_2 - \lambda_1 - \lambda_2 + \lambda_1 P_1 \left(\lambda_1 | \lambda_1 \right) + \lambda_2 P_1 \left(\lambda_2 | \lambda_2 \right) \right] \pi_1 \left(t | t_0 \right) +$$
$$+ \left[\alpha_2 + \lambda_2 - \lambda_2 P_1 \left(\lambda_2 | \lambda_2 \right) \right]$$

for $t_0 \to -\infty$.

Lemmas 1 and 2 yield the following theorem.

Theorem 1. *On time intervals (t_0, t_1) and $(t_k + T, t_{k+1})$, $k = 1, 2, \ldots$, the posterior probability $w(\lambda_1 | t)$ follows the following explicit formula:*

$$w \left(\lambda_1 | t \right) =$$
$$= \frac{w_1 [w_2 - w(\lambda_1 | t_k + T)] - w_2 [w_1 - w(\lambda_1 | t_k + T)] e^{-a(w_2 - w_1)(t - t_k - T)}}{w_2 - w(\lambda_1 | t_k + T) - [w_1 - w[\lambda_1 | t_k + T]] e^{-a(w_2 - w_1)(t - t_k - T)}}, \tag{11}$$

$$w_{1,2} = \frac{1}{2a} \left(\alpha_1 + \alpha_2 + \lambda_1 - \lambda_2 + 2\lambda_2 P_0 \left(\lambda_1 | \lambda_2 \right) \mp \sqrt{D} \right),$$
$$a = \lambda_1 - \lambda_2 - \lambda_1 P_0 \left(\lambda_2 | \lambda_1 \right) + \lambda_2 P \left(\lambda_1 | \lambda_2 \right), a \neq 0,$$
$$D = \left[\left(\lambda_1 - \lambda_2 \right) - \left(\alpha_1 + \alpha_2 \right) \right]^2 + 4\alpha_1 \left(\lambda_1 - \lambda_2 \right) + 4 \left[\alpha_1 \lambda_2 P_0 \left(\lambda_1 | \lambda_2 \right) + \right.$$
$$\left. \alpha_2 \lambda_1 P_0 \left(\lambda_2 | \lambda_1 \right) \right] + 4\lambda_1 \lambda_2 P_0 \left(\lambda_1 | \lambda_2 \right) P_0 \left(\lambda_2 | \lambda_1 \right),$$

where $t_k + T < t < t_{k+1}$ $(k = 0, 1, \ldots)$; $w \left(\lambda_1 | t_k + T \right)$ is defined further in assertion by formula (12), $k = 1, 2, \ldots$; $w \left(\lambda_1 | t_0 + 0 \right) = w \left(\lambda_1 | t_0 = 0 \right) = \pi_1$, where π_1 is defined in (10).

Let us consider semiinterval $(t_k, t_k + T]$, $k = 1, 2, \ldots$. On this semiinterval, the event takes place at the boundary point t_k, and there are no events on the semiinterval itself.

Assertion. The posterior probability $w(\lambda_1|t)$ on time semiinterval $(t_k, t_k + T]$, $k = 1, 2, ...$, is given by the following explicit formula:

$$w(\lambda_1|t) = \pi_1 + [w(\lambda_1|t_k + 0) - \pi_1] e^{-b(t-t_k)}, \tag{12}$$

$t_k < t \leq t_k + T$, $k = 1, 2, ...$; $w(\lambda_1|t_k + 0)$ is given by formula (7), $k = 1, 2, ...$; π_1 defined in (10); $b = \alpha_1 + \alpha_2 + \lambda_1 [1 - P_1(\lambda_1|\lambda_1)] + \lambda_2 [1 - P_1(\lambda_2|\lambda_2)]$.

This formulas let us construct the algorithm to compute the probability $w(\lambda_1|t)$ $(w(\lambda_2|t) = 1 - w(\lambda_1|t))$ and the algorithm to make a decision regarding the state of process $\lambda(t)$ at any time moment t:

(1) at time moment $t_0 = 0$, specify $w(\lambda_1|t_0 + 0) = w(\lambda_1|t_0 = 0) = \pi_1$;

(2) according to formula (11), for $k = 0$ compute the probability $w(\lambda_1|t)$ at any time moment t $(0 < t < t_1)$ where t_1 is the moment when the first event in the observed flow occurs;

(3) according to formula (11), for $k = 0$ compute $w(\lambda_1|t_1) = w(\lambda_1|t_1 - 0)$;

(4) increment k by one, and according to formula (7), for $k = 1$, compute the probability $w(\lambda_1|t_1 + 0)$ which is the initial value for $w(\lambda_1|t)$ in formula (12);

(5) according to formula (12), for $k = 1$ compute $w(\lambda_1|t)$ at any time moment t $(t_1 < t \leq t_1 + T)$ and the probability $w(\lambda_1|t_1 + T)$ is the initial value for $w(\lambda_1|t)$ on the next step of the algorithm;

(6) according to formula (11), for $k = 1$ compute $w(\lambda_1|t)$ ay any time moment t $(t_1 + T < t < t_2)$, where t_2 is the observation moment for the second event;

(7) according to formula (11), for $k = 1$ compute the probability $w(\lambda_1|t_2) = w(\lambda_1|t_2 - 0)$;

(8) go to step (4), repeat steps (4) – (8) for $k = 2$ and so on.

As we compute $w(\lambda_1|t)$, at any time moment t we can make a decision regarding the state of process $\lambda(t)$: if $w(\lambda_1|t) \geq w(\lambda_2|t)$ then we estimate $\hat{\lambda}(t) = \lambda_1$, otherwise $\hat{\lambda}(t) = \lambda_2$.

References

1. Kingman, J.F.C.: On Doubly Stochastic Poisson Process. Proceedings of Cambridge Philosophical Society 60(4), 923–930 (1964)
2. Basharin, G.P., Kokotushkin, V.A., Naumov, V.A.: On the Equivalent Substitutions Method for Computing Fragments of Communication Networks. Izv. Akad. Nauk USSR. Tekhn. Kibern. (6), 92–99 (1979)
3. Basharin, G.P., Kokotushkin, V.A., Naumov, V.A.: On the Equivalent Substitutions Method for Computing Fragments of Communication Networks. Izv. Akad. Nauk USSR Tekhn. Kibern. (1), 55–61 (1980)
4. Neuts, M.F.: A Versatile Markov Point Process. J. Appl. Probab. 16, 764–779 (1979)
5. Lucantoni, D.M.: New Results on the Single Server Queue with a Batch Markovian Arrival Process. Communications in Statistics Stochastic Models 7, 1–46 (1991)
6. Dudin, A.N., Klimenok, V.I.: Queueing Systems with Correlated Flows. Belarus Gos. Univ., Minsk (2000)
7. Apanasovich, V.V., Kolyada, A.A., Chernyavskii, A.F.: Statistical Analysis of Random Flows in a Physical Experiment. Universitetskoe, Minsk (1988)

8. Khazen, E.M.: Methods of Optimal Statistical Decisions and Optimal Control. Sovetskoe Radio, Moscow (1968)
9. Gortsev, A.M., Shmyrin, I.S.: Optimal Estimation of States of the Double Stochastic Flow of Events in the Presence of Measurement Errors of Time Instants. Automat. Remote Control 60(1, part 1), 41–51 (1999)

Modification of SSIM Metrics

Alexander Osokin and Dmitry Sidorov

National Research Tomsk Polytechnic University, Tomsk, Russia
OAO "Tomskoe pivo", Tomsk, Russia
osokin@vt.tpu.ru, rauco@mail.ru
http://www.tpu.ru/en

Abstract. Objective methods and metrics for assessing image quality
are more convenient, less expensive and time-consuming than subjective methods. There are a number of reference objective metrics that
provide a good correlation with perceived image quality, such as SSIM,
MS-SSIM, CW-SSIM, IW-SSIM, but they are slow and have sufficient
computational complexity. In this paper we proposed a simple modification for SSIM algorithms, in particular SSIM and MS-SSIM, which, for
a small deterioration in quality of results, can *significantly* reduce the
time of calculations.

Keywords: structural similarity (SSIM), multi-scale structural similarity (MS-SSIM), image quality assessment, perceptual quality, error sensitivity, image processing, experiment.

1 Introduction

Mathematical algorithms for assessing the quality of images are widely used in
development, testing and modification of various image processing algorithms:
lossy compression, separation and removal of noise, digitization and restoration
of old videotapes, lighting, etc. In such applications images are viewed by human beings and the correct method of quantifying image quality is through the
subjective evaluation. In practice subjective evaluation is usually too inconve-
nient, expensive and time-consuming. An objective image quality assessment, in
contrast to subjective evaluation, can greatly simplify and expedite the process
of evaluation. There are a number of reference algorithms (metrics) for image
quality assessment [2,7]. Reference metric in image processing is a function that
determines the distance from the distorted image to the reference, an original or
perfect image in the image space [2]. Some of these metrics are simple and easily
computed, such as PSNR, MSE, MSAD, but they do not provide a good correlation with perceived image quality and don't reflect the way that human beings
perceive images. Other metrics, like well known SSIM, MS-SSIM, CW-SSIM,
IW-SSIM allow to get more reliable and adequate results in the image assessment, but they are slow and have sufficient computational complexity [2-7,9].
However it is possible to solve the problem of high computational complexity
without significant loss in correlation with the subjective human quality assessment. We proposed a simple modification to some SSIM algorithms, in particular

A. Dudin et al. (Eds.): ITMM 2014, CCIS 487, pp. 351–355, 2014.

SSIM and MS-SSIM, which, for a small deterioration in quality of results, can significantly reduce the time of calculations.

2   SSIM and MS-SSIM

2.1   SSIM

According to work [5] SSIM metric is calculated as a degree of similarity of the corresponding square areas (windows) reference and distorted images:

$$Luminance\ L(x,y) = \frac{2\mu_x\mu_y + C_1}{\mu_x^2 + \mu_y^2 + C_1} \tag{1}$$

$$Contrast\ C(x,y) = \frac{2\sigma_x\sigma_y + C_2}{\sigma_x^2 + \sigma_y^2 + C_2} \tag{2}$$

$$Structure\ S(x,y) = \frac{\sigma_{x,y} + C_3}{\sigma_x\sigma_y + C_3}, \tag{3}$$

where x - reference (original) image, y - distorted image, μ_x^2 - weighted mean for a sliding window of M pixels of reference image $\mu_x^2 = \frac{1}{M}\sum_{i=1}^{M} w_i x_i$, σ_x^2 - standard deviation for reference image, calculated as follows $\sigma_x^2 = \frac{1}{M}\sum_{i=1}^{M} w_i(x_i - \mu_x)^2$, $\sigma_{x,y}$ - correlation coefficient between reference and distorted images $\sigma_{x,y} = \frac{1}{M}\sum_{i=1}^{M} w_i(x_i - \mu_x)(y_i - \mu_y)$, w={$w_i$ i=1,2,...,M} - normalized circular-symmetric Gaussian weighting coefficients with standard deviation σ=1.5, $K_1 = 0.01$, $K_2 = 0.0.3$, R=255, $C_1 = (K_1 R)^2$, $C_2 = (K_2 R)^2$, $C_3 = C_2/2$, size of the sliding window is 11 pixels (M=121).

Local $SSIM_L$ for a window is calculated using (1), (2) and (3):

$$SSIM_L(x,y) = L(x,y)C(x,y)S(x,y) \tag{4}$$

In practice, more required a single overall quality measure of the entire image, so the final SSIM index is calculated as a mean of (4):

$$SSIM(x,y) = \frac{1}{N}\sum_{i=1}^{N} SSIM_L(x,y) \tag{5}$$

where N - number of sliding windows that cover the image with step 1 pix.

2.2   MS-SSIM

Metric MS-SSIM [2, 4] is a kind of multi-scale SSIM metric and is calculated as a weighted mean of ratings SSIM, obtained for different scales of the reference and distorted images. Algorithm for calculate MS-SSIM can be formulated as follows:

1. at the current size (scale) of the images SSIM is computed using (5) and stored;

2. images are scaled down by 4;
3. Steps 1,2 repeated up to 6 times unless the size of scaled images is small-erthan the sliding window over each of the dimensions (height, width);
4. MS-SSIM is calculated as weighted average SSIM, saved on Step 1.

More details about MS-SSIM are in work [4].

It should be noted that due to the higher sensitivity of the human visual system to a slight change in brightness than in color, all quality assessments are typically calculated only in the luminance space (component) of images [7, 8].

During the implementation of SSIM and MS-SSIM metrics and testing we found that the requirement for the step of the sliding window in one pixel is not strict and can be changed. We expect that the increase of the sliding window step up to the it size (11 pix.) would have little impact on the quality assessment and will significantly reduce time for calculations. To confirm this we make a computer experiment.

3 Experiment

The goal of experiment is to determinate the adequacy (level of correlation with the MOS) and relative gain in time for quality assessment by modified and original metrics SSIM and MS-SSIM.

Input:

- TID2008 contains 25 reference images and 1700 distorted images (25 reference images x 17 types of distortions x 4 levels of distortions) [1];
- MOS for TID2008 as the results of 838 subjective experiments [1];
- implemented according to [4, 5] algorithms SSIM, MS-SSIM in C language with the ability to specify the step of sliding window (project compiled without use of any parallel calculations);
- step of the sliding window for the modified metrics 11 pix. (chosen by the authors), 1 pix. for the original metrics;
- ranking metrics in accordance with Spearman and Kendall correlation with MOS.

Hardware:

- CPU Intel Core i5-2500 (3,3 GHz);
- memory DDR III 10600 MB/s, 8 Gb;
- device for time measurement - integrated timer (measurement error is less than 10 ms).

Software:

- OS Microsoft Windows 7 SP1 (64-bit);
- Visual Studio 2010 compiler without code optimizations (default settings for C++ console project).

Boundary value of the sliding window step (11 pix.) was chosen by the authors due to the fact that the further increase of step will give more gain in speed but with completely poor quality assessments. Assessments generated by modified metrics begin to degrade and fluctuate within a small range (0.970,0.999), which shows no difference in the type and strength of distortions in the images.

The experimental results on TID2008 are shown in table 1.

Table 1. The experimental results for modified and original metrics SSIM and MS-SSIM

Metric	Step (pix.)	Spearman	Kendall	Calc. time (s.)
SSIM	1	0.795	0.595	75.222
	11	0.792	0.594	24.298
Relative change in %		-0.377	-0.168	-67.698
MS-SSIM	1	0.843	0.641	86.894
	11	0.839	0.637	27.464
Relative change in %		-0.474	-0.624	-68.393

According to results, the proposed modification for metrics SSIM and MS-SSIM based on the change of step of sliding window allows, without losing the quality assessments, to gain three or more times in reducing computation time.

4   Conclusion

Authors proposed a modification for metrics SSIM and MS-SSIM that can be successfully applied in the development, testing and configuring various image processing algorithms and methods. This modification provides the same level of image quality assessment as the original algorithm and allows to get a significant gain (more than a three times) in the rate calculation.

References

1. Ponomarenko, N.: Tampere image database 2008 TID2008, version 1.0, http://www.ponomarenko.info/tid2008.htm
2. Wang, Z., Bovik, A.C.: Modern image quality assessment, p. 157. Morgan & Claypool, NY (2006)
3. Wang, Z., Li, Q.: Information content weighting for perceptual image quality assessment. IEEE Transactions on Image Processing 20, 1185–1198 (2011)
4. Wang, Z., Simoncelli, E.P., Bovik, A.C.: Multi-scale structural similarity for image quality assessment. IEEE Asilomar Conf. on Signal, Systems and Computers, Pacific Grove, California (2003)
5. Wang, Z., Bovik, A.C., Sheikh, H.R., Simoncelli, E.P.: Image quality assessment: from error visibility to structural similarity. IEEE Transactions on Image Processing 13, 600–612 (2004)

6. Wang, Z., Simoncelli, E.P.: Translation insensitive image similarity in complex wavelet domain. In: IEEE Inter. Conf. Acoustic, Speech and Signal Processing, Philadelphia, vol. 2, pp. 673–676 (2005)
7. MSU Video Quality Measurement Tool, http://compression.ru/video/quality_measure/video_measurement_tool_en.html
8. Vatolin, D., Ratushnyak, A., Smirnov, M., Yukin, V.: Everything about the data compression, http://www.compression.ru/index_en.htm
9. Sidorov, D.: Image quality metrics, http://imq.vt.tpu.ru/indexeng.html

Queueing System $MAP/M/\infty$ with n Types of Customers*

Ekaterina Pankratova and Svetlana Moiseeva

National Research Tomsk State University,
Lenin Avenue. 36, 634050 Tomsk, Russia
pankate@sibmail.com,
smoiseeva@mail.ru

Abstract. The research of the queueing system with incoming MAP, n types of customers, infinite number of servers and exponential service time is proposed. Investigation of n-dimensional stochastic process that characterizes the number of busy servers for different types of customers is held by the method of initial moments. There are expressions for the characteristic function of the number of busy servers for different types of customers in the system $MAP/M/\infty$ under the asymptotic condition that service time infinitely grows equivalently to each type of customers.

Keywords: queueing system, Markovian arrival process, different types of customers, method of initial moments, asymptotic analysis.

1 Introduction

The research of the queuing system with infinite number of servers can be found in articles of A.V. Pechinkin [1–3], A.A. Nazarov, P. Abaev, R. Razumchik [4], B. D'Auria [5], D. Baum and L. Breuer [6, 7], J. Bojarovich L. Marchenko [8], E.A. van Doorn A.A Jagers [9], N.G. Duffield [10], C. Fricker and M. R. Jaïbi [11], E. Girlich [12], A. K. Jayawardene and O. Kella [13], M. Parulekar and A. M. Makowski [14] and others. Numerous studies of real flows in various subject areas, in particular, telecommunication flows and flows in economic systems led to the conclusion about the inadequacy of the classic models of substantial flows of random events to real data. There is an interest in investigation of flows, in which the customers are not identical and therefore require fundamentally different services. The queuing systems with heterogeneous devices include systems of parallel service, which can be found in articles of G.P. Basharin, K.E. Samuylov [15], A. Movaghar [16], M. Kargahi [17], J.A. Morrisson, C. Knessl [18], D.G. Down [19], N. Bambos, G. Michalidis [20] and others. In these works, all systems have a Poisson input and exponential service time. In the papers [21, 22], systems with parallel service of MMPP and renewal arrivals with paired customers are investigated. In this paper, we study a queueing system with MAP arrivals and

---

* This work is performed under the state order No. 1.511.2014/K of the Ministry of Education and Science of the Russian Federation.

A. Dudin et al. (Eds.): ITMM 2014, CCIS 487, pp. 356–366, 2014.

heterogeneous service. The main difference between the system in the paper from the previously considered ones is that when the customer comes in the system it is marked by i-th$(i = 1, \ldots, n)$ type in order to given probabilities. Service times for customers of different types has different stochastic parameters.

2   Statement of the Problem

Consider the queuing system with infinite number of servers of n different types and exponential service time. Incoming flow is a Markovian Arrival Process (MAP) with n types of customers. The underlying Markov chain $k(t)$ has a finite number of states K. $k(t)$ is determined by the matrix of infinitesimal characteristics Q, the set of non-negative integers λ_k and probabilities $d_{\nu k}$. At the time of occurrence of the event in this stream only one customer comes in the system. The type of incoming customer is defined as i-type with probability p_i $(i = 1, \ldots, n)$. It goes to the servers with appropriate type where it is servicing during a random time having an exponential distribution function with parameter μ_i corresponding to the type of the customer.

Set the problem of analysis of n-dimensional stochastic process $\{i_1(t), i_2(t), \ldots, i_n(t)\}$ of the number of busy servers of each type at the moment t. Incoming stream is not Poisson, therefore the n-dimensional process $\{i_1(t), i_2(t), \ldots, i_n(t)\}$ is non-Markov. Consider a $(n + 1)$-dimensional Markov process $\{k(t), i_1(t), i_2(t), \ldots, i_n(t)\}$ for which we can write the joint probability distribution $P\{k, i_1, i_2, \ldots, i_n, t\} = P\{k(t) = k, i_1(t) = i_1, i_2(t) = i_2, \ldots, i_n(t) = i_n\}$. Here $k(t)$ – the state of management Markov chain. The system of Kolmogorov differential equations for the probability distribution $P\{k, i_1, i_2, \ldots, i_n, t\}$ is the following:

$$\frac{\partial P(k, i_1, i_2, \ldots, i_n, t)}{\partial t} = \left(-\lambda_k - \sum_{l=1}^{n} i_l \mu_l\right) P(k, i_1, i_2, \ldots, i_n, t)+$$

$$+ \lambda_k p_1 P(k, i_1 - 1, i_2, \ldots, i_n, t) + \ldots + \lambda_k p_n P(k, i_1, i_2, \ldots, i_n - 1, t) + \quad (1)$$

$$+(i_1 + 1)\mu_1 P(k, i_1 + 1, i_2, \ldots, i_n, t) + \ldots + (i_n + 1)\mu_n P(k, i_1, i_2, \ldots, i_n + 1, t)+$$

$$+ \sum_{\nu \neq k} \{(1 - d_{\nu k}) P(\nu, i_1, i_2, \ldots, i_n, t) + d_{\nu k}\, (p_1 P(\nu, i_1 - 1, i_2, \ldots, i_n, t) + \ldots$$

$$+ p_n P(\nu, i_1, i_2, \ldots, i_n - 1, t))\} q_{\nu k}, \quad k = 1, \ldots, n.$$

The initial conditions have the form $P(k, 0, 0, \ldots, 0, t) = R(k)$, where $R(k)$ - stationary probability distribution of the Markov chain $k(t)$. We will find the solution of the system (1) during stationary operation of the system.

Introduce the characteristic function of the form:

$$H(k, u_1, \ldots, u_n) = \sum_{i_1=0}^{\infty} \ldots \sum_{i_n=0}^{\infty} e^{ju_1 i_1} \times \ldots \times e^{ju_n i_n} P(k, i_1, \ldots, i_n),$$

where $j = \sqrt{-1}$ – imaginary unit.

Using (1) write the system of differential equations for the characteristic function $H(k, u_1, \ldots, u_n)$

$$\sum_{l=1}^{n} \mu_l j \left(e^{-ju_l} - 1\right) \frac{\partial H(k, u_1, \ldots, u_n)}{\partial u_l} = \lambda_k \left(\sum_{l=1}^{n} p_l e^{ju_l} - 1\right) H(k, u_1, \ldots, u_n) +$$

$$+ \sum_{\nu=1, \nu \neq k}^{K} H(\nu, u_1, \ldots, u_n) \left[1 + \left(\sum_{l=1}^{n} p_l e^{ju_l} - 1\right) d_{\nu k}\right] q_{\nu k}, \qquad (2)$$

$$H(k, 0, \ldots, 0) = R(k), \ k = 1, \ldots, n.$$

Denote

• $\mathbf{H}(u_1, \ldots, u_n) = [H(1, u_1, \ldots, u_n), H(2, u_1, \ldots, u_n), \ldots, H(K, u_1, \ldots, u_n)]$ – row vector consisting of the characteristic functions of the random process $\{k(t), i_1(t), \ldots, i_n(t)\}$ for each state of the management Markov chain $k(t)$;

• $\mathbf{Q}$ – the matrix of infinitesimal characteristics with elements $q_{\nu k}$,

$$\nu = 1, \ldots, n, \ k = 1, \ldots, n;$$

• $\mathbf{\Lambda}$ – the diagonal matrix with elements $\lambda_k \, (k = 1, \ldots, K)$ on the main diagonal;

• $\mathbf{A}$ – Hadamard product of matrix $\mathbf{D}$ and $\mathbf{Q}$, that is the matrix of the elements $d_{\nu k} q_{\nu k}$, $\nu = 1, \ldots, K, \ k = 1, \ldots, K$;

• $\mathbf{B} = \mathbf{\Lambda} + \mathbf{A}$.

Taking into account (2), write vector-matrix equation for the vector of the characteristic function $\mathbf{H}(u_1, \ldots, u_n)$:

$$\sum_{l=1}^{n} \mu_l j \left(e^{-ju_l} - 1\right) \frac{\partial \mathbf{H}(u_1, \ldots, u_n)}{\partial u_l} = \mathbf{H}(u_1, \ldots, u_n) \left[\mathbf{Q} + \left(\sum_{l=1}^{n} p_l e^{ju_l} - 1\right) \mathbf{B}\right],$$

$$\mathbf{H}(0, \ldots, 0) = \mathbf{r} = [R(1), R(2), \ldots, R(K)]. \qquad (3)$$

The equation (3) will be considered as the basis for further research.

3   Method of Initial Moments

Theorem 1. *In the system $MAP|M|\infty$ with heterogeneous service, the average number fm_l of busy servers of the l-th type $(l = 1, \ldots, n)$ has the form:*

$$fm_l = \frac{p_l}{\mu_l} \mathbf{rBe}, \qquad (4)$$

where $\mathbf{e}$ – an identity column vector.

Theorem 2. *In the system $MAP|M|\infty$ with heterogeneous service, the second order moment of numbers sm_l of busy servers of the l-th type $(l = 1, \ldots, n)$ have the form:*

$$sm_l = p_l \mathbf{rB} \left\{\mathbf{I} + [\mu_l \mathbf{I} - \mathbf{Q}]^{-1} [\mu_l \mathbf{I} + 2p_l \mathbf{B}]\right\} \{2\mu_l \mathbf{I} - \mathbf{Q}\}^{-1} \mathbf{e}, \qquad (5)$$

where $\mathbf{I}$ – an identity matrix.

Theorem 3. *Correlation moment cm_{lg} of busy the l-th and the g-th type devices' number ($l = 1, \ldots, n$, $g = 1, \ldots, n$, $l \neq g$) in system $MAP|M|\infty$ with heterogeneous service has the form:*

$$cm_{lg} = (p_g \mathbf{fm}_l + p_l \mathbf{fm}_g)\mathbf{B}\left[(\mu_l + \mu_g)\mathbf{I} - \mathbf{Q}\right]^{-1} \mathbf{e}. \tag{6}$$

Then correlation coefficient has the form:

$$r(i_l, i_g) = \frac{cov(i_l, i_g)}{\sqrt{Di_l Di_g}} = \frac{cm_{lg} - fm_l fm_g}{\sqrt{Di_l Di_g}}, \ (l = 1, \ldots, n, \ g = 1, \ldots, n, \ l \neq g).$$

4   Method of the Asymptotic Analysis

4.1   Asymptotics of the First Order

We will solve the basis equation for the characteristic function (3) in the asymptotic condition that service time on appliances growths equivalently to each other $(\mu_l \to 0, l = 1, \ldots, n)$.
    Denote

$$\mu_1 = \varepsilon, \ \mu_2 = q\varepsilon, \ \ldots, \mu_n = q^{n-1}\varepsilon, \ u_1 = \varepsilon x_1, \ u_2 = q\varepsilon x_2, \ \ldots, u_n = q^{n-1}\varepsilon x_n,$$

$$\mathbf{H}(u_1, u_2, \ldots, u_n) = \mathbf{F}_1(x_1, x_2, \ldots, x_n, \varepsilon). \tag{7}$$

Taking into account (7) we can write (3) as

$$\sum_{l=1}^{n} j\left(e^{-jq^{l-1}x_l\varepsilon} - 1\right)\frac{\partial \mathbf{F}_1(x_1, \ldots, x_n, \varepsilon)}{\partial x_l} = \tag{8}$$

$$= \mathbf{F}_1(x_1, \ldots, x_n, \varepsilon)\left[\mathbf{Q} + \left(\sum_{l=1}^{n} p_l e^{jq^{l-1}x_l\varepsilon} - 1\right)\mathbf{B}\right].$$

Lemma 1

$$\lim_{\varepsilon \to 0} \mathbf{F}_1(x_1, \ldots, x_n, \varepsilon) = \mathbf{F}_1(x_1, \ldots, x_n) = \mathbf{r}\exp\left\{j\kappa \sum_{l=1}^{n} p_l x_l\right\}, \tag{9}$$

$\kappa = \mathbf{rBe}$, $\mathbf{e}$ − *an identity column vector.*

Proof. If $\varepsilon \to 0$ in (8), then obtain:

$$\mathbf{F}_1(x_1, \ldots, x_n) = \lim_{\varepsilon \to 0} \mathbf{F}_1(x_1, \ldots, x_n, \varepsilon).$$

Since $\mathbf{rQ} = 0$, then we look for $\mathbf{F}_1(x_1, \ldots, x_n)$ as

$$\mathbf{F}_1(x_1, \ldots, x_n) = \mathbf{r}\Phi_1(x_1, \ldots, x_n), \tag{10}$$

where $\Phi_1(x_1, \ldots, x_n)$ - the desired scalar function.

Let's multiply (8) on the identity vector-column $\mathbf{e}$:

$$\sum_{l=1}^{n} j \left(e^{-jq^{l-1}x_l \varepsilon} - 1 \right) \frac{\partial \mathbf{F_1}(x_1, \ldots, x_n, \varepsilon)}{\partial x_l} \mathbf{e} =$$

$$= \mathbf{F_1}(x_1, \ldots, x_n, \varepsilon) \left(\sum_{l=1}^{n} p_l e^{jq^{l-1}x_l \varepsilon} - 1 \right) \mathbf{Be}.$$

Expand exponents in the received equation into a Taylor series, substitute into it the vector-function $\mathbf{F_1}(x_1, \ldots, x_n)$ in the form (10) and let $\varepsilon \to 0$:

$$\sum_{l=1}^{n} q^{l-1} x_l \frac{\partial \Phi_1(x_1, \ldots, x_n)}{\partial x_l} = \Phi_1(x_1, \ldots, x_n) \left(\sum_{l=1}^{n} p_l j q^{l-1} x_l \right) \mathbf{rBe}.$$

Denote $\kappa = \mathbf{rBe}$ and obtain the partial differential equation for the function $\Phi_1(x_1, \ldots, x_n)$:

$$\sum_{l=1}^{n} q^{l-1} x_l \frac{\partial \Phi_1(x_1, \ldots, x_n)}{\partial x_l} = \Phi_1(x_1, \ldots, x_n) \left(\sum_{l=1}^{n} p_l j q^{l-1} x_l \right) \kappa.$$

Taking into account the initial condition $\Phi_1(0, \ldots, 0) = 1$ we obtain the following expression

$$\Phi_1(x_1, \ldots, x_n) = \exp \left\{ j\kappa \sum_{l=1}^{n} p_l x_l \right\}.$$

Thus,

$$\mathbf{F_1}(x_1, \ldots, x_n) = \mathbf{r} \exp \left\{ j\kappa \sum_{l=1}^{n} p_l x_l \right\}.$$

$\square$

Taking into account (10) and substitutions (7) we can write the asymptotic approximate equality ($\varepsilon \to 0$):

$$\mathbf{H}(u_1, \ldots, u_n) = \mathbf{F_1}(x_1, \ldots, x_n, \varepsilon) \approx \mathbf{F_1}(x_1, \ldots, x_n) = \mathbf{r} \exp \left\{ j\kappa \sum_{l=1}^{n} p_l x_l \right\}.$$

For the characteristic function of process $\{i_1(t), i_2(t), \ldots, i_n(t)\}$ denote

$$h_1(u_1, \ldots, u_n) = M e^{j \sum_{l=1}^{n} u_l i_l(t)} = \mathbf{H}(u_1, \ldots, u_n)\mathbf{e} = \exp \left\{ j r \mathbf{Be} \sum_{l=1}^{n} \frac{p_l}{\mu_l} u_l \right\}.$$

The $h_1(u_1, \ldots, u_n)$ will be called the asymptotics of the first order for the system $MAP|M|\infty$ with heterogeneous service.

Defenition 1. *The functions*

$$h_1^{(l)}(u_l) = M e^{j u_l i_l(t)} = h_1(0, \ldots, u_l, \ldots, 0) = \exp \left\{ j r \mathbf{Be} p_l \frac{u_l}{\mu_l} \right\}, \quad l = 1, \ldots, n,$$

$$(11)$$

will be called the asymptotics of the first order for the characteristic function of the busy servers of any type in system $MAP|M|\infty$ with heterogeneous service.

Consider the asymtotics of the second order for more accurate approximation.

4.2   Asymptotics of the Second Order

Consider function $\mathbf{H}(u_1, \ldots, u_n)$:

$$\mathbf{H}(u_1, \ldots, u_n) = \mathbf{H_2}(u_1, \ldots, u_n) exp \left\{ jr\mathbf{Be} \sum_{l=1}^{n} p_l \frac{u_l}{\mu_l} \right\}. \tag{12}$$

Using (12) in (3) obtain the expression for $\mathbf{H_2}(u_1, \ldots, u_n)$:

$$\sum_{l=1}^{n} \mu_l j (e^{-ju_l} - 1) \frac{\partial \mathbf{H_2}(u_1, \ldots, u_n)}{\partial u_l} = \tag{13}$$

$$= \mathbf{H_2}(u_1, \ldots, u_n) \left[\mathbf{Q} + \left(\sum_{l=1}^{n} p_l e^{ju_l} - 1 \right) \mathbf{B} + \kappa \sum_{l=1}^{n} p_l (e^{-ju_l} - 1) \mathbf{I} \right], \tag{14}$$

where $\kappa = \mathbf{rBe}$, $\mathbf{e}$ – the identity vector-column, $\mathbf{I}$ - the identity matrix.
   Substitute the folliwing in (13):

$$\mu_1 = \varepsilon^2, \ \mu_2 = q\varepsilon^2, \ldots, \mu_n = q^{n-1}\varepsilon^2, \tag{15}$$

$$u_1 = \varepsilon x_1, \ u_2 = \varepsilon q x_2, \ldots, u_n = \varepsilon q^{n-1} x_n, \mathbf{H_2}(u_1, \ldots, u_n) = \mathbf{F_2}(x_1, \ldots, x_n, \varepsilon)$$

and obtain:

$$\sum_{l=1}^{n} j\varepsilon (e^{-j\varepsilon q^{l-1} x_l} - 1) \frac{\partial \mathbf{F_2}(x_1, \ldots, x_n, \varepsilon)}{\partial x_l} = \tag{16}$$

$$= \mathbf{F_2}(x_1, \ldots, x_n, \varepsilon) \left[\mathbf{Q} + \left(\sum_{l=1}^{n} p_l e^{j\varepsilon q^{l-1} x_l} - 1 \right) \mathbf{B} + \kappa \sum_{l=1}^{n} p_l \left(e^{-j\varepsilon q^{l-1} x_l} - 1 \right) \mathbf{I} \right].$$

Theorem 4

$$\lim_{\varepsilon \to 0} \mathbf{F_2}(x_1, \ldots, x_n, \varepsilon) = \mathbf{F_2}(x_1, \ldots, x_n) = \mathbf{r} \exp \left\{ \kappa \sum_{l=1}^{n} p_l q^{l-1} \frac{x_l^2}{2} + \tag{17} \right.$$

$$\left. + \sum_{l=1}^{n} \sum_{s-1}^{n} p_l p_s \frac{x_l x_s}{q^{l-1} + q^{s-1}} f_l [\mathbf{B} - \kappa \mathbf{I}] \mathbf{e} \right\},$$

where $\kappa = \mathbf{rBe}, \mathbf{e}$ – the identity vector-column, and functions $f_l, l = 1, \ldots, n$ are defined by the following system of equations

$$f_l \mathbf{Q} + \mathbf{r} [\mathbf{B} - \kappa \mathbf{I}] = 0, \ l = 1, \ldots, n. \tag{18}$$

Proof. Desirable solution of the equation (16) should be like the following:

$$\mathbf{F_2}(x_1,\ldots,x_n,\varepsilon) = \varPhi_2(x_1,\ldots,x_n)\left\{\mathbf{r} + j\varepsilon\sum_{l=1}^{n}p_l q^{l-1}x_l f_l\right\} + O(\varepsilon^2). \qquad (19)$$

Using (19) in (16), obtain:

$$\sum_{l=1}^{n} j\varepsilon\left(e^{-j\varepsilon q^{l-1}x_l} - 1\right)\left[\frac{\partial\varPhi_2(x_1,\ldots,x_n)}{\partial x_l}\left(\mathbf{r} + j\varepsilon\sum_{s=1}^{n}p_s x_s f_s q^{s-1}\right) + \right.$$

$$\left. + \varPhi_2(x_1,\ldots,x_n)j\varepsilon p_l f_l q^{l-1}\right] =$$

$$= \varPhi_2(x_1,\ldots,x_n)\left[\mathbf{r} + j\varepsilon\sum_{l=1}^{n}p_l x_l f_l q^{l-1}\right]\left[\mathbf{Q} + \left(\sum_{l=1}^{n}p_l e^{j\varepsilon q^{l-1}x_l} - 1\right)\mathbf{B} + \right.$$

$$\left. + \kappa\sum_{l=1}^{n}p_l\left(e^{-j\varepsilon q^{l-1}x_l} - 1\right)\mathbf{I}\right] + O(\varepsilon^2),$$

hence taking into account $\mathbf{rQ} = 0$ may earn the following system of equations for the functions f_l, $l = 1,\ldots,n$ when $\varepsilon \to 0$:

$$f_l\mathbf{Q} + \mathbf{r}\left[\mathbf{B} - \kappa\mathbf{I}\right] = 0, \; l = 1,\ldots,n, \qquad (20)$$

which coincides with (18).

To get the form of the function $\varPhi_2(x_1,\ldots,x_n)$ sum all equations of the system (16) and expand exponents into a Taylor series:

$$\varepsilon^2\sum_{l=1}^{n}q^{l-1}x_l\frac{\partial\mathbf{F_2}(x_1,\ldots,x_n,\varepsilon)}{\partial x_l}\mathbf{e} = \mathbf{F_2}(x_1,\ldots,x_n,\varepsilon)\left\{j\varepsilon\sum_{l=1}^{n}p_l q^{l-1}x_l\left[\mathbf{B} - \kappa\mathbf{I}\right]\mathbf{e} + \right.$$

$$\left. + (j\varepsilon^2)\sum_{l=1}^{n}p_l\frac{(q^{l-1}x_l)^2}{2}\left[\mathbf{B} + \kappa\mathbf{I}\right]\mathbf{e}\right\} + O(\varepsilon^3).$$

Substitute into received expression (19):

$$\varepsilon^2\sum_{l=1}^{n}q^{l-1}x_l\frac{\partial\varPhi_2(x_1,\ldots,x_n)}{\partial x_l}\mathbf{re} = \varPhi_2(x_1,\ldots,x_n)j\varepsilon\sum_{l=1}^{n}p_l q^{l-1}x_l\mathbf{r}[\mathbf{B} - \kappa\mathbf{I}]\mathbf{e} + $$

$$+ \varPhi_2(x_1,\ldots,x_n)(j\varepsilon)^2\left\{\sum_{l=1}^{n}p_l\frac{(q^{l-1}x_l)^2}{2}\mathbf{r}[\mathbf{B} + \kappa\mathbf{I}]\mathbf{e} + \right. \qquad (21)$$

$$\left. + \sum_{l=1}^{n}\sum_{s=1}^{n}p_l q^{l-1}x_l p_s q^{s-1}x_s f_l[\mathbf{B} - \kappa\mathbf{I}]\mathbf{e}\right\} + O(\varepsilon^3).$$

Since $\mathbf{rBe} = \kappa$ that $\mathbf{r}[\mathbf{B} - \kappa\mathbf{I}]\mathbf{e} = 0$ and $\mathbf{r}[\mathbf{B} + \kappa\mathbf{I}]\mathbf{e} = 2\kappa$. Then taking into account $\mathbf{re} = 0$, divide both sides of the expression (21) by ε^2 and pass to the limit provided $\varepsilon \to 0$:

$$\sum_{l=1}^{n} q^{l-1} x_l \frac{\partial \Phi_2(x_1, \ldots, x_n)}{\partial x_l} = \tag{22}$$

$$= \Phi_2(x_1, \ldots, x_n) j^2 \left\{ \kappa \sum_{l=1}^{n} p_l (q^{l-1} x_l)^2 + \sum_{l=1}^{n} \sum_{s=1}^{n} p_l q^{l-1} x_l p_s q^{s-1} x_s f_l [\mathbf{B} - \kappa\mathbf{I}] \mathbf{e} \right\}.$$

Solution of the differential equation (22) corresponding to the initial condition $\Phi_2(0, \ldots, 0) = 1$ is the function $\Phi_2(x_1, \ldots, x_n)$ of the form:

$$\Phi_2(x_1, \ldots, x_n) = \tag{23}$$

$$= \exp \left\{ \kappa \sum_{l=1}^{n} p_l q^{l-1} \frac{(jx_l)^2}{2} + \sum_{l=1}^{n} \sum_{s=1}^{n} p_l p_s \frac{jx_l jx_s}{q^{l-1} + q^{s-1}} f_l [\mathbf{B} - \kappa\mathbf{I}] \mathbf{e} \right\}.$$

□

Taking into account the approximate equations of the form $\mathbf{H_2}(u_1, \ldots, u_n) = \mathbf{F_2}(x_1, \ldots, x_n, \varepsilon) \approx \mathbf{F_2}(x_1, \ldots, x_n) = \mathbf{r}\Phi_2(x_1, \ldots, x_n)$ and (15) write expression for the function $\mathbf{H_2}(u_1, \ldots, u_n)$:

$$\mathbf{H_2}(u_1, \ldots, u_n) = \mathbf{r} \cdot \exp \left\{ \kappa \sum_{l=1}^{n} p_l \frac{(ju_l)^2}{2\mu_l} + \sum_{l=1}^{n} \sum_{s=1}^{n} p_l p_s \frac{ju_l ju_s}{\mu_l + \mu_s} f_l [\mathbf{B} - \kappa\mathbf{I}] \mathbf{e} \right\}.$$

Then using (12) obtain:

$$\mathbf{H}(u_1, \ldots, u_n) = \mathbf{r} \cdot \exp \left\{ j\kappa \sum_{l=1}^{n} p_l \frac{u_l}{\mu_l} + \kappa \sum_{l=1}^{n} p_l \frac{(ju_l)^2}{2\mu_l} + \right.$$

$$\left. + \sum_{l=1}^{n} \sum_{s=1}^{n} p_l p_s \frac{ju_l ju_s}{\mu_l + \mu_s} f_l [\mathbf{B} - \kappa\mathbf{I}] \mathbf{e} \right\},$$

therefore, for the characteristic function of the random process $\{i_1(t), i_2(t), \ldots, i_n(t)\}$ obtain:

$$M e^{j \sum_{l=1}^{n} u_l i_l(t)} = \mathbf{H}(u_1, \ldots, u_n) \mathbf{e} = \tag{24}$$

$$= \exp \left\{ j\kappa \sum_{l=1}^{n} p_l \frac{u_l}{\mu_l} + \kappa \sum_{l=1}^{n} p_l \frac{(ju_l)^2}{2\mu_l} + \sum_{l=1}^{n} \sum_{s=1}^{n} p_l p_s \frac{ju_l ju_s}{\mu_l + \mu_s} f_l [\mathbf{B} - \kappa\mathbf{I}] \mathbf{e} \right\}.$$

The expression (24) will be called the asymptotics of the second order for the system $MAP|M|\infty$ with heterogeneous service.

5 Numerical Analysis

Consider the particular case: $n = 2$. Let

$$\mathbf{Q} = \begin{pmatrix} -0.1 & 0.1 \\ 0.9 & -0.9 \end{pmatrix}, \qquad \Lambda = \begin{pmatrix} 1 & 0 \\ 0 & 100 \end{pmatrix},$$

$$d_{\nu k} = 0, \qquad (\nu, k = 1, 2), \qquad p_1 = 0.4, \qquad p_2 = 0.6.$$

Table 1. Analysis of the range of applicability of asymptotic algorithms

μ_1	0.1	0.05	0.01	0.001
D_1	1326.640	2775.474	14409.703	145353.41
AD_1	1454.944	2909.888	14549.44	145494.4
$\Delta = \left\lvert \frac{D_1 - AD_1}{D_1} \right\rvert 100\%$	9%	4.8%	1%	0.1%

If we assume that the allowable relative error Δ should be less than 5%, the good convergence occurs when $\mu \leq 0.05$.

6 Conclusion

In this paper, we construct and investigate the mathematical model of the queuing system with the MAP arrivals and heterogeneous service. The main probabilistic characteristics are found. In particular, the first and the second order initial moments of the number of busy servers of different types are obtained. The system under consideration are studied using asymptotic analysis. Namely, the expression for the asymptotic of the first and the second order are obtained for the characteristic function of the busy servers of each type. The numerical analysis of the convergence of the main probabilistic asymptotic characteristics to exact ones is carried out.

References

1. Pechinkin, A.V.: Boundary of change of stationary queue in queuing systems with various service disciplines. In: Proc. of the Seminar "Problems of Stability of Stochastic Models", vol. 109, pp. 118–121. All-Union Scientific Research Institute for System Studies, Moscow (1985) (in Russian)
2. Pechinkin, A.V.: The inversion procedure with probabilistic priority in queuing system with extraordinary incoming flow. Stochastic processes and their applications. Mathematical research. Shtiintsa, Kishinev (1989) (in Russian)

3. Pechinkin, A.V., Sokolov, I.A., Chaplygin, V.V.: Stationary characteristics of multi-line queuing system with simultaneous failures of devices. Computer Science and Applications 1(2), 28–38 (2007) (in Russian)

4. Abaev, P., Pechinkin, A., Razumchik, R.: On Mean Return Time in Queueing System with Constant Service Time and Bi-level Hysteric Policy. In: Dudin, A., Klimenok, V., Tsarenkov, G., Dudin, S. (eds.) BWWQT 2013. CCIS, vol. 356, pp. 11–19. Springer, Heidelberg (2013)

5. Auria, B.D.: $M|M|\infty$ queues in semi-Markovian random environment. Queueing Systems 58(3), 221–237 (2008)

6. Baum, D.: The infinite server queue with Markov additive arrivals in space. In: Proc. of the International Conference on Probabilistic Analysis of Rare Events, Riga, pp. 136–142 (1999)

7. Baum, D., Breuer, L.: The Inhomogeneous $BMAP|G|\infty$ queue. In: Proc. of the 11th GI/ITG Conference on Measuring, Modelling and Evaluation of Computer and Communication Systems (MMB 2001), Aachen, pp. 209–223 (2001)

8. Bojarovich, J., Marchenko, L.: An open queueing network with temporarily non-active customers and rounds. In: Dudin, A., Klimenok, V., Tsarenkov, G., Dudin, S. (eds.) BWWQT 2013. CCIS, vol. 356, pp. 33–36. Springer, Heidelberg (2013)

9. Doorn, E.A., Jagers, A.A.: Note on the $GI|GI|\infty$ system with identical service and interarrival-time distributions. Journal of Queueing Systems 47, 45–52 (2004)

10. Duffield, N.G.: Queueing at large resources driven by long-tailed $M|G|\infty$-modulated processes. Queueing Systems 28(1-3), 245–266 (1998)

11. Fricker, C., Jaïbi, M.R.: On the fluid limit of the $M|G|\infty$ queue. Queueing Systems 56(3-4), 255–265 (2007)

12. Girlich, E., Kovalev, M.M., Listopad, N.I.: Optimal choice of the capacities of telecommunication networks to provide qoS-routing. In: Dudin, A., Klimenok, V., Tsarenkov, G., Dudin, S. (eds.) BWWQT 2013. CCIS, vol. 356, pp. 93–104. Springer, Heidelberg (2013)

13. Jayawardene, A.K., Kella, O.: $M|G|\infty$ with alternating renewal breakdowns. Queueing Systems 22(1-2), 79–95 (1996)

14. Parulekar, M., Makowski, A.M.: Tail probabilities for $M|G|\infty$ input processes: I. Preliminary asymptotics. Queueing Systems 27(3-4), 271–296 (1997)

15. Basharin, G.P., Samouylov, K.E., Yarkina, N.V., Gudkova, I.A.: A new stage in mathematical teletraffic theory. Automation and Remote Control 70(12), 1954–1964 (2009)

16. Movaghar, A.: Analysis of a Dynamic Assignment of Impatient Customers to Parallel Queues. Queueing Systems 67(3), 251–273 (2011)

17. Kargahi, M., Movaghar, A.: Utility Accrual Dynamic Routing in Real-Time Parallel Systems. Transactions on Parallel and Distributed Systems (TDPS) 21(12), 1822–1835 (2010)

18. Knessl, C.A., Morrison, J.: Heavy Traffic Analysis of Two Coupled Processors. Queueing Systems 43(3), 173–220 (2003)

19. Down, D.G., Wu, R.: Multi-layered round robin routing for parallel servers. Queueing Systems 53(4), 177–188 (2006)

20. Bambos, N., Michailidis, G.: Queueing Networks of Random Link Topology: Stationary Dynamics of Maximal Throughput Schedules. Queueing Systems 50(1), 5–52 (2005)

21. Ivanovskaya (Sinyakova), I., Moiseeva, S.: Investigation of the queuing system $MMP^{(2)}|M_2|\infty$ by method of the moments. In: Proc. of The Third International Conference, Problems of Cybernetics and Informatics, Baku, vol. 2, pp. 196–199 (2010)
22. Sinyakova, I., Moiseeva, S.: Investigation of queuing system $GI^{(2)}|M_2|\infty$. In: Proc. of the International Conference, Modern Probabilistic Methods for Analysis and Optimization of Information and Telecommunication Networks, Minsk, pp. 219–225 (2011)

Continuous Stochastic Dynamic Model for the Evolution of Polysemy and Sense Volume of Signs Ensembles of Natural Language and the Derivation of Their Synchronouos Distribution[*]

Vasily Poddubny[1] and Anatoly Polikarpov[2]

[1] National Research Tomsk State University, Tomsk, Russia
[2] Lomonosov Moscow State University, Moscow, Russia
vvpoddubny@gmail.com, anatpoli@mail.ru

Abstract. A continuous stochastic dynamic model for the evolution of polysemy and semantic volume of natural language signs is offered. The model is based on the assumption of the dissipative nature of the polysemy development of linguistic signs. On the basis of this model theoretical laws for synchronous (simultaneous) probability distributions for signs' ensembles are derived (age-and-polysemy, polysemy, age-and-sense-volume, sense-volume, frequency and frequency-rank distributions). Theoretically derived conclusions are compared with the corresponding empirical polysemy and sense volume distributions for lexical signs obtained from representative explanatory and frequency dictionaries of Russian and English.

Keywords: Evolution, linguistic sign, polysemy, semantic volume of a sign, stochastic mathematical model, explanatory and frequency dictionaries, identification of the model, the probability distributions.

1   Introduction and Basic Assumptions of the Model

It is believed that the frequency-rank distribution of signs of the natural language obeys to Zipf's law [1], [2], [3]. This law is expressed by the inverse power (i.e. hyperbolic) function of the Pareto distribution [4]. The graph of this function in the log-log coordinates is a straight line with a negative slope coefficient. However, the graph of the empirical distribution function of words of the natural language in the same coordinate system deviates from a straight line and is a concave function. It forces researchers to introduce amendments to the Zipf law, leading to its modifications in the form of Zipf–Mandelbrot law [5], [6]. Attempt at a theoretical justification of the Zipf–Mandelbrot law was undertaken by Yu. K. Krylov [7], M. A. Montemurro [8]. We propose an alternative approach to the

[*] The publication is prepared within the framework of the scientific project No. 14-14-70010 supported by the Russian Humanitarian Scientific Fund. This work is performed under the state order No. 1.511.2014/K of the Ministry of Education and Science of the Russian Federation.

A. Dudin et al. (Eds.): ITMM 2014, CCIS 487, pp. 367–376, 2014.

theoretical conclusion of the frequency-rank distribution of signs of the natural language based on the stochastic dynamic model of the polysemic evolution of signs ensembles.

The proposed mathematical model is based on the assumption of the dissipative nature of the linguistic sign polysemy development [9], [10]. In these investigations, we based on a discrete version of this model using simulation methods. In this paper, we present a continuous version of this model, using analytical methods. On this basis we undertake an attempt for theoretical derivation laws for polysemy, age-polysemy, frequency and age-frequency distributions for signs' ensembles.

The basic assumption of the model is that each linguistic sign at the moment of its birth has an individual limit of G – ability of a sign to generate (to acquire) some certain number of meanings through its life time. This ability called associative semantic potential (ASP) is gradually being implemented (wasted) in the course of use and corresponding development of any sign polysemy. At the same time the rate of generation for new meanings (usually the more abstract than later they are formed) assumed to be proportional still unspent part of ASP, whereby the rate of new meanings generation is gradually slowing down. At the same time, but with a delay of time τ_0, starts to flow a similar to signs generation process – the process of losses of some previously generated meanings. Because of relatively more specific nature of more initial meanings they are relatively less stable than the subsequent meanings. That is why the process of losses also is gradually slowing down.

Current polysemy of a sign at any time t of its life-cycle is expressed by the difference $x(t) = x_1(t) - x_2(t)$ between processes of new meanings gain $x_1(t)$ and loss of the previously acquired meanings $x_2(t)$. Continuous model suggests that these processes are continuous and subject to linear differential equations of the form:

$$\frac{dx_1(t)}{dt} = \frac{1}{\tau_1}(G - x_1(t)), \ x_1(t_0) = 1, \ t \geq t_0; \tag{1}$$

$$\frac{dx_2(t)}{dt} = \frac{1}{\tau_2}(G - x_2(t)), \ x_2(t_0 + \tau_0) = 0, \ t \geq t_0 + \tau_0, \tag{2}$$

where $\tau_1 = a_1/G$, $\tau_2 = a_2/G$ are inversely proportional to the ASP time constants of growth and decline of polysemy with coefficients of proportionality a_1 and a_2 respectively, and $\tau_1 \ll \tau_2$. Inequality $\tau_1 < \tau_2$ is required to ensure nonnegative values of polysemy $x(t)$. Value of $x_1(t)$ characterizes the maximum level of implementation of the sign's ASP, which is reached at a moment of time t. Correspondingly, value $x_2(t)$ characterizes respectively the minimum level of the sign's ASP, which is still remained unspent at time t.

The model suggests further that individual values of G and delay τ_0 – in a sign ensemble – should be independent exponentially distributed random variables with parameters (mathematical expectations) $\langle G \rangle$ and $\langle \tau_0 \rangle$ respectively.

Flood of signs birthes assumed to be stationary Poisson process with the intensity $\lambda = 1/\langle \tau \rangle$, so that the time intervals between neighboring occurrences of adjacent signs in the stream are assumed to be independent exponentially distributed random variables with parameter (expectation) $\langle \tau \rangle$. Moments of occurrence of events in a stationary Poisson flow have a uniform distribution.

Individual curve for the development of sign's polysemy x in time t (which is a curve for the polysemic life cycle of a sign) depending on G and τ_0 can be obtained by solving the above equations (1)-(2):

$$x(t) = (G - (G - 1)\exp(-G(t - t_0)/a_1)) \cdot 1(t \geq t_0) -$$
$$-G(1 - \exp(-G(t - t_0 - \tau_0)/a_2)) \cdot 1(t \geq t_0 + \tau_0), \quad x(t) \leq G, \quad (3)$$

where $1(\cdot)$ is an indicator for the fulfillment of the condition in brackets (this indicator takes the value 1, if the condition is satisfied, and the value 0, if otherwise). A typical course of the polysemy curve development for a particular linguistic sign is shown in Fig. 1. Number of meanings usually starts from 1, since each sign usually appears in a language having at least 1 meaning. Sum of lost meanings begins from 0, because we use continuous model for the polysemy development.

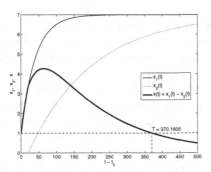

Fig. 1. Typical curve for an individual sign polysemy development ($G = 7$, $\tau_0 = 20$, $a_1 = 280$, $a_2 = 1260$), T is longevity of sign's life

As it can be seen, the curve of the polysemy development of a sign is a unimodal curve with a maximum of it located closer to the beginning of the evolution trajectory of a sign. So, polysemy of a sign first increases very fast, but with constant deceleration and then (after the beginning of the process of loosing of sign's initial meanings), reaches a maximum at some point. After that polysemy decreases until it reaches a value of 1 (it means a sign has the last meaning). Polysemy reduction to a level lower than 1 meaning means that a sign looses its last meaning and goes out of use. This occurs at the time point $t = t_0 + T$ where T is the longevity of a sign ($T \geq \tau_0$). If $t \leq t_0 + T$, the difference $t - t_0$ determines the age of a sign at time t. To find analytically longevity T of an individual linguistic sign is still difficult. It obeys analytically irresolvable transcendental equation:

$$(G - 1)\exp(-GT/a_1) - G\exp(-G(T - \tau_0)/a_2) = 1.$$

In the continuous model it is possible to suggest that, when linguistic signs reach a level of polysemy below 1, they do not extinct immediately, but remain in the language and live with exponentially decreasing level of polysemy up to 0 during increasing of length of age noticeably above T. This assumption has no significant impact on the results obtained, but greatly simplifies the procedures of theoretical drawing of polysemic and age-polysemic distributions. Therefore, then we shall not impose severe restrictions on the duration of the life of signs.

For the beginning point of time it is convenient to take the observation point, putting $t = 0$. For this reason we need to count the occurrence moments of linguistic signs in the stream in the opposite direction. As a result, we have $t \leq 0$, and the age of the sign will be equal $-t_0 \geq 0$. Now we denote by t the age of a sign. Formula (3) takes a simpler form, which expresses the dependence of the quantity sign's meanings of age t from its determinative values G and τ_0:

$$x(t) = (G - (G - 1)\exp(-Gt/a_1)) \cdot 1(t \geq 0) -$$
$$-G(1 - \exp(-G(t - \tau_0)/a_2)) \cdot 1(t \geq \tau_0) = x(t, G, \tau_0) \leq G. \qquad (4)$$

Since signs which conform to a higher level of implementation of the ASP have relatively more abstract meanings, each of them cover a wider sense area of meaning (have a greater sense volume) as compared to signs of lower level of ASP. And signs, having a large amount of meanings, should be used more often in the language of communication. Moreover, it can be assumed that the growth of frequency of use of the sign shall be directly proportional to its sense width (to the sum of sense volumes) of all signs' meanings. This, in particular, should be manifested in the fact that the overall semantic volume of the linguistic sign must be associated with its rank in the rank-frequency distribution of signs' textual use.

Natural to assume that the semantic scope of each of new meanings of each of next levels of implementation of ASP increases monotonically with the increasing level of it. This dependence can be represented by a power function of this level with an exponent $\mu - 1 > 0$. Then the total sense amount v of the interval from x_1 to x_2 for levels of implementation of the ASP to be expressed in the continuous model by the integral:

$$v = \mu \int_{x_1}^{x_2} x^{\mu-1}dx = x_2^\mu - x_1^\mu, \ \mu > 1.$$

Consequently, with regard to the expression (4), the linguistic sign of the age t having an actual polysemy defined by the difference between "plus" and "minus" semantic derivation $x_1(t)$ and $x_2(t)$, is characterized by semantic volume

$$v(t) = (G - (G - 1)\exp(-Gt/a_1))^\mu \cdot 1(t \geq 0) -$$
$$-(G(1 - \exp(-G(t - \tau_0)/a_2)))^\mu \cdot 1(t \geq \tau_0) = v(t, G, \tau_0) \leq G^\mu, \ \mu > 1. \quad (5)$$

When $\mu = 1$, this formula coincides with the formula (4) for polysemy $x(t)$. Therefore, formula (5) can be used as a universal formula describing sign polysemy (with $\mu = 1$), as well as its semantic volume (at $\mu > 1$). And, consequently, the formula (5) can be used also for describing the frequency features of the linguistic sign – taking into account that its sense volume should be directly related to the frequency of its use in the process of communication.

2 Theoretical Conclusion on Age-Polysemy, Polysemy, Age-Sense-Volume, Age-Frequency, Frequency and Rank-Frequency Distributions of Units in Sign Ensembles

Conditional density distribution of the sense volume v for an individual sign depending on the time t, associative semantic potential G and delay τ_0 for fixed values of these parameters is represented by Dirac delta function:

$$p(v|t, G, \tau_0) = \delta(v - v(t, G, \tau_0)) \cdot 1(v \leq G). \tag{6}$$

With regard to (5), the expression (6) takes the following form:

$$p(v|t, G, \tau_0) = \delta(v - (G - (G - 1)\exp(-Gt/a_1))^\mu \cdot 1(t \geq 0) +$$
$$+(G - G\exp(-G(t - \tau_0)/a_2))^\mu \cdot 1(t \geq \tau_0)) \cdot 1(v \leq G). \tag{7}$$

Averaging this distribution according to τ_0 and G distributions, one can get the conditional distribution density $p(v|t)$ for the sense volume (and for the polysemy) of signs of the same age. Integrating polysemy according v ranging from k to $k + 1$, it is possible to obtain the probability distribution $P(k|t)$ for integer values of the sense volume (and the polysemy) $k = 1, 2, \ldots$ of signs of the same age. More convenient, however, to calculate not the density distribution, but the cumulative distribution function $F(v|t)$, using it further to calculate $P(k|t)$. Averaging $p(v|t, G, \tau_0)$ according to an exponential distribution $p(\tau_0) = \exp(-\tau_0/\langle\tau_0\rangle)$ of the delay within the range $0 \leq \tau_0 < \infty$ is carried out without difficulty and leads to the formula:

$$p(v|t, G) = e^{-t/\langle\tau_0\rangle}\delta(v - v_1) + \frac{a_2 e^{-t/\langle\tau_0\rangle}}{G^2\langle\tau_0\rangle\mu} \cdot$$
$$\cdot \frac{1(v_1 - v_2 \leq v \leq v_1)}{(v_1 - v)^{1-1/\mu}(1 - (v_1 - v)^{1/\mu}/G)^{\frac{a_2}{G\langle\tau_0\rangle}+1}}, \tag{8}$$

where indicated:

$$v_1 = (G - (G - 1)\exp(-Gt/a_1))^\mu, \quad v_2 = (G - G\exp(-Gt/a_2))^\mu. \tag{9}$$

Conditional cumulative distribution function can also be calculated easily:

$$F(v|t, G) = \int_0^v p(\theta|t, G)d\theta = 1(v \geq v_1 - v_2) -$$
$$-e^{-t/\langle\tau_0\rangle}\frac{1(v_1 - v_2 \leq v \leq v_1)}{(1 - (v_1 - v)^{1/\mu}/G)^{\frac{a_2}{G\langle\tau_0\rangle}}}, \tag{10}$$

It is easy to see that $F(v|t, G) = 0$, if $v < v_1 - v_2$, and $F(v|t, G) = 1$, if $v > v_1$. And in the interval between these values the function $F(v|t, G)$ increases monotonically from 0 to 1 with increasing of v.

Formula (10) has a relatively simple form due to the fact that the integration is being done from 0, but not from 1, i.e. on the assumption that a sign may have polysemy lesser than 1, but not being formally out of use. However, the interpretation of the probability distribution of the polysemy for an ensemble of signs is necessary to consider that reducing polysemy to a value less than 1, the sign actually means going out of use.

Unfortunately, this averaging according G of the conditional distribution function fails to be carried out analytically. The distribution function $F(v|t)$ remains in an integral form:

$$F(v|t) = 1 - \int_1^\infty 1(v \geq v_1 - v_2) \exp(-(G-1)/\langle G \rangle) dG/\langle G \rangle -$$

$$-e^{-t/\langle \tau_0 \rangle} \int_1^\infty \frac{1(v_1 - v_2 \leq v \leq v_1) \exp(-(G-1)/\langle G \rangle) dG/\langle G \rangle}{(1 - (v_1 - v)^{1/\mu}/G)^{\frac{a_2}{G\langle \tau_0 \rangle}}}, \tag{11}$$

and can be calculated only numerically.

The difference $P(k|t) = F(k+1|t) - F(k|t)$, $k = 1, 2, \ldots$, at $\mu = 1$ determines the theoretical age-polysemic distribution for values of polysemy, but at $\mu > 1$ determines sense-volume distribution of units in a signs' ensemble.

We assume Poisson nature for the flow of words births in language life. Intervals of time τ_i between adjacent signs in the flow of them are thus statistically independent exponentially distributed random variables with mathematical expectation $\langle \tau \rangle$. The moment of occurrence $t_{0i} = \sum_{j=1}^i \tau_j$ of each next sign in the flow is determined by the Erlang distribution, corresponding with an order number of signs in the flow. It is possible to show that the arithmetic mean of Erlang distributions – with an infinite increase of the number of words in the language – tends to an uniform distribution at each time interval length $\langle \tau \rangle$. Thus, the time of occurrence t_{0i} of linguistic signs can be regarded as realizations of uniformly distributed random variables within any desired finite time interval.

The density distribution of polysemy at any time in an ensemble of n signs can be expressed by the arithmetic average of density for the distributions of terms involved in the averaging. For the received zero observation time, the above is equivalent to averaging of the distribution $F(v|t)$ within the uniform distribution of t on the whole semiaxis of ages:

$$F(v) = \lim_{T \to \infty} \int_0^T F(v|t) dt. \tag{12}$$

The difference $P(k) = F(k+1) - F(k)$, $k = 1, 2, \ldots$, at $\mu = 1$ determines the theoretical distribution of integer values of polysemy in an ensemble of linguistic signs, regardless of their age. While the difference at $\mu > 1$ determines theoretical distribution of their sense volumes. Distribution (12) is an irrespectable one for the age of signs.

In the assumption adopted here, the frequencies of use of signs in acts of communication (including, for example, a specific set of texts) are proportional to their sense volumes. Moreover, integer values of semantic volumes for signs may simply be identified with the absolute frequencies of use of signs in a communications field, taking the proportionality factor equal to 1. Then the theoretical distribution $P(k)$ of integer values of k for sense volume of a sign (at $\mu > 1$) expresses the share of signs in some language dictionary with absolute frequency k of their use in acts of communication. Thus, $P(k)$ (at $\mu > 1$) becomes the theoretical frequency distribution for the ensemble of signs (distribution of shares for signs with the frequency of use k).

3   Identification of the Model

To compare the empirical and theoretical distributions (obtained by the use of the proposed evolutionary model), it is necessary to identify the mathematical model with the real empirical data obtained from representative explanatory, frequency and historical dictionaries of a language under consideration. The proposed model is characterized by six parameters: $\langle G \rangle$, $\langle \tau \rangle$, $\langle \tau_0 \rangle$, a_1, a_2, μ. However, the parameter $\langle \tau \rangle$ is not explicitly included in the theoretical distribution, and the parameter $\langle \tau_0 \rangle$ can be set equal to 1, selecting it as the unit of time, and therefore – the unit of sign age. There are remained three parameters that define the shape of the probability distribution of polysemy: $\langle G \rangle$, a_1, a_2. These parameters can be determined according to the test of language dictionaries, using the least squares method for suitably chosen criterion. Parameter μ can then also be determined by use of the least squares method comparing theoretically drawn frequency distributions with the empirical one obtained from reliable frequency dictionaries. But someone can define all four parameters $\langle G \rangle$, a_1, a_2, μ at once, using the least squares method, comparing theoretical and empirical frequency distributions. In this case, the model identification is performed using only the frequency as the most objective kind of data for basing on it. One can also use a weighted criterion that assigns all available theoretical and empirical distributions – the frequency of signs' use as well as polysemic distribution. Finally, if necessary, one can associate a model unit of time with a real time, using empirical data on the age distribution of signs in language.

For the comparison of the theoretical and empirical polysemy and frequency distributions, changing in a very wide range (up to several orders of magnitude), it is convenient to choose the mean square deviation of their logarithms ratio from 1 as the criterion of their proximity:

$$\epsilon^2(q) = \sum_{k=1}^{k_{\max}} \left(\frac{\ln P_e(k)}{\ln P(k|q)} - 1 \right)^2, \quad q^* = \arg\min_q \epsilon^2(q), \tag{13}$$

where q^* is an estimation of the parameter vector $q = (\langle G \rangle, a_1, a_2)$, $q = \mu$ or $q = (\langle G \rangle, a_1, a_2, \mu)$, obtained by the least squares method (13), $P(k|q)$ is the theoretical distribution, depending on the parameter vector q, $P_e(k)$ is some empirical distribution obtained from the corresponding dictionary.

For identification and comparison of theoretical models with empirical distributions we used representative dictionaries of Russian and English languages. Russian is presented by "Dictionary of Pushkin's Language" (1958-1961) (explanatory and frequency) and English – by the explanatory dictionary Webster's Collegiate (9th ed.) After the identification of the model there were obtained values of the parameters, as it is shown in Table. 1 (accepted value $\langle \tau_0 \rangle = 1$). During the procedure we used a sequence identification – first, using the parameters $\langle G \rangle$, a_1, a_2, and then – the parameter μ.

Table 1. Model Parameters

Dictionaries	$\langle G \rangle$	a_1	a_2	μ (for x)	μ (for v)
Pushkin	2.18	0.122	36.5	1	2.86
Webster's	3.28	6.12	183	1	–

4   Comparative Analysis of Theoretical and Empirical Polysemy and Age-Polysemy Distributions

Fig. 2 shows diagrams calculated using the formula (11) of the conditional distribution $P(k|t)$ of polysemy and of sense volumes for an ensemble of signs, depending on their age at the parameters corresponding to the Pushkin dictionary. Age t of signs is present in units $\langle \tau_0 \rangle$.

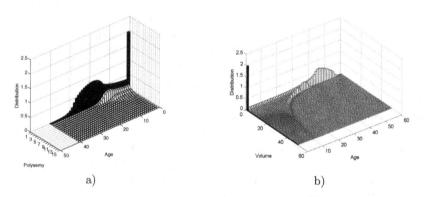

a)   b)

Fig. 2. Theoretical a) age-polysemy distribution and b) age-sense-volumes distribution with parameters of the Pushkin dictionary

It is seen from Fig. 2a) that the most of lexical signs in the theoretical distribution are nonpolysemic (having a polysemy "1" – see "far wall" on the chart). This group includes most of the signs (majority of them are rather young, but some of them are long-living). It is evident that signs of old ages are leaving language (signs of age, exceeding 35 units in the accepted system of measuring time are practically nonexistent, their probability is close to zero). Much less longevity is typical for signs with a polysemy 2. Relatively higher polysemy is inherent for signs which are not too young and not too old. Maximum polysemy falls on signs of some "ripe age" – in the region of several units of simulated time.

Fig. 2b) demonstrates quite curious behavior of the semantic volumes of signs distribution, depending on their age.

Fig. 3 presents theoretical and empirical distributions for the polysemy parameters of Pushkin and Webster dictionaries.

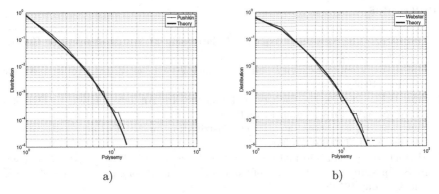

a)   b)

Fig. 3. Theoretical and empirical polysemy distributions with parameters a) of the Pushkin dictionary and b) that of Webster's dictionary

One can see a good agreement between the calculated theoretical and empirical polysemy distributions for two dictionaries.

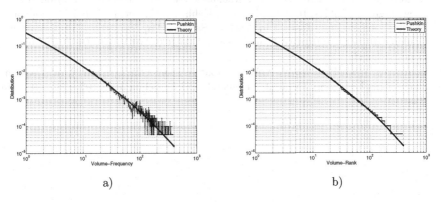

a)   b)

Fig. 4. Theoretical and empirical a) frequency distributions and b) frequency-rank distributions of words in dictionary of Pushkin

Fig. 4 presents the theoretical distribution of word frequencies (functionally equal to the distribution of their sense volumes) and the empirical frequency and frequency-rank distributions of words from the dictionary of Pushkin.

It is seen that theoretically obtained frequency distribution of words (which is the same – the sense volume of words) agrees well with the empirical frequency and frequency-rank distributions.

This allows theoretically calculate and investigate not only the frequency and frequency-rank distributions of sign ensembles, but also their polysemy distributions.

References

1. Zipf, G.K.: The Meaning-Frequency Relationship of Words. Journal of General Psychology 33, 251–256 (1945)
2. Zipf, G.K.: Human Behavior and the Principle of Least Effort. Addison-Wesley, Reading (1949)
3. Zipf, G.K.: The Psycho-Biology of Language. In: An Introduction to Dynamic Philology. MIT Press, Cambridge (1965)
4. Newman, M.E.J.: Power Laws, Pareto Distributions and Zipf's Law. Contemporary Physics 46(5), 323–351 (2005)
5. Mandelbrot, B.: An Informational Theory of the Statistical Structure of Language. In: Symposium on Applications of Communications Theory, pp. 486–500. W. Jackson, London (1953)
6. Mandelbrot, B.: Information Theory and Psycholinguistics: A Theory of Words Frequencies. In: Lazafeld, P., Henry, N. (eds.) Readings in Mathematical Social Science. MIT Press, Cambridge (1966)
7. Krylov, Y.K.: A Markov Model for the Evolution of Lexical Ambiguity. Journal of Quantitative Linguistics 2(1), 19–26 (1995)
8. Montemurro, M.A.: Beyond the Zipf-Mandelbrot Law in Quantitative Linguistics. Physica A: Statistical Mechanics and its Applications 300, 567–578 (2001)
9. Poddubny, V.V., Polikarpov, A.A.: Stochastic Dynamic Model of Evolution of Language Sign Ensembles. In: Obradovic, I., Kelih, E., Kohler, R. (eds.) Methods and Applications of Quantitative Linguistics – Selected Papers of the 8th International Conference on Quantitative Linguistics (QUALICO 2012), pp. 69–83. Academic Mind, Belgrade (2013)
10. Poddubny, V.V., Polikarpov, A.A.: Dissipative Stochastic Dynamic Model of Language Signs Evolution. Computer Research and Modeling 3(2), 103–124 (2011) (in Russian)

Formation of Probabilistic Distributions
of RSE by Associative Functions

Daria Semenova and Natalia Lukyanova

Institute of Mathematics and Computer Science
of Siberian Federal University
79 Svobodny pr., 660041 Krasnoyarsk, Russia
{dariasdv,nata00sfu}@gmail.com
http://www.sfu-kras.ru/

Abstract. One of the most important, complex and significant tasks
for science in general and for specific areas of application of probability,
such as economics and statistics, is to develop methods for the determination, study and statistical evaluation of the dependency structure of
complex distributions of large dimension. In the paper a new approach of
the description of probabilistic distributions of random sets of events by
means of the device of associative functions is offered. The feature of this
approach is that for definition of probabilistic distribution of a random
set of events it is enough to know N probabilities of events and a type of
associative function, whereas for definitions of probabilistic distribution
of any random set of events it is necessary to set 2^N of probabilities.

Keywords: Random set of events (RSE), probabilistic distribution, associative function.

1   Introduction

Each set of N events is characterized by a set of probabilities 2^N which play the
same role for a random set of events and which the probability distribution plays
for a random variable with a finite set of values. Distribution of a random set of
events is a convenient mathematical tool to describe all variants of interaction
between elements. It is obvious that the more events have been contained in the
selected set of events, the more complex structures of probabilistic dependence
of events it possesses. Nowadays a collection of probabilistic (eventological) distributions which completely the define structure of dependence of a random set
of events has been studied and described [1]. The most complete study of the
distributions is written in [1].

The key role is played by three basic structures of dependences such as embedded, least intersecting and independently-pointwise [1], [2], [3].

In this work the new approach of formation of probabilistic distributions of
random sets of events by means of the device of associative functions is offered.
The feature of this approach is that for definition of probabilistic distribution
of a random set from N events it is enough to know only N probabilities of

A. Dudin et al. (Eds.): ITMM 2014, CCIS 487, pp. 377–386, 2014.

events and a type of associative function, whereas for definitions of probabilistic distribution of any random set of events it is necessary to set 2^N of probabilities.

1.1 Random Set of Events

Consider a probability space $(\Omega, \mathcal{F}, \mathbf{P})$. Let $\mathfrak{X} \subset \mathcal{F}$ be a finite set of events chosen from algebra $\mathcal{F}$ of that space. Let designate $N = |\mathfrak{X}|$.

Definition 1. *Random set of events[1] (RSE) on a set of the chosen events $\mathfrak{X}$ is decided on probability space as a random element of*

$$K : \Omega \to 2^{\mathfrak{X}}$$

on values from measurable space $\left(2^{\mathfrak{X}}, 2^{2^{\mathfrak{X}}}\right)$, where $2^{\mathfrak{X}}$ is power set $\mathfrak{X}$, $2^{2^{\mathfrak{X}}}$ is algebra of all its subsets.

The random set of K maps any $\omega \in \Omega$ in $2^{\mathfrak{X}}$, i.e $K(\omega \in \Omega) \in 2^{\mathfrak{X}}$. And this mapping is measurable, it means that for any $A \in 2^{2^{\mathfrak{X}}}$ there is a preimage of $K^{-1}(A) \in \mathcal{F}$ such that $\mathbf{P}(A) = \mathbf{P}(K^{-1}(A))$.

Definition 2. *Probabilistic distribution of RSE of K which has been set on a finite set of the chosen events $\mathfrak{X} \subseteq \mathcal{F}$ can be presented several equivalent distributions of the probabilities generated by a set of events $\mathfrak{X}$ [1]:*

– *probabilistic distribution of the I-st sort is an array from 2^N probabilities of a type*

$$\left\{ p(X) = \mathbf{P}(K = X) = \mathbf{P}\left(\left(\bigcap_{x \in X} x \right) \cap \left(\bigcap_{x \in X^c} x^c \right) \right), \quad X \subseteq \mathfrak{X} \right\};$$

– *probabilistic distribution of the II-nd sort is an array from 2^N probabilities of a type*

$$\left\{ p_X = \mathbf{P}(X \subseteq K) = \mathbf{P}\left(\bigcap_{x \in X} x \right), \quad X \subseteq \mathfrak{X} \right\}.$$

Probabilistic distribution of RSE contains the comprehensive information about all types of probabilistic dependencies of events from this set. The structure of probabilistic dependencies includes types of dependence between events, which make couples of events, the three of events and more powerful subsets of events of this set. If the power of a considered set of events is equal to N, it is possible to imagine 2^N as classical types of dependencies between events of this set, i.e. it is exactly equal to number of set's subsets.

Probabilistic distributions of the I-st and the II-nd sorts are connected with a help of the Möbius inversion formulas[1].

$$p_X = \sum_{Y \in 2^{\mathfrak{X}}:\, X \subseteq Y} p(Y), \quad p(X) = \sum_{Y \in 2^{\mathfrak{X}}:\, X \subseteq Y} (-1)^{|Y|-|X|} p_Y. \tag{1}$$

---

[1] Further, the abbreviation *RSE* will be used for the convenience.

Further, let formulate necessary and sufficient conditions of legitimacy of probabilistic distribution of RSE.

Necessary and Sufficient Conditions of Legitimacy of Probabilistic Distribution of RSE. *Let say that the random set of events of K possesses legitimate probabilistic distribution if for probabilistic distribution of the I-st sort the following conditions are satisfied:*

- $0 \leq p(X) \leq 1, X \subseteq \mathfrak{X}$,
- $\sum\limits_{X \subseteq \mathfrak{X}} p(X) = 1$.

Necessary Condition of Legitimacy of Probabilistic Distribution of the II-nd Sort. *Legitimate probabilistic distribution of the II-nd sort $\{p_X, X \subseteq \mathfrak{X}\}$ of RSE of K on $\mathfrak{X}$ satisfies to system from 2^N inequalities of Frechet:*

$$p_X^- \leq p_X \leq p_X^+,$$

where

$$p_X^- = \max\left\{0, 1 - \sum_{x \in X}(1 - p_x)\right\}$$

— *Frechet's lower bound,*

$$p_X^+ = \min_{x \in X} p_x$$

— *Frechet's upper bound.*

Notice that only probabilistic distribution of the I-st sort satisfies to a ratio of a probabilistic normalization. This results from the fact that probabilities of events of the I-st sort $\mathbf{P}\left(\left(\bigcap\limits_{x \in X} x\right) \cap \left(\bigcap\limits_{x \in X^c} x^c\right)\right)$ unlike to probability of events of the II-nd sort $\mathbf{P}\left(\bigcap\limits_{x \in X} x\right)$ are not crossed and form a partition the spaces of elementary outcomes. In other words, if the distribution of the I-st sort is defined, a legitimate distribution of the II-nd sort by Möbius inversion formulas will be always received. Vice versa is not always truly, for example, the set from 2^N numbers $\mathbf{P}(X)$, $X \subseteq \mathfrak{X}$ from $[0, 1]$ satisfying to Frechet's bounds, does not always define legitimate distribution of RSE. It demands additional research in each case.

In work [1] the convenient tool of the analysis of structures of event dependence – covariance – was offered. *Covariance* of two events $x, y \in \mathfrak{X}$ is defined as the variable

$$\text{Kov}_{xy} = \mathbf{P}(x \cap y) - \mathbf{P}(x)\mathbf{P}(y),$$

which is equal zero when these events are independent; it is greater than zero when they occur together more frequently (are statistically attracted), and it is less than zero when they occur together more rarely (are statistically repelled) than in independent situations. Covariance of events serves as a measure of an additive deviation of events from an independent situation. For a set of events $X \subseteq \mathfrak{X}$ ary covariance is determined by a formula [1]:

$$\text{Kov}_X = \mathbf{P}\left(\bigcap_{x \in X}\right) - \prod_{x \in X} \mathbf{P}(x), \qquad X \subseteq \mathfrak{X}. \tag{2}$$

1.2   Associative Functions

Classes of associative functions [4], in particular, triangular norms [5], [6], [7] and copulas [8], [9] are widely used in contemporary theories of uncertainty.

In this paper the following definition of associative functions is used.

Definition 3. *Associative function in the theory of RSE*

$$\text{AF} : [0, 1]^2 \to [0, 1]$$

is defined as a two-place function satisfying to the following properties:

A1. Boundary conditions

$$\text{AF}(a, 0) = \text{AF}(0, a) = 0,$$
$$\text{AF}(a, 1) = \text{AF}(1, a) = a, \tag{3}$$

$a \in [0, 1].$
A2. Monotonicity

$$\text{AF}(a_1, b_1) \le \text{AF}(a_2, b_2), \tag{4}$$

when $a_1 \le a_2$, $b_1 \le b_2$.
A3. Commutativity, i.e. for all a, $b \in [0, 1]$

$$\text{AF}(a, b) = \text{AF}(b, a).$$

A4. Associativity, i.e. for all a, b, $c \in [0, 1]$

$$\text{AF}(\text{AF}(a, b), c) = \text{AF}(a, \text{AF}(b, c)). \tag{5}$$

A5. condition of Lipschitz's continuity

$$\text{AF}(c, b) - \text{AF}(a, b) \le c - a, \quad a \le c, \quad a, b, c \in [0, 1]. \tag{6}$$

Note that the properties A1-A4 corresponds to the definition t-norm [6]. Thus, under the associative function it is understood as a continuous t-norm satisfying the Lipschitz's condition or, equivalently, associative, commutative copula[4], [8].

2 Recurrent Approach of Constructing of Probabilistic Distributions of Random Sets of Events by Associative Functions

Calculation methods of copulas and triangular norms [4], [6], [7], [8] are applicable to probabilistic distributions of RSE. It is offered to consider probabilities of events (their number coincides with a power of a basic set) as arguments of associative function. Properties of associative function allow to receive probabilistic distributions with the set structure of dependence. Let proceed to describe the method.

Input:

- set of events $\mathfrak{X}$, $|\mathfrak{X}| = N$;
- N probabilities of events p_x, $x \in \mathfrak{X}$;
- associative function of $\mathrm{AF}_\alpha(a, b)$, where α is vector of parameters of function.

Output: probabilistic distribution of the II-nd sort $\{p_X, X \subseteq \mathfrak{X}\}$, of a set of events $\mathfrak{X}$.

Main Idea: probabilities of the intersection of sets of events p_X are defined by a recurrence relation at known probabilities of events $p_x = \mathbf{P}(x)$, $x \in \mathfrak{X}$

$$p_{xy} = \mathbf{P}\left(x \bigcap y\right) = \mathrm{AF}\left(p_x, p_y\right),$$

$$p_{xyz} = \mathbf{P}\left(x \bigcap y \bigcap z\right) = \mathrm{AF}\left(p_x, \mathrm{AF}(p_y, p_z)\right) = \mathrm{AF}\left(p_x, \mathbf{P}\left(y \bigcap z\right)\right),$$

$$p_X = \mathbf{P}\left(\bigcap_{x \in X} x\right) = \mathrm{AF}\left(p_x, \mathbf{P}\left(\bigcap_{y \in X \setminus \{x\}} y\right)\right), \qquad X \subseteq \mathfrak{X}.$$

The offered recurrent approach allows to define a RSE through probabilistic distribution of the II-nd sort. Probabilities of events and a type of associative function act as input parameters. In works [4], [5], [6], [8] it is shown that associative function satisfies to Frechet's bounds. Thus, this method allows to receive $(2^N - N - 1)$ probabilities of the II-nd sort which satisfy to Frechet's bounds and which do not suffice to a totality of probabilities of the II-nd sort. However, these distributions can turn out illegitimate. Therefore, for each family of associative functions it is necessary to define sufficient conditions of legitimacy of probabilistic distribution.

2.1 Families of Associative Random Sets of Events

Let $\mathfrak{X}$ be a set of the chosen events and let probabilities of events $p_x = \mathbf{P}(x)$ be fixed. A general view of probabilities of the II-nd sort p_X is received by the recurrent method and sufficient conditions of legitimacy of probabilistic distribution of the random set determined by the following functions AF are found:

- $\text{AF}(a, b) = a \cdot b;$
- $\text{AF}(a, b) = \min\{a, b\};$
- $\text{AF}(a, b) = \max\{a + b - 1, 0\};$
- $\text{AF}(a, b) = \dfrac{ab}{a + b - ab}.$

Notice that the first three functions define basic structures of dependences [2], [3] RSE, namely independently-pointwise, embedded and least intersecting sets of events.

Further, results of work of a recurrent approach in the form of theorems will be formulated for the listed above associative functions.

Theorem 1. *Associative function* $\text{AF}(a, b) = a \cdot b$ *defines an associative RSE with a legitimate probabilistic distribution of the II-nd sort*

$$p_X = \prod_{x \in X} p_x, \qquad X \subseteq \mathfrak{X}.$$

An associative random set of events with the function of the type $\text{AF}(a, b) = a \cdot b$ *is an independently-pointwise RSE with a probabilistic distribution of the I-st sort*

$$p(X) = \boldsymbol{P}(K = X) = \prod_{x \in X} \boldsymbol{P}(x) \prod_{x \in X^c} \boldsymbol{P}(x^c), \ X \in 2^{\mathfrak{X}}.$$

Theorem 2. *Associative function* $\text{AF} = \min\{a, b\}$ *defines an associative RSE with a legitimate probabilistic distribution of the II-nd sort*

$$p_X = \min_{x \in X} p_x.$$

An associative RSE with the function of the type $\text{AF} = \min\{a, b\}$ *is random set of embedded events[2].*

Theorem 3. *Associative function* $\text{AF}(a, b) = \max\{a + b - 1, 0\}$ *defines*

1. *a random set with nonintersecting structure of dependency with a legitimate probabilistic distribution if the probabilities of the events* $p_x = \boldsymbol{P}(x) > 0$, $x \in \mathfrak{X}$ *satisfy the inequality*

$$\sum_{x \in \mathfrak{X}} p_x \leq 1;$$

2. *a RSE with a legitimate probabilistic distribution if the probabilities of the events* $p_x = \boldsymbol{P}(x) > 0$, $x \in \mathfrak{X}$ *satisfy the inequalities*

$$|\mathfrak{X}| - 1 \leq \sum_{x \in \mathfrak{X}} p_x \leq |\mathfrak{X}|.$$

---

[2] If every two events from X are embedded, then X is called as a set embedded events [1], or the set of random events with embedded structure [1].

Theorem 4. *Let the probabilities of events $p_x = \mathbf{P}(x) > 0$, $x \in \mathfrak{X}$ be given. Then associative function* $\mathrm{AF}(a,b) = \dfrac{ab}{a+b-ab}$ *defines an associative RSE with a legitimate probabilistic distribution of the II-nd sort:*

$$p_X = \frac{\prod\limits_{x \in X} p_x}{\sum\limits_{Y \subseteq C_X^{|X|-1}} \left[\prod\limits_{x \in Y} p_x \right] - (|X|-1) \cdot \prod\limits_{x \in X} p_x}, \quad |X| > 1,\ X \in \mathfrak{X},$$

where $C_X^{|X|-1} = \{Y : Y \subseteq X, |Y| = |X| - 1\}$.

Theorem 5. *For associative RSE which is defined by the function*

$$\mathrm{AF}(a,b) = \frac{ab}{a+b-ab},$$

all ary covariance Kov_X *are non-negative and have the form*

$$\mathrm{Kov}_X = \prod_{x \in X} p_x \cdot \left[\frac{1}{\sum\limits_{Y \subseteq C_X^{|X|-1}} \left[\prod\limits_{x \in Y} p_x \right] - (|X|-1) \cdot \prod\limits_{x \in X} p_x} - 1 \right],$$

for all $X \subseteq \mathfrak{X}$, $|X| > 1$.

In these theorems the families of associative RSE were considered which probabilistic distribution of the II-nd sort is completely defined by probabilities of events and a type of associative functions.

2.2   Associative RSE of Frank and Ali-Mikhail-Haq

In the theory of associative functions [4], [6] an one-parameter function has been well researched and studied. Further, in the paper let consider the one-parameter family of Frank and the one-parameter family of Ali-Mikhail-Haq.

Family of Ali-Mikhail-Haq. In the paper the family of Ali-Mikhail-Haq is considered and the general formula of probability of the II-nd sort is received. Proven following theorem allows to obtain always legitimate probabilistic distributions of RSE by narrowing the area of constraints on the values of the parameter α.

Theorem 6. *Let the probability of events $p_x = \mathbf{P}(x)$, $x \in \mathfrak{X}$ satisfy the inequalities $0 < p_x < 1$. Then associative function*

$$\mathrm{AF}_\alpha(a,b) = \frac{ab}{1 - \alpha(1-a)(1-b)}, \quad \alpha \in [0,1]$$

defines family of associative random sets of Ali-Mikhail-Haq with probabilistic distribution of the II-nd sort for all $X \subseteq \mathfrak{X}$, $|X| > 1$

$$p_X = \frac{\displaystyle\prod_{x \in X} p_x}{(1-\alpha)^{|X|-1} + \displaystyle\sum_{k=1}^{|X|-1} \alpha^k \cdot \left[\displaystyle\sum_{Y \subseteq C_X^k} \left[(-1)^{k-|Y|} \cdot \delta_k(Y) \cdot \prod_{x \in Y} p_x \right] - \prod_{x \in X} p_x \right]},$$

where $C_X^k = \{Y : Y \subseteq X, |Y| = k\}$,
$$\delta_k(Y) = \begin{cases} 1, & k = 1, \ \ k = |X| - 1; \\ k - |Y| + 1, & 1 < k < |X| - 1. \end{cases}$$

Consequence. *For associative distribution of Ali-Mikhail-Haq*

– *for $\alpha = 1$ obtain a random set defined associative function*

$$\mathrm{AF}(a, b) = \frac{ab}{a + b - ab};$$

– *for $\alpha = 0$ obtain an independently-pointwise random set which is determined by the associative function*

$$\mathrm{AF}(a, b) = a \cdot b.$$

For a family of Ali-Mikhail-Haq $\alpha \in [0, 1]$, respectively, the covariance is always positive. Thus, the distribution of Ali-Mikhail-Haq allows to model events which are statistically attracted.

Family of Frank. In 1979 was introduced in [10] the one-parameter family of functions of Frank that are associative.

$$Frank(x_1, ..., x_n, \alpha) = -\frac{1}{\alpha} \ln \left(1 + \frac{(e^{-x_1 \cdot \alpha} - 1) \cdot ... \cdot (e^{-x_n \cdot \alpha} - 1)}{(e^{-\alpha} - 1)^{n-1}} \right),$$

$(x_1, \cdots, x_n) \in [0, 1]^n$, $\alpha \in (-\infty, +\infty) \setminus \{0\}$.

In this paper it is proposed to introduce an associative family of Frank of random sets of events. Theorem about the form of probabilistic distributions of the I-st and II-nd sorts for the family of Frank has been formulated and proved.

Theorem 7. *Let the probability of events is $p_x = \mathbf{P}(x) > 0$, $x \in \mathfrak{X}$, then the associative function*

$$\mathrm{AF}_\alpha(a, b) = Frank(a, b, \alpha) = -\frac{1}{\alpha} \ln \left(1 + \frac{(e^{-\alpha \cdot a} - 1)(e^{-\alpha \cdot b} - 1)}{(e^{-\alpha} - 1)} \right),$$

where $\alpha \in (-\infty; \infty) \setminus \{0\}$, defines an associative random set of Frank

– *with the probabilistic distribution of the II-nd sort:*

$$p_X = -\frac{1}{\alpha} \ln \left(1 + \frac{\prod\limits_{x \in X} (e^{-\alpha p_x} - 1)}{(e^{-\alpha} - 1)^{|X|-1}} \right), X \subseteq \mathfrak{X},$$

– *with the probabilistic distribution of the I-st sort:*

$$p(X) = \ln \prod_{Y \supseteq X} \left(1 + \frac{\prod\limits_{x \in Y} (e^{-\alpha p_x} - 1)}{(e^{-\alpha} - 1)^{|Y|-1}} \right)^{\frac{(-1)^{|Y|-|X|+1}}{\alpha}}.$$

Associative random set of Frank will have a legitimate probability distribution, if all probabilities $p_x, x \in \mathfrak{X}$ and parameter $\alpha \neq 0$ satisfy the following system from $2^{|\mathfrak{X}|}$ inequalities

$$1 \leq \prod_{Y \supseteq X} \left(1 + \frac{\prod\limits_{x \in Y} (e^{-\alpha p_x} - 1)}{(e^{-\alpha} - 1)^{|Y|-1}} \right)^{\frac{(-1)^{|Y|-|X|+1}}{\alpha}} \leq e, \qquad X \subseteq \mathfrak{X}.$$

Theorem 8. *For an associative random set of Frank all ary covariance* Kov_X, $X \subseteq \mathfrak{X}$, $|X| > 1$, *have the form*

$$\mathrm{Kov}_X = -\frac{1}{\alpha} \ln \left(\left(1 + \frac{\prod\limits_{x \in X} (e^{-\alpha \cdot p_x} - 1)}{(e^{-\alpha} - 1)^{|X|-1}} \right) \cdot e^{\alpha \prod\limits_{x \in X} p_x} \right), X \subseteq \mathfrak{X}.$$

The sign of the parameter $\alpha \in (-\infty; +\infty) \setminus \{0\}$ determines the sign of covariance.

3   Conclusion

One of the most important, complex and significant tasks for science in general and for specific areas of application of probability, such as economics and statistics, is to develop methods for determination, study and statistical evaluation of the dependency structure of complex distributions of large dimension.

Distribution of a RSE is a convenient mathematical tool to describe all variants of interaction between elements. It should be noted that probabilistic distribution of a RSE is defined by $2^{|\mathfrak{X}|}$ a set of parameters.

According to statistical data only probabilities of monoplet of events $p_x = \mathbf{P}(x \in K)$, $x \in \mathfrak{X}$ ($|\mathfrak{X}|$ parameters), as a rule, are available to us in practice. In the paper the new approach of modeling of probabilistic distributions of a RSE by means of the device of associative functions is offered. This approach allows to reduce the dimensionality of the problem from $2^{|\mathfrak{X}|}$ to $|\mathfrak{X}|$ parameters.

In work two new families of random sets of events Frank and Ali-Mikhail-Haq are entered by means of the device of associative functions. For these families the following theorems are proved:

- about a probabilistic distribution of the RSE of II-nd sort;
- about the form of the ary covariances;
- about the sufficient condition for the legitimacy of a probabilistic distribution of II-nd sort.

The offered method does not apply for universality, however, it allows to obtain input data in the form of probabilistic distributions of a RSE and their characteristics for a number of models of the statistical systems which possess a complex structure of dependences and interrelations.

Acknowledgments. Authors express sincere gratitude to their friends and colleagues: V.V. Bykova, I.V. Baranova, E.E. Goldenok and A.A. Novoselov whose effective cooperation with was the reason of emergence of the main results of this work. Special gratitude is expressed to the teacher prof. O.Yu. Vorobyev who has stimulated the research of authors in the theory of sets of events for many years.

References

1. Vorobyev, O.Y.: Eventology. Siberian Federal University, Krasnoyarsk (2007) (in Russian)
2. Vorobyev, O.Y., Goldenok, E.E.: Structural set-analysis of dependences of random events. ICM of RAS, Krasnoyarsk (2002) (in Russian)
3. Vorobyev, O.Y., Fomin, A.Y.: Set-regressional analysis of dependences of random events in statistical systems. ICM of RAS, Krasnoyarsk (2004) (in Russian)
4. Alsina, S., Frank, M.J., Scheiveizer, B.: Associative functions Triangular Norms and Copulas. World Scientific Publishing Co. Pte. Ltd. (2006)
5. Klement, E.P., Mesiar, R., Pap, E.: Triangular norms. Kluwer Academic Pub., Boston (2000)
6. Klement, E.P., Mesiar, R.: Logical, Algebraic, Analytic and Probabilistic Aspects of Triangular Norms. Kluwer, Dordrecht (2005)
7. Mesiar, R.: On some constructions of new triangular norms. Mathware and Some Computing 2, 39–45 (1995)
8. Nelsen, R.B.: An Introduction to Copulas, 2nd edn. Springer Science+Business Media, Inc., New York (2006)
9. Embrechts, P., Lindskog, F., McNeil, A.J.: Modelling dependence with copulas and applications to risk management. In: Rachev, S.T. (ed.) Handbook of Heavy Tailed Distributions in Finance. Elsevier/North-Holland, Amsterdam (2003)
10. Frank, M.J.: On the simultaneous associativity of $F(x,y)$ and $x + y - F(x,y)$. Aequationes Math. 19, 194–226 (1979)
11. Schweizer, B., Sklar, A.: Probabilistic metric spaces. North-Holland, New York (1983)
12. Goldenok, E.E.: Economic eventology and eventological economics. KGTEI, Krasnoyarsk (2010) (in Russian)
13. Nguyen, H.T.: An Introduction to Random Sets. Taylor & Francis Group, LLC (2006)
14. Shiryaev, A.N.: Probability-1, Moscow (2004) (in Russian)
15. Orlov, A.I.: Non-numerical statistics, Moscow (2004) (in Russian)

The Reduction of the Multidimensional Model of the Nonlinear Heat Exchange System with Delay

Aleksandr Shilin and Viktor Bukreev

National Research Tomsk Polytechnic University, 634050, Russia, Tomsk
{shilin,bukreev}@tpu.ru

Abstract. A method of reducing of a multi-dimensional model of the complex non-linear heat exchange system (HES) with delay based on structural changes in equilibrium points and approximation of delay function by the end of inertial units is presented. Simulation results confirming the adequacy of the process of reduced and original models, as well as their compliance with the real data of the experimental control object are demonstrated.

Keywords: complex heat-exchange system, equilibrium state, structural transformation, reduction of multidimensional models.

1 Introduction

In modern systems of intelligent management of complex equipment there are often used optimal control algorithms providing real-time management of production processes. One of the well-approved approaches for the synthesis of such algorithms is the theory of linear systems, which assumes a description of non-linear processes and objects in terms of linearized models [1]. The main obstacle to the use of these methods is a significant degree of differential equations describing the behavior of a complex multi-dimensional control object with the necessary range of accuracy. The most preferred variant of priori mathematical models allowing to record the control laws in an analytical form, and submit the results of the analysis of dynamic processes in a convenient form, is second order differential equations. For example, in the relay control systems in sliding mode control methods for nonlinear second-order systems are well-developed [2], [3].

There are several approaches to reduction of dimensional models of high dimension [4]: the use of linearized matrix properties when converting to block matrix according to division into independent blocks ; the use of the coefficients properties of low interconnection; aggregating the matrix elements; the models selection according to the frequency hierarchy of submatrices; separation in time or frequency.

1.1 Problem Formulation

We will consider the problem of reduction of non-linear mathematical model of a distributed heat-exchange system with maximum accounting its features. We

A. Dudin et al. (Eds.): ITMM 2014, CCIS 487, pp. 387–396, 2014.

assume interval stationary of HES element parameters and determinate nature of interrelated thermal processes. It is appropriate to present the approximation procedure in several stages [4]: decomposition of the original model [5]; formation of several reduced models [6]; structural transformation in equilibrium state points and leading to a simpler form; checking the approximate model for the adequacy of the complete model or the real process of high order.

2   The Original Non-linear Model in the Space of State Variables

As an illustrative example here is the heat exchange system (Fig. 1), characterized by significant non-linear properties [7]. Notation: 1 heat exchanger, 2 circulating pumps, 3 control valve with AC motor 4, 5 temperature sensors and a microprocessor controller (MC).

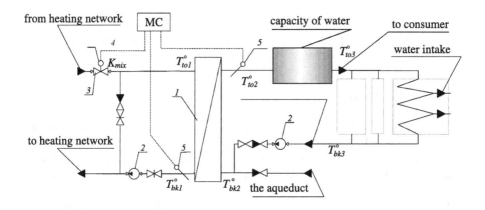

Fig. 1. The technological scheme structure of the heat exchange system

Salient features of HES with high-performance heat exchangers are not only the non-linearity of the three-point relay control, but also the delay in the formation channels of control actions and the flow of heat transfer in a distributed pipe network of the secondary circuit. In addition, significant disturbance on the characteristics and parameters of the heat exchange system is made by a periodic flow of cold water into the secondary loop of the HES to compensate for the irreplaceable HES discharge.

Arbitrary arrangement of risers-of couplers of the HES secondary circuit and their different distances from the heat exchanger causes a variable delay which makes a significant impact on the dynamics of thermal processes in the primary circuit.

Under certain admissions for thermal processes taking place in contours of HES, the original non-linear model of the system in the space of state variables can be represented by the following differential equations:

$$\begin{cases} \dfrac{dK_{mix}(t)}{dt} = (k_{mx} - K_{mix}(t)) \cdot \dfrac{k_h}{T_{vlv}} \cdot u(t) \\[2mm] \dfrac{dT_{to1}^{\circ}(t)}{dt} = \dfrac{(T_1^{\circ} - T_{bk1}^{\circ}(t)) \cdot K_{mix}(t) + T_{bk1}^{\circ}(t) - T_{to1}^{\circ}(t)}{T_{mix}} \\[2mm] \dfrac{dT_{to2}^{\circ}(t)}{dt} = \dfrac{k_{exc} \cdot T_{to1}^{\circ}(t) + (1 - k_{exc}) \cdot T_{bk2}^{\circ}(t) - T_{to2}^{\circ}(t)}{T_{exc}} \\[2mm] \dfrac{dT_{bk1}^{\circ}(t)}{dt} = \dfrac{k_{exc} \cdot T_{bk2}^{\circ}(t) + (1 - k_{exc}) \cdot T_{to1}^{\circ}(t) - T_{bk1}^{\circ}(t)}{T_{exc}} \\[2mm] \dfrac{dT_{to3}^{\circ}(t)}{dt} = \dfrac{T_{to2}^{\circ}(t) - T_{to3}^{\circ}(t)}{T_{cp}} \\[2mm] \forall i = 1..n \rightarrow \left\{ \forall j = 1..m \rightarrow \left\{ \dfrac{dT_{bkz(i,j)}^{\circ}(t)}{dt} = \dfrac{T_{bkz(i,j-1)}^{\circ}(t) - T_{bkz(i,j)}^{\circ}(t)}{\tau_{z(i)}/m} \right\} \right\} \\[2mm] \dfrac{dT_{bk2}^{\circ}(t)}{dt} = \dfrac{k_{cw} \cdot T_{cw}^{\circ}(t) + (1 - k_{cw}) \cdot T_{bk3}^{\circ}(t) - T_{bk2}^{\circ}(t)}{T_{cw}} \end{cases} \quad (1)$$

$$\forall i = 1..n \rightarrow T_{bkz(i,0)}^{\circ}(t) = (1 - k_{cl}) \cdot T_{to3}^{\circ}(t) + k_{cl} \cdot T_{rm}^{\circ}$$

$$T_{bk3}^{\circ}(t) = \sum_{i=1}^{n} \left(k_{zi} \cdot T_{bkz(i,m)}^{\circ}(t) \right), \sum_{i=1}^{n} k_{zi} = 1$$

where $K_{mix}(t)$ - the coefficient of coolant mixing in the external circuit to the mixing unit; k_{mx} and k_h - coefficient characterizing the nonlinear properties of the mixing process; T_{vlv} - the time constant of the electric control valve; $u(t)$ - electric valve control action that takes one of discrete values $u \in (-1, 0, +1)$; T_{to1}° - the coolant temperature at the inlet HES in the external circuit; T_1° - the coolant temperature coming out of the backbone network; $T_{mix}, T_{exc}, T_{cp}, T_{cw}$ - respectively, constants mixing time of the valve in the heat exchanger, in the intermediate storage device, in the input node of cold water; T_{bk1}° - the coolant temperature at the outlet of the external circuit of HES; T_{to2}° - the coolant temperature at the outlet of the internal circuit of HES; T_{bk2}° - the coolant temperature at the inlet of the internal circuit of HES; k_{exc} - the coefficient of heat exchange efficiency; T_{to3}° - the coolant temperature at the outlet of the intermediate storage, which is located in the internal circuit; T_{bk3}° - return temperature before unit mixing with cold water; $\tau_{z(i)}$ - the time delay of the transport carrier in the secondary circuit (i riser-branch); $T_{bkz(i,j)}^{\circ}$ - the equivalent temperature in inertial units to be used for the approximation of the transport delay; k_{cw} - the coefficient of cold water influence on the coolant in the internal circuit; k_{cl} - the coefficient of the coolant cooling in the internal circuit; n - the number of secondary circuits (riser-branch) in HES; m - the number of inertial units approximating transport delay.

Obviously, the order of the system of differential equations (1) will be determined by $(6 + n \cdot m)$. For example, for a system with two risers $n = 2$ and branches of inertial delay line units equal to $m = 5$ the number of equations of the system will be 16. The coefficient k_{zi} means relative proportion of the i-th heat flow throughout the entire volume of the coolant flow q_2, and characterizes the distribution of the i-th flow on risers-branches.

3 The Reduction of the Original Non-linear Multidimensional Model

The procedure of the original nonlinear model HES converting assumes finding the equilibrium points, which can be calculated by solving the system (1) with priori known parameters of the object and control $u(t) = 0$.

As a result of this calculation with the required accuracy we can find steady values of the following variables state of the heat exchange system:

$$\left[K_{mix}^0, T_{to1}^{o0}, T_{to2}^{o0}, T_{bk1}^{o0}, T_{to3}^{o0}, \forall i = 1..n \to \left\{ \forall j = 1..m \to \left\{ T_{bkz(i,j)}^{o0} \right\} \right\}, T_{bk3}^{o0} \right] \quad (2)$$

where the superscript "0" denotes the state variables belong to the fields of steady state. Further, using the calculated values (2) we write equations that reflect dynamic processes in a neighborhood of steady state:

$$\begin{cases}
x_1(t) = K_{mix}(t) - K_{mix}^0 \\
x_2(t) = T_{to1}^\circ(t) - T_{to1}^{o0} \\
x_3(t) = T_{to2}^\circ(t) - T_{to2}^{o0} \\
x_4(t) = T_{bk1}^\circ(t) - T_{bk1}^{o0} \\
x_5(t) = T_{to3}^\circ(t) - T_{to3}^{o0} \\
\forall i = 1..n \to \left\{ \forall j = 1..m \to \left\{ x_{5+(i-1)\cdot m+j}(t) = T_{bkz(i,j)}^\circ(t) - T_{bkz(i,j)}^{o0} \right\} \right\} \\
x_{6+n\cdot m}(t) = T_{bk3}^\circ(t) - T_{bk3}^{o0}
\end{cases} \quad (3)$$

Defining the $x(t) = [x_1(t), x_2(t), .., x_{6+n\cdot m}]^T$, the linearized model can be written in a vector-matrix form:

$$\dot{x} = \begin{vmatrix}
0 & 0 & 0 & 0 & 0 & 0 & \cdot & 0 & 0 \\
a_{2,1} & a_{2,2} & 0 & a_{2,4} & 0 & 0 & \cdot & 0 & 0 \\
0 & a_{3,2} & a_{3,3} & 0 & 0 & 0 & \cdot & 0 & a_{3,6+nm} \\
0 & a_{4,2} & 0 & a_{4,4} & 0 & 0 & \cdot & 0 & a_{4,6+nm} \\
0 & 0 & a_{5,3} & 0 & a_{5,5} & 0 & \cdot & 0 & 0 \\
0 & 0 & 0 & 0 & A_{in1} & A_1 & \cdot & 0 & 0 \\
\cdot & \cdot & \cdot & \cdot & \cdot & \cdot & & \cdot & \cdot \\
0 & 0 & 0 & 0 & A_{inn} & 0 & \cdot & A_n & 0 \\
0 & 0 & 0 & 0 & 0 & A_{out1} & \cdot & A_{outn} & a_{6+nm,6+nm}
\end{vmatrix} \cdot x + \begin{vmatrix} b_1 \\ 0 \\ 0 \\ 0 \\ 0 \\ 0 \\ \cdot \\ 0 \\ 0 \end{vmatrix} \cdot u \quad (4)$$

where the matrices $A_1, A_2, .., A_n$ represent an approximation of the transport delay of the coolant in riser-branches by inertial units:

$$\forall i = 1..n \to A_i = \begin{vmatrix}
ai_{1,1} & 0 & \cdot & 0 \\
ai_{2,1} & ai_{2,2} & \cdot & 0 \\
0 & 0 & \cdot & ai_{m,m}
\end{vmatrix};$$

where $\forall j = 1..m \to ai_{j,j} = -m/\tau_{zi}, ai_{j,j-1} = m/\tau_{zi}$ the input vectors-columns A_{ini}^T of dimension m are calculated as follows:

$$\forall i = 1..n \to A_{ini}^T = [(1 - k_{cl}) \cdot m/\tau_{zi}, 0, .., 0]$$

the output vectors-lines A_{outi} of the dimension m are defined by the expression:

$$\forall i = 1..n \rightarrow A_{outi} = [0, .., 0, (1 - k_{cw}) \cdot k_{ci}/T_{cw,}]$$

where $\sum_{i=1}^{n} k_{ci} = 1$; the coefficient $b_1 = \left(k_{mx} - K_{mix}^0\right) \cdot \frac{k_h}{T_{vlv}}$.

Let us consider in more details the peculiarities of a heat exchange system as a control object. As it is known from the description of the object its nonlinear properties are reflected in the coefficients $a_{2,1}, a_{2,4}, b_1$. The remaining elements of the matrix HES parameters are stationary coefficients, the components $A_1, A_2, .., A_n$ are determined by transport delay with different values within certain limits.

The traditional problem of regulating in heat exchange systems is to stabilize the temperature $T_{to3}^{\circ}(t)$ of the coolant at the outlet of the intermediate tank, which in terms of the taken denotation corresponds to a state variable $x_5(t)$. Further, assuming that the $a_{2,1}, a_{2,4}, b_1$ coefficients of the linearized model are stationary points in the equilibrium state, the transfer functions can be used for structural analysis of the object mathematical model. In addition, note the assumption that is made on the analysis of the functioning of the control object. This assumption is as following: the minimum time of transport delay in heat exchange systems is next larger than the mixing time constant, therefore the impact of the return coolant in the HES is considered as an external disturbance on the closed loop control. In the absence of the influence of cold water on the coolant in the inner-loop heating systems, i.e. $k_{cw} = 0$, this disturbance will be stationary and it can be used in the mathematical model (4) in the form of a fixed coefficient.

In case of significant effect of cold water on the heat exchange system, which leads to the inequality coefficient $k_{cw} > 0$, the disturbance takes the form of time-dependent function, which in the mathematical model (4) is appropriate to distinguish as a separate term. Denoting the disturbance as a symbol $v(t)$, we can reduce the dimension of the mathematical model to the fifth order:

$$\dot{x} = \begin{vmatrix} 0 & 0 & 0 & 0 & 0 \\ a_{2,1} & a_{2,2} & 0 & a_{2,4} & 0 \\ 0 & a_{3,2} & a_{3,3} & 0 & 0 \\ 0 & a_{4,2} & 0 & a_{4,4} & 0 \\ 0 & 0 & a_{5,3} & 0 & a_{5,5} \end{vmatrix} \cdot x + \begin{vmatrix} b_1 \\ 0 \\ 0 \\ 0 \\ 0 \end{vmatrix} \cdot u + \begin{vmatrix} 0 \\ 0 \\ q_3 \\ q_4 \\ 0 \end{vmatrix} \cdot v(t) \tag{5}$$

where $q_3 = a_{3,6+nm}, q_4 = a_{4,6+nm}$

Next, using the Laplace transformation in the point of the equilibrium state, we will write the linearized model:

$$\begin{cases} X_1(s) = \frac{b_1 \cdot U}{s} \\ X_2(s) = \frac{a_{2,1} \cdot X_1 + a_{2,4} \cdot X_4}{s - a_{2,2}} \\ X_3(s) = \frac{a_{3,2} \cdot X_2 + q_3 \cdot V_3}{s - a_{3,3}} \\ X_4(s) = \frac{a_{4,2} \cdot X_2 + q_4 \cdot V_4}{s - a_{4,4}} \\ X_5(s) = \frac{a_{5,3} \cdot X_3}{s - a_{5,5}} \end{cases} \tag{6}$$

With the notation of the functional blocks

$$W_1(s) = \frac{-a_{2,1} \cdot b_1 \cdot a_{2,2}^{-1}}{s}, W_2(s) = \frac{-a_{3,2} \cdot a_{3,3}^{-1}}{1 - a_{2,2}^{-1} \cdot s}, W_3(s) = \frac{1}{1 - a_{3,3}^{-1} \cdot s},$$

$$W_4(s) = \frac{-a_{2,4} \cdot a_{4,4}^{-1}}{1 - a_{4,4}^{-1} \cdot s}, W_5(s) = \frac{-a_{5,3} \cdot a_{5,5}^{-1}}{1 - a_{5,5}^{-1} \cdot s}, \tag{7}$$

$$K_{v3} = -\frac{q_3}{a_{3,3}}, K_{v4} = -\frac{q_4}{a_{4,4}}, K_{2,4} = \frac{a_{4,2} \cdot a_{3,3}}{a_{3,2} \cdot a_{4,4}};$$

the mathematical model (6) can be represented as a block diagram (Fig. 2):
After conversion (dotted line marked shifts directions of adders) we obtain the

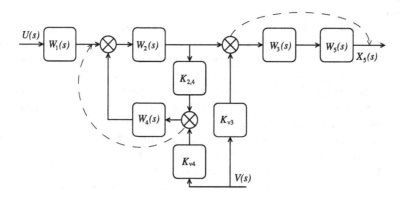

Fig. 2. The block diagram of the linearized model (7)

system shown in Figure 3.

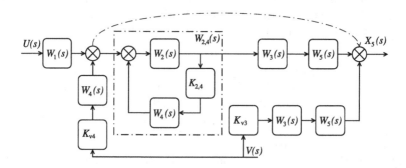

Fig. 3. The converted block diagram of the linearized model (7)

Under the conditions of HES functioning let us consider some assumptions that allow quite adequately to convert a block diagram (Fig. 3) in order to get an equivalent transfer function. Thus the equivalent element with transfer function $W_{2,4}(s)$ is assumed to be stable, because the static open-loop transfer coefficient is less than one:

$$W_2(0) \cdot k_{2,4} \cdot W_4(0) = \frac{a_{4,2} \cdot a_{2,4}}{a_{4,4} \cdot a_{2,2}} = (1 - K_{mix}^0) \cdot (1 - k_{exc}) < 0.2 \qquad (8)$$

where the coefficient $k_{exc} = 0.9$ (based on the practical experience of heat exchange systems maintenance). In addition, using Vieta theorem, we can write the following approximation:

$$\frac{(1 - T_2 \cdot s)}{(1 - T_2 \cdot s) \cdot (1 - T_1 \cdot s) - 0.2} \approx \frac{(1 - 0.2)^{-1}}{(1 - T_1 \cdot s)} \qquad (9)$$

This assumption is transformed to the ratio of the roots of the characteristic equation, which allows to write the conditions:

$$T_{min} < T_1 < T_{max}, \frac{T_{max}}{T_{min}} \approx \left(1 + 2\sqrt{1 - \frac{4 \cdot T_1 \cdot T_2 \cdot 0.8}{(T_1 + T_2)^2}}\right) < 2 \qquad (10)$$

Execution of the inequalities (10) presents the measure of inaccuracy of log-magnitude of the open loop not more than 6 dB.

Thus, the transfer function unit $W_{2,4}$ with a positive feedback, with the assumptions noted above, can be written as:

$$W_{2,4}(s) = \frac{(-a_{3,2}/a_{3,3}) \cdot (1 - (a_{4,2} \cdot a_{2,4})/(a_{4,4} \cdot a_{2,2})^{-1})}{1 + (-a_{2,2} \cdot k_{tt}(K_{mix}^0) \cdot s)},$$
$$\forall K_{mix}^0 \in (0..1) \vee (k_{exc} > 0.9) \rightarrow k_{tt}(K_{mix}^0) \in ((\sqrt{2})^{-1}..\sqrt{2}) \qquad (11)$$

where $k_{tt}(K_{mix}^0)$ - the function that characterizes the change in the time constant of the object.

Next, the object is divided into the following parts: the unit of control signal delay $W_z(s)$, the integrating part which is a unit of the equivalent transfer function $W_i(s)$ of the electric control valve, the inertial part which is an aperiodic link $W_o(s)$ of the first-order transfer function of the thermal object, the transfer function $W_v(s)$ of the disturbance signal $V(s)$. As a result of transformations, we obtain the final block diagram (Fig. 4), representing the linearized model graphically (6). The corresponding transfer functions are defined by the following equations:

$$W_z(s) \cdot W_i(s) \cdot W_o(s) = W_1(s) \cdot W_{2,4}(s) \cdot W_3(s) \cdot W_5(s),$$
$$W_v(s) = (K_{v3} + K_{v4} \cdot W_4(s) \cdot W_{2,4}(s)) \cdot W_3(s) \cdot W_5(s) \qquad (12)$$

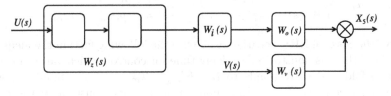

Fig. 4. The final block diagram

The blocks are distributed in such a way that the inertial link with a maximum time constant to be present in the unit $W_o(s)$, the other two units are replaced by the delay link (Fig. 3).

As a result of the transformation we obtain more convenient for the synthesis of closed-loop control transfer function of the electric control valve, including the main parameters of the heat exchanger system:

$$W_i(s) = \left(\frac{b_1 \cdot a_{3,2} \cdot a_{2,1} \cdot a_{5,3}}{a_{3,3} \cdot a_{2,2} \cdot a_{5,5}} \right) \cdot \left(1 - \frac{a_{4,2} \cdot a_{2,4}}{a_{4,4} \cdot a_{2,2}} \right)^{-1} \cdot s \tag{13}$$

After the substitution of physical quantities, which are the parameters of HES, this expression takes on the following form:

$$W_i(s) = \frac{k_h \cdot k_{exc} \cdot (k_{mx} - K^0_{mix}) \cdot (T^\circ_1 - T^{\circ 0}_{bk1}(k_{cw}))}{(1 - (1 - k_{exc}) \cdot (1 - K^0_{mix})) \cdot (T_{vlv} \cdot s)} \tag{14}$$

In the research of complex objects with delay a very important requirement is a preliminary assessment of the main parameters that have a significant impact on the stability of the closed-loop control system. A valid transfer coefficient and time constants in the inner loop heat exchange systems that define the retarded reaction of control to the disturbance can be such parameters for the object in question.

For the specific HES performance with known interval values of the constituents parameters, in particular, the transfer coefficient k_g of the control valve, the coefficient k_cw of cold water influence on the coolant in the internal contour of the heat exchange system we can rather accurately estimate the range of variation of the static coefficient of the actuator transfer function $W_i(s)$.

The determining factor in assessing of the delay in the HES control channel is the ratio of the time constant T_{cp} of the fluid mixing in the storage container (if it exists in the HES) and time constant T_{mix} of the fluid mixing at the valve of the system. It is clear that in the absence of the storage container the delay duration in the control channel will be determined only by the time constant T_{mix}. Let us write the transfer functions $W_o(s)$ and $W_z(s)$ the coefficients of which are largely determined by the ratio of the time constants data:

$$\begin{cases} T_{cp} > T_{mix} \begin{cases} W_o(s) = \frac{1}{1 - a_{5,5}^{-1} \cdot s} = \frac{1}{1 + T_{exc} \cdot s}, \\ W_z(s) = exp((k_{tt}(K^0_{mix}) \cdot a_{2,2}^{-1} + a_{3,3}^{-1}) \cdot s) \\ = exp((k_{tt}(K^0_{mix}) \cdot T_{mix} + T_{exc}) \cdot s) \end{cases} \\ T_{cp} < T_{mix} \begin{cases} W_o(s) = \frac{1}{1 - k_{tt}(K^0_{mix}) \cdot a_{2,2}^{-1} \cdot s} = \frac{1}{1 + k_{tt}(K^0_{mix}) \cdot T_{mix} \cdot s}, \\ W_z(s) = exp((a_{5,5}^{-1} + a_{3,3}^{-1}) \cdot s) = exp((T_{cp} + T_{exc}) \cdot s) \end{cases} \end{cases} \tag{15}$$

To solve the tasks of HES research we can use the following fact: in the inequality $T_{cp} > T_{mix}$ there is nonstationary delay time for control, which varies not more than two fold and corresponds to $(k_{tt}(K^0_{mix}) \cdot T_{mix} + T_{exc})$. Similarly, when performing inequality $T_{cp} < T_{mix}$ nonstationary time constant of the object $(k_{tt}(K^0_{mix}) \cdot T_{mix})$ also changes not more than two fold.

4 The Nonlinear HES Model of the Second Order with Delay in Controlling

Using the equations (15) without disturbance, with the notation $y_1(t) = x_5(t)$, $y_2(t)$ - output unit with the transfer function $W_i(s)$

$$k_g(s) = \frac{k_h \cdot k_{exc} \cdot (k_{mx} - K^0_{mix}) \cdot (T^\circ_1 - T^{\circ 0}_{bk1}(k_{cw}))}{(1 - (1 - k_{exc}) \cdot (1 - K^0_{mix})) \cdot T_{vlv}}$$
$$T_{cp} > T_{mix} \left\{ T_o = T_{cp}, \tau_z = (k_{tt}(K^0_{mix}) \cdot T_{mix} + T_{exc}) \right.$$
$$T_{cp} < T_{mix} \left\{ T_o = k_{tt}(K^0_{mix}) \cdot T_{mix}, \tau_z = (T_{cp} + T_{exc}) \right. \tag{16}$$

we can write the system of differential equations of the second order

$$\begin{cases} \dot{y}_1(t) = \frac{y_2(t) - y_1(t)}{T_o(t,y)}, \\ \dot{y}_2(t) = k_g(t,y) \cdot u(t - \tau_z) \end{cases} \tag{17}$$

The canonical representation of the system of differential equations in Frobenius form after changing variables $z_1(t) = y_1(t)$, $z_2(t) = \dot{y}_1(t)$ becomes:

$$\begin{cases} \dot{z}_1(t) = z_2(t), \\ \dot{z}_2(t) = (k_g(t,z)/T_o(t,z)) \cdot u(t - \tau_z) - T^{-1}_o(t,z) \cdot z_2(t) \end{cases} \tag{18}$$

4.1 The Results of the Simulation

The adequacy of the resulting model (18) is confirmed by comparing the s-shaped curves of the speed-up after the numerical simulation of the transition process which conforms to conditions of the experiment [8]. The experiment was performed in the same conditions for each model, the parameters of the model (18) were calculated by the formula (16) on the base of the parameters of the initial model (1). The simulation was performed in the C language; the source code is available on the web resource. [9]

Fig. (5a) illustrates transients: full red line T°_{to3} for the model (1); green dotted line $y_1(t) = z_1(t)$ for the model (18), and curve T°_{sto3} from the sensor output of the current heat exchange system - the blue dotted line. Some difference between

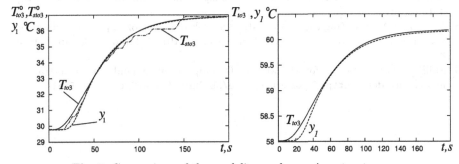

Fig. 5. Comparison of the modeling and experiment outcomes

the models at the beginning of the transition process can be explained by the fact that in the system of equations (18) several inertial links are replaced by a single delay element.

Fig. 5b shows the laws T°_{to3} and $y_1(t) = z_1(t)$ for both models. They correspond to the same conditions and values of the parameters in the control valve stem position $h = 0.7$. The results of modeling and experimental research imply a high degree of adequacy of the reduced model to the real object.

Conclusion

A solution to the problem of approximation of a complex nonlinear mathematical model of heat transfer delay system to a nonlinear system of differential equations of the second order allows us to use modern methods of relay control. For the dimensional model reduction we used the method of decomposition of the linearized model at the steady state point, where the basic coefficients are given in a general form which allows us to get the non-linear law of the reduced model coefficients at an arbitrary point of equilibrium. The results of numerical simulations confirm the adequacy of the reduced model and real heat exchange system.

References

1. Klyuev, A.S., Kolesnikov, A.A.: Optimization of automatic speed control systems, p. 240. Energoizdat, Moscow (1982)
2. Khalil Hassan, K.: Nonlinear Systems, 3rd edn., p. 750. Prentice Hall (2002) ISBN 0-13-067389-7
3. Shilin, A.A., Bukreev, V.G.: Investigation of the three-position relay temperature controller in a sliding mode. Reoprts of TUSUR, Russian Tomsk 2(1), 251–257 (2012)
4. Buslenko, N.P.: Modeling of complex systems, p. 360. Nauka, Moscow (1978)
5. Anderson, B.D.O., Liu, Y.: Controller reduction: concepts and approaches. IEEE Trans. Automat. Control. 34, 802–812 (1989)
6. Moore, B.C.: Principal component analysis in linear systems: controllability, observability and model reduction. IEEE Trans. Automat. Control. AC-26, 17–32 (1981)
7. Shilin, A.A., Bukreev, V.G., Koykov, K.I.: Mathematical model of a nonlinear system with delay. Devices and Systems. Management, Monitoring, Diagnostics (6), 3–10 (2013)
8. Pan'ko, M.A.: Selecting the mathematical models of a controlled piece of equipment from experimental data. Thermal Engineering 53(10), 782–786 (2006) ISSN 0040 6015
9. Shilin, A.A.: Materials for the article: Decomposition of nonlinear heat transfer model with delay, http://portal.tpu.ru/SHARED/s/SHILIN/tethiss/Tab4

Robust Semiparametric Regression Estimates

Valery Simakhin and Oleg Cherepanov

Kurgan State University, Kurgan, Russia
sva_full@mail.ru,
ocherepanov@inbox.ru

Abstract. The paper deals with the new semiparametric regression estimates for the different level of *a priori* data. The estimates are based on the weighted maximum likelihood method. The investigations show the estimates are effective for symmetrical and asymmetrical outliers and adaptive to outliers and distribution type.

Keywords: regression, semiparametric, robust, adaptive.

1 Problem Formulation

Let us consider a classical problem of regression of the form

$$y = r(x, \Theta) + \varepsilon, \tag{1}$$

where X is a random variable with distribution function F_1 and density f_1; $r(x, \Theta)$ is the regression function defined to within the parameter $\Theta = (\theta_1, \ldots, \theta_k)$; ε is a random variable independent of X with distribution function F_2 and symmetric density f_2. Let $Z = (X, Y)$ be a random variable with distribution function G and density g. Consider that G is unknown and belongs to a certain class Γ of supermodels. It is required to estimate the vector of the parameters Θ on the sample $(x_i, y_i), i = 1 \ldots N$.

As is well known, the least squares method (LSM), and the maximum likelihood method (MLM) are conventionally used to solve this problem. In case of presence of outliers along the x and y axes, various method of truncation and fitting of robust regression algorithms [1]-[5] are used. Unfortunately, many robust regression estimates, being robust on supermodels, can have very low efficiency in different concrete situations G. As is well known, the robustness and efficiency are inconsistent [4]. In this regard, various adaptive estimates are applied. One of the approaches that allows efficient adaptive estimates to be obtained is the weighed maximum likelihood method (WMLM) [6].

2 Weighted Maximum Likelihood Method

Estimates of the coefficients of the regression function by the weighted maximum likelihood method for problem (1) are found from estimation equations of the form

A. Dudin et al. (Eds.): ITMM 2014, CCIS 487, pp. 397–405, 2014.

$$\sum_{i=1}^{N} \psi_j(x_i, y_i) = 0, j = 1, \ldots, k, \tag{2}$$

$$\psi_j(x, y) = \frac{\partial}{\partial \theta_j} \log f_2(\varepsilon) f_1^{l_1}(x) f_2^{l_2}(\varepsilon)$$

$$\varepsilon = y - r(x, \Theta_N)$$

where $\Psi = (\psi_1, \ldots, \psi_k)$ is the estimation vector function, $l = (l_1, l_2)$ is the vector of radical parameters defining the robustness and efficiency of the estimates It is easy to note that at $l_1 = l_2 = 0$, we obtain maximum likelihood estimate (MLE), at $l_1 = l_2 = 0.5$, we obtain the radical estimates (RE) [4], and at $l_1 = l_2 = 1$, we obtain estimates with maximum stability [4]. The radical parameter l_1 is responsible for the degree of *soft* truncation in x and the radical parameter l_2 is responsible for the degree of *soft* truncation in y. It can be shown [6] that for the generalized Tukey error model, the equation of type (2) corresponds to the weighted maximum likelihood equations. By minimization of the integral quadratic regression error with respect to the vector of radical parameters, we derive the regression estimate adaptive with respect to outliers both along the and y axes. Effective regression algorithms are largely determined by *a priori* data on the form of distributions F_1 and F_2. For adaptation to a given uncertainty type, nonparametric methods based on Rosenblatt-Parzen estimates are used in the present work.

Bellow we consider efficient regression estimates for the following semiparametric models:

1. The form of distribution $F_1(x, \theta_1)$ is known to within the vector θ_1 of the parameters to be estimated (parametric level), and the form of distribution $F_2(\varepsilon)$ is unknown (nonparametric level). As an estimate of the distribution density f_2, we take advantage of the symmetrized Rosenblatt-Parzen estimate of the form

$$f_{2N}(\varepsilon) = \frac{1}{2Nh_{2N}} \sum_{i=1}^{N} \left(K\left(\frac{\varepsilon - \varepsilon_i}{h_{2N}}\right) + K\left(\frac{\varepsilon + \varepsilon_i}{h_{2N}}\right) \right), \tag{3}$$

where $K(x)$ is the kernel function, h_{2N} is the bandwidth parameter, N is the sample size. According to Eqs. (2) and (3), the semiparametric estimate of regression coefficients is defined by the system of equations of the form

$$\sum_{i=1}^{N} \frac{\partial r(x, \Theta)}{\partial \theta_t}\Big|_{\Theta=\Theta_N} f_{2N}^{l_2-1}(\varepsilon_i) f_1^{l_1}(x_i) \sum_{j=1}^{N} \phi\left(\frac{\varepsilon_i + \varepsilon_j}{h_{2N}}\right) K\left(\frac{\varepsilon_i + \varepsilon_j}{h_{2N}}\right) = 0 \tag{4}$$

$$\varepsilon_i = y_i - r(x_i, \Theta_N)$$

$$\phi(x) = \frac{1}{K(x)} \frac{dK(x)}{dx}$$

As a result, we obtain the semiparametric regression estimate independent of the from of distribution F_2.

2. The form of distribution $F_1(x)$ is unknown (nonparametric level), and the form of distribution $F_2(\varepsilon, \theta_2)$ is known to within the vector θ_2 of the parameters to be estimated (parametric level). As an estimate of density $f_1(x)$, we take advantage of the classical Rosenblatt-Parzen estimate:

$$f_{1N}(x) = \frac{1}{Nh_{1N}} = \sum_{i=1}^{N} K_1\left(\frac{x - x_i}{h_{1N}}\right), \tag{5}$$

where $K_1(x)$ is the kernel function, h_{1N} is the bandwidth parameter. Substitution Eq. (5) into Eq. (2) we derive the system of equations of the form

$$\sum_{i=1}^{N} \frac{\partial}{\partial \theta_j} \log f_2(\varepsilon_i) f_{1N}^{l_1}(x_i) f_2^{l_2}(\varepsilon_i), = 0, j = 1, \ldots, k. \tag{6}$$

Let us now derive estimates of the regression parameters for typical distributions of residuals with different degree of tail stretching:

(a) Let the random variable ε obey the generalized normal distribution of the fourth degree $f_2(\varepsilon)$; then estimation equations assume the following form:

$$\sum_{i=1}^{N} \left.\frac{\partial r(x, \Theta)}{\partial \theta_j}\right|_{\Theta=\Theta_N} \varepsilon_i^3 f_{1N}^{l_1}(x_i) f_2^{l_2}(\varepsilon_i), = 0, j = 1, \ldots, k.$$

(b) Let the random variable ε be a random variable with normal distribution $f_2(\varepsilon)$; then estimation equations assume the following form

$$\sum_{i=1}^{N} \left.\frac{\partial r(x, \Theta)}{\partial \theta_j}\right|_{\Theta=\Theta_N} \varepsilon_i f_{1N}^{l_1}(x_i) f_2^{l_2}(\varepsilon_i), = 0, j = 1, \ldots, k.$$

(c) Let ε obey the Laplace distribution $f_2(\varepsilon)$; then estimation equations assume the following form

$$\sum_{i=1}^{N} \left.\frac{\partial r(x, \Theta)}{\partial \theta_j}\right|_{\Theta=\Theta_N} Sign(\varepsilon_i) f_{1N}^{l_1}(x_i) f_2^{l_2}(\varepsilon_i), = 0, j = 1, \ldots, k.$$

(d) Let ε obey the Cauchy distribution $f_2(\varepsilon)$; then estimation equations assume the following form

$$\sum_{i=1}^{N} \left.\frac{\partial r(x, \Theta)}{\partial \theta_j}\right|_{\Theta=\Theta_N} \varepsilon_i f_{1N}^{l_1}(x_i) f_2^{l_2+1}(\varepsilon_i), = 0, j = 1, \ldots, k.$$

3. The form of distributions F_1 and F_2 are unknown. As estimates of the distribution density we take advantage of the Rosenblatt-Parzen estimates described by Eqs. (3) and (5). As a result, we derive semiparametric estimates

of the regression parameters independent of distributions of random variable X and ε of the form

$$\sum_{i=1}^{N} \frac{\partial r(x, \Theta)}{\partial \theta_t}\bigg|_{\Theta=\Theta_N} f_{2N}^{l_2-1}(\varepsilon_i) f_{1N}^{l_1}(x_i) \sum_{j=1}^{N} \phi\left(\frac{\varepsilon_i + \varepsilon_j}{h_{2N}}\right) K\left(\frac{\varepsilon_i + \varepsilon_j}{h_{2N}}\right) = 0 \quad (7)$$

$$\varepsilon_i = y_i - r(x_i, \Theta_N)$$

$$\phi(x) = \frac{1}{K(x)} \frac{dK(x)}{dx}$$

Equations (4), (6), and (7) describe algorithms of finding the semiparametric estimates of the regression parameters for problem (1) with different levels of *a priori* data on distributions.

Algorithms (4), (6), and (7) can be investigated theoretically based on a study of properties of conditional functional [7]. This is an independent problem.

3 Modeling

The efficiency of adaptive semiparametric WMLM regression estimates (4) and (7) was investigated on the class of symmetric distributions of residues consisting of generalized normal distribution of the fourth degree, normal distribution, Laplace distribution, and Cauchy distribution. The following outlier models were used:

– Model of outliers asymmetric in y:

$$f_2(\varepsilon) = (1 - p)g_2(\varepsilon) + pg_2(\varepsilon - \alpha);$$

– Model of outliers symmetric in y:

$$f_2(\varepsilon) = (1 - p)g_2(\varepsilon) + pg_2(\varepsilon/\lambda);$$

– Model of outliers asymmetric in x:

$$f_1(x) = (1 - p)g_1(x) + pg_1(x - b).$$

Semiparametric regression estimates (4) and (7) were compared with estimates by the maximum likelihood estimates (MLE), the radical estimates (RE), the estimates with maximum stability (EMS), the least squares estimates (LSE) and the least modules estimates (LME).

The relative efficiency of estimation was defined as

$$\epsilon = \frac{V}{V_{opt}},$$

where V is the integral variation of the estimate and V_{opt} is the minimum integral variation of the estimate among considered estimates.

To estimate the integral variance of the regression estimate and the radical parameters, the naive bootstrap method [8] was used.

Investigations were performed for the linear model of the regression function

$$y = x + 3.$$

Let us introduce following designations: SM is the supermodel; ASPE1 is the adaptive semiparametric regression estimate (4); ASPE2 is the adaptive semiparametric regression estimate (7).

3.1 Generalized Normal Distributions of the Fourth Degree

The efficiency of the adaptive semiparametric regression estimates by WMLM was evaluated for the following supermodels:

1. Pure sample:
$$f(x, \varepsilon) = N(x, 0, 1) GND_4(\varepsilon, 0, 1.767). \tag{8}$$

2. Outliers symmetric in y:
$$f(x, \varepsilon) = N(x, 0, 1)(0.9 GND_4(\varepsilon, 0, 1.767) + 0.1 GND_4(\varepsilon, 0, 5.391)). \tag{9}$$

3. Outliers asymmetric in y:
$$f(x, \varepsilon) = N(x, 0, 1)(0.9 GND_4(\varepsilon, 0, 1.767) + 0.1 GND_4(\varepsilon, 8, 1.767)). \tag{10}$$

4. Outliers asymmetric in x:
$$f(x, \varepsilon) = (0.9 N(x, 0, 1) + 0.1 N(x, 8, 1)) GND_4(\varepsilon, 0, 1.767). \tag{11}$$

Table 1. Efficiency of estimates on distributions (8)-(11)

Estimates	MLE	LSE	LME	EMS	RE	ASPE1	ASPE2
SM (8)	1.000	0.645	0.339	0.556	0.690	0.741	0.741
SM (9)	0.264	0.679	0.535	0.691	0.809	1.000	1.000
SM (10)	0.005	0.036	0.188	0.493	0.654	1.000	1.000
SM (11)	0.045	0.029	0.034	0.755	0.954	1.000	0.770

3.2 Normal Distribution

The efficiency of adaptive semiparametric WMLM regression estimates was evaluated on the following supermodels:

1. Pure sample:
$$f(x, \varepsilon) = N(x, 0, 1)N(\varepsilon, 0, 1). \tag{12}$$

2. Outliers symmetric in y:

$$f(x, \varepsilon) = N(x, 0, 1)(0.9N(\varepsilon, 0, 1) + 0.1N(\varepsilon, 0, 3)). \tag{13}$$

3. Outliers asymmetric in y:

$$f(x, \varepsilon) = N(x, 0, 1)(0.9N(\varepsilon, 0, 1) + 0.1N(\varepsilon, 8, 1)). \tag{14}$$

4. Outliers asymmetric in x:

$$f(x, \varepsilon) = (0.9N(x, 0, 1) + 0.1N(x, 8, 1))N(\varepsilon, 0, 1). \tag{15}$$

Table 2. Efficiency of estimates on distributions (12)-(15)

Estimates	MLE	LSE	LME	EMS	RE	ASPE1	ASPE2
SM (12)	1.000	1.000	0.700	0.369	0.647	0.983	0.983
SM (13)	0.730	0.730	0.701	0.437	0.753	1.000	1.000
SM (14)	0.015	0.015	0.143	0.352	0.629	1.000	1.000
SM (15)	0.014	0.014	0.014	0.317	0.644	1.000	0.867

3.3 Laplace Distribution

The efficiency of adaptive semiparametric WMLM regression estimates was evaluated on the following supermodels

1. Pure sample:
$$f(x, \varepsilon) = N(x, 0, 1)L(\varepsilon, 0, 0.7144). \tag{16}$$

2. Outliers symmetric in y:

$$f(x, \varepsilon) = N(x, 0, 1)(0.9L(\varepsilon, 0, 0.7144) + 0.1L(\varepsilon, 0, 2.1432)). \tag{17}$$

3. Outliers asymmetric in y:

$$f(x, \varepsilon) = N(x, 0, 1)(0.9L(\varepsilon, 0, 0.7144) + 0.1L(\varepsilon, 8, 0.7144)). \tag{18}$$

4. Outliers asymmetric in x:

$$f(x, \varepsilon) = (0.9N(x, 0, 1) + 0.1N(x, 8, 1))L(\varepsilon, 0, 0.7144). \tag{19}$$

Table 3. Efficiency of estimates on distributions (16)-(19)

Estimates	MLE	LSE	LME	EMS	RE	ASPE1	ASPE2
SM (16)	1.000	0.680	1.000	0.732	0.907	0.893	0.893
SM (17)	1.000	0.600	1.000	0.707	0.916	0.837	0.837
SM (18)	0.322	0.012	0.322	0.166	0.362	1.000	1.000
SM (19)	0.011	0.009	0.011	0.628	1.000	0.928	0.879

3.4 Cauchy Distribution

The efficiency of adaptive semiparametric WMLM regression estimates was evaluated on the following supermodels:

1. Pure sample:
$$f(x, \varepsilon) = N(x, 0, 1)C(\varepsilon, 0, 0.2605). \tag{20}$$

2. Outliers symmetric in y:
$$f(x, \varepsilon) = N(x, 0, 1)(0.9C(\varepsilon, 0, 0.2605) + 0.1C(\varepsilon, 0, 7815)). \tag{21}$$

3. Outliers asymmetric in y:
$$f(x, \varepsilon) = N(x, 0, 1)(0.9C(\varepsilon, 0, 0.2605) + 0.1C(\varepsilon, 8, 0.2605)). \tag{22}$$

4. Outliers asymmetric in x:
$$f(x, \varepsilon) = (0.9N(x, 0, 1) + 0.1N(x, 8, 1))C(\varepsilon, 0, 0.2605). \tag{23}$$

Table 4. Efficiency of estimates on distributions (20)-(23)

Estimates	MLE	LSE	LME	EMS	RE	ASPE1	ASPE2
SM (20)	1.000	0.040	0.377	0.235	0.595	0.974	0.974
SM (21)	1.000	0.036	0.379	0.291	0.658	0.835	0.835
SM (22)	1.000	0.002	0.099	0.254	0.605	0.873	0.873
SM (23)	0.324	0.0001	0.001	0.115	0.414	1.000	0.939

3.5 Discussion of the Results of Modeling

Figures 1-4 show some results of implementation of the algorithms considered above on an example of a linear regression function with outliers of different types. The linear regression function of the form $y = x + 3$ on supermodels (12) - (15) was considered for $p = 0.1, \alpha = 8, \lambda = 3, b = 7, N = 30; X$ and ε obeyed standard normal distribution). In Figs. 1- 4, we have used following designations:

1 is the least squares estimate;
2 is the least modules estimate;

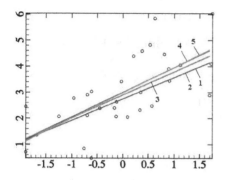

Fig. 1. Plot of regression estimates on distribution (12)

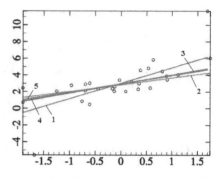

Fig. 2. Plots of regression estimates on distribution (13)

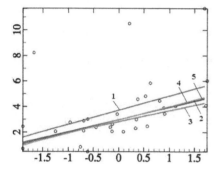

Fig. 3. Plots of regression estimates on distribution (14)

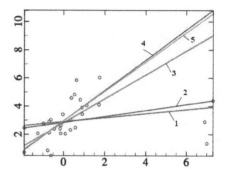

Fig. 4. Plots of regression estimates on distribution (15)

3 is the estimate with maximum stability;
4 is the adaptive parametrical estimate by WMLM [6];
5 is the adaptive semiparametric estimate (4).

Figure 1 shows that without outliers, all methods do not lead to biases of the regression estimates. In Figs. 1-3, the LSE have significant biases of unknown nature, and as demonstrated Tables (1)-(4), low efficiency. Figure 4 shows that in the presence of outliers along the x axis, the LSM estimates are almost inoperative, and the recommended robust EMS and LME estimates have considerable biases and low efficiency (see Table 2). Regression estimates (4) proposed in this work effectively cope with outliers along the and y axes.

Our analysis of Tables 1-4 allows us to conclude the following:

1. In the presence of outliers, the recommended robust EMS and LME estimates have low efficiency practically on all distributions.
2. In the presence of outliers, the LSE and MLE estimates have low efficiency on all distributions and are even inoperative on some distributions.

3. Radical estimates corresponding to estimates at the Hellinger distance have intermediate efficiency.
4. Results of modeling demonstrated that the WMLM estimates have highest efficiency for different distribution and outlier types.

Conclusions

The regression problem has a long history, and methods of its solving (LSE and LME) are traditional. Since the 70s of the XX century, considerable attention has been paid to a search for robust (stable) regression algorithms [1]-[5]. Investigations have demonstrated that these approaches provide robust (rough by efficiency) algorithms. The approach based on WMLM allows new effective algorithms (4), (6), and (7) of solving the classical regression problem based on *a priori* data on all elements of the mathematical model of problem formulation to be considered. The conjunction of all *a priori* data sets yields a final efficient algorithm for finding regressions.

References

1. Huber, P.J.: Robust Statistics. Mir, Moscow (1981)
2. Hampel, F.R., Ronchetti, E.M., Rousseeuw, P.J., Stahel, W.A.: Robust Statistics. Mir, Moscow (1989)
3. Tsypkin, Y.Z.: Principles of Information Theory of Identification. Nauka, Moscow (1984)
4. Shurygin, A.M.: Applied Statistics. Robustness. Estimates. Prediction. Financy i Statistika, Moscow (2000)
5. Shulenin, V.P.: Mathematical Statistics. Part 3. Robust Statistics. Publishing House of Scientific and Technology Literature, Tomsk (2012)
6. Simakhin, V.A.: Robust Nonparametric Estimates. LAMBERT Academic Publishing, Germany (2011)
7. Vasiliev, V.A., Dobrovidov, A.V., Koshkin, G.M.: Nonparameteric estimation of stationary sequences distributions. Nauka, Moscow (2004)
8. Efron, B.: Nonconventional Methods of Multidimensional Statistical Analysis. Financy i Statistika, Moscow (1998)

Application of a Bypass Pipeline
during Pressure Control

Dmitry Starikov, Evgeny Rybakov, and Evgeny Gromakov

Institute of Cybernetics, National Research Tomsk Polytechnic University
Lenina Avenue, 30, 634050, Tomsk, Russia
dstarikov@me.com

Abstract. Relevant problem of modern oil transportation is being solved by automatic control systems. In the article an algorithm of pressure control is described. Application of bypass pipeline with valve can improve quality of regulation system. Simulation of research is given in the article.

Keywords: Pressure control, bypass, energy efficient, regulation.

1   Introduction

Main pipeline pumps are difficult technical constructions and play crucial part in oil pipeline transportation. Some of them are intended for oil supply form buster pumps to main pipeline. And others are used for energy losses replenishment during pressure control and also for pipeline hydrodynamic separation on given in draft sectors to provide pumping and hydra impact effects localization in the main pipeline. The main problem of oil transportation is pressure maintenance set by regulatory requirements. The recent time tendency of pressure regulation is controlling pressure by pump rotating speed changing.

To provide desired operating mode main pump stations include serially connected pumps which are controlled with frequency adjustable motors with high power consuming. This power depends on an oil supply volume in pipeline Q and value of pressure H:

$$P_p = \frac{QHg\rho}{\eta_p \eta_{el} \eta_{fc}}, \tag{1}$$

where g and ρ - acceleration of gravity and oil density;
$\eta_p, \eta_{el}, \eta_{fc}$ - efficiency of a pump, power suppliers and frequency converter;
And it can reach up to megawatt of consuming.

The main aim of this work is automatic pressure control system (APCS) increasing which can decrease main pump energy consumption in dynamic and stabilization mode. Stationary or trigger modes are being considered in comparative calculations of energy consuming in frequency adjustable main pumps. But it is not counted that pump is in a control loop and therefore the actuator of this loop will consume power to overcome internal resistance caused by huge weight

A. Dudin et al. (Eds.): ITMM 2014, CCIS 487, pp. 406–414, 2014.

of motor shaft. Because of bad quality regulation (oscillations and overshooting) that losses may be significant. This fact follows from pump motion equation:

$$n_s = \frac{GD^2}{375}\frac{d\omega}{dt} = M_p - M_s,$$ (2)

where GD^2 - the turning moment of the pump;
$M_p = \frac{3I^2R^2}{s}$ - moment of a pump motor;
n_s - synchronous speed of rotating;
I - loop current;
R - loop active resistance;
s - slipping;
M_s - moment of resistance on motor shaft.

Internal losses caused by continuous pump acceleration and deceleration are proportional to motor acceleration during transient, turing and motor moment. The last one is significant for huge pumps.

At the same time energy losses in dynamic mode in a valve are less than in a pump. This fact is caused by lower lag of valves during moving from the point.

That is why simultaneous use of frequency controlled pump and throttling component is an attractive solution of pressure control problem. In this way it is possible to implement power economic consumption system.

For decreasing of internal moment value it is necessary to provide smooth frequency changing of pump power and fast regulation of valve motor. But eigenfrequency of fast loop must be higher than pump loop. To find a solution it is possible to use Splint-technology of oil supply from the pump with two pipelines: general and bypass. On both pipelines there should be installed two controlled valves with different purposes. A low diameter bypass valve has high speed. And a general pipeline valve has normal and typical (at the moment) characteristics. It has a lower speed.

And in this case a bypass valve loop can suppress high frequency influences of ACS dynamic and release pump loop from that influence regulation. But pump loop should suppress low frequency parts of dynamic. And in steady and quasi-steady modes the opening degree of a general pipeline valve should be maximum.

In real pressure control system lag of pump and valve differ by an order. Big currents supplying provide relatively small pump acceleration time. The special system of smooth frequency changing is used for excluding hydra impacts during acceleration. At the same time it is impractical to use that system during pressure stabilization in main oil pipeline. Generally valve lag is around 200-300 seconds. And this values of eigenfrequency must be changed appropriately.

It may be solved by adding low frequency filter in pump loop and choosing quite fast valve moving. Filter is an aperiodic link which can be written as:

$$W_f(s) = \frac{k_f}{T_f + 1},$$ (3)

where k_f - direct transfer coefficient;
T_f - time constant(lag).

If big value of a pump filter lag will be chosen it will be possible to provide (by an algorithm) smooth speed changing of a pump. This way the necessity of big currents using for switching pump mode will disappear and as a result unnecessary power losses will be excluded.

The typical scheme of a system Pump Station - Pipeline is shown in Figure 1.

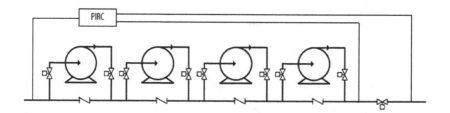

Fig. 1. Typical oil transport scheme

Fig. 1 shows that the scheme involves the presence of at least two loops ACS (including possible variations). Will continue to be considered a combination of control algorithms: Split range control and Parallel control.

The main goal is to choose an algorithm (scheme) under control optimality criterion relative to a functional energy intensity (the choice of the type of control depending on the amount of energy used).

Described earlier technology of pressure regulation with help of valves in general and bypass pipelines can be presented in Fig. 2.

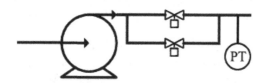

Fig. 2. Bypass pressure regulation scheme

2   Modeling

Research model valve and pump was developed in MatLAB Simulink. Block diagram of the general valve is shown in Fig. 3.

Figure 4 shows the second valve in bypass pipeline.

It is clear that valves (Fig. 3, 4) have different value of time lag an as a result different operating time. In Fig. 5 pump model is presented.

Block diagram of the pipeline is shown in Figure 6.

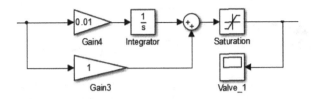

Fig. 3. General pipeline valve

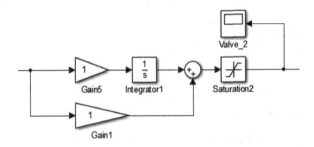

Fig. 4. Bypass pipeline valve

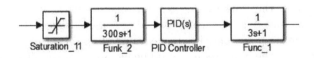

Fig. 5. Model of a main pump

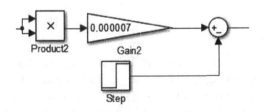

Fig. 6. Pipeline model

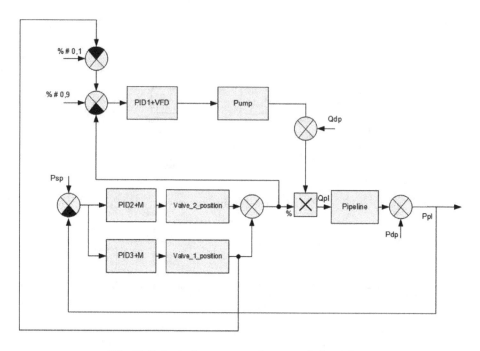

Fig. 7. Principal structure of a control algorithm

The proposed algorithm for controlling the pressure of the structure is presented in Fig. 7.

Suppose that in the initial state general pipeline valve is opened (opening degree $\%_{op}=0,9$) and bypass valve is closed (opening degree $\%_{cl}=0,1$). In this case pump operating point matches to set value of an oil supply (Fig 8).

In this scheme in the high-frequency influence pressure stabilization loop (by-pass valve) counteracts initially. Simultaneously pump regulation loop changes oil supply to return movement of general valve to initial conditional. This recovery is provided by set point in pump loop opening degree of general valve.

3   Research

Figure 9 represents a model of a set of the system in Simulink. Transients chats of described models during step influence at 15th second of pump acceleration up to 15,6 MPa are shown on Figures 10-13. There is a moment of disturbance appearing and its neutralization by a APCS with help of two valves an pump[1]. Both valves starts fast that is why pressure changing trend has smooth bend which exclude abrupt rotating speed and big power consuming. Moreover it is clear that pump is helping control during pressure falls. It is clear that influence

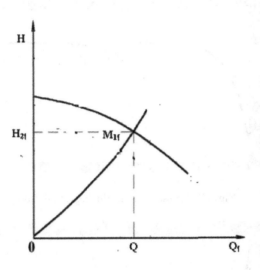

Fig. 8. Pump operating point during control

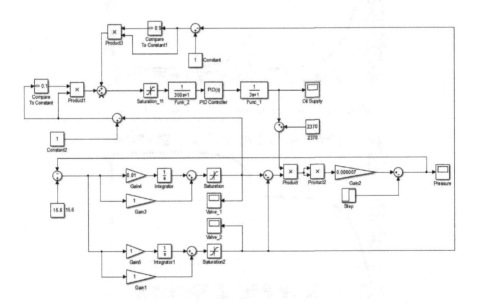

Fig. 9. Pump operating point during control

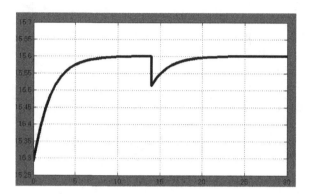

Fig. 10. Pressure trend during simulation

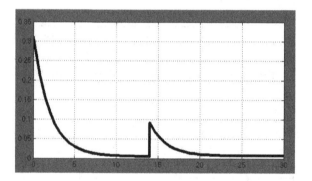

Fig. 11. Trend of a bypass valve opening degree

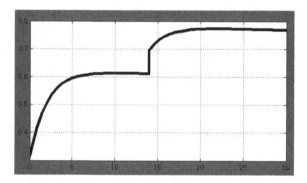

Fig. 12. Opening degree trend of a general valve

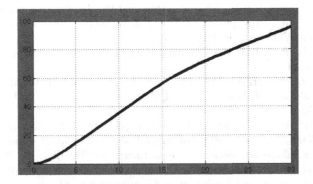

Fig. 13. Changing of an oil supply by a pump

is appearing on 14th second (as it set) when setpoint (15,6 MPa) is being reached. Main valve tends to change opening degree to set required oil supply on condition pressure stabilization. At the same time fast bypass valve starts to change supply value. After transient oil pressure in pipeline is 15,6 MPa. During counter-influence minor changes of pump supply are being observed and it makes pump control more effective in terms of energy consumption.

4 Conclusion

During modeling the fact that bypass valve application can reduce energy losses in pump control of pressure was obtained. Figure 13 shows significant result of regulation. It is possible to recognize the deflection of a pump operating characteristic from the straight line. According to this fact energy losses may be reduced. Moreover we obtained transient operating without oscillation and overshoot. The inertia torque component of the actuator shaft can be readily reduced by controlling pressure with help of two valves: general and bypass.

The structure of oil supply APCS was proposed. It includes main and fast valves. It provides reaction to fast and slow pressure disturbances in a main oil pipeline. Reducing power losses of a frequency controlled pump in dynamic modes for pressure stabilization is being reached by smooth rotating speed change by slow frequency restructuring. The pump control loop of an oil supply allows continuously monitor pump operating point in static mode. The loops of valve opening degree recovery (appropriate to setpoints of OD) provide main pipeline opening and necessary bypass line closing after transient. Made parametric reconfigures during modeling show APCS tuning ease on transients without oscillations and overshooting.

References

1. Starikov, D.P., Rybakov, E.A., Gromakov, E.A.: Minimization of pump energy losses in dynamic automatic control of pressure in the main oil pipeline. In: FCICS 2014, Beijing (2014)

2. Roffel, B., Betlem, B.H.L.: Advanced Practical Process Control. Springer (2004)
3. Carlos, A.: Smith, Principles and Practice of Automatic Process Control, 2nd edn., p. 563. John Wiley & Sons, Inc. (2006)
4. Phillips, C.L., Parr, J.M.: Parr Feedback Control Systems, 5th edn., p. 774. Prentice Hall PTR (2011)
5. Harnefors, L., Nee, H.-P.: Model-Based Current Control of AC Machines Using the Internal Control Model. Method IEEE Transactions on Industry Applications 34(1), 133–141 (1998)
6. Corriou, J.P.: Process Control: Theory and applications. Springer (2004)
7. Ma, Z., Wang, S.: Energy efficient control of variable speed pumps in complex building central air-conditioning systems. Energy and Buildings 41, 197–205 (2009)
8. Pedersen, G.K., Yang, Z.: Efficiency Optimization of a Multi-pump Booster system. In: Proc. of Genetic and Evolutionary Computation Conference (GECCO 2008), Atlanta, Georgia, USA, July 12-16, pp. 1611–1618 (2008)
9. Perez, M.A., Cortes, P., Rodriguez, J.: Predictive control algorithm technique for multilevel asymmetric cascaded h- bridge inverters. IEEE Transactions on Industrial Electronics 55(12), 4354–4361 (2008)
10. Reeves, D.: Study on improving the energy efficiency of pumps. European Commission (2001)
11. Shiels, S.: Optimizing centrifugal pump operation. World Pumps, 35–39 (2001)
12. Miesner, T.O., Leffler, W.L.: Oil & Gas Pipelines in Nontechnical Language, p. 357. PennWell Corp. (2006)
13. Yang, Z., Borsting, H.: Energy Efficient Control of a Boosting System with Multiple Variable-Speed Pumps in Parallel. In: 49th IEEE Conference on Decision and Control, Atlanta, Georgia, USA, December 15-17, pp. 2198–2203 (2010)
14. Wang, Z., Duan, R., Xu, X.: Model Identification of Hydrostatic Center Frame Control System based on MATLAB. Journal of Networks (6), 1322–1328 (2013)
15. Popov, D.N., Sosnovskii, N.G.: Strukturnyi metod modelirovaniia na EVM nestatsionarnykh protsessov v sistemakh s lopastnymi nasosami (Structural method of computer simulations of nonstationary processes in systems with vane pumps). Nauchno-tekhnich. konferentsiia 4-go Mezhdunarodnogo foruma PCVEXPO?2005 Nasosy. Effektivnost' i ekologiia.: tez. dokl (Scientific-technical conference of the 4-th International forum PCVEXPO?2005 Pumps. Efficiency and ecology: theses of reports), Moscow, pp. 13-14 (2005)
16. Popov D.N. Osobennosti dinamiki upravliaemykh sistem s nasosami (Features of dynamics of controllable systems with pumps). Mezhdunar. nauchnotekhnich. ECOPUMP .RU 2007. Effektivnost' i nasosnogo oborudovaniia: tez. dokl. Scientific-Technical Conference ECOPUMP .RU 2007. Effectiveness and ecological properties of pump equipment: theses of reports, pp. 37–38. Bauman MSTU Publ., Moscow (2007)
17. Kluev, A.S., Lebedev, A.T., Kluev, S.A., Tovarov, A.G.: Naladka sredstv avtomatizacii I avtomaticheskih system upravleniya, p. 368. Energoatomizdat, Moscow (1989)
18. Arshenevsky, N.N., Pospelov, B.B.: Perehodnye process v krupnih nasosnih stanciyah, p. 112. Energia, Moscow (1980)
19. Popov, D.N., Sosnovsky, N.G.: Avtomaticheskoe regulirovanie davleniya na vhode v magistralnii nasos pri vikluchenii elektoprivoda. ENTI Nauka I obrazovanie (2013)

Network Society: Aggregate Topological Models

Alexei Tikhomirov[1], Alexandr Afanasyev[2], Nikolay Kinash[2],
Andrey Trufanov[2], Olga Berestneva[3], Alessandra Rossodivita[4],
Sergey Gnatyuk[5], and Rustem Umerov[5]

[1] INHA University,
Nam-gu,Yonghyun-dong, 253, Incheon, Republic of Korea
[2] Irkutsk State Technical University,
Lermontovastr. 83, 664074 Irkutsk, Russia
[3] National Research Tomsk Polytechnic University,
Lenin Avenue 30, 634050 Tomsk, Russia
[4] Ospedale Civile di Legnano,
via Papa Giovanni Paolo II - C.P. 3 - 20025 Legnano, Italy
[5] National Aviation University,
Prosp Kosmonavta Komarova 1, 03058 Kyiv, Ukraine
alexei-tikhomirov@hotmail.com, aad@istu.edu,
{kinash_family,ogb2004}@mail.ru, troufan@istu.edu,
a_rossodivita@yahoo.it, s.gnatyuk@nau.edu.ua,
rustem.amdy.umerov@gmail.com

Abstract. An innovative approach for analysis of "network society" with its large- scale and multicomponent features has been proposed. A new network model - a model of so-called aggregate networks has been developed as a key tool for such analysis. These aggregate structures topologically are not identical in their global and local scales, and thus distinguished from canonical large-scale networks. It was elicited that aggregate network entities have significant features in their topological vulnerability in comparison with canonical ones. This is crucial for building resilient constructions of the network society. Also some additional distinctions for the concepts of "network" and "graph" have been formulated.

Keywords: network society, complex networks, ontologies, aggregate structures, topological vulnerability.

1 Introduction

A predominant way of organizing a network of modern society, is reflected in the phenomenon, which is called network society. This means that society, although composed of many separate individuals, groups, and communities becomes organized and linked together increasingly, forming a large-scale networks. It should be noted that a common definition of network society does not exist. A number of researchers focuse their interest on political components of the society, others pay attention to social or technological constituents of the society [1-3]. However,

A. Dudin et al. (Eds.): ITMM 2014, CCIS 487, pp. 415–421, 2014.

defining network society all recognize the Internet as a valuable communication tool that covers the planet, enabling individuals and groups to establish links, surpassing interaction environments and systems that had been used before. Currently, the theory of complex networks, which is associated with the name of A.L.Barabashi have been used successfully to analyze a variety of complex systems.

The last two years a specific scientific interest have been paid to complex networks in their interdependent [4], multiplex [5], composite [6] and large-scale interpretations [7]. Nevertheless, it has been typical that most famous publications demonstare results of studying networks without thorough attention to their scale, assuming that their topology does not change significantly with the size of the network. The paper [8] gave a suggestion that a network while having become a large-scale acquires other topological characteristics than it had before. Naturally , the size of such networks is the matter of discussion to clarify what the critical size of such networks is .

Meanwhile, methods of mathematics with its highly specific language and even significant results remain inaccessible to researchers from other domains and a wide range of users. Going beyond graph theory as a branch of discrete mathematics, the complex networks approach with its substantive part, realistic models, effective calculation tools and impressive visualization might be considered as a solid platform to further interdisciplinary practice.

In general, it is of sense to set a problem of clarification of the "network society" phenomenon through construction and study of adequate network models with taking into account huge sizes and special topological properties characteristic of contemporary social formations.

2   Methods

Authors approach that we called a Comprehensive Network Lace , CNL (see, [9]) has been applied as a basic method to study the problem. This approach enables thorough and many-sided description and analysis of complex systems. Related to this approach, necessary network models are constructed on the basis of specific ontologies [10] to promote cross-cutting seamless solutions. At the same time our approach provides a platform for international and interdisciplinary cooperation in development of a network society

3   Findings

The authors of this work disclose that a common narrow perception of a "network society" concept, focusing on social and media entities only, is somewhat apart from many modern infrastructures, which provide material and humanitarian life of citizens (energy, transport, food supply, water supply and sewage, educational, scientific, medical, and other technological and organizational networks).

The current study of "network society" has been based on a general concept of network model construction for complicated systems by means of ontologies.

We define the term ontology as most complete formal description with specifications of a network society. CNL composed by its complex networks as a method contrary to graphs has no many pathologies , that make difficulties to work with. Here, it is of sense to declare explicitly the distinctions between theory of complex networks theory and that of graphs . First, in defining a network, which we formulate as a set of homogeneous elements and their homogeneous relations of elements, thus that there not few connections but many, and besides the set evolves and has its own ontogeny. Second, graph theory is just only one of the tools for complex networks (might be most effective) along with other math ones: as linear algebra, probability theory, fractal geometry and engineering: as high-performance computing, optimal control, etc., as well as economic and humanitarian means.

Along with the development of the mathematical formalism for CNL, in this paper we investigate network models, reflecting the network society phenomenon. These structures have not only millions of nodes, but otherwise inhomogeneous topologies which are different to small structure topologies. It is reasonable to call the networks with such properties not large-scale (or massive), but aggregate. A rendering of aggregate network made by means of **graph-tool** [11] is presented by Figure 1.

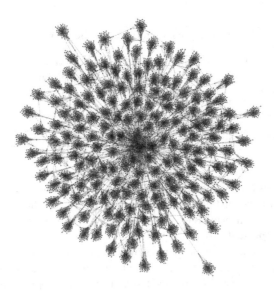

Fig. 1. Aggregate network

Sample aggregate network of a society is modeled on the following algorithm:

1. It has been generated a Barabasi network having $NG=$ 17,000 nodes with an indicator γ_G , which is equal to 1 and with a connectivity which is a random chosen uniformly on the range from 3 to 24 for each newly created node of the network (the coefficient $m \sim U$ (3,24)).

2. For all the nodes of the network it has been created a community of size NL = 100nodes , according to Barabasi model, but with a random γ_L, chosen uniformly on the interval from 1 to 3 for the community L. The coefficient m ∼ U (3, 15) for each community.
3. For the resulting network of N = 1,700,000 nodes it has been calculated an integer $I = (0.03 * N)$. Then it has been created I links between randomly selected nodes. Thus, the aggregate network in contrast to a large-scale one not only brings together a huge number of nodes (typically millions), but the topology of its inter-cluster links (global) is different from the connection topology of nodes within a cluster (local).

Choosing an adequate N_L value was in line with that which is lower than a certain number of Dunbar [12], N_G must be one that a total number of nodes N corresponds to the sizes of real well-known networks *skitter* [13] and *flickr* [14], which allow an adequate comparison of all these networks. Additionally, as comparison it has been generated a synthetic traditional large-scale network (synthetic) with uniform Barabasi topology (γ_L=γ_G=1.04, *m* ∼ *U(2,3)*).

To generate a network and perform calculations a network analysis tool *igraph* [15] has been used . Some key characteristics of the networks are shown in the Table1.

Table 1. Key characteristics of the networks

Network	Node Number	Link Number	Maximal Connectivity	Average Connectivity	Giant Cluster Size
aggregate	1,700,000	13,952,907	4438	16	1,700,000
skitter	1,696,415	11,095,298	35450	13	1,694,616
flickr	1,715,255	15,555,042	18137	18	1,624,992
syntetic	1,700,000	3,399,997	118001	4	1,700,000

Some differences are noticeable not only in the maximum and average connectivities but also in the distribution of this parameter for the aggregate network and the canonical (syntetic) one, (see Figure 2.). All the plots were performed with *matplotlib pyplot* [16]

One of the most important properties of network structures is their topological robustness. In order to identify features of the sample aggregate network that models a network society the prepared structures - aggregate and synthetic, as well as the structures *skitter* and *flickr* have been tested with the removal of nodes. Relative size of giant cluster G has been considered as a sensitive network vulnerability metrics.

The results of calculations of changes in relative size of giant cluster G versus number of removed nodes (proportion to the initial number of nodes in the networks) are shown in Figure 3 and 4 has been revealed the following. Firstly changes in G for the aggregate network and homogeneous canonical (synthetic) one differ sharply. Really observed networks *skitter* and *flickr* demonstrate an

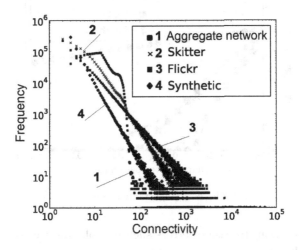

Fig. 2. Connectivity distribution

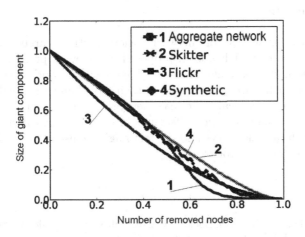

Fig. 3. Error vulnerability

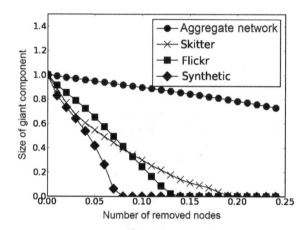

Fig. 4. Attack vulnerability

intermediate position. In case of errors such notable differences in the changes of G was not found.

Secondly, exploring the tails of the vulnerability curves for errors , it is noticeable that in the case of huge removal of nodes ($> 60\%$), the aggregate network is considerably weaker than the others in robustness.

The following selection strategy for node removal has been applied : a) random and b) connectivity-based in descending order, which corresponds to errors and targeted attacks, respectively.

4   Conclusions

A new approach for analysis of "network society" with its large- scale and multicomponent features that has been proposed in the work promotes to prepare a new network model - a model of so-called aggregate network. These aggregate networks topologically are not identical in their global and local scales, and thus distinguished from canonical large-scale networks. It was found that aggregate network structures have significant features in their topological vulnerability in comparison with canonical ones and the fact might be useful in building resilient structures of the network society.

References

1. Barney, D.D.: The Network Society, p. 216. Polity Press, Malden (2004)
2. Castells, M.: Informationalism, networks, and the network society: a theoretical blueprint. Edward Elgar, Northampton (2004)
3. van Dijk, J.: Outline of a multilevel theory of the network society, http://www.utwente.nl/gw/vandijk/research/network_theory/ network_theory_plaatje/a_theory_outline_outline_of_a/

4. Li, W., Bashan, A., Buldyrev, S.V., Stanley, H.E., Havlin, S.: Cascading Failures in Interdependent Lattice Networks: The Critical Role of the Length of Dependency Links. Physical Review Letters PRL 108, 228702 (2012)
5. Gomez-Gardenes, J., Reinares, I., Arenas, A., Floria, L.M.: Evolution of Cooperation in Multiplex Networks. Sci. Rep. 2, 620 (2012), http://www.ncbi.nlm.nih.gov/pmc/articles/PMC3431544/pdf/srep00620.pdf
6. Frivolt, G.: Analysis of Massive Networks. In: Bielikova, M. (ed.) IIT.SRC, pp. 35–40 (2005)
7. Perumal, S., Basu, P., Guan, G.: Minimizing Eccentricity in Composite Networks via Constrained Edge Additions. In: MILCOM 2013, San Diego, CA (November 2013), http://www.ir.bbn.com/~pbasu/pubs/milcom2013-composite.pdf
8. Leskovec, J.: Dynamics of large networks. PhD Dissertation. Carnegie Mellon University. Technical report CMU-ML-08-111 (2008)
9. Aminova, M., Rossodivita, A., Tikhomirov, A., Trufanov, A.: Comprehensive network lace (how to govern the world). Proceedings of Free Economic Society of Russia 148, 190–207 (2011), http://www.iuecon.org/2011/148/20VEOR_PRINT.pdf
10. Frye, L., Cheng, L., Heflin, J.: An Ontology-Based System to Identify Complex Network Attacks. In: First IEEE International Workshop on Security and Forensics in Communication Systems, part of IEEE International Conference on Communications 2012, Ottawa, Canada (2012), http://swat.cse.lehigh.edu/pubs/frye12a.pdf
11. Graph-tool, http://graph-tool.skewed.de/
12. Dunbar, R.I.M.: Neocortex size as a constraint on group size in primates. Journal of Human Evolution 22(6), 469–493 (1992)
13. Skitter, http://konect.uni-koblenz.de/networks/as-skitter
14. Flickr, http://konect.uni-koblenz.de/networks/flickr-links
15. Igraph, http://igraph.org/python/
16. Matplotlib Pyplot, http://matplotlib.org/api/pyplot_api.html

On Evaluation of Discrete States of Hidden Markov Chain under Uncertainty Conditions

Vasily Vasilyev and Alexander Dobrovidov

Trapeznikov Institute of Control Sciences,
Russian Academy of Sciences, 117997, Moscow, Russia
{evil.vasy,dobrovidov}@gmail.com

Abstract. In this paper, we develop methods of nonlinear filtering and interpolation of an unobservable Markov chain with a finite set of states. This Markov chain controls coefficients of AR(p) model. Using observations generated by AR(p) model we have to estimate the state of Markov chain in the case of an unknown probability transition matrix. To solve this problem we construct a system of equations with respect to the posterior probability of Markov states. According to the idea of empirical Bayes approach we represent these equations in the form independent of the unknown transition matrix. The resulting equations are solved using nonparametric kernel procedures by dependent observations. Comparison of proposed non-parametric algorithms with the optimal methods in the case of the known transition matrix is carried out by simulating.

Keywords: hidden Markov chain, autoregressive model, nonlinear filtering and interpolation, multivariate kernel density estimation.

Introduction

In practice it is often required to know the state of the system described by some stochastic differential or difference equation with a given set of values of coefficients. Each given set of equation coefficients values corresponds to some system state. The problem arises to estimate the system state at the specified time from observations of the system output. Switching of states is convenient to describe by Markov chain (ϑ_n) with a finite set of states. In this case, usually say that the coefficients of the observed random process equation is governed by the Markov chain.

There is the well known optimal procedure (maximization of posterior probabilities) for estimating these states by observations of the random process. However, in applications there exists a great number of examples (such as radio and sonar) where it is impossible to construct the transition probability matrix and hence use a standard methods of posterior analysis. In this paper, we propose methods of estimation of these states when transition matrix is unknown. These methods are founded on nonlinear optimal filtering and interpolation equations. These equations can be represented in the form independent of the unknown transition probability matrix. Such representation is a consequence of applying

A. Dudin et al. (Eds.): ITMM 2014, CCIS 487, pp. 422–431, 2014.

of empirical Bayes approach to the problem under consideration. The resulting equations contain some unknown statistics which are restored by using non-parametric kernel estimation techniques by dependent observations. Empirical Bayes approach with the nonparametric estimation was widespread in solving problems under uncertainty (see, for instance, [1–4]). However, authors had not come across with a case where this approach would be applied to a sequence of dependent random variables, i.e. to processes. Application of this approach to stationary processes reveals broad opportunities for solving filtering, interpolation and prediction problems under unknown distributions of the desired signal. The theoretical description of such approach can be found in [5]. In this paper, it is proposed a more effective modifications of algorithms developed in [5] which lead to more accurate results.

In the first section, system model is represented. In the second section, we consider optimal filtering in the case of known transition matrix and propose non-parametric filtering under unknown one. The third section is devoted to multivariate kernel density estimation. Interpolation methods of state estimation with unknown probability characteristics are developed in the fourth section. The comparison of optimal and non-parametric procedures are performed in the last section.

1   System Model

Let (ϑ_n, X_n) be a two-component process, where (ϑ_n) is unobservable process and (X_n) is observable one, $n \in \overline{1, N}$, $N \in \mathbb{N}$; (ϑ_n) controls coefficients of (X_n). Let (ϑ_n) be a stationary Markov chain with M discrete states and transition matrix $\|p_{i,j}\|$, $p_{i,j} = \mathsf{P}\{\vartheta_n = j \mid \vartheta_{n-1} = i\}$. The process (X_n) is described by the autoregressive model of order p:

$$X_n = \mu(\vartheta_n) + \sum_{i=1}^{p} a_i(\vartheta_n)(X_{n-i} - \mu(\vartheta_n)) + b(\vartheta_n)\xi_n \ , \tag{1}$$

where $\{\xi_n\}$ are i.i.d. random variables with the standard normal distribution, $\mu, a_i, b \in \mathbb{R}$ are coefficients controlled by the process (ϑ_n).

As a quality measure for the proposed methods we use mean risk with a simple loss function L:

$$L(\vartheta_n, \hat{\vartheta}_n) = \begin{cases} 1, & \vartheta_n \neq \hat{\vartheta}_n \ , \\ 0, & \vartheta_n = \hat{\vartheta}_n \ , \end{cases} \tag{2}$$

where $\hat{\vartheta}_n = \hat{\vartheta}_n(X_1^n)$ is an estimator of ϑ_n and $X_1^n = X_1, X_2, \ldots, X_n$.

For the risk function with the loss function (2) the optimal estimator is

$$\hat{\vartheta}_n = \arg\max_m \mathsf{P}\{\vartheta_n = m \mid X_1^n\} \ , \tag{3}$$

where $P\{\vartheta_n = m \mid X_1^n\}$ is a posterior probability with respect to a σ-algebra, generated by r.v. X_1^n. Its realization will be denoted by

$$\mathsf{P}\{\vartheta_n = m \mid X_1^n = x_1^n\} = w_n(m \mid x_1^n) = w_n(m) \ . \tag{4}$$

2   Filtering

2.1   Optimal Filtering

In filtering we estimate the value of unobservable ϑ_n using observable values x_1^n. Probability $w_n(m)$ satisfies the recurrent Stratonovich's equation [5]

$$w_n(m) = \frac{f_m(x_n)}{f(x_n \mid x_1^{n-1})} \sum_{j=1}^{M} p_{j,m} w_{n-1}(j) \,, \tag{5}$$

$$f_{m,n} = f_m(x_n) = f(x_n \mid x_{n-p}^{n-1}, \vartheta_n = m) \,. \tag{6}$$

Using (1) the conditional density function $f_m(x_n)$ is calculated as:

$$f_m(x_n) = \mathcal{N}\left(x_n,\ \mu(m) + \sum_{i=1}^{p} a_i(m)(x_{n-i} - \mu(m)),\ b^2(m)\right) \,. \tag{7}$$

Conditional density $f(x_n \mid x_1^{n-1})$ is obtained by summing of (5) over m:

$$f(x_n \mid x_1^{n-1}) = \sum_{m=1}^{M}\left(f_m(x_n) \sum_{j=1}^{M} p_{j,m} w_{n-1}(j)\right) \,. \tag{8}$$

If transition matrix $\|p_{i,j}\|$ is known, then all elements in (5) are defined and the solution of the problem is found, otherwise we offer methods in the next section.

2.2   Non-parametric Filtering

In this method we overcome uncertainty in the $\|p_{i,j}\|$. For that we analyse $f(x_n \mid x_1^{n-1})$. We assume that process (ϑ_n, X_n) is α-mixing, then instead of $f(x_n \mid x_1^{n-1})$ we use "truncated" approximation $f(x_n \mid x_{n-\tau}^{n-1})$, $\tau \in \overline{1, n-1}$. Then (5) can be rewritten as

$$w_n(m) \approx \frac{f_m(x_n) f(x_{n-\tau}^{n-1})}{f(x_{n-\tau}^n)} \sum_{j=1}^{M} p_{j,m} w_{n-1}(j) \,, \tag{9}$$

$$w_n(m) = \frac{f_m(x_n)}{f(x_{n-\tau}^n)} u_n(m) \,, \tag{10}$$

where the introduced variable $u_n(m)$ does not depend on x_n. Then we sum of (10) over m and get:

$$f(x_{n-\tau}^n) = \sum_{m=1}^{M} f_m(x_n) u_n(m) \,. \tag{11}$$

To calculate $u_n(m)$ it is necessary to find $M - 1$ more equations, which we can obtain by differentiating and integrating (11) with respect to x_n.

Therefore obtained system of equations in matrix form is

$$\mathbf{F}_n u_n = b_n ,\tag{12}$$

where

$$\mathbf{F}_n = \begin{pmatrix} 1 & 1 & \cdots & 1 \\ f_{1,n} & f_{2,n} & \cdots & f_{M,n} \\ f_{1,n}^{(1)} & f_{2,n}^{(1)} & \cdots & f_{M,n}^{(1)} \\ \vdots & \vdots & \ddots & \vdots \\ f_{1,n}^{(M-2)} & f_{2,n}^{(M-2)} & \cdots & f_{M,n}^{(M-2)} \end{pmatrix} ,\tag{13}$$

$$u_n = \begin{pmatrix} u_n(1) \ u_n(2) \ u_n(3) \ \dots \ u_n(M) \end{pmatrix}^T ,\tag{14}$$

$$b_n = \begin{pmatrix} f(x_{n-\tau}^{n-1}) \ f^{(0)}(x_{n-\tau}^n) \ \dots \ f^{(M-2)}(x_{n-\tau}^n) \end{pmatrix}^T .\tag{15}$$

We estimate unknown elements of b_n by using multivariate kernel density estimators from Sect. 4.

2.3 Adaptive Non-parametric Filtering

Note that solution of (12) is dependent on the properties of the matrix $\mathbf{F}$. If the matrix $\mathbf{F}$ is near singular then the solution is diverged. This difficulty could be overcome using following method.

Consider system of equations (12) for $M = 2$:

$$\begin{cases} u_n(1) + u_n(2) = f(x_{n-\tau}^{n-1}) \\ f_{1,n}u_n(1) + f_{2,n}u_n(2) = f(x_{n-\tau}^n) \end{cases} .\tag{16}$$

We add another compatible equation obtained by differentiating with respect to x_n the second equation:

$$f_{1,n}' u_n(1) + f_{2,n}' u_n(2) = f'(x_{n-\tau}^n) .\tag{17}$$

To find u_n we solve one of the 3 systems

$$\mathbf{F}_{n,1}u_n = b_{n,1} : \begin{pmatrix} 1 & 1 \\ f_{1,n} & f_{2,n} \end{pmatrix} \cdot \begin{pmatrix} u_n(1) \\ u_n(2) \end{pmatrix} = \begin{pmatrix} f(x_{n-\tau}^{n-1}) \\ f(x_{n-\tau}^n) \end{pmatrix} ,\tag{18}$$

$$\mathbf{F}_{n,2}u_n = b_{n,2} : \begin{pmatrix} f_{1,n} & f_{2,n} \\ f_{1,n}' & f_{2,n}' \end{pmatrix} \cdot \begin{pmatrix} u_n(1) \\ u_n(2) \end{pmatrix} = \begin{pmatrix} f(x_{n-\tau}^n) \\ f'(x_{n-\tau}^n) \end{pmatrix} ,\tag{19}$$

$$\mathbf{F}_{n,3}u_n = b_{n,3} : \begin{pmatrix} 1 & 1 \\ f_{1,n}' & f_{2,n}' \end{pmatrix} \cdot \begin{pmatrix} u_n(1) \\ u_n(2) \end{pmatrix} = \begin{pmatrix} f(x_{n-\tau}^{n-1}) \\ f'(x_{n-\tau}^n) \end{pmatrix} .\tag{20}$$

For solving we propose to choose the system with the best-conditioned matrix for each u_n using the following test

$$I_n = \arg \min_{i=1,2,3} \kappa(F_{n,i}) ,\tag{21}$$

where $\kappa(A) = \|A\| \cdot \|A^{-1}\|$ is a condition number, $\|A\| = \sup_{x \neq 0} \frac{\|A\bar{x}\|_2}{\|\bar{x}\|_2}$.
Choosing of necessary matrix is represented in Fig. 1.

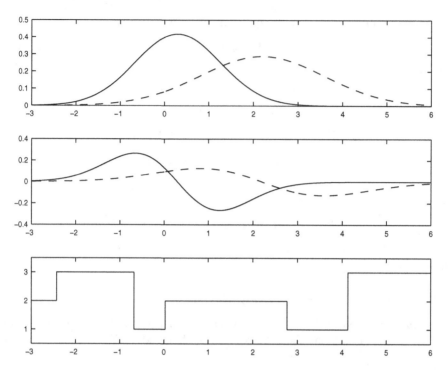

Fig. 1. From top to bottom: 1 – densities $f_{1,n}(solid)$ and $f_{2,n}(dashed)$; 2 – derivatives $f'_{1,n}(solid)$ and $f'_{2,n}(dashed)$; 3 – index I_n

3 Multivariate Kernel Density Estimation

Let us introduce partial density derivatives

$$f^{(r)}(y) = \frac{\partial^r f(y)}{\partial y_1^{r_1} \partial y_2^{r_2} \ldots \partial y_d^{r_d}} \,, \tag{22}$$

where $y = (y_1, y_2, \ldots, y_d) \in \mathbb{R}^d$, $d \in \mathbb{N}$; $r_1, r_2, \ldots, r_d$ are orders of partial derivatives and $r = r_1 + r_2 + \ldots + r_d$.

Let us introduce sample vector $Y_i = (Y_{i1}, Y_{i2}, \ldots, Y_{id})$, $i = 1, 2, \ldots, N'$ drawn from $f(y)$. Then nonparametric estimator of $f^{(r)}(y)$ is

$$\hat{f}^{(r)}(y) = \frac{1}{N'} \sum_{i=1}^{N'} \prod_{j=1}^{d} \frac{1}{h_{j,N'}^{1+r_j}} K^{(r_j)} \left(\frac{y_j - Y_{ij}}{h_{j,N'}} \right) \,, \tag{23}$$

where $K(\cdot)$ is a kernel, $h_{j,N}$ is called the bandwidth. Properties and choosing kernel and bandwidth see in [6], [7]. We use Gaussian kernel

$$K(u) = \frac{1}{\sqrt{2\pi}} e^{-\frac{1}{2}u^2} \,, \quad u \in \mathbb{R} \,, \tag{24}$$

with corresponding derivative with respect to u

$$K'(u) = -\frac{u}{\sqrt{2\pi}}e^{-\frac{1}{2}u^2} = -u \cdot K(u) \ . \tag{25}$$

For bandwidth $h_{j,N}$ we use estimator "Rule of thumb" proposed in [7]:

$$\hat{h}_{j,N} = \left(\frac{4}{d + 2r_j + 2}\right)^{1/(d+2r_j+4)} \hat{\sigma}_j N^{-1/(d+2r_j+4)} \ , \tag{26}$$

where $\hat{\sigma}_j$ is a sample standard deviation of realization $y_{1j}, y_{2j}, \ldots, y_{Nj}$.

The sample vector $Y_i = (Y_{i1}, Y_{i2}, \ldots, Y_{id})$ is constructed using process realization $X_1, X_2, \ldots X_N$ according the following rule:

$$(Y_{i1}, Y_{i2}, \ldots, Y_{id}) = (X_i, X_{i+1}, \ldots X_{i+d-1}) \ . \tag{27}$$

Then the resulting vector sequence is $(X_i, X_{i+1}, \ldots X_{i+d-1})_{i=1}^{N'}$, $N' = N - d + 1$. To estimate value of a density $f(x_{n+a}^{n+b}) = f(x_{n+a}, x_{n+a+1}, \ldots x_{n+b})$, $n \in \overline{1, N'}$, $a, b \in \mathbb{Z}$, $a \le b$, $d = b - a + 1$ we use sample sequence $(x_i, x_{i+1}, \ldots x_{i+d-1})_{i=1}^{N'}$ without vector x_{n+a}^{n+b}. Then the resulting sequence $(x_i, x_{i+1}, \ldots x_{i+d-1})_{i=1, i \neq n+a}^{N-d+1}$ has $N - d$ elements. Therefore, estimator of density $f(x_{n+a}^{n+b})$ has the form

$$\hat{f}(x_{n+a}^{n+b}) = \frac{1}{N-d} \sum_{\substack{k=1-a, \\ k \neq n}}^{N-b} \prod_{j=a}^{b} \frac{1}{h_{n+j,N-d}} K\left(\frac{x_{n+j} - x_{k+j}}{h_{n+j,N-d}}\right) \ , \tag{28}$$

$$h_{n+j,N-d} = \left(\frac{4}{d+2}\right)^{1/(d+4)} \hat{\sigma}_j (N - d)^{-1/(d+4)} \ . \tag{29}$$

In this case $(r_{n+a}, r_{n+a+1}, \ldots, r_{n+b}) = (0, 0, \ldots 0)$.

For example, the estimator of $f(x_{n-\tau}^n)$ is

$$\hat{f}(x_{n-\tau}^n) = \frac{1}{n-\tau-1} \sum_{\substack{k=1+\tau, \\ i \neq n}}^{n} \prod_{j=-\tau}^{0} \frac{1}{h_{n+j,n-\tau-1}} K\left(\frac{x_{n+j} - x_{k+j}}{h_{n+j,n-\tau-1}}\right) \tag{30}$$

with $(r_{n-\tau}, r_{n-\tau+1}, \ldots r_n) = (0, 0, \ldots, 0)$.

The estimator of the derivative $f^{(1)}(x_{n-\tau}^n)$ is

$$\hat{f}^{(1)}(x_{n-\tau}^n) = \frac{1}{n-\tau-1} \sum_{\substack{k=1+\tau, \\ k \neq n}}^{n} \prod_{j=-\tau}^{0} \frac{1}{h_{n+j,n-\tau-1}^{1+r_{n+j}}} K^{(r_{n+j})}\left(\frac{x_{n+j} - x_{k+j}}{h_{n+j,n-\tau-1}}\right) \ , \tag{31}$$

where $(r_{n-\tau}, r_{n-\tau+1}, \ldots r_n) = (0, 0, \ldots, 1)$.

4 Interpolation

4.1 Optimal Interpolation

In interpolation we estimate the value of unobservable ϑ_n, $n \in \overline{1,N}$ using observable values x_1^N. To get optimal interpolation estimator one need to use markov property of sequence (ϑ_n, X_n)

$$f(\theta_1^N, x_1^N) = f(\theta_1, x_1) \prod_{n=2}^N g(\theta_n, x_n \mid \theta_{n-1}, x_{n-1}) , \qquad (32)$$

where (θ_n, x_n) is a realization of (ϑ_n, X_n) and $g(\cdot \mid \cdot)$ is a transition density of process (ϑ_n, X_n). We use the formula (32) to calculate a posterior probability $\pi_n(\theta_n \mid x_1^N)$:

$$\pi_k(\theta_n \mid x_1^N) = \frac{1}{f(x_1^N)} f(\theta_n, x_1^N) = \frac{1}{f(x_1^N)} f(\theta_n, x_1^n, x_{n+1}^n) = \qquad (33)$$

$$= \frac{1}{f(x_1^N)} f(x_1^n) w_n(\theta_n \mid x_1^n) f(x_{n+1}^N \mid \theta_n, x_1^n) = \qquad (34)$$

$$= \frac{1}{f(x_1^N)} f(x_1^n) w_n(\theta_n \mid x_1^n) f(x_{n+1}^N \mid \theta_n, x_n) . \qquad (35)$$

It is necessary to find all elements in latter equation, which depend on x_n. According to total probability formula

$$f(x_1^n) w_n(\theta_n \mid x_1^n) = \sum_{\theta_{n-1}=1}^M f(\theta_{n-1}, x_1^{n-1}) g(\theta_n, x_n \mid \theta_{n-1}, x_{n-1}) . \qquad (36)$$

For dynamical model AR(1)

$$g(\theta_n, x_n \mid \theta_{n-1}, x_{n-1}) = p(\theta_n \mid \theta_{n-1}) f(x_n \mid x_{n-1}, \theta_n) \qquad (37)$$

holds, thus we get

$$f(x_1^n) w_n(\theta_n \mid x_1^n) = f(x_n \mid x_{n-1}, \theta_n) \sum_{\theta_{n-1}=1}^M p(\theta_n \mid \theta_{n-1}) f(\theta_{n-1}, x_1^{n-1}) = \qquad (38)$$

$$= f(x_1^{n-1}) f(x_n \mid x_{n-1}, \theta_n) p(\theta_n \mid x_1^{n-1}) , \qquad (39)$$

where only the second factor depends on x_n. Then we consider the third factor of (35):

$$f(x_{n+1}^N \mid \theta_n, x_n) = \sum_{\theta_{n+1}=1}^M f(x_{n+2}^N \mid \theta_{n+1}, x_{n+1}) g(\theta_{n+1}, x_{n+1} \mid \theta_n, x_n) = \qquad (40)$$

$$= \sum_{\theta_{n+1}=1}^M f(x_{n+2}^N \mid \theta_{n+1}, x_{n+1}) p(\theta_{n+1} \mid \theta_n) f(x_{n+1} \mid x_n, \theta_{n+1}) = \qquad (41)$$

$$= \sum_{\theta_{n+1}=1}^M f(x_{n+2}^N, \theta_{n+1}, x_{n+1}) \frac{p(\theta_{n+1} \mid \theta_n)}{p(\theta_{n+1})} \frac{f(x_{n+1} \mid x_n, \theta_{n+1})}{f(x_{n+1} \mid \theta_{n+1})} \qquad (42)$$

Here only $f(x_{n+1} \mid x_n, \theta_{n+1})/f(x_{n+1} \mid \theta_{n+1})$ depends on x_n. Because of this dependence we can not find equations for optimal interpolation estimator in the case of dynamical systems. However, if model of observations is static, i.e. $X_n = \varphi(\vartheta_n, \xi_n)$ with φ is known Borel function and ξ_n is independent noise, then conditional density $f(x_{n+1} \mid x_n, \theta_{n+1}) = f(x_{n+1} \mid \theta_{n+1})$ does not depend on x_n. Using last assumption we present the formula (35) to

$$\pi(\vartheta_n = m \mid x_1^N) = \pi_n(m) = \frac{1}{f(x_1^N)} f(x_1^n) w_n(m \mid x_1^n) f(x_{n+1}^N \mid m, x_n) = \quad (43)$$

$$= \frac{f(x_1^{n-1})}{f(x_1^N)} f(x_n \mid m) p(m \mid x_1^{n-1}) \sum_{\theta_{n+1}=1}^{M} f(x_{n+2}^N, \theta_{n+1}, x_{n+1}) \frac{p(\theta_{n+1} \mid m)}{p(\theta_{n+1})} = \quad (44)$$

$$= \frac{f(x_1^{n-1}) f(x_{n+1}^N)}{f(x_1^N)} \cdot \frac{f_m(x_n) v_n(m) \widetilde{v}_n(m)}{p_n(m)} , \quad (45)$$

where

$$v_n(m) = \sum_{j=1}^{M} p_{j,m} w_{n-1}(j) , \quad \widetilde{v}_n(m) = \sum_{j=1}^{M} p_{j,m}^+ \widetilde{w}_{n+1}(j) , \quad (46)$$

$$p_{j,m}^+ = \mathsf{P}\{\vartheta_n = m \mid \vartheta_{n+1} = j\} , \quad p_n(m) = \mathsf{P}\{\vartheta_n = m\} . \quad (47)$$

Filtering probability w_n is defined in (5) and $\widetilde{w}_n$ comply with recurrent equation

$$\widetilde{w}_n(m) = \frac{f_m(x_n)}{f(x_n \mid x_{n+1}^N)} \sum_{j=1}^{M} p_{j,m}^+ \widehat{w}_{n+1}(j) . \quad (48)$$

If $\|p_{i,j}\|$ is known then all variables in (45) could be calculated and the problem could be solved.

4.2   Non-parametric Interpolation

As in non-parametric filtering we overcome the uncertainty in $\|p_{i,j}\|$ by introducing new variable $z_n(m)$ that does not depend on x_n and includes unknown $p_{i,j}$. Then (45) can be represented like

$$\pi_n(m) = \frac{f_m(x_n) z_n(m)}{f(x_{n-\tau}^{n+\tau})} . \quad (49)$$

If we sum of latter equation over m, we get:

$$f(x_{n-\tau}^{n+\tau}) = \sum_{m=1}^{M} f_m(x_n) z_n(m) . \quad (50)$$

To calculate $z_n(m)$ it is necessary to find $M - 1$ more equations, which we can obtain by differentiating and integrating (50) with respect to x_n. The resulting system of linear equations in $z_n(m)$ can be represented as

$$\mathbf{F}_n z_n = c_n , \quad (51)$$

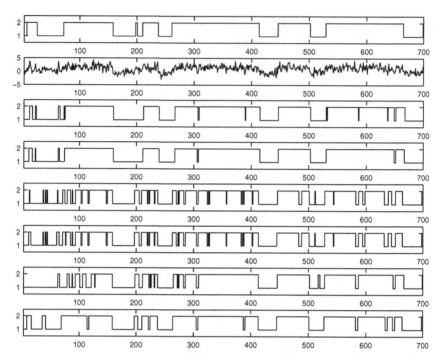

Fig. 2. From top to bottom: 1 – unobservable θ_n; 2 – observable x_n; 3, 4 – estimator $\hat{\vartheta}_n$ by optimal filtering and interpolation; 5, 6 – estimator $\hat{\vartheta}_n$ by non-parametric filtering and interpolation; 7, 8 – estimator $\hat{\vartheta}_n$ by adaptive non-parametric filtering and interpolation

where $\mathbf{F}_n$ is defined in (13) and

$$z_n = \big(z_n(1)\, z_n(2)\, z_n(3) \ldots z_n(M)\big)^T \ , \tag{52}$$

$$c_n = \big(f(x_{n-\tau}^{n-1}, x_{n+1}^{n+\tau})\, f^{(0)}(x_{n-\tau}^{n+\tau}) \ldots f^{(M-2)}(x_{n-\tau}^{n+\tau})\big)^T \ . \tag{53}$$

To estimate unknown multivariate densities and its derivatives of column c_n we use kernel estimator (23). Convergence condition of kernel estimators by dependent observations one can find in [5].

4.3 Adaptive Non-parametric Interpolation

This method is similar to the adaptive non-parametric filtering. For solving we propose to choose the system with the best-conditioned matrix for each z_n.

5   Example

Let the Markov chain (ϑ_n) have two states $(M = 2)$ and the transition matrix

$$\|p_{i,j}\| = \begin{pmatrix} 0.98 & 0.02 \\ 0.02 & 0.98 \end{pmatrix} . \tag{54}$$

Consider the AR(1) model with coefficients $\mu \in \{0, 1.5\}$, $a_1 \in \{0.3, 0.2\}$, $b \in \{0.9539, 0.9797\}$. Coefficients b are calculated using Yule-Walker [8].

The volume of generated sample is 700. Results are presented in Fig. 2 and sample mean errors after 50 repeated experiments in Table 1.

Table 1. Sample mean errors

	Filtering	Interpolation
Optimal	0.0838	0.0738
Non-parametric	0.1510	0.1454
Adaptive non-parametric	0.1338	0.1090

Conclusion

Developed nonparametric methods for nonlinear estimation of the hidden Markov chain states with the unknown transition probability matrix belong to a class of unsupervised stochastic algorithms. To obtain more accurate results in the future we suppose to choose the unconstrained matrix bandwidth for kernel nonparametric estimates of multivariate densities entering in the algorithms.

References

1. Singh, R.S.: Empirical Bayes estimation in Lebesgue-exponential families with rates near the best possible rate. Annals of Statistics 7, 890–902 (1979)
2. Pensky, M.: A general approach to nonparametric empirical Bayes estimation. Statistics 29, 61–80 (1997)
3. Pensky, M.: Empirical Bayes estimation of a scale parameter. The Mathematical Methods of Statistics 5, 316–331 (1996)
4. Pensky, M., Singh, R.S.: Empirical Bayes estimation of reliability characteristics for an exponential family. The Canadian Journal of Statistics 27, 127–136 (1999)
5. Dobrovidov, A.V., Koshkin, G.M., Vasiliev, V.A.: Non-parametric models and statistical inference from dependent observations. Kendrick Press, USA (2012)
6. Turlach, B.A.: Bandwidth Selection in Kernel Density Estimation: A Review. In: CORE and Institut de Statistique, pp. 23–493 (1993)
7. Chacon, J.E., Duon, T., Wand, M.P.: Asymptotics for General Multivariate Kernel Density Derivative Estimators. Centre for Statistical and Survey Methodology, University of Wollongong, Working Paper 08-09, p. 30 (2009)
8. Box, G., Jenkins, G.: Time series analysis: Forecasting and control. Holden-Day, San Francisco (1970)

Growing Network: Nonlinear Extension
of the Barabasi-Albert Model

Vladimir Zadorozhnyi and Evgeniy Yudin

Omsk State Technical University, Russia
zwn@yandex.ru,
udinev@asoiu.com

Abstract. We introduce a model of random graphs that follows a nonlinear preferential attachment rule and give a recursive formula to determine the vertex degree distribution for the graphs of this model. We demonstrate how the nonlinear preferential attachment model can be calibrated so that the generated graphs will require vertex degree distributions.

Keywords: networks, random graphs, nonlinear preferential attachment rule, structural properties.

1   Introduction

At present the Barabasi-Albert (BA) model for generating random graphs [1] is widely spread when modeling large network structures like social networks, web networks and the Internet [2–4]. The BA model starts with a small "seed" graph and increases by subsequent addition of new vertices one at a time, each with $m = const$ incident edges. A new vertex with its incident edges that is added to the graph will be named a graph differential (GD). When growing the BA graph, free ends of the GD edges attach to any vertices of the graph. The probability p_i of attaching to vertex i depends on degree k_i:

$$p_i = \frac{k_i}{\sum\limits_j k_j} \, . \tag{1}$$

The infinite addition of new GDs leads to growing the infinite BA graph, which is scale-free because its vertex degree distribution is asymptotically power-law (a scale-free distribution). The exact expression for stationary vertex degree distribution of the infinite BA graph was first given in [5]:

$$Q_k = \frac{2m(m+1)}{k(k+1)(k+2)}, \quad k = m, m+1, ..., \tag{2}$$

where Q_k is the probability that a randomly chosen vertex has degree k. From Eq. (2) it follows that $Q_k \sim 2m(m+1)k^{-3}$, and $Q_k \propto k^{-3}$ as $k \to \infty$. By the construction of the BA graph, the average degree of the graph is $\langle k \rangle = 2m$.

A. Dudin et al. (Eds.): ITMM 2014, CCIS 487, pp. 432–439, 2014.

It can also be derived from Eq. (2). Asymptotic power-law vertex degree distribution of the BA graph is only consistent with node degree distribution of some real networks [4]. And in cases when node degree distributions of networks are close to power-law, other characteristics such as clustering coefficient or network diameter can differ from the corresponding characteristics of the BA graphs [2, 3].

2   Structural Characteristics of the BA Graphs

Consider such structural properties of the BA graphs as vertex degree correlation and clustering coefficient [6, 7]. Let the BA graph be realized as a directed graph. This modification helps us to focus on the difference in the characteristics of the two ends of edges. This difference occurs while attaching an edge of a GD because only one end of the edge chooses a vertex to attach to it. A GD of the directed BA graph is a vertex with m outgoing directed edges. The ends of these edges attach to the existing vertices according to the preferential attachment rule (1).

The stationary joint probability distribution $\{Q_{l,k}\}$ of vertex degrees l and k of a random directed edge of the BA graph can be represented in the general case as a recursive equation [8]:

$$Q_{l,k} = \begin{cases} 0, \quad l \geq m, & k = m, \\ \frac{2}{(m+2)(2m+3)}, & l = m, \quad k = m+1, \\ \frac{2(m+1)+k(k^2-1)}{k(k+1)(k+m+2)}Q_{l,k-1}, & l = m, k \geq m+2, \\ \frac{(k-1)Q_{l,k-1}+(l-1)Q_{l-1,k}}{(l+k+2)}, & l \geq m+1, k \geq m+1. \end{cases} \tag{3}$$

The calculation of matrix $\mathbf{Q} = \| Q_{l,k} \|$ starts with the completion of its leftmost column (column $k = m$) by zero values and calculation of the upper line (line $l = m$): firstly, element $Q_{m,m+1}$ of this line is calculated by the formula (3), then elements $Q_{m,k}$ such that $k \geq m + 2$ are calculated. After completing the upper line, elements of the following lines are calculated left to right line by line. As a special case, result (3) contains the following solution derived in [6] for the BA model with $m = 1$ in the closed form:

$$Q_{lk} = \frac{4(k-1)}{l(l+1)(l+k)(l+k+1)(l+k+2)} + \tag{4}$$
$$+ \frac{12(k-1)}{l(l+k-1)(l+k)(l+k+1)(l+k+2)}, \quad l, k \geqslant 1.$$

The marginal distributions for the degrees of vertices incident to directed edges are found in [8]:

$$Q_{k,*} = \frac{2m(m+1)}{k(k+1)(k+2)}, \quad Q_{*,k} = \frac{2(k-m)(m+1)}{k(k+1)(k+2)}, \quad l, k \geqslant m, \tag{5}$$

where $Q_{k,*}(Q_{*,k})$ is the probability that the initial vertex of the edge (the terminal vertex of the edge) has the degree k. From Eq. (5) it follows that the average degree of the initial vertex is $2m$ and the average degree of the terminal vertex is infinite. The degrees of vertices incident to directed edges are dependent variables.

We also find the following expression for the clustering coefficient C of the finite BA graph:

$$C \sim \frac{(m-1)(\ln N)^2}{8} \frac{1}{N} + (m-1)c_m \frac{\ln N}{N}, \qquad (6)$$

where N is the number of vertices in the graph, c_m is a coefficient that can be evaluated when the large BA graph is generating. So, for $m = 2, 3, ..., 8$ values c_m are approximately $0.5, 0.35, 0.275, 0.18, 0.12, 0.10$ and 0.07 respectively. Found in [7] the shifted for m estimate

$$C \sim \frac{m}{8} \frac{(\ln N)^2}{N}$$

is close to our estimation (6).

3   Nonlinear Preferential Attachment Graphs

By analogy with the BA model, we study a more general model for generating nonlinear preferential attachment graphs (the NPA graphs). To grow the NPA graph we use the weight function $f = f(k)$ (preference function, weight) defined for integer k where $g \leq k \leq M$, $(g \geq 1, M \leq \infty)$. The weights $f_k = f(k)$ must be non-negative values. The probability p_i of attaching to vertex i depends on the degree k_i of this vertex such that

$$p_i = \frac{f(k_i)}{\sum_j f(k_j)}. \qquad (7)$$

In addition, the NPA graphs have stochastic GDs, i.e. each GD is a vertex with a random number x of incident edges. The random variable x has discrete probability distribution $\{r_k\}$, where the probability $r_k = P\{x = k\} \geq 0$, $g \leq k \leq h$ $(g \geq 1, h \leq M)$ and $\sum_{k=g}^{h} r_k = 1$. The average number of GD edges is $m = \langle x \rangle = \sum_{k=g}^{h} k r_k < \infty$. Thus, the NPA model can be calibrated by varying f and $\{r_k\}$ parameters.

In [9] the existence condition of a stationary NPA graph where $M + 1 \geq 2m$ is found. If $M + 1 = 2m$, then a pseudo-lattice is generated, which is an infinite random graph where the vertex degree is $2m = M + 1$ with probability one.

We propose the following modification of the NPA model. When an attachment to the next vertex is unrealizable (due to a possible lack of vacant positions when M is finite), the GD is queued, and the next GD is generated.

The vertex degree distribution for the NPA graph is expressed recursively:

$$Q_g = \frac{r_g \langle f \rangle}{\langle f \rangle + m f_g}, \qquad Q_k = \frac{r_k \langle f \rangle + m f_{k-1} Q_{k-1}}{\langle f \rangle + m f_k}, \qquad k \geq g+1, \qquad (8)$$

where $\langle f \rangle = \sum f_k Q_k$ is the average preference of a vertex. The reasoning used in [9] in the derivation of Eq. (8) is similar to the reasoning used in [8] in the derivation of Eq. (3). If $x \equiv m = g$ (i.e. $r_g = 1$), then a stochastic GD reduces to a constant GD (a vertex with m incident edges). In addition, if $f_k = k$, then the NPA graph reduces to the BA graph, and in Eq. (7) we have $\langle f \rangle = \langle k \rangle = 2m$, $f_g = g = m$,

$$Q_g = \frac{2}{2+m}, \qquad Q_k = \frac{k-1}{2+k} Q_{k-1}, \quad k \geq g+1.$$

From here, the formula (2) is derived easily by induction on k for the BA graph.

The direct use of recursion (8) for the numerical calculation of probabilities Q_k is generally complicated by the fact that if m, r_k and f_k are given, then the average weight $\langle f \rangle$ is unknown. However, it can be derived from the equation that is obtained by substitution of Eq. (8) in equation $\langle k \rangle = \sum_{k \geqslant g} k Q_k = 2m$. Such calculation of $\langle f \rangle$ is easily implemented in a spreadsheet environment by the following procedure.

1. Form columns of values k, f_k, r_k and cell m.
2. Enter any value $a > 0$ as the initial approximation of $\langle f \rangle$ in a separate cell.
3. Form a column of probabilities Q_k (initial approximation $\{Q_k\}$) with the reference to the cell a as a parameter $\langle f \rangle$ by recursive formulas (8).
4. Enter in separate cells:
   – the formula that calculates $\langle f \rangle$ as the sum of productions of the columns k and Q_k,
   – the formula that calculates $\langle f \rangle$ as the sum of productions of the columns f_k and Q_k.
5. With the help of the service Goal Seek, find the required a for which cell $\langle k \rangle$ equals $2m$. There exists a unique value of a [9]. With the value of a given, cell $\langle f \rangle$ will automatically be equal to the value of a. As a result, we obtain the required values $\langle f \rangle$ and Q_k.

We can also solve the inverse task that is synthesis (or calibration) of the model by the formulas (8). If distribution $\{\tilde{Q}_k\} = \tilde{Q}_g,...,\tilde{Q}_{M+1} > 0$, $(\sum Q_k = 1)$ and the average degree $\langle k \rangle = 2m \geq 2g$ are given, then we can find parameters $\{r_k\} = r_g,...,r_h$ and $\{f_k\}$ such that graph with the required vertex degree distribution $\{Q_k\} = \{\tilde{Q}_k\}$ is generated. From Eq. (7) it follows that the preference function f for the NPA model can be determined with an accuracy to a multiplicative factor. That allows to impose the condition $\langle f \rangle = m$ and to determine $\{f_k\}$ from Eq. (8) as:

$$f_k = \begin{cases} \frac{r_g}{\tilde{Q}_g} - 1, & k = g, \\ \frac{\tilde{Q}_{k-1}}{\tilde{Q}_k} f_{k-1} + \frac{r_k}{\tilde{Q}_k} - 1, & k = g+1, ..., M, \\ 0, & k = M+1 \ (where \ M < \infty). \end{cases} \tag{9}$$

If $\tilde{Q}_k$ are empirical estimates, which accuracy decreases with increasing k, the calculation of a few values f_k at a small k by the formulas (9) is recommended.

For the remained k, it is necessary either to smooth the sequence f_k derived with the use of the estimated $\tilde{Q}_k$, or to replace estimates $\tilde{Q}_k$ in Eq. (9) by their smoothed estimates. Distribution of $\{r_k\}$ in Eq. (9) can be chosen arbitrarily from the constraint region defined by the following conditions:

$$0 \leqslant r_i \leqslant 1, \quad i = \overline{g, h}, \tag{10}$$

$$\sum_{i=g}^{h} r_i = 1, \tag{11}$$

$$\sum_{i=g}^{h} i r_i = m, \tag{12}$$

$$\sum_{i=g}^{k} r_i \geqslant \sum_{i=g}^{k} \tilde{Q}_i, \quad k = \overline{g, h}. \tag{13}$$

Conditions (10) and (11) characterize the properties of probabilities r_k, condition (12) follows from $\langle x \rangle = m$, which provides the required average degree $\langle k \rangle = 2m$, condition (13) provides non-negative preferences $\{f_k\}$ and is calculated by Eq. (9).

Let $F_Q(t)$ denote the cumulative distribution function (DF) of vertex degree k: $F_Q(t) = \sum_{k \leq t} Q_k$. Let $F_r(t)$ denote the DF of number x of the edges in a stochastic GD: $F_r(t) = \sum_{i \leq t} r_i$. Functions $F_Q(t)$ and $F_r(t)$ are determined uniquely by their values at the points $t \in \{g, g+1, ...\}$. Conditions (11) and (12) in terms of DF take the form of $F_r(h) = 1$ and $F_r(t) \geq F_Q(t)$. For the chosen h these conditions generally satisfy many different functions $F_r(t)$. But the values of h can also be changed. A lower bound for h is derived from condition (12) given in the form of $S_r = \sum_{i=0}^{h-1} [1 - F_r(i)] = m$ and condition (13) in the form of $F_r(t) \geq F_Q(t)$:

$$S_Q = \sum_{i=0}^{h-1} [1 - F_Q(i)] \geqslant m. \tag{14}$$

Based on the above, we propose a method to find the distributions $\{r_i\}$ that satisfy the constraint region (10)-(13). This method is illustrated in Figure 1 and may be described by the following sequence of steps.

Step 1. Determine the smallest integer h satisfying inequation (14).

Step 2. Assume $F_r(i) = F_Q(i)$, $i = g, ..., h - 1$; $F_r(h) = 1$.

Step 3. Calculate S_Q from Eq. (14) and $\Delta = S_Q - m$. If $\Delta = 0$, then go to Step 5.

Step 4. Separating Δ on supplements to the values $F_r(g), ..., F_r(h-1)$, increase them so that equality $S_r = m$ is true for resulting DF $F_r(t)$.

Step 5. Assume $r_i = F_r(i) - F_r(i - 1)$, $i = g, ..., h$.

The Step 4 can be completed in different ways. One of the examples is to increase $F_r(h - 1)$ by $\Delta = S_Q - m$ and assume $F_r(h) = 1$ (Fig. 1). In that

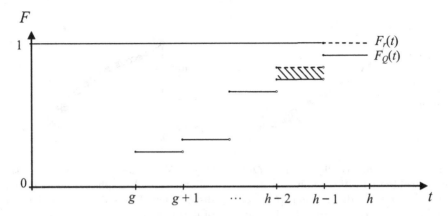

Fig. 1. The proposed method to build DF $F_r(t)$ (dashed line) based on the DF $F_Q(t)$ (solid line). When $g \leq t < h - 1$ assume $F_r(t) \equiv F_Q(t)$. Let S_Q denote the area of the region bounded by the plot DF $F_q(t)$, straight $F = 1$ and vertical lines $t = 0$ and $t = h$. According to expression (14) $S_Q \geq m$. The height and the area of the shaded rectangle is $\Delta = S_Q - m$. Since h is the smallest integer satisfying in equation (14), then $\Delta \leq 1 - F_Q(h - 1)$. And since $\Delta = S_Q - m$, then $S_Q - \Delta = m$, i.e. that part of the area S_Q, this part is located above the plot of DF $F_r(t)$ is equal to m. Therefore $\langle x \rangle = m$, and all the conditions (10)–(13) are satisfied.

case, Step 5 gives $r_g = Q_g$, ..., $r_{h-2} = Q_{h-2}$ and $r_{h-1} = Q_{h-1} + \Delta$, $r_h = 1 - F_Q(h - 1) - \Delta$. Another possible way to complete Step 4 is to increase $F_r(g)$ so that $S_r = m$, which can be obtained by increasing $F_r(g + 1)$, $F_r(g + 2)$, ..., $F_r(g)$ so that DF $F_r(t)$ is non-decreasing. In this case, several consequent probabilities r_{g+1}, r_{g+2}, ... can become equal to zero. If the model is calibrated on the basis of nonnegative weights (9), which are calculated by $\{\tilde{Q}_k\}$ (for the given r_k), and conditions $M + 1 \geq 2m$ and $\langle k \rangle \equiv 2m \geq 2g$ are satisfied, then the NPA model has the required distribution $\{\tilde{Q}_k\} = \{Q_k\}$ and also $\langle f \rangle = m$.

A "seed" graph can be a fully connected graph where vertex degrees equal g. The calibration of the model by empirical node degree distribution of a network can be considered as a solution of a network identification task. Fig. 2 reflects the quality of the solution of an identification task for the Internet autonomous systems network based on data [10], which describe $22,961$ nodes and $48,436$ edges. The calibrated model parameters are $r_1 = 0.342$, $r_2 = 0.432$, $r_3 = 0.096$, $r_4 = 0.13$, $m = r_1 + 2r_2 + 3r_3 + 4r_4 = 2.014$; $f_1, ..., f_4 = 0.0017, 0.0245, 0.0999$, 2.5303, and $f_k = 0.8603k$ for $k \geq 5$. Analogically, vertex degree distribution of the NPA graph is presented in Fig. 3 (solid line), where the model of the graph was calibrated by the empirical degree distribution of the network of movie actors (markers). We used the published on the Internet data [11] about the network with 511,416 non-isolated nodes and 1,463,331 edges. Nodes of this network are actors, and two nodes have a common edge if the corresponding actors have acted in a movie together. The calibrated model parameters $r_1, ..., r_8 > 0$, $f_1, ...,$

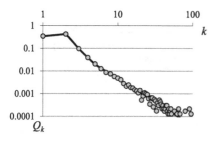
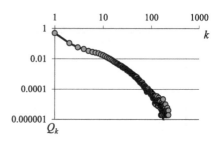

Fig. 2. Vertex degree distribution $\{\tilde{Q}_k\}$ of the calibrated NPA graph (markers) and node degree distribution $\{Q_k\}$ of the Internet structure at the level of autonomous systems (solid line)

Fig. 3. Vertex degree distribution of the calibrated NPA graph and nodes degree distribution of the collaboration network of movie actors

f_{10} are established, and if $k > 10$, the general weight function takes the form of $f_k = 4.429 \ln(k)$. The fact that the preference function in this model is logarithmical can be explained by human subjective sensation (in this case – the actor's fame), which is proportional to the logarithm of the stimulus intensity.

The required vertex degree distribution can be obtained by the NPA model in which GDs have a constant number of edges $m = const$. We give detailed recommendations on the calibration of the NPA model with examples in [9]. For example, the general formula for the implementation of a triangular degree distribution is derived for a GD with a constant number of edges ($x \equiv g = m$). In particular, for $m = 3$ the model with weigths $f_3, ..., f_8 = 15, \frac{13}{2}, \frac{10}{3}, \frac{3}{2}, 1, \frac{1}{2}$ (and $f_k = 0$ for $k < 3$ and $k > 8$) implements the graph with triangular degree distribution $Q_3, ..., Q_9 = \frac{1}{16}, \frac{2}{16}, \frac{3}{16}, \frac{4}{16}, \frac{3}{16}, \frac{2}{16}, \frac{1}{16}$. There we also give useful examples and accurate theoretical relations.

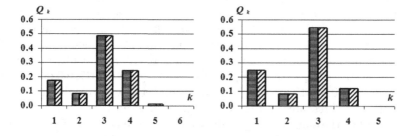

Fig. 4. Node degree distributions of Pennsylvania (left) and Omsk (right) transport network

4   Conclusions

Thus, we have described new results of the analysis of the BA graphs and a nonlinear preferential attachment model to research large network structures. The NPA model allows to grow graphs so that the probability that GD edges attach to nodes is not necessarily in the proportion to the degrees of the nodes. Fig. 4 shows the node degree distributions of transport networks (horizontal hatching) of Pennsylvania and Omsk city. The vertex degree distributions of the random graphs of size 100 thousand vertices grown by calibrated generators are marked by diagonal hatching.

The exact formula of the degree distribution found for the NPA graphs confirms previously obtained results and presents them as a special case. The proposed methods for calibration of the NPA models are simple in practical application. The use of the accelerated methods [9] of generating preferential attachment graphs allows to generate the NPA graphs that contain hundreds of thousands of vertices. The above results allow solving the task of structural identification, modeling, and optimization of large networks effectively. The calibrated NPA models can be used for simulation, forecasting and optimization of processes occurring or planned to be implemented in the networks, or for testing different algorithms and programs dealing with large graphs.

The authors would like to thank the Omsk State Technical University and O.A. Saveleva for help with translating the article in English. This work is supported by Russian Foundation for Basic Research under grants 12-07-00149-a and 14-01-31551-mol-a.

References

1. Barabasi, A.-L., Albert, R.: Science 286, 509 (1999)
2. Barabasi, A.-L., Albert, R.: Rev. Mod. Phys. 74, 47 (2002)
3. Newman, M.: Rev. Mod. Phys. 45, 167 (2003)
4. Clauset, A., Shalizi, C.R., Newman, M.: Rev. Mod. Phys. 51, 661 (2009)
5. Dorogovtsev, S.N., Mendes, J.F.F., Samukhin, A.: Phys. Rev. Lett. 85, 4633 (2000)
6. Krapivsky, P.L., Redner, S.: Phys. Rev. E 63, 066123 (2001)
7. Klemm, K., Eguiluz, V.M.: Phys. Rev. E 65, 057102 (2002)
8. Zadorozhnyi, V.N., Yudin, E.B.: Automation and Remote Control 73, 702 (2012)
9. Zadorozhny, V.N.: Control Sciences 6, 2 (2011)
10. Newman, M.: A symmetrized snapshot of the structure of the internet at the level of autonomous systems, reconstructed from bgp tables posted by the university of oregon route views (2006), http://wwwpersonal.umich.edu/mejn/netdata/
11. Barabasi, A.-L., Albert, R.: Actor network data (1999), http://www.nd.edu/networks/resources/actor/actor.dat.gz

Stability and Unloading Cost of Time-Sharing Dual-Server Systems in Random Environment*

Andrei Zorine

N.I. Lobachevsky State University of Nizhni Novgorod
23 Prospekt Gagarina, 603950, Nizhni Novgorod, Russia
zoav1602@gmail.com

Abstract. We investigate a time-sharing queueing process with readjustments. Conflicting input flows are formed in a random external environment with two states. The input flows are Poisson with intensities dependent on environment's states. There are two parallel homogeneous servers in the system. Service and readjustment durations have exponential laws of probability distribution. A continuous-time denumerable multidimensional Markov chain is defined to describe the dynamics of the servers, the sizes of the queues and the environment state. The customers sojourn cost during the period of reduction of the number of customers in the system is chosen as a performance metric. Numerical study in case of two input flows and a class of priority and threshold service algorithms is conducted.

Keywords: Time-sharing queueing system with readjustments, parallel servers, random environment, continuous-time Markov chain ergodicity condition, Chung functionals, cost of unloading a queueing system.

1 Introduction

The service of conflicting input flows in the class of time-sharing algorithms is a well-known problem in queueing theory [1–12]. The algorithms perform dividing the total service time of a customer into small quanta, while between quanta the customers are sent back to waiting queues. In the first place the interest in such systems was caused by modeling certain information processing operations in computer control systems. As a rule, there are multiple classes of customers present in real-world queueing systems. Customers of different classes have different probability distribution functions for service duration. In a paper by G.P. Klimov [2] concerning the optimal control of multi-class customers according to the criterion of minimal expected sojourn cost per time unit it was established for the first time that priority indices should be assigned to customer

---

* This work was fulfilled as a part of State Budget Research and Development program No. 01201456585 "Mathematical modeling and analysis of stochastic evolutionary systems and decision processes" of N.I. Lobachevsky State University of Nizhni Novgorod and was supported by State Program "Promoting the competitiveness among world's leading research and educational centers".

A. Dudin et al. (Eds.): ITMM 2014, CCIS 487, pp. 440–449, 2014.

classes. Moreover, an algorithm to compute the indices was given. Other works cited above allowed readjustments (setup times, orientations, etc.) of the server, as well as branching input flows of secondary customers generated by customers just serviced. At the same time in the majority of researches the input flows were assumed stationary with no after-effect. In [9–12] a time-sharing queueing system was studied whose input flows were modulated by a random external environment synchronized with the server. After each service cycle a readjustment cycle occurred. The objective of control was the expected sojourn cost per work cycle or readjustment cycle for all customers present in the system during this cycle.

Study of time-sharing algorithms in parallel servers was started by several authors (see [7, 8, 13] for references there). The treatment there is in discrete time. A heuristic priority-index rule, based on Klimov's solution to the single-server model was applied to control a network of a finite number of parallel homogeneous servers in [7, 8]. In [13] a mathematical model for a double-server queueing system with input flows modulated by an external random environment with control within a class of time-sharing algorithms with readjustments of the servers was considered. The model took form of a multi-dimensional discrete-time Markov chain with several denumerable components. Using an iterative-dominating method a sufficient condition for stationarity was found. The method has already been used by a number of authors to study discrete-time Markov queueing models [4, 5, 11, 14].

The aim of this article is to establish a continuous-time model for the exponential time-sharing dual-server system in random environment and then to formulate an optimization problem of unloading the system at smallest cost and to report conclusions based on numerical experiments. We also extend the iterative-dominating method in order to obtain a sufficient condition for the ergodicity of the continuous-time Markov process.

2   The Markov Process and Its Ergodicity Condition

The following controlled queueing system is studied. A finite number $m < \infty$ of input flows Π_1, Π_2, ..., Π_m enter. The flows are formed in a random external environment with two states $e^{(1)}$, $e^{(2)}$. In the state $e^{(k)}$, $k = 1, 2$, the flow Π_j, $j = 1, 2, \ldots, m$, is Poisson with intensity $\lambda_j^{(k)} > 0$. In each state of the external environment the input flows are independent. Customers of the flow Π_j are kept in a queue O_j of infinite capacity. Customers are chosen for service in order of arrivals. There are two homogeneous servers in the system, Σ_1 and Σ_2. Each server has $n = 2m + 1$ states named $\Gamma^{(0)}$, $\Gamma^{(1)}$, ..., $\Gamma^{(2m)}$. A server is in the state $\Gamma^{(r)}$, $1 \leqslant r \leqslant m$ if it serves a customer from O_r. After the state $\Gamma^{(r)}$ the server switches to the state $\Gamma^{(m+r)}$. In the state $\Gamma^{(m+r)}$ no customers are serviced by the server and a cycle of server readjustment and control takes place. When a server readjustment terminates the server either chooses next customer to service if there are any, according to the following rule. Let $h(\cdot)$ be a fixed mapping of the nonnegative integer lattice $X = \{0, 1, \ldots\}^m$ onto $\{0, 1, \ldots, m\}$

with constraints: $h(x) = j$, $x \in X$ implies $x_j > 0$ and only the zero vector $y^{(0)} = (0, 0, \ldots, 0) \in X$ is mapped into 0. Then, given that upon a readjustment cycle termination the sizes of queues are described with a nonzero vector $x \in X$, the service starts for a customer from the queue O_j where $j = h(x)$. The customer is taken out from the queue to the server. When the queues are empty upon the readjustment cycle termination, the server switches to the state $\Gamma^{(0)}$ waiting for a new arrival. Upon arrival the service of the new customer starts by the idle server, and the state of the server becomes $\Gamma^{(j)}$ if the first arrival occurred from Π_j. If upon arrival both servers are in the state $\Gamma^{(0)}$ then the server Σ_1 begins service. Durations of service cycles and readjustment cycles are independent with exponential probability distributions. The expected duration of the state $\Gamma^{(r)}$ equals $\beta_r > 0$, $1 \leqslant r \leqslant 2m$. External random environment is synchronized with the servers. Changes in the environmental state may occur only at cycle termination epochs. The probability of transition from the state $e^{(k)}$ to the state $e^{(l)}$ equals $a_{k,l} \in (0, 1)$, k, $l = 1, 2$. So the continuous-time random process describing the changes in the random environment is not a Markov process. After service the customer is either redirected into the queue O_r for repeated service with probability $p_{j,r}$ or leaves the queueing system with probability $p_{j,0} = 1 - \sum_{r=1}^{m} p_{j,r}$. We assume that every customer leaves the queueing system after a finite number of service cycles with positive probability. Thus, besides the primary input flows there are input flows of secondary customers and the resulting input flows have complex probabilistic structure. Finally, the sojourn cost per unit time for single customer in O_j is given and it equals c_j.

All random objects are considered on a probability space $(\Omega, \mathfrak{F}, \mathbf{P})$ where Ω is a set of elementary outcomes, $\mathfrak{F}$ is a σ-algebra of events $A \subset \Omega$, $\mathbf{P}$ is a probability on $\mathfrak{F}$. We shall need random variables $\chi(\omega, t) \in \{e^{(1)}, e^{(2)}\}$ — the state of the random external environment at time $t \geqslant 0$, $\Gamma(\omega, t) \in \Gamma^2 = \{\Gamma^{(0)}, \Gamma^{(1)}, \ldots, \Gamma^{(2m)}\} \times \{\Gamma^{(0)}, \Gamma^{(1)}, \ldots, \Gamma^{(2m)}\}$ — the servers' states at time t, $\kappa_j(t)$ — the number in O_j at time t, and random vector

$$\kappa(\omega, t) = (\kappa_1(\omega, t), \kappa_2(\omega, t), \ldots, \kappa_m(\omega, t)), \quad t \geqslant 0.$$

As usual in probability theory, the argument ω of random objects will be omitted.

Since the future of the process

$$\{(\Gamma(t), \kappa(t), \chi(t)); t \geqslant 0\} \tag{1}$$

after time t is determined only by its state at time t, the termination moments of the cycles taking place at time t, future arrivals after time t and future states of the external environment after time t, random process (1) is a Markov process given an initial distribution, with the state space

$$S = \{(\Gamma^{(0)}, \Gamma^{(0)}, y^{(0)}, e^{(k)}) : k = 1, 2\}$$
$$\cup \{(\Gamma^{(r)}, \Gamma^{(0)}, y^{(0)}, e^{(k)}) : r = 1, 2, \ldots, 2m; k = 1, 2\}$$
$$\cup \{(\Gamma^{(0)}, \Gamma^{(r)}, y^{(0)}, e^{(k)}) : r = 1, 2, \ldots, 2m; k = 1, 2\}$$
$$\cup \{(\Gamma^{(r)}, \Gamma^{(r')}, x, e^{(k)}) : r, r' = 1, 2, \ldots, 2m; x \in X; k = 1, 2\}.$$

Markov process (1) is time-homogeneous. Denote by $q(r_1, r_2, x, k; r_1', r_2', w, l)$, $(\Gamma^{(r_1)}, \Gamma^{(r_2)}, x, e^{(k)}) \in S$, $(\Gamma^{(r_1')}, \Gamma^{(r_2')}, w, e^{(l)}) \in S$, the transition rates for Markov process (1). Put $q(r_1, r_2, x, k) = -q(r_1, r_2, x, k; r_1, r_2, x, k)$. Finally, let $\delta_{s,s'}$ be the Kronecker's delta and set $y^{(s)} = (\delta_{1,s}, \delta_{2,s}, \ldots, \delta_{m,s}) \in X$, $s = 1, 2, \ldots, m$.

The proof of the next theorem follows the usual technique of considering the process state changes over small amounts of time, Δt and obtaining Chapman – Kolmogorov differential equations. Set $\lambda_+^{(k)} = \lambda_1^{(k)} + \lambda_2^{(k)} + \ldots + \lambda_m^{(k)}$, $k = 1, 2$.

Theorem 1. *Let $r, r' = 1, 2, \ldots, 2m$, $s, s' = 0, 1, \ldots, m$, $x \in X$, $j = 1, 2, \ldots, m$, $k, l = 1, 2$. The transition rates for Markov process (1) are*

$$q(0, 0, y^{(0)}, k) = \lambda_+^{(k)},$$

$$q(r, 0, y^{(0)}, k) = q(0, r, y^{(0)}, k) = \beta_r^{-1} + \lambda_+^{(k)},$$

$$q(r, r', x, k) = \beta_r^{-1} + \beta_{r'}^{-1} + \lambda_+^{(k)},$$

$$q(0, 0, y^{(0)}, k; j, 0, y^{(0)}, k) = q(r, 0, y^{(0)}, k; r, j, y^{(0)}, k)$$

$$= q(0, r, y^{(0)}, k; j, r, y^{(0)}, k) = q(r, r', x, k; r, r', x + y^{(j)}, k) = \lambda_j^{(k)},$$

$$q(r, 0, y^{(0)}, k; m + r, s, y^{(0)}, l) = q(0, r, y^{(0)}, k; s, m + r, y^{(0)}, l)$$

$$= q(r, r', x, k; m + r, r', x + y^{(s)}, l) = q(r', r, x, k; r', m + r, x + y^{(s)}, l)$$

$$= a_{k,l}\beta_r^{-1} p_{r,s}, \qquad 1 \leqslant r \leqslant m,$$

$$q(r, 0, y^{(0)}, k; 0, 0, y^{(0)}, l) = q(0, r, y^{(0)}, k; 0, 0, y^{(0)}, l)$$

$$= q(r, r', x, k; j, r', x, l) = q(r', r, x, k; r', j, x, l)$$

$$= a_{k,l}\beta_r^{-1} \qquad h(x) = j, \; m + 1 \leqslant r \leqslant 2m.$$

The remaining transition rates equal zero.

It follows from the form of the transition intensities that Markov process (1) has only stable non-absorbing states and a conservative infinitesimal matrix.

When $x = (x_1, x_2, \ldots, x_m) \in X$ and $v = (v_1, v_2, \ldots, v_m)$ is a real or complex vector we write $v^x = v_1^{w_1} \cdot v_2^{w_2} \cdots v_m^{x_m}$ for short, assuming $0^0 = 1$. In the remaining of this section we assume that for $r, r' = 1, 2, \ldots, 2m$, $j = 1, 2, \ldots, m$, $k = 1, 2$, and $x \in X$, the series

$$\frac{\beta_r + \beta_{r'}}{\beta_r + \beta_{r'} + \lambda_+^{(k)} \beta_r \beta_{r'}}$$

$$\times \sum_{\substack{w \in X \\ h(w) = j}} v^{w-x} \frac{(x_1 + x_2 + \cdots + x_m)!}{x_1! x_2! \cdots x_n!} \prod_{s=1}^{m} \left(\frac{\lambda_s^{(k)} \beta_r \beta_{r'}}{\beta_r + \beta_{r'} + \lambda_+^{(k)} \beta_r \beta_{r'}} \right)^{w_s - x_s}$$

depends on x in a finite number of forms. For example, this is the case for priority algorithms. Let $\vartheta_1, \vartheta_2, \ldots, \vartheta_m$ be the solution to the system

$$\vartheta_j = \sum_{r=1}^{m} p_{j,r} \vartheta_r + \beta_j, \qquad j = 1, 2, \ldots, m.$$

It can be shown that the solution exists since every customer leaves the queueing system after a finite number of services. Since $\beta_j > 0$ we have $\vartheta_j > 0$, $j = 1, 2, \ldots, m$. Set $\beta_- = \min\{\beta_1, \ldots, \beta_m\}$, $\beta_+ = \max\{\beta_1, \ldots, \beta_m\}$, $\bar{\beta}_- = \min\{\beta_{m+1}, \ldots, \beta_{2m}\}$, $\bar{\beta}_+ = \max\{\beta_{m+1}, \ldots, \beta_{2m}\}$, $\vartheta_- = \min\{\vartheta_1, \ldots, \vartheta_m\}$, $\vartheta_+ = \max\{\vartheta_1, \ldots, \vartheta_m\}$, $\bar{\vartheta} = \max\{\vartheta_1 - \beta_1, \ldots, \vartheta_m - \beta_m\}$, $\beta = (\beta_1, \ldots, \beta_m)$, $\lambda^{(k)} = (\lambda_1^{(k)}, \ldots, \lambda_m^{(k)})^T$, $Q = (p_{j,r})_{j,r=\overline{1,m}}$, $\rho^{(k)} = \beta(I_m - Q^T)^{-1}\lambda^{(k)}$, $k = 1, 2$, I_m is the size m identity matrix.

Theorem 2. *Inequality*

$$\max\left\{\rho^{(l)}\beta_+\bar{\beta}_+ + (a_{l,1}\rho^{(1)} + a_{l,2}\rho^{(2)})\frac{\beta_+^3 + \bar{\beta}_+^3}{\beta_- + \bar{\beta}_-} : l = 1, 2\right\} < \vartheta_- - \bar{\vartheta}. \quad (2)$$

is sufficient for the existence of the stationary probability distribution of Markov process (1).

Proof. Under the assumptions on Markov process (1) its trajectories are piecewise constant and have only discontinuities of the first kind. Let $\tau_0 = 0$ and τ_1, $\tau_2, \ldots$ be the jump instants of (1). Consider the embedded Markov chain

$$\{(\Gamma_i, \kappa_i, \chi_i); i = 0, 1, \ldots\} \quad (3)$$

where $\Gamma_i = \Gamma(\tau_i + 0)$, $\kappa_i = \kappa(\tau_i + 0)$, $\chi_i = \chi(\tau_i + 0)$. Observe that

$$\sup_{r,r',x,k} q(r, r', x, k) < \infty. \quad (4)$$

By virtue of Theorems 5.4.4, 5.4.4 in [15], inequality (4) and positive recurrence of all states of Markov chain (3) imply positive recurrence of Markov process (1) as well as existence of its unique stationary probability distribution. Let us prove that all the states of Markov chain (3) are positive recurrent. The jump instants $\{\tau_i; i = 0, 1, \ldots\}$ are generated both by new arrivals or by server operation terminations. Set $\hat{\Gamma}_0 = \Gamma_0$, $\hat{\kappa}_0 = \kappa_0$, $\hat{\chi}_0 = \chi_0$ and let

$$\{(\hat{\Gamma}_i, \hat{\kappa}_i, \hat{\chi}_i); i = 0, 1, \ldots\} \quad (5)$$

be Markov chain (3) sampled at an operation termination by either server. An operation termination instant $\theta_i \in \{0, 1, \ldots\}$ can be identified from the past and present values of Γ_i component of (3). Then Strong Markov property ensures that stochastic sequence (5) is a Markov chain. Denote by

$$\sigma = \inf\{i \geqslant 1 : \Gamma_i = (\Gamma^{(0)}, \Gamma^{(0)}), \kappa_i = y^{(0)}, \chi_i = e^{(1)}\},$$
$$\hat{\sigma} = \inf\{i \geqslant 1 : \hat{\Gamma}_i = (\Gamma^{(0)}, \Gamma^{(0)}), \hat{\kappa}_i = y^{(0)}, \hat{\chi}_i = e^{(1)}\}$$

the steps when the first return to the state $(\Gamma^{(0)}, \Gamma^{(0)}, y^{(0)}, e^{(1)})$ occurs for Markov chains (3) and (5) correspondingly. Obviously, $\sigma \leqslant \theta_{\hat{\sigma}}$. Consider an event

$$\{\omega : \hat{\Gamma}_0 = (\Gamma^{(0)}, \Gamma^{(0)}), \hat{\kappa}_0 = y^{(0)}, \hat{\chi}_0 = e^{(1)}, \hat{\sigma} = 2i_0\}$$

which occurs if and only if i_0 service cycles and i_0 readjustment cycles have terminated. In this case no more than i_0 primary customers may have entered the queueing system. Each primary customer invokes exactly three jumps (its arrival, service termination and readjustment termination), while each secondary customer brings only two jumps (service termination and readjustment termination). Hence $\theta_{\hat{\sigma}} \leqslant 3i_0$. So, on the event $\{\omega\colon \hat{I}_0 = (\Gamma^{(0)}, \Gamma^{(0)}), \hat{\kappa}_0 = y^{(0)}, \hat{\chi}_0 = e^{(1)}\}$ the inequality $\sigma \leqslant 3\hat{\sigma}$ is established. By virtue of Theorem 3 in [13], Markov chain (5) is positive recurrent, hence

$$\mathbf{E}(\hat{\sigma}|\{\omega\colon \hat{I}_0 = (\Gamma^{(0)}, \Gamma^{(0)}), \hat{\kappa}_0 = y^{(0)}, \hat{\chi}_0 = e^{(1)}\}) < \infty.$$

By definition, $\Gamma_0 = \hat{I}_0$, $\kappa_0 = \hat{\kappa}_0$, $\chi_0 = \hat{\chi}_0$. Therefore

$$\mathbf{E}(\sigma|\{\omega\colon \Gamma_0 = (\Gamma^{(0)}, \Gamma^{(0)}), \kappa_0 = y^{(0)}, \chi_0 = e^{(1)}\})$$
$$\leqslant \mathbf{E}(3\hat{\sigma}|\{\omega\colon \hat{I}_0 = (\Gamma^{(0)}, \Gamma^{(0)}), \hat{\kappa}_0 = y^{(0)}, \hat{\chi}_0 = e^{(1)}\}) < \infty.$$

The state $(\Gamma^{(0)}, \Gamma^{(0)}, y^{(0)}, \chi^{(1)})$ turns out to be positive recurrent with respect to Markov chain (3) as well. The theorem is proven.

3   Chung Functionals for Computation of Unloading Cost

Let $S = \tilde{S}_- \cup \tilde{S}_0 \cup \tilde{S}_+$ be a partition of S into nonempty disjoint sets. We call $\tilde{S}_+$ a final set and $\tilde{S}_-$ a taboo set. Let $c(\gamma, x, e^{(k)}) = c(\Gamma^{(r)}, \Gamma^{(r')}, x, e^{(k)}) \geqslant 0$, $\gamma = (\Gamma^{(r)}, \Gamma^{(r')})$, be a sojourn cost for Markov process (1) per unit time. Let

$$\eta(\omega) = \inf\{t \geqslant 0\colon (\Gamma(\omega, t'), \kappa(\omega, t'), \chi(\omega, t')) \notin \tilde{S}_-, 0 \leqslant t' \leqslant t,$$
$$(\Gamma(\omega, t), \kappa(\omega, t), \chi(\omega, t)) \in \tilde{S}_+\}$$

be the first passage time of the final set $\tilde{S}_+$ with taboo set $\tilde{S}_-$. Put $\Omega(r, r', x, k) = \{\omega\colon \Gamma(0) = (\Gamma^{(r)}, \Gamma^{(r')}), \kappa(0) = x, \chi(0) = e^{(k)}\}$. Then the cost $\zeta(\omega)$ of taboo first passage is defined as

$$\zeta(\omega) = \int_0^{\eta(\omega)} c(\Gamma(\omega, t), \kappa(\omega, t), \chi(\omega, t))\, dt.$$

The mean cost of taboo first passage from the state $(\Gamma^{(r)}, \Gamma^{(r')}, x, e^{(k)}) \in \tilde{S}_0$ is defined as $\mathbf{E}(\zeta|\Omega(r, r', x, k) \cap \{\omega\colon \eta < \infty\})$. The probability $f(r, r', x, k) = \mathbf{P}(\{\omega\colon \eta < \infty\}|\Omega(r, r', x, k))$ is called a taboo-probability (of taboo first passage) [16]. Numerical evaluation of taboo probabilities and mean costs of taboo first passage can be carried out by the solving two sets of linear equations. Proofs are similar to those in [17].

Theorem 3. *Taboo-probabilities $\{f(r, r', x, k) \colon (\Gamma^{(r)}, \Gamma^{(r')}, x, e^{(k)}) \in \tilde{S}_0\}$ are a solution of the linear algebraic system*

$$q(r, r', x, k)f(r, r', x, k) = \sum_{(\Gamma^{(s)}, \Gamma^{(s')}, w, e^{(l)}) \in \tilde{S}_+} q(r, r'x, k; s, s', w, l)$$

$$+ \sum_{\substack{(\Gamma^{(s)}, \Gamma^{(s')}, w, e^{(l)}) \in \tilde{S}_0 \\ (s, s', w, l) \neq (r, r', x, k)}} f(s, s', w, l)q(r, r', x, k; s, s', w, l), \quad (\Gamma^{(r)}, \Gamma^{(r')}, x, e^{(k)}) \in \tilde{S}_0,$$

$$(6)$$

obtained by successive approximations with zero initial conditions. The conditional expected costs of taboo first passage have the form

$$\mathbf{E}(\zeta | \Omega(r, x, k) \cap \{\omega \colon \eta < \infty\})$$

$$= G(r, r', x, k)(f(r, r', x, k))^{-1} \quad \text{for } f(r, r', x, k) \neq 0, \quad (7)$$

where the numbers $G(r, r', x, k)$, $(\Gamma^{(r)}, \Gamma^{(r')}, x, e^{(k)}) \in \tilde{S}_0$, are solutions of the linear algebraic system

$$q(r, r', x, k)G(r, r', x, k) = c(\Gamma^{(r)}, \Gamma^{(r')}, x, e^{(k)})f(r, r', x, k)$$

$$+ \sum_{\substack{(\Gamma^{(s)}, \Gamma^{(s')}, w, e^{(l)}) \in \tilde{S}_0 \\ (s, s', w, l) \neq (r, r', x, k)}} G(s, s', w, l)q(r, r', x, k; s, s', w, l), \quad (\Gamma^{(r)}, \Gamma^{(r')}, x, e^{(k)}) \in \tilde{S}_0,$$

$$(8)$$

obtained by successive approximations with zero initial conditions.

We are interested in computation of the unloading cost for the time-sharing dual-server queueing system. Let us define the sojourn cost per time unit in the state $(\Gamma^{(r)}, \Gamma^{(r')}, x, e^{(k)})$ by

$$c(\Gamma^{(r)}, \Gamma^{(r')}, x, e^{(k)}) = x_1 c_1 + x_2 c_2 + \ldots + x_m c_m.$$

Thus defined, the sojourn cost is independent either of the external environment state or of the server state, and it depends on the sizes of the queues only. Further, let the state space X of the queues be partitioned into disjoint nonempty subsets X_0, X_+, X_-, $X_0 \cup X_+ \cup X_- = 0$. The sets are interpreted as the set of permissible sizes of the queues, as the set of desirable sizes of the queues, and the set of prohibited (taboo) sizes of the queues correspondingly. The representation $X = X_- \cup X_0 \cup X_+$ of the queues sizes state space leads to a partition $S = \tilde{S}_- \cup \tilde{S}_0 \cup \tilde{S}_+$ of the state space of random process (1) with

$$\tilde{S}_- = \{(\gamma, x, e^{(k)}) \colon x \in X_-\}, \quad \tilde{S}_0 = \{(\gamma, x, e^{(k)}) \colon x \in X_0\},$$

$$\tilde{S}_+ = \{(\gamma, x, e^{(k)}) \colon x \in X_+\}.$$

For instance, the partition of the state space S can be generated by sets

$$X_+ = \{x \colon x_j \leqslant N_j, j = 1, \ldots, m\}, \quad X_- = \{x \colon x_j > N_j' \text{ for some } j = 1, \ldots, m\}, \tag{9}$$

with some positive integers N_j, N_j', $j = 1, 2, \ldots, m$. A switching function $h(\cdot)$ given, the conditional expected cost of taboo first passage for any initial state in $\tilde{S}_0$ can be computed by means of equations (6), (8), and (7). Choice between switching functions $h_1(\cdot)$ and $h_2(\cdot)$ for a partition of the state space S of process (1) can be carried out by comparison of corresponding values in (7). A single objective function can be obtained, for instance, by a convolution

$$J(h, \tilde{S}_+, \tilde{S}_0, \tilde{S}_-) = \frac{1}{|\tilde{S}_0|} \sum_{(\Gamma^{(r)}, \Gamma^{(r')}, x, e^{(k)}) \in \tilde{S}_0} G(r, r', x, k)(f(r, r', x, k))^{-1} \tag{10}$$

of objective functions (7). Now the problem can be stated as follows: given a partition $S = \tilde{S}_- \cup \tilde{S}_0 \cup \tilde{S}_+$ solve the minimization problem

$$J(h_0, \tilde{S}_-, \tilde{S}_0, \tilde{S}_+) = \inf_{h(\cdot)} J(h, \tilde{S}_+, \tilde{S}_0, \tilde{S}_-). \tag{11}$$

For numerical experiments a queueing system with $m = 2$ input flows was considered. A program in Octave programming language [18] was developed to generate and solve systems of linear equations (6), (8). The program uses sparse representation for big matrices and the default Octave solver. We consider a more narrow class of switching functions $h(\cdot)$ than that in (11). We chose the following switching functions for the study. If $(x_1, x_2) \neq (0, 0)$ let $h_{\max}(x_1, x_2) = 1$ for $x_1 \geqslant x_2$ and $h_{\max}(x_1, x_2) = 2$ for $x_1 < x_2$, $h_{\mathrm{thr},1}(x_1, x_2) = 1$ for $x_2 = 0$ or $x_1 \geqslant a$ and $h_{\mathrm{thr},1}(x_1, x_2) = 2$ for $x_2 > 0$ and $x_1 < a$, $h_{\mathrm{thr},2}(x_1, x_2) = 2$ for $x_1 = 0$ or $x_2 \geqslant a$ and $h_{\mathrm{thr},2}(x_1, x_2) = 1$ for $x_1 > 0$ and $x_2 < a$. The switching function $h_{\max}$ performs the "fair" service of the longest queue, $h_{\mathrm{thr},j}$ corresponds to the threshold policy with the threshold value $a > 0$ for the queue O_j. When $a = 1$ the switching function $h_{\mathrm{thr},j}$ implements a priority service and assigns the higher priority to the queue O_j.

Lemma 1. *For a partition of type (9) and any of the switching functions $h_{\max}$, $h_{\mathrm{thr},1}$, and $h_{\mathrm{thr},2}$ the taboo-probabilities in (6) are strictly positive.*

Proof. It is sufficient to find for every state in $\tilde{S}_0$ a path with positive probability which leads to $\tilde{S}_+$ without visiting $\tilde{S}_-$. It follows from Theorem 1 that process (1) can leave each state in $\tilde{S}_0$ due to a change in a server state. Assuming that only this kind of jumps of process (1) takes place, any path in $\tilde{S}_0$ leading to $\tilde{S}_+$ can be depicted by a path in X_0 leading to X_+ if one considers only the changes in the sizes of the queues. If $h(x_1, x_2) = 1$ then the path may have a segment only from (x_1, x_2) to $(x_1 - 1, x_2)$. If $h(x_1, x_2) = 2$ then the path may have a segment only from (x_1, x_2) to $(x_1, x_2 - 1)$. Typical paths for different positions of the point (x_1, x_2) and different switching functions are shown in Fig. 1. From this geometric interpretation we may conclude that there is a path from any point in X_0 to X_+ for the switching functions under consideration. The claim is proven.

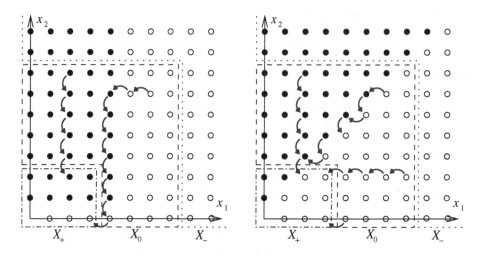

Fig. 1. Switching functions: $h_{\mathrm{thr},1}$ for $a = 5$ (on the left) and $h_{\max}$ (on the right). A white circle means switching to service of O_1, a black circle means switching to service of O_2.

In our numerical experiments we observed that smaller values of the objective functions (10) were obtained for $h_{\mathrm{thr},j}$ for some $a \leqslant N_j$, $j = 1, 2$, and never for $h_{\max}$. As an example, see Table 1. We searched for a quasi-optimal switching function amongst $h_{\max}$, $h_{\mathrm{thr},1}$ $(1 \leqslant a \leqslant N_1')$, and $h_{\mathrm{thr},2}$ $(1 \leqslant a \leqslant N_2')$. The parameters for the example are: $a_{1,1} = 0.5$, $a_{2,2} = 0.8$, $\lambda_1^{(1)} = 0.3$, $\lambda_1^{(2)} = 0.6$, $\lambda_2^{(1)} = 0.45$, $\lambda_2^{(2)} = 0.15$, $p_{1,1} = 0$, $p_{1,2} = 0.05$, $p_{2,1} = 0.02$, $p_{2,2} = 0.01$, $\beta_1 = 0.375$, $\beta_2 = 0.5$, $\beta_3 = 0.125$, $\beta_4 = 0.25$, $c_1 = c_2 = 1$, $N_1 = 3$, $N_1' = 7$, $N_2 = 2$, $N_2' = 7$. For these parameters condition in Theorem 2 is fulfilled.

Table 1. Average unloading cost $J_h = J(h; \tilde{S}_+, \tilde{S}_0, \tilde{S}_-)$ for different switching functions (see parameters in the text). The smallest value is in bold face.

$h_{\max}$	$h_{\mathrm{thr},1}$						$h_{\mathrm{thr},2}$					
	$a=1$	$a=2$	$a=3$	$a=4$	$a=5$	$a=7$	$a=1$	$a=2$	$a=3$	$a=4$	$a=5$	$a=7$
8.745	7.556	**7.505**	7.527	7.831	12.129	12.831	12.690	10.427	8.617	10.070	9.482	8.443

References

1. Takács, L.: A single server queue with feedback. The Bell Syst. Tech. J. 42, 487–503 (1963)
2. Klimov, G.P.: Time-sharing service systems. I. Theor. Probab. Appl. XIX(3), 558–576 (1974)
3. Kitaev, A.Y., Rykov, V.V.: A service system with a branching flow of secondary customers. Avtom. i Telemech. 9, 52–61 (1980)

4. Fedotkin, M.A.: Optimal control for conflict flows and marked point processes with selected discrete component. I. Liet. Mat. Rinkinys. 4(28), 783–794 (1988)
5. Fedotkin, M.A.: Optimal control for conflict flows and marked point processes with selected discrete component. II. Liet. Mat. Rinkinys. 1(29), 148–159 (1989)
6. Chao, X.: On Klimov's model with two job classes and exponential processing times. J. Appl. Probab. 3(30), 716–724 (1993)
7. Glazebrook, K.D., Niño-Mora, J.: Scheduling multiclass queueing networks on parallel servers: Approximate and heavy-traffic optimality of Klimov's priority rule. In: Burkard, R.E., Woeginger, G.J. (eds.) ESA 1997. LNCS, vol. 1284, pp. 232–245. Springer, Heidelberg (1997)
8. Glazebrook, K.D.: An analysis of Klimov's problem with parallel servers. Math. Meth. Oper. Res. 58, 1–28 (2003)
9. Fedotkin, M.A., Zorine, A.V.: Analysis of time-sharing processes. Vestn. of NNSU (Mat. ser.) 1, 18–28 (2003)
10. Fedotkin, M.A., Zorine, A.V.: Analysis and optimization of time-sharing processes, functioning in a random environment. Vestn. of NNSU (Mat. ser.) 1, 92–103 (2004)
11. Fedotkin, M.A., Zorine, A.V.: Optimization of Control of Doubly Stochastic Nonordinary Flows in Time-Sharing Systems. Autom. and Remot. Control. 7(66), 1115–1124 (2005)
12. Zorine, A.: On ergodicity conditions in a polling model with Markov modulated input and state-dependent routing. Queueing Syst. 76, 223–241 (2014)
13. Zorine, A.: Time-sharing service in parallel in random environment. In: Proceedings of the International Conference "Distributed Computer and Communication Networks: Control, Computation, Communications", pp. 265–272. Technosphera, Moscow (2013)
14. Projdakova, E.V.: Probability properties of output flows in a priority queueing system. Vestn. of NNSU 5(2), 190–196 (2012)
15. Kannan, D.: An introduction to stochastic processes. North Holland, New York (1979)
16. Fedotkin, M.A.: Algebraic properties of distributions for Chung functionals of homogeneous Markov chains with a countable set of states. Sov. Math., Dokl. 17, 350–353 (1976)
17. Zorine, A.V.: Minimization of unloading cost for an exponential time-sharing queueing process. Tomsk State Univer. J. of Contr. and Comp. Sci. 4(17), 55–63 (2011)
18. Eaton, J.W., Bateman, D., Hauberg, S.: GNU Octave Manual Version 3. Network Theory, Ltd. (2008)

Author Index